HANDBOOK ON GEOGRAPHIES OF TECHNOLOGY

RESEARCH HANDBOOKS IN GEOGRAPHY

Series Editor: Susan J. Smith, *Honorary Professor of Social and Economic Geography* and *The Mistress of Girton College, University of Cambridge, UK*

This important new *Handbook* series will offer high quality, original reference works that cover a range of subjects within the evolving and dynamic field of geography, emphasising in particular the critical edge and transformative role of human geography.

Under the general editorship of Susan J. Smith, these *Handbooks* will be edited by leading scholars in their respective fields. Comprising specially commissioned contributions from distinguished academics, the *Handbooks* offer a wide-ranging examination of current issues. Each contains a unique blend of innovative thinking, substantive analysis and balanced synthesis of contemporary research.

Handbook on Geographies of Technology

Edited by

Barney Warf

Department of Geography, University of Kansas, USA

RESEARCH HANDBOOKS IN GEOGRAPHY

Edward **Elgar**
PUBLISHING

Cheltenham, UK • Northampton, MA, USA

Published by
Edward Elgar Publishing Limited
The Lypiatts
15 Lansdown Road
Cheltenham
Glos GL50 2JA
UK

Edward Elgar Publishing, Inc.
William Pratt House
9 Dewey Court
Northampton
Massachusetts 01060
USA

A catalogue record for this book
is available from the British Library

Library of Congress Control Number: 2016953920

This book is available electronically in the **Elgar**online
Social and Political Science subject collection
DOI 10.4337/9781785361166

ISBN 978 1 78536 115 9 (cased)
ISBN 978 1 78536 116 6 (eBook)

Typeset by Servis Filmsetting Ltd, Stockport, Cheshire
Printed and bound by CPI Group (UK) Ltd, Croydon, CR0 4YY

Contents

Contributors

Ravi Baghel, University of Heidelberg, Germany

Michael Batty, Centre for Advanced Spatial Analysis, University College London, UK

Ryan E. Baxter, Penn State Institutes for Energy and Environment, Pennsylvania State University, USA

Thomas Birtchnell, University of Wollongong, Australia

M.J. Blair, Department of Geography and Planning, Queen's University, Canada

Laura Cabral, Centre Universitaire de Formation en Environnement et Développement Durable, Université de Sherbrooke, Canada

Kirby E. Calvert, Department of Geography, University of Guelph, Canada

Min Chen, Department of Geography, Nanjing Normal University, China

Julie Cidell, Department of Geography and GIS, University of Illinois, USA

Jonathan C. Comer, Department of Geography, Oklahoma State University, USA

Daphne Comfort, The Business School, University of Gloucestershire, Cheltenham, UK

Scott W. Cunningham, Department of Multi-Actor Systems, Delft University of Technology, The Netherlands

Martin Dodge, Department of Geography, University of Manchester, UK

Andrew R. Goetz, Department of Geography and the Environment, University of Denver, USA

Aaron Golub, Portland State University, USA

Alana Grech, Department of Environmental Science, Macquarie University, Australia

David Hillier, Centre for Police Sciences, University of South Wales, Pontypridd, UK

Adelheid Holl, Institute of Public Goods and Policies, Consejo Superior de Investigaciones Científicas, Spain

Jordan P. Howell, Department of Geography and Environment, Rowan University, USA

Aaron Johnson, Portland State University, USA

Peter Jones, The Business School, University of Gloucestershire, Cheltenham, UK

Aharon Kellerman, Department of Geography and Environmental Studies, University of Haifa, Israel

Lado Kurdgelashvili, Center for Energy and Environmental Policy, University of Delaware, USA

Linna Li, The University of Hong Kong

Hui Lin, Institute of Space and Earth Information Science, The Chinese University of Hong Kong, Hong Kong

Ramon Lobato, Swinburne University of Technology, Australia

Becky P.Y. Loo, The University of Hong Kong, Hong Kong

Antonio López Peláez, Department of Social Work, Faculty of Law, National Distance Education University, Spain

Edward Louie, School of Public Policy, Oregon State University, USA

Sophia Maalsen, Faculty of Architecture, Design and Planning, University of Sydney, Australia

Warren E. Mabee, Department of Geography and Planning, Queen's University, Canada

Jeff D. Makholm, National Economic Research Associates, USA

Jessica McLean, Department of Geography and Planning, Macquarie University, Australia

Marcus Nüsser, University of Heidelberg, Germany

Gabriel Popescu, Indiana University South Bend, USA

Ruth Rama, Institute of Economics, Geography and Demography, Consejo Superior de Investigaciones Científicas, Spain

Paul L. Robertson, Australian Innovation Research Centre, University of Tasmania, Australia

Jean-Paul Rodrigue, Department of Global Studies and Geography, Hofstra University, USA

Mark W. Rosenberg, Department of Geography and Planning, Queen's University, Canada

Barry Solomon, Department of Social Sciences, Michigan Technological University, USA

Jamie D. Stephen, Department of Geography and Planning, Queen's University, Canada

Daniel Sui, Department of Geography, Ohio State University, USA

Govinda Timilsina, Development Research Group, World Bank, USA

Natalie Waldbrook, Business Technology Management/Health Studies, Wilfrid Laurier University (Brantford Campus) and Innovation Policy Lab, Munk School of Global Affairs, University of Toronto, Canada

Barney Warf, Department of Geography, University of Kansas, USA

Thomas A. Wikle, Department of Geography, Oklahoma State University, USA

Catherine Wilkinson, Edge Hill University, UK

1. Introduction: geography, technology, society
Barney Warf

> Any technology sufficiently advanced is indistinguishable from magic.
> Arthur C. Clarke

Few phenomena play a more important role in our economies, societies, and daily lives as technology. Much, if not most, of the world's populations live in technologically rich – if not technologically saturated – environments. Human beings have, of course, used technologies of one sort or another for as long as there have been human beings: fire, stone axes, digging sticks, boomerangs, fishing hooks, bows and arrows, adzes, and countless other devices to hunt, farm, and make goods. Indeed, technological prowess was one of the keystones to the emergence of the planet's first superspecies (Ambrose 2001). Technologies are integral to making our products, cleaning up our messes, fighting our wars, moving us around, and building our cities, landscapes, and social structures. Technologies shape how we think about and act in the world: they do not simply reflect societies, they also constitute them. From the individual body to the global economy, technologies are ubiquitous, inescapable, and surrounded by clouds of hope, fear, dreams and, often, unrealistic expectations.

Not surprisingly, there exist considerable popular confusion and misunderstanding about technologies. Technologies are not simply 'things' – machines, robots, airplanes – but *systems* that enmesh people, objects, knowledge, techniques, procedures, and places into a seamlessly integrated whole. Some equate 'technology' with advanced machinery – computers, nuclear weapons, and space flight. Yet a technology, in the simplest and broadest definition, is but a means of converting inputs into outputs; technological change involves the growth of output per unit input (e.g. labor hour or hectare of land) or, conversely, reduced inputs per unit of output. Technologies can be primitive or amazingly complex, used to enhance human and environmental wellbeing or to surveil, harm or kill people.

Since the dawn of capitalism, and particularly the Industrial Revolution, technological change, grounded in theoretical science and applied engineering, has accelerated at exponential rates, raising productivity levels, moving people, goods, and information ever more quickly across the Earth's surface, allowing us to communicate more easily, entertaining us, and making daily life immeasurably safer, cleaner, and more convenient. Not surprisingly, technological change has captured the popular imagination: think, for example, of the first flight of the airplane in 1903, or Neil Armstrong landing on the moon in 1969. Typically, important new technologies are greeted with breathless enthusiasm, and their long-term effects are greatly over-estimated (recall that nuclear power in the 1950s was going to lead to free electricity). Technological change is widely heralded as being synonymous with progress, national or regional competitiveness, and a solution to pressing social dilemmas.

Arguably the most common and pernicious myth about technology is that of technological determinism (Staudenmaier and John 1985; Smith and Marx 1994), a term widely attributed to Thorsten Veblen. In this reductionist view, technological change acquires the aura of some omnipotent, external, asocial actor whose power drives all other changes. Technology acts, society reacts. All other domains – the social, political, and cultural – are reduced to secondary analytical importance. There is, simply put, a one-way line of causality, one that denies the historical and geographical contingency with which technologies are produced, adopted, and have effects. Technological determinists range from famed historian Lynn White (1966), who focused on the impacts of the stirrup on medieval European warfare, to noted columnist and author Thomas Friedman (2005), who proudly accepted the label in his best-selling book *The World is Flat*. Marxism too exhibits aspects of this line of thought (Bimber 1990).

Given the speed and depth with which technological change has progressed, it is admittedly difficult to avoid falling into this trap. The advent of sophisticated microelectronics instruments has unleashed so many changes that contemporary life is inconceivable without their fruits, including the Internet and cellular or mobile phones. Yet technological determinism is a fatally flawed, and thus widely rejected, ideology. Technological determinism frequently offers an unwarranted optimism, the notion that new technologies will inevitably offset diminishing returns or resolve environmental crises, when the evidence indicates otherwise (Huesemann and Huesemann 2011). More importantly, technologies are always and inevitably social products (Bijker et al. 1987). Their design and purpose emanate from concrete historical circumstances; they are, in short, created to address particular problems. Embedding technologies in their social contexts allows us to appreciate the complexity and unevenness of innovation and technological adoption, the power relations and politics that accompany it, and the differential effects as costs and benefits are borne by different classes, genders, ethnicities, and regions. Far from being inevitable, new technologies can be resisted (e.g. the Luddites). To approach technologies in any other way is to reify technological change, to assign it an autonomous status it does not deserve, to make it into a teleological force in which politics and culture play no role. Viewed in this way, technological relations and social relations are deeply intertwined. Rather than a one-way causality, it is more productive to view this relationship as simultaneously determinant.

Wresting our gaze away from the traditional economic focus on technology, cultural critics have pointed to its countless social, cultural and ideological effects (e.g. Green 2001). The printing press, for example, facilitated widespread literacy, the rise of nationalism, the Protestant Reformation, and the Enlightenment (Eisenstein 1979). Neil Postman (1985, 1992) similarly laments the role of television on consciousness and, more broadly, how discourses of scientific progress marginalize other ways of knowing the world. In the same vein, critics of the Internet argue that it is having profound effects on attention spans and the ability to concentrate (Carr 2010). In short, technologies are every bit as much cultural and political as they are economic in nature.

Another serious but widespread myth about technology is that it is only a force for good. Given that Western capitalism has benefited enormously from rapid and continuous technological change, this view is not altogether unexpected. For many, technological change is intimately wrapped with broader notions of social progress. Yet even a casual glance at the evidence reveals that technologies can be used against people as well as for them. Military technologies come to mind, such as the potential of nuclear weapons to annihilate whole societies, whereas drones raise serious questions about the legality of targeted assassinations (see Chapter 16 this volume). Likewise, the Internet can be used for surveillance. There is, in short, nothing inherently good or evil about technologies: their effects are contingent, dependent on the intentions of those who use them and the power relations that enable or constrain their deployment. Moreover, new technologies frequently have unintended consequences (Tenner 1997).

There are numerous superb histories of technology that portray in depth the multiple ways in which technologies arose, their movements across and within cultures, and their innumerable social, economic, and scientific consequences. World histories abound (Pacey 1991; Cardwell 1995; McClellan and Dunn 2006; Headrick 2009; Friedel 2010), while others focus only on the United States (Pursell 1995). Influential historian William McNeill (1982) focused on the role of military technology during and since the medieval era, while Headrick (1981, 1988) detailed how technologies enabled European imperialism. David Landes's (1993) magisterial *The Unbound Prometheus* still stands as the definitive history of technological change during the Industrial Revolution. At a very different spatial scale, authors such as Cowan (1983) reveal how technologies have reshaped the meaning of housework, and not entirely in ways that liberate women. Many other histories can be found easily. This vast corpus of work serves to show how technologies are deeply, inevitably *social* in nature, that they are wrapped up in relations of power and culture, and that their effects vary enormously over time and space: historicizing technology is the antidote to technological determinism.

Technologies have clear implications for gender relations (see Chapter 3), both reflecting and shaping the power differences between men and women. Traditionally, machinery was a man's world, and men enjoyed disproportionate advantages from things such as automobiles (Oldenziel 1999). The Internet is used by more men than women in many countries. Yet, as an insightful stream of feminist research has illustrated, it is not enough to point out the differential uses and effects of technologies. Rather, jettisoning dichotomies such as male/female or human/non-human has led feminists to theorize technologies in new and creative ways (Haraway 1991; Wajcman 2010).

Economists have long celebrated technological change as a major driver – if not the driver – of productivity growth and rising standards of living (Helpman 1998; Archibugi and Filippetti 2015). In this view, the dynamism of market-based economies unleashes round upon round of Schumpeterian 'creative destruction' as firms innovate and adopt new technologies. This process is widely held to have given the West a decisive advantage over other parts of the world, as argued by Jared Diamond in his hugely popular but controversial book *Guns, Germs, and Steel* (1997), a discrepancy that accelerated in the 19th century (Allen 2012) and still accounts for global differences in growth rates today (Fagerberg 1994).

There are, of course, also multiple, complex and contingent geographies of technology, just as there is a geography of everything else. Vast literatures have been dedicated to the subject. Entire regions are named after specific technologies (Silicon Valley, Steel Belt). The global expansion of capitalism and the forging of a world-system were integrally intertwined with the acceleration of technological change (Hugill 1993). Historically and at the present moment, technologies are bound up in geopolitics, including the Cold War (Hecht 2011). The invention and adoption of new technologies are intermingled with the uneven geographies of science, as Livingstone's (2003) careful analysis of Enlightenment science illustrates. Geographers study technology from several conceptual perspectives, although Science and Technology Studies (STS) has become perhaps the dominant mode (Jasanoff et al. 1995; Truffler 2008; see Chapter 4 this volume). STS attempts to overcome traditional empiricist interpretations of technology by embedding it within shifting networks of people, practices, and power, emphasizing the contingent nature of scientific discovery, innovation, and adoption. Much geographical work has focused on which places are innovative, and which are not, and the reasons that underpin these differentials (Fagerberg 2006). Technological innovation is highly uneven, typically concentrated in large cities; density, it appears, is key to the social production of creativity (Boschma 2005; Gordon and McCain 2005). Knowledge spillovers represent a kind of technological diffusion in this regard. Indeed, because technologies diffuse unevenly over time and space, diffusion has been a core geographic concern (Rogers 2003; Robertson and Patel 2007; Robertson and Jacobson 2011; see Chapter 2 this volume). The impacts of technologies are unevenly felt: for example, labor-saving agricultural technologies may enhance productivity in temperate grasslands environments in the developed world but increase unemployment in tropical environments in the developing world. Others focus on how transportation and communications technologies lead to massive time–space compression and the creation of new geographies of centrality and peripherality (Kirsch 1995; Warf 2008).

The discipline of geography is also, of course, shaped by and in turn a producer of technologies. One collection of essays, *Geography and Technology* (Brunn et al. 2004), is more focused on technology's impacts on the discipline of geography rather than the geographies of technological change in society at large. Earlier generations relied on maps, globes, and compasses, which enabled the exploration and conquest of the globe (McDonald and Withers 2016). Geographical information systems (GIS), or more broadly, geospatial technologies that include remote sensing and global positioning systems, have been an extremely important example of the discipline's contributions to technological change, revolutionizing not only academic geography but also applied fields such as marketing and urban planning.

The *Handbook on Geographies of Technology* is an attempt to provide meaningful insights into a series of technologies, both old and new, that generate important social and spatial repercussions. The focus of this volume is not so much geography as a discipline but on how key technologies have been deployed to shape the world at large. Its goal is to elucidate the multiple and complex means by which technologies come into being, their social uses and misuses, how they shape landscapes and social formations, and the ideologies and politics that swirl in their wake. Obviously, given the plethora of changes that have occurred over the last few decades, it cannot hope to cover all relevant technologies. For example, missing from this volume (among others) are discussions of wind energy, nuclear energy, fusion energy, lasers, and submarines; alas, too few geographers study these topics. Geographic Information Systems have received so much attention elsewhere that they are not addressed here.

SKETCH OF THIS VOLUME

The volume is divided into seven sections, one of which is conceptual in nature while the others are concerned with a cluster of related technologies. In Part I, three approaches to understanding geography and technology are proffered. Chapter 2, by Paul L. Robertson, focuses on technological diffusion and transfer, a long-standing concern for geographers. Robertson analyzes this issue at several scales, ranging from individual organizations to the global economy. Far from a simple linear path from science to development to diffusion, he shows that the process is much more complex and path-dependent, involving the uneven movement of different types of knowledge, external returns and spillovers, outsourcing, and differential ability to incorporate new techniques. At the social level, rates and patterns of diffusion reflect different national propensities to innovate, the size and level of integration of networks of firms and individuals, and the presence or absence of industrial clusters. International movements of knowledge are even more complex, with complicated distributions for its export and import that function with varying degrees of effectiveness, including foreign direct investment.

In Chapter 3, Jessica McLean, Sophia Maalsen, and Alana Grech turn to the question of gender and technology. Various feminist perspectives highlight how technologies are embedded in the power relations that form the core of gender relations, an important means of noting that technologies are much more than simply objects. Opportunities for women in technologically advanced fields have traditionally been limited. Moreover, feminism helps to overcome simple dichotomies such as human/machine that have long underpinned masculinist understandings, and open the door to relational and post-human understandings. They conclude with a case study of Destroy the Joint, a feminist online group, to assess feminist geographical research in cyberspace.

The fourth chapter, by Jordan P. Howell, summarizes the literature on STS, perhaps the most popular mode for theorizing science and technology today within the social sciences. Born of the post-structural turn that celebrates positionality, embodiment, and relational interpretations – particularly the work of Bruno Latour – STS emphasizes networks of actors (both human and non-human) in the production of scientific knowledge. Howell critically summarizes the origins and evolution of STS, its leading journals, and major conceptual debates, including Actor-Network Theory. This approach profoundly socializes science, leading Howell to examine related issues such as the influence of industry and the state on the construction of scientific knowledge, as well as the public's understanding and science education. He concludes by pointing to the geographic implications of this line of thought.

Part II addresses a series of computational technologies. As capitalism has become ever more information-intensive in nature, a process manifested in the steady, inexorable rise of services the world over, technologies to collect, process, and transmit information have grown accordingly. Martin Dodge, in Chapter 5, delves into the reciprocal relations between software and space: so pervasive has code become that contemporary geographies are inconceivable without it. Code turns the world into algorithms and databases, foregrounding some issues and backgrounding others. Dodge penetrates the taken-for-granted nature of software to explore the discourses that surround it, how it animates ever-larger legions of objects to give them almost lifelike qualities. His geographic exploration notes how code is embedded in a hierarchy of phenomena ranging from individual objects to coded infrastructures and processes. The final sections delineate code in spaces such as the home to the surveilled self.

Chapter 6, by Daniel Sui, offers a comprehensive look at location-based services (LBS), those that deploy users' spatial locations to provide individually tailored outcomes. As networked devices become increasingly common, the LBS industry has grown in size and influence. Sui summarizes the technical aspects of LBS, including RFID tags, and then turns to key applications. For individuals, LBS not only offers convenient information, but can also be used to track children or people with dementia. For businesses, LBS has become central to the so-called 'sharing economy' (e.g. Uber) as well as marketing and geofencing to delineate specified areas digitally. Governments also use LBS, such as for emergency management or to deploy citizens as sensors. Sui also looks at concerns about LBS such as privacy, inequality, and environmental sustainability.

In Chapter 7, Michael Batty, Hui Lin and Min Chen describe the geographic dimensions of virtual reality. As the real and the virtual worlds become more intertwined, virtual reality has become ever more sophisticated and lifelike, engaging users interactively. The chapter traces the history of virtual reality systems, and notes the various types such as standalone and networked systems. The primary focus is on virtual reality representations of cities, although they also discuss virtual geographic environments. Virtual reality systems have become commonplace, and are widely used in planning and other applications. Finally, the chapter turns to how the virtual and real worlds can be blended as virtual data are projected back into the world, such as with augmented reality.

Part III concerns communications technologies, arguably the most dynamic sector of contemporary capitalism. The ongoing aftermath of the microelectronics revolution, computers, and the digitization of information has been so unprecedented that it is almost impossible to document these changes in their entirety. In Chapter 8, Barney Warf describes fiber optics – by far the most important telecommunications medium in the world, forming the core technology that underpins the Internet as well as electronic funds transfer systems. Warf summarizes the history of fiber optics and situates it within the contemporary information-intensive global economy. He points to the urban implications of fiber, and maps the world's major systems that emerged over the last three decades. Finally, the chapter turns to some of the impacts of the massive global boom in fiber capacity, including the dot-com crash, excess capacity, and the steady erosion of the satellite industry.

Today, roughly 50% of the world's population uses the Internet, perhaps the defining technology of our historical moment. Chapter 9, by Aharon Kellerman, notes how the Internet came to be, and the primary types of applications, including mobile Internet usage. He emphasizes that the Internet is deeply geographical, including the location of users and the screens that allow them access. The spatiality of the Internet is also evident in the movement of information through that medium, including the widespread use of open code. The impacts of the Internet on physical space – making life safer, faster, and more convenient for many – also speak to its geographic nature. Kellerman also writes of the Internet as action space, in which it substitutes for physical movements. Finally, he notes that the Internet is inevitably shaped by local cultures; abstract as cyberspace may appear, it is not independent of the physical and social realities that it reflects and in turn affects.

Radio is such a long-standing technology that it may appear unworthy of attention;

geographers have written remarkably little about it, preferring to study visual media. Yet as Catherine Wilkinson shows in Chapter 10, the soundscapes of radio are important in several ways. She offers a brief history of radio, from its infancy in the 1920s to the explosion in usage in the 1960s, when transistors made it portable. Today radio is an intimate part of everyday life, a major source of news and entertainment. Traditionally, the geographies of radio were bound by the transmission capacities of stations: it has long been primarily a community medium, and she stresses that it helped to forge 'imagined communities' at that scale. In the digital age, the spatiality of radio has undergone a sustained transformation, including podcasts, which greatly expanded the medium's spatial reach, creating complex new sonic geographies.

Chapter 11 concerns satellites, which have had a series of economic, military, and discursive implications. Here, Barney Warf defines the oft-confused terms concerning satellites and Earth stations, then turns to the history of the technology. Much of the chapter is concerned with the international regulation of geostationary satellites, a story that traces the rise and demise of the International Satellite Organization (Intelsat) and several regional competitors. As neoliberalism has reshaped telecommunications, like everything else, Intelsat's power has eroded, and private satellite operations have risen in importance. Finally, Warf notes the powerful impacts of fiber optics on the satellite industry and the hopes presented by low-orbiting satellites that service the world's mobile phones.

Cellular or mobile phones have become increasingly ubiquitous worldwide: 70% of the planet now owns one. Jonathan C. Comer and Thomas A. Wikle summarize this technology in Chapter 12. Far from being simply devices for talking, smart mobile phones allow Internet access, photography, video, and other applications. The impacts of mobile phone adoption are monumental. They note that it has diminished the importance of physical distance, a common consequence of telecommunications. More people than ever before can now communicate over long distances and search for information, a process that has blurred the boundaries between public and private spaces. The chapter traces the evolution of the cellular concept and the global diffusion of mobile telephony, mapping its growth over time and space. They also explore the factors that lead to cell phone adoption, paying particular attention to the developing world.

In Chapter 13, Ramon Lobato addresses the changing nature of television, not a new technology to be sure but surely one of the most influential. The digital revolution thoroughly altered the landscapes of television, as witnessed by the rise of Netflix, which he uses to explore contemporary geographies of the medium. Noting that television involves a bundle of technologies, he also cautions that the medium is embedded in multiple geographies simultaneously: the individual viewer, the infrastructure, flows of culture across borders, and so forth. The streaming infrastructure that makes Netflix possible has changed how people watch TV. The chapter also explores the changing distribution of content distribution, which has altered the relationship between programming and place. Finally, he turns to television platform spaces, the interface between users and their screens, in which complex algorithmic structures become intertwined with viewers' consciousness.

In Part IV, five transportation technologies are examined. Some, such are railroads, are relatively old, while others, such as drones, are products of the 21st century. Capitalism has long sought to conquer space by means of more rapid movements of people and goods, a process Harvey (1982) famously attributed to the constant need to minimize the turnover rate of capital and produce successive new 'spatial fixes'. Initiating this section is Chapter 14, by Aaron Golub and Aaron Johnson, who write about automobility, or the geographies created by the world's one billion cars. The world today would be unthinkable without the automobile, which shapes cities, production, consumption, trade, and everyday life in countless ways that vary greatly by class, gender, ethnicity, and place. It is a major consumer of energy and producer of CO_2. Few innovations can rival it in importance. Drivers are enmeshed in complex systems of automobility that greatly transcend driver and car, but form, as Golub and Johnson note, assemblages of people, things, ideas, and power. They trace the history of automobility, how it varied over time, and then proceed systematically to uncover the various systems that enter into its making, such as government policies, household behavior, and planners and developers. They also explore the infrastructures, including global flows of petroleum, which are essential to the mobility enjoyed by so many. Finally, they offer a useful summary of the externalities imposed by driving, including fatalities, air pollution, health impacts, and social inequality. They conclude by speculating on the nature of an auto-free future.

Aviation is the aerial equivalent to automobility. In Chapter 15, Andrew R. Goetz notes the historical development of this technology, which saw the Wright brothers' first flight eventually evolve into the Concorde. The changing regulatory framework that governs air travel also receives scrutiny, as does air freight. Goetz also examines conceptual issues pertaining to this industry, such as its role in time–space convergence (or compression) and globalization. Next he turns to the impacts of deregulation and the rise of low-cost carriers, which increased competition and gave rise to the familiar hub-and-spoke pattern we see today. Finally, Goetz examines recent trends in aviation and the associated geographies that accompany them, as assessed by airlines and airports.

Drones have recently surfaced as one of the most ominous – yet simultaneously promising – technologies. In Chapter 16, Thomas Birtchnell studies the role these machines play in military and civilian life, their definition, history, and much-debated role in conflicts, where they have revolutionized warfare. Yet drones have wide non-military uses as well, such as delivering cargo, nature conservation (e.g. keeping an eye on poachers), and emergency management. Concerns about privacy and safety loom large in this context. Many researchers also use drones, which have, among other things, facilitated the growth of volunteered geographic information.

Since the Industrial Revolution, railroads have been an important form of transportation within and among cities, albeit one often overlooked by geographers. Chapter 17, by Linna Li and Becky P.Y. Loo, explicates the dynamics of this technology at several spatial scales. Recent years have witnessed a railroad renaissance, including high-speed trains. Li and Loo's chapter examines the global distribution of railroads, then delves into their geographical implications, such as increased regional integration. The governance and financing of rail systems vary considerably among nations, as does their integration with other forms of transportation.

Shipping moves most of the world's goods. In Chapter 18, Jean-Paul Rodrigue notes that this ancient technology has been utterly modernized since the advent of containerization in the mid-20th century, which dramatically reduced shipping costs. The geographies of shipping networks reflect both the shifting landscapes of global capitalism and physical constraints (e.g. the Malacca Straits). Enormous undertakings such as the Suez and Panama Canals are also testimony to capitalism's incessant need to remake landscapes to accelerate the movement of capital, goods, and people. Rodrigue notes that the push for economies of scale has led to stunningly large 'post-Panamax ships' capable of carrying vast quantities of cargo, further driving down costs. Finally, he turns to ports and the multiple ways they have been woven into their hinterlands, adopted automation, and cultivated supply chains.

Part V concerns itself with a series of technologies related to the production and use of energy in different forms. Absolutely essential to the functioning of advanced divisions of labor, energy technologies have grown in diversity and complexity over time. In Chapter 19, by Kirby E. Calvert, Jamie D. Stephen, M.J. Blair, Laura Cabral, Ryan E. Baxter, and Warren E. Mabee, biofuels are given due consideration. An important alternative to fossil fuels, biofuels utilize portions of animal feed, food, and pulp production that otherwise would go to waste. Liquid biofuels include bioethanol and biodiesel. Using evolutionary economic geography, their chapter draws attention to the changing supply chains of biorefining as a means of revealing how economies and environments presuppose one another. They proceed in three steps: first, by examining the pathways of biofuels and products in the production process; second, by examining the implications of biorefining in light of regional development and land use; and three, undertaking an empirical survey of existing patterns of biorefining.

Dams are the focus of Chapter 20, in which Marcus Nüsser and Ravi Baghel shine light on the 45,000 projects that have fragmented half of the world's major rivers, with profound ecological and economic effects. They classify these hydroscapes and unearth how they were produced historically, which typically involved constellations of power and often bitter disputes. Beyond the dam-building industry, with legions of contractors and engineers, national governments were often involved, viewing dams as signs of modernization, as well as international entities such as the World Bank. Dams are often geopolitically important, as when they restrict flows of water between countries. Rich in examples, Nüsser and Baghel's chapter also touches on related issues such as neoliberalism and climate change.

Fracking, or the exploitation of shale gas reserves, has become one of the most contentious energy-related issues in the world. New technologies have made once-unprofitable fields open to exploitation. In Chapter 21, by Peter Jones, Daphne Comfort, and David Hillier, fracking in the United Kingdom is explored in depth, a case study that illuminates the technology and politics of the procedure in many places. They situate British fracking within changing manifolds of global energy supply and demand as well as wider debates about energy security. They also explain the technical dimensions en route to understanding why many regions have adopted fracking. In the British context, they focus on potential shale gas reserves. The environmental risks are explored at length, from local footprints to climate change. They also discuss fracking's poor reputation and why so many people are fearful of it, which has resulted in heated opposition. Such controversial processes invite government regulation and planning, which they also summarize.

Geothermal energy, the topic of Chapter 22 by Edward Louie and Barry Solomon, has become an attractive alternative to fossil fuels. The authors summarize the literature on this topic, including a variety of environmental, land use, and regulatory issues, then move on to pressing conceptual debates. Is geothermal energy renewable? Is it clean? Is it sustainable? Next they address geographic issues pertaining to this energy source, including its role in a variety of uses such as electricity generation, noting that there remain underutilized sources.

Julie Cidell's chapter (23) on Leadership in Energy and Environmental Design (LEED) buildings is apropos of geographic work on energy conservation. She provides a history of these 'green' buildings, then examines four dimensions: their spatial distribution, the economics of implementing and maintaining them in light of the extra costs incurred, the social aspects (their valuation and uses) and their environmental facets (just how green are they?).

Pipelines are another essential, and efficient, feature of the energy landscape, particularly for natural gas. In Chapter 24, Jeff D. Makholm notes that, while the technology does not vary much among regions, the institutional environment that surrounds them certainly does. First, Makholm addresses pipeline costs and their ties to the energy markets they serve. Next he delves into the technologies of these natural monopolies with significant barriers to entry, in which pressure and distance figure prominently. Third, he turns to market problems of pipelines, whose capital is immobile despite shifting resource patterns and are the topic of government regulation. Frequently pipelines are protected from competition, leading to odd pricing systems. In short, while pipelines may appear simple, or as he notes, not romantic, they lie at the core of complex systems of markets, governments, and geopolitics.

Another alternative to fossil fuels is solar energy, which recently has grown rapidly in popularity. Govinda Timilsina and Lado Kurdgelashvili, in Chapter 25, examine the dynamics of solar energy in depth. Government subsidies are the norm. They begin by charting the evolution of solar energy technologies from their modest beginnings as a way to cook food and heat water to the gradual adoption of solar heaters in a variety of countries. They note its use in electricity generation and explosive growth of photovoltaics. The popularity of solar has, not surprisingly, often fluctuated in inverse proportion to the price of fossil fuels. Next they turn to the evolution of markets for this technology, notably China, the world's largest producer of solar equipment. The largest single use is for heating in residential homes. They also look at various national policies to encourage the growth of solar energy, some adopted with an eye toward climate change, which have led to a precipitous decline in the cost of this technology.

In Part VI, three manufacturing technologies are explored. Just-in-time (JIT) delivery systems have been a hallmark of post-Fordist production, and are explored by Ruth Rama and Adelheid Holl in Chapter 26. They note that, in contrast to most technologies explored in this volume, JIT is a 'soft' technology that consists of procedures and processes. Japanese in origin, it has become widely deployed. They examine its applicability in other contexts, unpacking the issue of whether its adoption is spatially homogeneous or not. Next they turn to the question of whether JIT promotes the clustering of firms, in part because vertically disintegrated production complexes deploy it extensively. Finally, they compare the adoption of JIT with that of other technologies, such as CAD/CAM systems.

Few technologies capture the popular imagination as much as robots, a term that dates back to 1917. In Chapter 27, Antonio López Peláez covers every feature of robots, from Isaac Asimov's three laws to their contemporary use in eldercare. Since the 1950s, industrial robots have grown widely in the number and importance of their applications, particularly with the advent of the microprocessor. In manufacturing, they have contributed greatly to the decline in the demand for labor. Service robots assist people (e.g. cleaning) but do not manufacture goods. Military robots are revolutionizing warfare. He also explores conceptual issues swirling around robots: few phenomena so poignantly illustrate the possibility of post-human life. Political debates also revolve around robots; while some envision emancipatory possibilities, others see them as a threat to the labor force.

Geographies of the extremely small – nanotechnology – are the subject of Chapter 28, by Scott W. Cunningham. The ability to manipulate matter at the molecular level holds great promise for material science and industrial chemistry, with broad applications in production, health care, biotechnology, and environmental management. Research in this area is funded by both private and public organizations, and universities play a key role. Globally, advanced economies invest the most and are likely to reap the greatest benefits of nanotechnology, and within some countries, such as the United States, emerging nanodistricts are unfolding. Because the industry is in its infancy, the long-term impacts are unclear.

In Part VII, three technologies in the life sciences are addressed. Barney Warf, in Chapter 29, focuses on the biotechnology industry, the molecular and genetic modification of living organisms. He traces its history, from beer making to cloning. Next he turns to its impacts, including the contentious issue of genetically modified organisms (GMOs), perhaps biotech's most famous product, as well as biofuels and uses in manufacturing and health care (e.g. gene therapy). Third, he examines the regulatory impacts at the global, national, and local scales. The fourth part unearths the economic geography of biotech districts, the life sciences' equivalent of new industrial spaces.

New technologies in health care – as described by Mark W. Rosenberg and Natalie Waldbrook in Chapter 30 – are viewed through two perspectives: how geographers have taken them up in their research, and how these technologies are creating new health care landscapes. In the first view, GIS has become instrumental in mapping diseases, understanding various populations and their contexts, and in health care planning (including emergency responses), all of which are facilitated by the rise of national health databases. In the second view, innovations such as telemedicine and virtual care are redefining how health care is provided and to whom; they also focus on the implications for understanding the health geographies of the elderly.

Finally, in Chapter 31 Gabriel Popescu examines biometrics, the digital measurement of individual's unique characteristics to ascertain their identity (e.g. with facial and fingerprint recognition technology). From iPhones to airports to daycare centers, biometrics have been evermore widespread. Understandably, the technology has aroused fear, suspicion and opposition, often over concerns regarding privacy. Popescu summarizes the technicalities of biometrics and critically discusses the ramifications. There are clear geographical implications from this manner of digitally scripting the body, including the changing meaning of borders (i.e. airports) and the ability of the state to restrict mobility.

REFERENCES

Allen, R. 2012. Technology and the great divergence: global economic development since 1820. *Explorations in Economic History* 49(1): 1–16.

Ambrose, S. 2001. Paleolithic technology and human evolution. *Science* 291(5509): 1748–1753.

Archibugi, D. and A. Filippetti (eds). 2015. *The Handbook of Global Science, Technology and Innovation*. Chichester: Wiley.
Bijker, W., T. Hughes and T. Pinch. 1987. *The Social Construction of Technological Systems: New Directions in the Sociology and History of Technology*. Cambridge, MA: MIT Press.
Bimber, B. 1990. Karl Marx and the three faces of technological determinism. *Social Studies of Science* 20(2): 333–351.
Boschma, R. 2005. Proximity and innovation: a critical assessment. *Regional Studies* 39(1): 61–74.
Brunn, S., S. Cutter and J.W. Harrington (eds). 2004. *Geography and Technology*. Dordrecht: Kluwer.
Cardwell, D. 1995. *Wheels, Clocks, and Rockets: A History of Technology*. New York: W.W. Norton.
Carr, N. 2010. *The Shallows: What the Internet is Doing to Our Brains*. New York: Norton.
Cowan, R. 1983. *More Work for Mother: The Ironies of Household Technology from the Open Hearth to the Microwave*. New York: Basic Books.
Diamond, J. 1997. *Guns, Germs, and Steel*. New York: W.W. Norton.
Eisenstein, E. 1979. *The Printing Press as an Agent of Change*. New York: Cambridge University Press.
Fagerberg, J. 1994. Technology and international differences in growth rates. *Journal of Economic Literature* 32(3): 1147–1175.
Fagerberg, J. 2006. *The Oxford Handbook of Innovation*. Oxford: Oxford University Press.
Friedel, R. 2010. *A Culture of Improvement: Technology and the Western Millennium*. Cambridge, MA: MIT Press.
Friedman, T. 2005. *The World is Flat: A Brief History of the 21st Century*. New York: Picador.
Gordon, I. and P. McCain. 2005. Innovation, agglomeration, and regional development. *Journal of Economic Geography* 5(5): 523–543.
Green, L. 2001. *Technoculture: From Alphabet to Cybersex*. Crows Nest: Allen & Unwin.
Haraway, D. 1991. *Simians, Cyborgs, and Women: The Reinvention of Nature*. London: Free Association Press.
Harvey, D. 1982. *The Limits to Capital*. Chicago, IL: University of Chicago Press.
Headrick, D. 1981. *The Tools of Empire: Technology and European Imperialism in the Nineteenth Century*. New York: Oxford University Press.
Headrick, D. 1988. *The Tentacles of Progress: Technology Transfer in the Age of Imperialism, 1850–1940*. New York: Oxford University Press.
Headrick, D. 2009. *Technology: A World History*. Oxford: Oxford University Press.
Hecht, G. 2011. *Entangled Geographies: Empire and Technopolitics in the Global Cold War*. Cambridge, MA: MIT Press.
Helpman, E. (ed.). 1998. *General Purpose Technologies and Economic Growth*. Cambridge, MA: MIT Press.
Huesemann, M. and J. Huesemann. 2011. *Technofix: Why Technology Won't Save Us or the Environment*. Gabriola Island, BC: New Society Publishers.
Hugill, P. 1993. *World Trade since 1431: Geography, Technology and Capitalism*. Baltimore, MD: Johns Hopkins University Press.
Jasanoff, S., G. Markle, J. Petersen and T. Pinch (eds). 1995. *Handbook of Science and Technology Studies*. Thousand Oaks, CA: Sage.
Kirsch, S. 1995. The incredible shrinking world? Technology and the production of space. *Environment and Planning D: Society and Space* 13(5): 529–555.
Landes, D. 2003. *The Unbound Prometheus: Technological Change and Industrial Development in Western Europe from 1750 to the Present*. Cambridge: Cambridge University Press.
Livingstone, D. 2003. *Putting Science in its Place: Geographies of Scientific Knowledge*. Chicago, IL: University of Chicago Press.
McClellan, J. and H. Dunn. 2006. *Science and Technology in World History: An Introduction*. 2nd ed. Baltimore, MD: Johns Hopkins University Press.
McDonald, F. and C. Withers (eds). 2016. *Geography, Technology and Instruments of Exploration*. London: Routledge.
McNeill, W. 1982. *The Pursuit of Power: Technology, Armed Force, and Society since A.D. 1000*. Chicago, IL: University of Chicago Press.
Oldenziel, R. 1999. *Making Technology Masculine: Men, Women, and Modern Machines in America*. Amsterdam: Amsterdam University Press.
Pacey, A. 1991. *Technology in World Civilization: A Thousand-year History*. Cambridge, MA: MIT Press.
Postman, N. 1985. *Amusing Ourselves to Death: Public Discourse in the Age of Show Business*. New York: Viking.
Postman, N. 1992. *Technopoly: The Surrender of Culture to Technology*. New York: Knopf.
Pursell, K. 1995. *The Machine in America: A Social History of Technology*. Baltimore, MD: Johns Hopkins University Press.
Robertson, P. and D. Jacobson (eds). 2011. *Knowledge Transfer and Technology Diffusion*. Cheltenham, UK and Northampton, MA, USA: Edward Elgar Publishing.

Robertson, P. and P. Patel. 2007. New wine in old bottles: technological diffusion in developed economies. *Research Policy* 36(5): 708–721.

Rogers, E. 2003. *Diffusion of Innovations.* New York: Free Press.

Smith, M. and L. Marx (eds). 1994. *Does Technology Drive History? The Dilemma of Technological Determinism.* Cambridge, MA: MIT Press.

Staudenmaier, S.J. and M. John. 1985. *Technology's Storytellers: Reweaving the Human Fabric.* Cambridge, MA: MIT Press.

Tenner, E. 1997. *Technology and the Revenge of Unintended Consequences.* New York: Random House.

Truffer, B. 2008. Society, technology, and region: contributions from the social study of technology to economic geography. *Environment and Planning A* 40(4): 966–985.

Wajcman, J. 2010. Feminist theories of technology. *Cambridge Journal of Economics* 34(1): 143–152.

Warf, B. 2008. *Time–Space Compression: Historical Geographies.* London: Routledge.

White, L. 1966. *Medieval Technology and Social Change.* New York: Oxford University Press.

PART I

CONCEPTUAL ISSUES

2. Technological diffusion in local, regional, national and transnational settings
Paul L. Robertson

Technological diffusion lies at the root of modern economic growth. If knowledge had not spread across industries and nations from its original locations, the unprecedented changes in productivity and living standards of the past 250 years that have favourably affected most of the world, including nations that still remain relatively underdeveloped today, would never have occurred. There would nevertheless have been advances in particular industries, as have happened throughout recorded history, but they would not have led to the cumulative transformation that since the Industrial Revolution has seen not only the rise of new industries, but tremendous advances in the oldest and most traditional sectors such as agriculture and mining.

Diffusion and knowledge transfer have not been assured, however, because they involve complicated processes that are sometimes beyond the ability of humans to manage quickly and efficiently. Technological factors obviously underlie diffusion, but adverse geographical, social, political and economic elements can block progress even when it is technically feasible. As a result, it is not surprising that technological transformation has remained uneven and even mysterious in some respects, but this only further emphasises the need to improve our understanding in order to allow greater proportions of humanity to be supported by technological change – or at least by its most beneficial aspects.

In this chapter, diffusion is analysed in order of increasing aggregation. In the next section, the basics of diffusion and technology transfer are outlined and change is examined at the individual and firm levels. The third part introduces geographical and social features of knowledge transfer to demonstrate the effects of proximity and industrial concentration on diffusion. The fourth section looks at the mechanisms that affect knowledge transfer at the global level, especially between developed and developing economies, followed by some concluding remarks in the final section.

ORGANISATION-LEVEL DIFFUSION

The Relationship Between Innovation and Diffusion

Technological diffusion occurs when an existing technological artefact or concept is used for a different purpose, by a different person or organisation, or in a different location than it has been previously used. In a traditional linear model (Godin 2006), diffusion is presented as following from innovation – from the development of *totally new* concepts or machinery:

Science → Development → Innovation → Diffusion

17

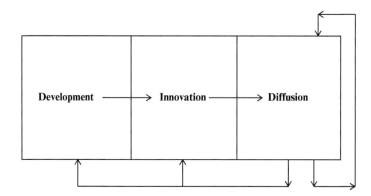

Figure 2.1   Paths of diffusion

In this formulation, potential adopters evaluate an innovation (Rogers 2003) on the basis of criteria such as the new product's relative advantage in comparison to existing alternatives and to those that might appear in the near future, its compatibility with existing technical and social frameworks, the innovation's complexity as it affects ease of adoption and use, the ability of a potential adopter to test the innovation before making a commitment (its trialability), and the extent to which the effects of potential adoption can be observed in advance. Depending on their individual assessments, would-be users will then adopt the innovation at various rates, some immediately, some fairly quickly, others after a considerable lag, and still others not at all (Rogers 2003).

However, the route followed by diffusion is often less straightforward. Far from being separate, innovation and diffusion can coincide, with diffusion leading to additional rounds of innovation as part of a protracted process that involves the repeated reuse of ideas and artefacts that have been employed for extended periods, perhaps decades, in other contexts. Moreover, diffusion can foster diffusion, as potential new uses are identified, not from the original innovation, but from subsequent reuses (Figure 2.1).

Consequently, innovation has multiple meanings. Of these the most restrictive, and one of the less useful from the standpoint of measuring its impact, is that an innovative object, concept or procedure must be 'new to the world', as is implied in the linear model (Kline and Rosenberg 1986). When an innovation is developed to deal with a contingency that is so narrow that it has no other uses, its impact on the wider environment will be muted. Many innovations have wider ranges of applicability, however, which allow them to be introduced with appropriate modifications into numerous environments where their effects proliferate and can eclipse their original use as they also become 'new to the industry' or 'new to the firm'. The innovations with the most extensive ramifications are termed General Purpose Technologies (GPTs) (Lipsey et al. 2005; Helpman 1998). Lipsey et al. (2005, 3) define GPTs as 'pervasive technologies that . . . transform a society's entire set of economic, social and political structures', and have been able to identify only around 25 in the last 11,000 years. Of these, the earliest were the domestication of plants and animals around 8000–9000 BCE and the most recent was the development of the Internet and of bio- and nano-technologies. The remainder include three forms of materials and material processing, six means of transportation and three ways of generating power.

Innovation and Economic Performance

Notwithstanding their undoubted importance, GPTs account for only a small proportion of the impact of technological advances in which the diffusion process has been one of the major drivers of economic change. There have been many narrower, but still impressive, examples of diffusion that, like GPTs, have been complicated, initially dependent on existing innovations, and that also acted as a spur to further innovation in other fields. As a result of their snowballing effects, and despite the current popularity of disruptive innovation (Christensen 1997), it is the diffusion and reuse of incremental non-general purpose technologies that are responsible for much of the growth in both developed and developing economies, especially since further uses increase the pay-off to the research and development activities that underpin new-to-the-world innovations, validating past expenditures on R&D and encouraging new investments that generate ongoing innovation cycles (Robertson et al. 2003).

Modern industrial economies are products of history that mix new elements with others from the past. Branches of agriculture, the oldest industry, and of microelectronics, one of the newest, both contribute enormous value to modern economies, alongside a wide diversity of industries in fields like construction and the extraction of natural resources, and in other types of manufacturing which may be high- or low-technology,[1] or more likely some combination of both (Robertson and Patel 2007). In recent decades, these traditional mainstays of economic performance have been overtaken by the miscellany of activities known as services that now comprises 70 per cent or more of the output of modern economies in North America, Europe and East Asia (Randhawa and Scerri 2015). Services industries include high-income fields dependent on high levels of skills and knowledge such as medicine, information and computer technology, and financial services; cleaning and other very traditional and unskilled activities; and a broad array of pursuits of intermediate levels of sophistication in areas including distribution and transportation.

Important new-to-the-world innovations generally originate in only a few manufacturing and service industries, including, in recent decades, electronics and, to a lesser extent, finance, but the use of these innovations extends much further as they are adopted in one form or another in other sectors. Although only perhaps 5 per cent of output and employment in even the most innovative economies originate in high-technology industries (Robertson and Patel 2007), the bulk of the influence of these industries derives from the use of high-tech products throughout developed and, to a reduced but still important extent, developing economies.

Diffusion as Knowledge Transfer

The importance of linkages between high-technology and low- and medium-technology (LMT) industries is long established in economic growth models (Hirschman 1958; Rostow 1960). These connections between sectors are best viewed as a form of knowledge transfer (Hirsch-Kreinsen 2015; Jensen et al. 2007) in which different people or groups learn how to do new things – including planning, design and implementation – that they could not have done previously but that others could have. They accomplish this through the transmission between parties of knowledge that is embodied, codified or tacit (Ancori et al. 2000; Johnson et al. 2002).

Embodied Technology

Embodied technology is an important vehicle for diffusion that occurs when innovative components and ideas are embedded in improved and more productive equipment that is purchased by LMT firms, generally in mature industries that Pavitt (1984) has termed 'supplier dominated' because they (are alleged to) engage in little development on their own and rely instead on producers of equipment to give them access to innovations. In many cases, embodied technology may include two stages of diffusion as the equipment suppliers have themselves adopted components such as electronic controls that originated in other firms or industries. Pavitt (1984) singles out cost cutting as the major motive for purchasing improved machinery, but in some circumstances established firms can also use better equipment to gain other strategic advantages by enhancing the range and quality of their own outputs.

Codified and Tacit Knowledge

Codified knowledge generally appears in written form, although oral codification is also possible; the acquisition of tacit knowledge, by contrast, may depend more on experiential learning to grasp the significance of events or processes that have not (yet) been codified. Learning innovative knowledge 'is seldom automatic – the idea of effortless "knowledge transfer" is normally misleading and a "prepared mind" helps a lot' (Jensen et al. 2007, 681), an observation that holds as well for reinvention, relearning and reconfiguration, which are also important aspects of diffusion and knowledge transfer. The possession of a high level of 'absorptive capacity' (Cohen and Levinthal 1989, 1990), which endows individuals and organisations with an ability both to learn and to understand the implications of new knowledge, can therefore be a substantial advantage in knowledge transfer because it permits the use of experiential learning not only to acquire but also to improve upon codified knowledge. People with prior involvement with particular classes of concepts and artefacts are at an advantage because they can approach new and somewhat familiar constructs from different angles than are likely to be open to people who have no prior mental models to apply when they confront something different. Participation in formal R&D is often cited as a good way for an organisation to build absorptive capacity because, in addition to experiential learning, it involves immersion in relevant technical literature (Cohen and Levinthal 1989, 1990), but for many organisations, particularly small ones, other ways of acquiring knowledge can be useful in working out ways of applying existing but new-to-the-firm techniques to solve problems. These are summarised by Lundvall and Johnson (1994) as know-what, know-why, know-how and know-who.

Know-what, know-why and know-how are all valuable when making analogies that allow problems to be solved in new ways by using techniques, or modifications of techniques, borrowed from other spheres. As was recognised centuries ago by Adam Smith (1937[1776]), people who are familiar with the advantages and disadvantages of certain products or ways of doing things are more likely than the uninitiated to be alert to possible improvements and, in some cases, to have greater incentives to implement changes. From the demand side, this can lead to innovation by analogy as solutions to similar problems elsewhere are applied to new uses (Franke et al. 2014; Enkel and Gassmann 2010; Kalogerakis et al. 2010).

Although thorough familiarity with procedures within one's own firm or industry can be indispensable, it can also be confining (Granovetter 1973) if important solutions arise from outside an immediate context. R&D and the preparation that goes into mastering a topic help to sidestep the problem, but similar if less formal paths are also open to organisations that do not have and perhaps could not afford R&D facilities (Huang et al. 2011). Alertness to innovations that have generated solutions to similar problems in different circumstances can lead to valuable, even radical, changes despite considerable cognitive distance between industries (Enkel and Gassmann 2010). From the supply side, open innovation can also help to diffuse knowledge across areas that might otherwise be cognitively distant. A great deal of attention is devoted to people with problems who are in search of solutions, but solution-holders can find it similarly difficult to locate others who can use their innovations (Robertson 1998). This notion is recognised by Chesbrough (2003) in his work on Open Innovation. Although 'inward' Open Innovation, in which firms search widely for solutions, has been researched thoroughly in the past decade, Chesbrough's original works were aimed as well at firms that do not bother to commercialise innovations that they have developed but do not meet their immediate needs, or – as in the case of Xerox PARC – at firms that are unimaginative in how they attempt to commercialise them (Chesbrough and Rosenbloom 2002; Chesbrough 2003). This suggests that diffusion can be improved by firms that are willing to disseminate their discoveries as widely as possible and to allow potential adopters to provide their own visions of what they want to do and how they intend to proceed.

External Returns and Spillovers

Diffusion and knowledge transfer are not necessarily deliberate because knowledge is inherently hard to confine and may 'leak' from its originators to others. The results, which Marshall (1920) called 'external returns' and are now known as 'spillovers', are controversial because they can have variable consequences. '[S]pillovers occur when someone's actions affect anyone else in either a positive or negative way and this effect is not [fully] *paid for* (in the case of a benefit) or [fully] *compensated* (in the case of a cost)' (Bureau of Industry Economics 1994, 7, emphasis in original). While access to cheap knowledge can benefit the recipients, and indeed society as a whole, it can also reduce the incentive to engage in innovative explorations if leakages diminish the returns to development activities to a level that does not cover their costs in an economic sense that includes a reasonable profit as well as the amounts invested. These concerns can be overstated (Langlois and Robertson 1996), and in any case spillovers do occur regularly when people acquire knowledge, sometimes only in snippets, that allows them to solve problems or otherwise improve their operations. This may involve geographic proximity, as discussed earlier, but this is not always necessary, especially in the age of the Internet. Agents with sufficient absorptive capacity, acquired through learning-by-doing and learning-by-using as well through R&D, can employ it to reengineer, and even improve on, existing innovations on the basis of knowledge that is inadvertently made publicly available in legitimate sources or through industrial espionage (Chen 2009).

Diffusion, Implementation and Further Innovation

Interaction between diffusion and innovation is heightened at the implementation stage. Both embodied and non-embodied diffusion frequently involve introducing change into existing frameworks – into contexts with rules and procedures that can be intricate and inflexible. From the standpoint of the adopters, these are incremental rather than radical innovations, associated with relatively minor modifications to products and processes, as when a piece of equipment is replaced because of obsolescence. Despite their incremental nature, the overall influence on productivity stemming from these changes to the 90 per cent or so of most modern economies that are classified as LMT is vital (Robertson and Patel 2007; Hirsch-Kreinsen et al. 2006). As a result, ways of overcoming barriers to implementing change in LMT sectors can remove major obstacles by decreasing the costs and time required to innovate.

To achieve compatibility between incremental innovations and existing plant and equipment and organisational frameworks, LMT firms may need access to knowledge on ways to *adapt* and *integrate* innovations that were originally intended for different purposes or to be used in different contexts. Capabilities that promote *adaptability* are required when a piece of equipment (or a concept or organisational form) developed for one purpose is used for another. For example, a common type of machine tool may need to be refined when used in a situation that requires tighter than normal tolerances or when it is applied to a different material. In such a case, diffusion demands not only that a machine be used in another way, but also that new knowledge be brought to bear. This knowledge can come from internal or external sources as suppliers may make the adaptations to secure new customers, or buyers may make changes themselves because they have inside knowledge of their operations that is too difficult to communicate or that they do not want to share for reasons of confidentiality (Robertson et al. 2012).

Integrative capabilities, on the other hand, may be needed to achieve compatibility between a new artefact or concept and an existing array of equipment or organisational forms and procedures. Prevailing patterns of balance and flow can be upset by introducing an innovative idea or piece of equipment. When this happens, a choice arises between discarding current arrangements or foregoing the innovation unless some means can be found of resolving the differences. As it is generally expensive to get rid of a whole range of equipment or to reorganise drastically, this sets a high standard for the performance necessary to justify upgrading a single item (Rogers 2003). Consequently, improved methods of adjustment between the old and the new can facilitate innovation. For instance, when a new machine works at a different pace than its predecessor, it might not fit efficiently into an existing production process if ways cannot be found of altering the new machine, the existing machines with which it is to be used, or both. These adjustments could entail physical modifications, but they are also likely to involve organisational changes in how machinery and workers are deployed in relation to each other. As with adaptive capabilities, integrative changes may therefore involve the creation of new knowledge, leading to additional incremental innovation that can then be further diffused (Robertson et al. 2012).

The adaptive and integrative capabilities associated with implementation therefore involve all of the types of knowledge identified by Lundvall and Johnson (1994). However, although know-what, know-why and know-how are central to problem solving, know-who has a special role in technology transfer, especially in LMT organisations with limited R&D capabilities. Absorptive capacity can be a great help in managing innovation, but it is also expensive to develop and it may not be sensible for organisations to acquire deep knowledge that will be seldom used (Winter 2003). Organisations that know-who can overcome at least part of the need for internal absorptive capacity by drawing on the expertise of others with relevant knowledge that can be tapped without the innovative organisation having to finance a full range of learning needed to gain knowledge that they might never need again and, in any case, quite possibly could not afford to acquire. Outsourcing knowledge acquisition by hiring consultants is a long-standing way of avoiding overinvestment and is a very useful means of knowledge transfer when consultants are able through analogy to apply learning gained from their work with other clients (Franke et al. 2014; Kalogerakis et al. 2010). Heavy dependence on outsourcing can be dangerous, however, as consultants and other outsiders cannot be expected to know an organisation's business in the same depth as its own managers. This means that innovative firms need to retain control over the introduction and use of innovations (Brusoni et al. 2001). Accordingly, the transfer of existing knowledge and the generation of new knowledge when adapting and integrating often entail co-development between the adopter and suppliers or consultants (Appleyard 2003; Edvardsson et al. 2010).

INSTITUTIONAL, GEOGRAPHICAL AND SOCIAL INFLUENCES ON DIFFUSION

The context in which diffusion takes place can be an important factor in determining the extent and spread of knowledge transfer because knowledge is situated both socially (Nidumolu et al. 2001) and geographically. The presence of strong or weak ties (Granovetter 1973) between problem-holders and solution-holders (Robertson 1998) not only influences who one associates with, but it can also lead to variations in the vocabulary used in codification and even in the basic approaches or mindsets employed in technology transfer (Nidumolu et al. 2001), creating relatively smooth channels between some parties and building solid barriers between others.

National Systems of Innovation

At a macroeconomic level, national systems of innovation (NIS) are among the greatest sources of impact on diffusion. The NIS framework, which was developed by evolutionary economists such as Lundvall (1992) and Nelson (1993), may be defined by either narrow or broad criteria (Lundvall et al. 2009). Narrowly, the main characteristics of an NIS are expressed through the interaction in a national context of research and development and other scientific and technical activities to generate new knowledge, either abstractly or in the form of physical products. As this emphasis on R&D effectively excludes many of the innovative activities of both developed and developing nations, however, the broader definition that Christopher Freeman proposes, that an NIS is '[t]he network of institutions in the public- and private-sectors whose activities and interactions initiate, import, modify and diffuse new technologies' (Christopher Freeman, quoted in Lundvall et al. 2009, 4) is more useful when considering technology transfer.

The variety of institutions involved in an NIS is broad, taking in firms, government agencies and independent research laboratories (Nelson 1993). Some of these are developers, others are users, and others provide finance. In many cases, the same institution, whether private or public, can play two roles or even all three. The geographical basis of an NIS is both explicit and artificial as it reflects political boundaries which, although they help to define institutional arrangements, may be of less importance from the standpoint of economic factors such as natural resource endowments or the location of markets.

Regional and Sectoral Systems of Innovation

The NIS literature overstates the degree of homogeneity of national institutional structures for innovation and diffusion (Malerba 1993), as well as underplaying some types of relationships that contribute to economic performance in general and to diffusion in particular. To deal with variations in the effectiveness of institutions on both regional and transnational bases, two other types of innovation systems have attracted attention. Regional innovation systems (RIS) centre on social and other relationships in given localities (Asheim et al. 2011; Cooke 1992). They involve networks between firms and also, in common with NIS, private and public institutions, but unlike clusters (Porter 1990; Delgado et al. 2014), an RIS is not confined to a single sector (Asheim et al. 2011). While some observers (Malerba 1993), emphasise differences between NIS and subnational groupings, others (Freeman 2002) believe that regional groups have historically coalesced into NIS.

Sectoral innovation systems (SIS) (Malerba 2002, 2004) are a second alternative way of grouping activities leading to innovation and diffusion. In Malerba's words (2002, 248), 'a sectoral system of innovation and production is a set of new and established products for specific uses and the set of agents carrying out market and non-market interactions for the creation, production and sale of those products'. An SIS, therefore, is similar to a cluster in terms of its stress on a single industry or group of related industries, but without a geographical emphasis.

Regional Diffusion in Industrial Districts, Clusters and Regional Innovation Systems

The related concepts of industrial districts, clusters and regional systems of innovation all underline the importance of geographical concentration for learning and technology diffusion. Industrial districts (IDs) first featured in the work of Alfred Marshall in the late nineteenth and early twentieth centuries (Whitaker 1975; Marshall 1920). Marshall found a strong tendency for firms in the same and closely related industries, such as shoemaking and machinery manufacturing, to locate in close proximity. In the latter part of the twentieth century, the industrial district framework began to be applied again (Becattini et al. 2009), particularly to sections of Italy where mature manufacturing industries had achieved international advantages in 'socio-territorial entities characterized by the active presence of both a community of people and a population of firms in one naturally and historically bounded area [with] a dominant industrial activity' (Giacomo Becattini, quoted in Porter and Ketels 2009, 172). Some IDs are characterised by cooperative competition, or 'co-option' in which resources that do not hold special competitive advantages to firms are sourced cooperatively while firms keep control over aspects that affect their competitive abilities. Clusters and RIS are similar in many respects to IDs and the terms are often used interchangeably despite attempts to distinguish between the concepts (Porter and Ketels 2009; Asheim et al. 2011). Although Italian industrial districts have generally been populated by small- and medium-sized firms, for example, it is by no means clear that this was necessary to Marshall's formulation as some of the shipbuilding firms in important centres before 1914 (e.g. Glasgow, Newcastle and Belfast; Pollard and Robertson 1979) were unquestionably large in terms of employment and the capital employed, as were many textile firms in major centres in Lancashire and Yorkshire. The degree of government involvement and market connections has also been used to discriminate between clusters, IDs and RIS, but again it seems clear that both government intervention and the extent of market and non-market relationships among firms is high in all three categories (Porter and Ketels 2009; Asheim et al. 2011; Becattini et al. 2009).

What is certain, however, is that the presence of clusters, IDs and RIS facilitates knowledge transfer among individuals and firms. Marshall's initial research led him to undertake considerable empirical research and eventually to conclude that proximity resulted in enhanced learning possibilities for individuals and also in substantial spillovers among firms, leading to accelerated innovation and diffusion. In relation to training and apprentices, he famously wrote that '[t]o use a mode of speaking which workmen themselves use, the skill required for their work "is in the air, and children breathe it as they grow up"' (Whitaker 1975, 197). The case of spillovers is more intricate (Langlois and Robertson 1996). When, as in Italy, proximity is associated with small firms and high degrees of vertical specialisation, concentrations of firms encourage the diffusion of technologies that are developed locally and those that are imported from outside the region. In the ceramic tile sector, for instance, the world-class performance of the Sassuolo district in the Modena Province of Emilia-Romagna can be traced to close relationships between tile makers and machinery manufacturers and other suppliers who have worked together to create innovative products and production processes. As levels of appropriability have been low, however, improvements have been imitated by other suppliers, eventually giving tile manufacturers a choice of 20–30 different models of machinery to choose from and ensuring price competition (Russo 1985). For technologies imported from beyond an ID, local packaging companies, for example, have been shown to depend on 'focal firms' with exceptional levels of absorptive capacity to identify opportunities that have then spread throughout the district at a faster rate than they are transmitted to outside firms (Munari et al. 2011).

The relative ease of knowledge transfer in IDs, clusters and RIS is in large part a result of close social relationships that allow personal and sometimes informal exchanges. Firms and their bosses and workers possess social capital and are socially embedded (Granovetter 1985; Grabher 1993) in their local environments as well as in broader settings such as sectoral systems of innovation. People with similar work interests who also know each other in other contexts – as neighbours, church-goers or fellow drinkers or diners in a pub or café – can talk problems over and exchange ideas. Equally importantly, workers can change employers if they are not happy (Bagnasco 2009), taking with them their knowledge of how things are done in their former firms. Similar, although variable results, have also been found for spillovers and labour mobility in clusters (Iammarino and McCann 2006; Lundmark and Power 2010). Finally, when social mobility is also high, as in many Italian IDs in the twentieth century, workers can move upwards and downwards as well as between firms, sometimes shifting from employees to employers and back again as markets expand or contract or as tastes change (Becattini et al. 2009; Brusco and Paba 2014; Brusco 1982; Paci 1991).

Taken together, the main effect of geographic concentrations, therefore, is to facilitate and accentuate the operation of diffusion mechanisms that are well known in other settings. Even though spatial proximity is often not necessary, especially as electronic communications improve (Casali and Robertson 2011), clusters, industrial districts and RIS may be useful in promoting knowledge transfer and improving productivity.

Other Types of Knowledge Communities

Communities of practice (CoPs) (Lave and Wenger 1992; Wenger 1998; Wenger et al. 2002) are groups of individuals, usually in direct contact with each other, who develop preferred methods for analysing and performing tasks, and who look to each other for answers as questions arise (Wenger 1998, 45):

> Over time, collective learning results in practices that reflect both the pursuit of our enterprises and the attendant social relations. These practices are thus the property of a kind of community created over time by the sustained pursuit of a shared enterprise. It makes sense, therefore, to call these kinds of communities *communities of practice*. (Emphasis in original)

As the name implies, a CoP comprises groups of people who perform similar activities, although a practitioner may belong to more than one community: all of the surgeons may belong to a hospital-wide community in relation to infection-control activities, but the heart, brain and thoracic surgeons may have their own local communities when wielding their scalpels within the same hospital. Owing to their close personal contacts, within these groupings members discuss problems and develop formal or informal rules for going about their work that may be distinctive even within their wider professions.

Projects

Even when CoPs are informal, in the sense that people may not be conscious that they belong to a more-or-less hermetic group of practitioners, they have a degree of stability as they are ongoing organisations whose memberships evolve. In contrast, the execution of projects, which is the form in which a high proportion of diffusion takes place, can involve hybrid structures that bring members of different CoPs together on an ad hoc basis. When it is necessary to modify equipment and organisational procedures to implement an incremental change, the services of several groups may be called upon – suppliers, who have a more profound knowledge of the new artefact; customers, who understand an existing set-up and have a good instinct for the ramifications of change; and consultants, who may know what is going on across a range of industries and be in a good position to suggest useful analogies (Ruuska and Vartiainen 2005; Ajman and Koskinen 2008; Ajman et al. 2009). Yet while a project organisation can open access to a wider range of knowledgeable talent than a CoP can command (Wu et al. 2015), multi-dimensional teams also face impediments that might not affect members of a CoP.

This relates especially to confidentiality as suppliers and consultants can be banned from sharing knowledge developed with other customers and buyers may be reluctant to risk their intellectual property by allowing outsiders access to their operations (Miozzo et al. 2015). Thus in some circumstances the use of ad hoc project teams may be constraining, forcing duplication in development efforts, driving up implementation costs and discouraging knowledge transfer and new-to-the-firm innovation without offering useful access to weak ties.

CROSS-COUNTRY KNOWLEDGE DIFFUSION

The diffusion of knowledge on an international level is perhaps the largest single influence on differences in living standards across nations, as well as being a major contributor to technological change, particularly in less developed countries. It is also more complex than sectoral, and especially than regional, diffusion because the factors underpinning knowledge transfer often vary far more on a global basis than within individual regions. In fact, this high level of variance can itself be a driver of international knowledge diffusion when it reflects different resource endowments that create profitable opportunities for firms to relocate their operations. One consequence is that, even though high-technology innovations gain more attention, some of the most important knowledge transfers in recent decades have involved mature technologies that are well codified and can be assimilated relatively easily in less developed nations with factor costs lower than those in the major industrialised economies.

Diffusion in History

Enormous increases in productivity since the eighteenth century, following a millennium of very slow and erratic change, were rooted in important cumulative technological changes that centred initially on very limited mechanisation, primarily in textiles, and improvements in power generation, including the use of coke for iron production and the development of efficient stationary steam engines (Maddison 2007; Landes 2003). Even during the Industrial Revolution, however, these innovations, including General Purpose Technologies, diffused quite slowly both within and across nations (Crafts and O'Rourke 2013). In 1870, a century after James Watt developed the separate condenser, water power was still dominant in the United Kingdom and stationary steam engines were used primarily in textile production, as they had been in 1800 (Musson 1976). As late as the 1960s, modern industrialisation was argued to be confined to a handful of countries in Western Europe as well as the United States (Denison 1967), although the list of relatively high-income nations also included so-called 'countries of recent settlement', most of which had been British colonies. A few years later, Denison did acknowledge the development of Japan (Denison and Chung 1976). Since then, a number of East Asian nations have progressed greatly, but even today, economic power at an international level is discussed in terms of 'Group of 7' or 'Group of 20' nations or of the OECD (although per capita incomes vary considerably in the latter two).

NIS and International Diffusion

Economic historians have long recognised the importance of technology for development (Rostow 1960; Gerschenkron 1962; Landes 2003), but economic growth theory was not equipped to deal with the role of innovation and diffusion for much of the twentieth century because technology was assumed to be an exogenous publicly available variable (Fagerberg 1994). Nevertheless, analyses of 'catching-up' and of the 'technology gap' between developed and developing countries have attracted significant attention in recent decades (Abramovitz 1986; Allen 2012; Crafts and O'Rourke 2013; Fagerberg and Verspagen 2002), especially in the framework of National Systems of Innovations. The narrow definition of NIS, with its emphasis on the roles of R&D and science and technology in promoting innovation, is inadequate because, even in the most advanced economies, a substantial proportion of innovation involves the adoption, perhaps with modifications, of new techniques and machinery that have been diffused from other places and other uses, and this holds most strongly in developing economies where almost none of the significant innovations potentially available would have originated locally. Therefore, it is best to rely on the broader definition of an NIS, with its strong institutional emphasis, because it applies more generally when analysing the diffusion that triggers international technology transfer.

Institutions and 'Technology Clubs'

Recent studies focus on the part played by national institutions in diffusion, especially in terms of absorptive capacity. Even though technological knowledge is acquired and used at the firm level, elements such as the provision of public education and other types of infrastructure help to determine the ability of organisations to recruit technologically and scientifically skilled personnel and to deal successfully with transportation and communications issues. Castellacci and Archibugi (2008) have used a selection of indicators to identify levels of knowledge creation, acquisition and deployment for 131 countries at the end of the twentieth century. On the basis of factor analysis, they reduce eight technological indicators to two factors covering technological infrastructure and human skills, which together underlie national absorptive capacity, and 'creation and diffusion of codified knowledge'.[2] They then use cluster analysis to isolate three 'technology clubs', or groups of countries with similar endowments of the two factors. The smallest club is dubbed Advanced and corresponds to highly industrialised economies.[3] A somewhat larger group of Followers includes economies that were moderately developed over the period, and the third group comprises Marginalised economies with poor infrastructures and low incidences of innovation. Their comparison of results for 1990 and 2000 shows only limited movement between categories, although it was always upwards. From our standpoint, however, the most interesting finding is that, while the Followers as a group seemed to be converging with the Advanced nations, the Marginalised nations were not as successful in closing the gap to Followers, leading to an increased tendency for innovative knowledge, as measured by patents and scientific articles, to originate in the more Advanced economies and for the Marginalised nations to be ever more dependent on diffusion (Castellacci and Archibugi 2008). Although this does not mean that the Marginalised economies have not been learning in absolute terms, they have been retreating further from the constantly moving technological frontier during the period, possibly eroding their ability to benefit from diffusion even more.

Nonetheless, this may not be a cause for alarm. Castellacci and Archibugi (2008, 1671) refer to the seeming polarisation of innovation in Advanced economies and of diffusion in the Marginalised ones as a 'vicious international division of labour', but a high dependence on diffusion has been a consistent practice in developing economies (Fagerberg and Verspagen 2002), which are likely to find the absorption and mobilisation of existing knowledge to be a more rapid, reliable and cost effective way of promoting growth than creating new knowledge would be. Moreover, rather than being a developmental dead end, diffusion can be a necessary platform for innovation in developing economies. The creation of technological capabilities is quite possibly a non-linear and cumulative process in which the establishment of R&D facilities and other indicators of innovation are built on the twin foundations of absorptive capacity and adequate financing. As small firms grow and their marketing and other ties improve, their ability to develop and exploit innovations increases (as indeed does their ability to exploit diffused knowledge). The mechanistic view of economic development that was popular in the 1950s has long been discredited, but even though development remains slow and uneven, membership in the wrong Technology Club is not necessarily a sign that a nation is permanently condemned to stagnation and backwardness.

Mechanisms for Importing Knowledge

Building absorptive capacity is only one of the prerequisites to successful diffusion. A willingness on the part of the current 'owners' of knowledge to share and the availability of sufficient investment capital may also be crucial (Dahmén 1989). Deficiencies in these areas help to explain why development continues to be so uneven despite widespread improvements in absorptive capacity among developing or Marginalised countries (Castellacci and Archibugi 2008). A recent framework presented by Castellacci and Natera (2015) provides a more comprehensive way of approaching international diffusion by dividing the concept of a National Innovation System into the two components of a socio-institutional system (based on social cohesion, education and human capital, and political institutions) and a techno-economic system (based on innovation and technological capabilities, openness, and infrastructures). As in the Castellacci and Archibugi (2008) model, elements such as R&D that are posited to contribute most strongly to innovation are separated from the aspects of absorptive capacity that underpin diffusion, but Castellacci and Natera (2015) also highlight additional elements that enhance the ability of firms and individuals in particular nations to gain access to and make use of existing scientific and technical knowledge.

In general, 'openness' refers to government policies that influence the ability of people within a country to tap into external knowledge. By restricting foreign direct investment (FDI) and foreign trade, governments can inadvertently choke off important routes to diffusion because these are two of the main channels for acquiring international knowledge. As part of receiving foreign financial aid, recipients of FDI are able to access both tacit and codified technical knowledge and to tap a panoply of types of complementary knowledge in areas such as marketing. Investing firms provide not only equipment but also managerial and technical assistance that reflect earlier learning-by-doing and learning-by-using in other environments. The second main aspect of openness, international trade, can also be a major boost to technological advance by increasing the availability of modern techniques that are embodied in imported equipment. Exports are beneficial in broadening markets and providing economies of scale that again encourage moves from traditional to more modern technologies.

In practice, the relative importance of FDI and open international trade in promoting diffusion is disputed, but both are clearly important in particular cases. Seck (2012) argues that, despite a sizeable contribution of FDI to development, R&D spillover gains from Advanced to Marginalised countries derived from imports are even greater. Belitz and Mölders (2016) have also found that the knowledge stocks of developing nations benefit substantially through importation of high-technology goods. Iammarino and McCann (2015), on the other hand, contend that multinational enterprises (MNEs):

> are today the largest source of technology generation, transfer, and diffusion in the world. . . . MNE access to a broad variety of sources of new knowledge, both intra- and inter-firm, provides immense opportunities to acquire new competitive advantages for both the firm itself and all the actors involved in its networks.

Spinoffs from FDI

However, a more dynamic analytical approach is needed to measure the outcomes arising from international knowledge diffusion, which are likely to vary from recipient to recipient depending on absorptive capacity, factor costs and the size of the economies concerned. Through FDI, MNEs can provide Follower and Marginalised nations with major advances from their current technological positions and substantial learning opportunities. The destinations of FDI have generally been segmented on the basis of their absorptive capacity, which affects the ability of host nations to assimilate new technologies (Alvarez and Marin 2013; Constantini and Liberati 2014). Until recently, investment in facilities to build high-tech products has generally been made by firms domiciled in Advanced countries to countries at similar levels of development, in part because of the relatively high skill levels and learning capabilities of their workforces, but also because, in the case of expensive products, high-income countries provide better markets. Nevertheless, the extensive migration of manufacturing firms that intensified at the end of the twentieth century has represented a very substantial diffusion of technological knowledge to less developed regions. In many cases, including automobiles and consumer electronics, the industries that were transferred were highly sophisticated even though their technologies were mature. Furthermore, as in industrial districts, clusters and RIS, movements of labour trained by MNEs can further diffuse knowledge internally in developing economies. Despite Fordist production practices, workers in the host countries become familiar with the production and assembly of complex goods and are introduced to learning routines that can be applied in other sectors. This has been accentuated when local firms have been groomed as suppliers by MNE investors. The outcome has been a pronounced, even if sometimes slow, diffusion of knowledge within host economies through labour mobility. The resulting impetus to local entrepreneurship has been further aided, perhaps, by the way FDI has accustomed consumers in developed economies to buying products from regions in South and East Asia and Latin America. In addition, for some of the countries that have been able to participate, the spread of mature technologies to developing economies has facilitated further moves up the technological scale. This has involved not only the assembly of more advanced products such as smart phones and flat-screen televisions, but has also led to important domestic R&D activities in countries such as South Korea, Taiwan and China that until the late twentieth century were regarded as severely underdeveloped. In electrical products and electronics, for example, China has moved quite quickly from being a producer of simple electrical products such as fans to a country that is making important breakthroughs in cellular phone technology (Guo 2011; Long and Laestadius 2011).

Thus the transfer of knowledge from mature industries can bring important (at times crucial) benefits to less developed nations. This is especially true when learning opportunities are included in the calculations. However, as Castellacci and Natera (2015) argue, diffusion is complex and may be dominated by social and political factors. Because of the substantial risks that can arise, FDI is not an attractive alternative when there is political instability, corruption or social resistance to foreign involvement. These factors may be deep-rooted and beyond the control not only of MNEs, but also of governments in potential recipient countries. As a consequence, trade may be a more practical route for international diffusion in some cases.

CONCLUSION

Since the Industrial Revolution, technological diffusion has been a vital influence on economic performance at all levels, but it is not a homogeneous phenomenon. Instead, it is accomplished through complex webs of personal and collective relationships. The processes involved are far from seamless and often fail, even within relatively uniform and stable environments, because of conflicting geographical, social, economic and political institutions. Obstacles can be especially severe in regions without the financial, legal, educational and other institutional structures that have traditionally co-evolved with modern industrialisation. Nevertheless, there does not appear to be a single route to successful economic development. Countries such as South Korea and Singapore have succeeded despite quite different views on FDI, and China has (so far) been able grow impressively over a period of decades, overcoming[4] institutional weaknesses that would be lethal in most highly developed economies.

Because of its importance to personal and national wellbeing, encouraging diffusion should be a central concern of governments as well as of citizens in all of their roles. New-to-the-world innovation is treated favourably through tax breaks, patent legislation and other policies, which can be justified in part by its seminal role in change, but the amount of attention that it receives is disproportionate in comparison to the assistance provided for the new-to-the-sector and new-to-the-firm innovation that diffusion enables. A better balance is needed between protecting the investments accruing to investors and realising the broader social rewards that flow from knowledge transfer. To some degree, this can achieved through governmental subsidisation, but this can lead to perverse outcomes if the public pays private individuals for innovations that are then largely denied to the community as a whole. To overcome these issues and to get better value from diffusion at all levels of aggregation, the social and economic institutions that are the foundations of innovation and diffusion should be continually rethought and refashioned to meet new needs.

NOTES

1. In official OECD studies, levels of technology (high, medium and low) are assigned on the basis of the percentage of an industry's revenues that is spent on R&D. As a guide to the technological sophistication of the technology employed in an industry, however, these figures can be misleading since, for example, newly developed goods may be produced with older machinery. More importantly, R&D figures fail dismally in measuring the technological impact of an industry since, when embodied in products sold to others (a major form of diffusion), technological advances can substantially improve the performance of the purchasing firm without figuring in its R&D expenditure.
2. The five indicators comprising technological infrastructure and human skills are telephone penetration, electricity consumption, tertiary science and engineering enrolment, mean years of schooling, and a nation's literacy rate. For the second factor, creation of codified knowledge is measured by patents per capita and production of scientific articles, while diffusion is measured by Internet penetration (Castellacci and Archibugi 2008).
3. Membership in the three Clubs is, however, somewhat mysterious. For example, Iceland is included in the small group of Advanced nations while France, Belgium and Austria are placed in the Followers group along with, among others, Jamaica, Paraguay and Turkmenistan.
4. Or papering them over for now.

REFERENCES

Abramovitz, M. 1986. Catching up, forging ahead, and falling behind. *Journal of Economic History* 46: 386–406.
Ajman, M. and K. Koskinen. 2008. Knowledge transfer in project-based organizations: an organizational culture perspective. *Project Management Journal* 39: 7–15.
Ajman, M., T. Kekäle and K. Koskinen. 2009. Role of organisational culture for knowledge sharing in project environments. *International Journal of Project Organisation and Management* 1: 358–374.
Allen, R. 2012. Technology and the great divergence: global economic development since 1820. *Explorations in Economic History* 490: 1–16.
Alvarez, I. and R. Marin. 2013. FDI and technology as levering factors of competitiveness in developing countries. *Journal of International Management* 19: 232–246.
Ancori, B., A. Bureth and P. Cohendet. 2000. The economics of knowledge: the debate about codification and tacit knowledge. *Industrial and Corporate Change* 9: 255–287.
Appleyard, M. 2003. The influence of knowledge accumulation on buyer-supplier codevelopment projects. *Journal of Product Innovation Management* 20: 356–373.
Asheim, B., H. Smith and C. Oughton. 2011. Regional innovation systems: theory, empirics and policy. *Regional Studies* 45: 875–891.
Bagnasco, A. 2009. The governance of industrial districts. In G. Becattini, M. Bellandi and L. De Propris (eds), *A Handbook of Industrial Districts*. pp. 216–228. Cheltenham, UK and Northampton, MA, USA: Edward Elgar Publishing.
Becattini, G., M. Bellandi and L. De Propris (eds). 2009. *A Handbook of Industrial Districts*. Cheltenham, UK and Northampton, MA, USA: Edward Elgar Publishing.
Belitz, H. and F. Mölders. 2016. International knowledge spillovers through high-tech imports and R&D of foreign-owned firms. *Journal of International Trade and Economic Development* 25: 590–613.
Brusco, S. 1982. The Emilian model: productive decentralisation and social integration. *Cambridge Journal of Economics* 6: 167–184.
Brusco, S. and S. Paba. 2014. Towards a history of the Italian industrial districts from the end of World War II to the nineties. DEMB Working Paper, Dipartimento di Economia Marco Biagi, Università degli studi di Modena e Reggio Emilia.
Brusoni, S., A. Prencipe and K. Pavitt. 2001. Knowledge specialization, organizational coupling, and the boundaries of the firm: why do firms know more than they make? *Administrative Science Quarterly* 46: 597–621.
Bureau of Industry Economics. 1994. *Beyond the Innovator: Spillovers from Australian R&D*. Canberra: Australian Government Printing Service.
Casali, G. and P. Robertson. 2011. Pre-packaging entrepreneurship? The Gold Coast marine precinct. *Australasian Journal of Regional Studies* 17: 352–383.
Castellacci, F. and D. Archibugi. 2008. The technology clubs: the distribution of knowledge across nations. *Research Policy* 37: 1659–1673.

Castellacci, F. and J.M. Natera. 2015. The convergence paradox: the global evolution of national innovation systems. In D. Archibugi and A. Filippetti (eds), *The Handbook of Global Science, Technology and Innovation.* Chichester: Wiley.

Chen, L.-C. 2009. Learning through informal local and global linkages: the case of Taiwan's machine tool industry. *Research Policy* 38: 527–535.

Chesbrough, H. 2003. *Open Innovation: The New Imperative for Creating and Profiting from Technology.* Boston, MA: Harvard Business School Press.

Chesbrough, H. and R. Rosenbloom. 2002. The role of the business model in capturing value from innovation: evidence from Xerox Corporation's technology spin-off companies. *Industrial and Corporate Change* 11: 529–555.

Christensen, C. 1997. *The Innovator's Dilemma: When New Technologies Cause Great Firms to Fail.* Boston, MA: Harvard Business School Press.

Cohen, W. and D. Levinthal. 1989. Innovation and learning: the two faces of R&D. *Economic Journal* 99: 569–596.

Cohen, W. and D. Levinthal. 1990. Absorptive capacity: a new perspective on learning and innovation. *Administrative Science Quarterly* 35: 128–152.

Constantini, V. and P. Liberati. 2014. Technology transfer, institutions and development. *Technological Forecasting and Social Change* 88: 26–48.

Cooke, P. 1992. Regional innovation systems: competitive regulation in the new Europe. *Geoforum* 23: 365–382.

Crafts, N. and K. O'Rourke. 2013. Twentieth century growth. Working Paper, Department of Economics, University of Warwick, Cage Online Working Paper Series, http://wrap.warwick.ac.uk/59305.

Dahmén, E. 1989. 'Development blocks' in industrial economics. In B. Carlsson (ed.), *Industrial Dynamics: Technological, Organizational, and Structural Changes in Industries and Firms.* Boston, MA: Kluwer.

Delgado, M., M. Porter and S. Stern. 2014. Clusters, convergence, and economic performance. *Research Policy* 43: 1785–1799.

Denison, E., assisted by J. Poullier. 1967. *Why Growth Rates Differ: Postwar Experience in Nine Western Countries.* Washington, DC: Brookings Institution.

Denison, E. and W.K. Chung. 1976. *How Japan's Economy Grew So Fast: The Sources of Postwar Expansion.* Washington, DC: Brookings Institution.

Edvardsson, B., A. Gustafsson, P. Kristensson and L. Witell. 2010. Customer integration in service innovation. In F. Gallouj and F. Djellal (eds), *The Handbook of Innovation and Services: A Multi-disciplinary Perspective.* pp. 301–317. Cheltenham, UK and Northampton, MA, USA: Edward Elgar Publishing.

Enkel, E. and O. Gassmann. 2010. Creative imitation: exploring the case of cross-industry innovation. *R&D Management* 40: 256–270.

Fagerberg, J. 1994. Technology and international differences in growth rates. *Journal of Economic Literature* 32: 1147–1175.

Fagerberg, J. and B. Verspagen. 2002. Technology-gaps, innovation-diffusion and transformation: an evolutionary approach. *Research Policy* 31: 1291–1304.

Franke, N., M. Poetz and M. Schreier. 2014. Integrating problem solvers from analogous markets in new product ideation. *Management Science* 60: 1063–1081.

Freeman, C. 2002. Continental, national and sub-national innovation systems – complementarity and economic growth. *Research Policy* 31: 191–211.

Gerschenkron, A. 1962. *Economic Backwardness in Historical Perspective.* Cambridge, MA: Harvard University Press.

Godin, B. 2006. The linear model of innovation: the historical construction of an analytical framework. *Science, Technology and Human Values* 31: 639–667.

Grabher, G. (ed.). 1993. *The Embedded Firm: On the Socioeconomics of Industrial Networks.* London: Routledge.

Granovetter, M. 1973. The strength of weak ties. *American Journal of Sociology* 78: 1360–1380.

Granovetter, M. 1985. Economic action and social structure: the problem of embeddedness. *American Journal of Sociology* 91: 481–510.

Guo, Y.-H. 2011. How low- and medium-technology industries in developing countries compete with multinationals: lessons from China's home electronics sector. In P. Robertson and D. Jacobson (eds), *Knowledge Transfer and Technology Diffusion.* pp. 211–237. Cheltenham, UK and Northampton, MA, USA: Edward Elgar Publishing.

Helpman, E. (ed.). 1998. *General Purpose Technologies and Economic Growth.* Cambridge, MA: MIT Press.

Hirsch-Kreinsen, H. 2015. Patterns of knowledge use in 'low-tech' industries. *Prometheus* 33: 67–82.

Hirsch-Kreinsen, H., D. Jacobson and P. Robertson. 2006. 'Low-tech' industries: innovativeness and development perspectives – a summary of a European research project. *Prometheus* 24: 3–21.

Hirschman, A. 1958. *The Strategy of Economic Development.* New Haven, CT: Yale University Press.

Huang, C., A. Arundel and H. Hollanders. 2011. How firms innovate: R&D, non-R&D, and technology adoption. Paper presented at the DIME Final Conference, 6–8 April, Maastricht.

Iammarino, S. and P. McCann. 2006. The structure and evolution of industrial clusters: transactions, technology and knowledge spillovers. *Research Policy* 35: 1018–1036.

Iammarino, S. and P. McCann. 2015. Multinational enterprises innovation networks and the role of cities. In D. Archibugi and A. Filippetti (eds), *The Handbook of Global Science, Technology and Innovation*. Chichester: Wiley.

Jensen, M., B. Johnson, E. Lorenz and B.-Å. Lundvall. 2007. Forms of knowledge and modes of innovation. *Research Policy* 36: 680–693.

Johnson, B., E. Lorenz and B.-Å. Lundvall. 2002. Why all this fuss about codified and tacit knowledge? *Industrial and Corporate Change* 11(2): 245–262.

Kalogerakis, K., C. Lüthje and C. Herstatt. 2010. Developing innovations based on analogies: experience from design and engineering consultants. *Journal of Product Innovation Management* 27: 418–436.

Kline, S. and N. Rosenberg. 1986. Innovation and the chain-linked model. In R. Landau and N. Rosenberg (eds), *The Positive Sum Strategy*. pp. 275–305. Washington, DC: National Academies Press.

Landes, D. 2003. *The Unbound Prometheus: Technological Change and Industrial Development in Western Europe from 1750 to the Present*. Cambridge: Cambridge University Press.

Langlois, R. and P. Robertson. 1996. Stop crying over spilt knowledge: a critical look at the theory of spillovers and technical change. University of Connecticut, Department of Economics Working Papers.

Lave, J. and E. Wenger. 1992. *Situated Learning: Legitimate Peripheral Participation*. Cambridge: Cambridge University Press.

Lipsey, R., K. Carlaw and C. Bekar. 2005. *Economic Transformations: General Purpose Technologies and Long-Term Economic Growth*. Oxford: Oxford University Press.

Long, V. and S. Laestadius. 2011. New patterns in knowledge transfer and catching up: Chinese R&D in ICT. In P. Robertson and D. Jacobson (eds), *Knowledge Transfer and Technology Diffusion*. pp. 238–259. Cheltenham, UK and Northampton, MA, USA: Edward Elgar Publishing.

Lundmark, M. and D. Power. 2010. Labour market dynamics and the development of the ICT cluster in the Stockholm region. In C. Karlsson (ed.), *Handbook of Research on Innovation and Clusters: Cases and Policies*. Cheltenham, UK and Northampton, MA, USA: Edward Elgar Publishing.

Lundvall, B.-Å. (ed.). 1992. *National Systems of Innovation: Towards a Theory of Innovation and Interactive Learning*. London: Pinter.

Lundvall, B.-Å. and B. Johnson. 1994. The learning economy. *Journal of Industry Studies* 1: 23–42.

Lundvall, B.-Å., J. Vang, K. Joseph and C. Chaminade. 2009. Innovation system research and developing countries. In B.-Å. Lundvall, K. Joseph, C. Chaminade and J. Vang (eds), *Handbook of Innovation Systems and Developing Countries: Building Domestic Capabilities in a Global Setting*. pp. 1–20. Cheltenham, UK and Northampton, MA, USA: Edward Elgar Publishing.

Maddison, A. 2007. *Contours of the World Economy, 1–2030 AD: Essays in Macro-Economic History*. Oxford: Oxford University Press.

Malerba, F. 1993. The national system of innovation: Italy. In R. Nelson (ed.), *National Innovation Systems: A Comparative Analysis*. pp. 230–259. New York: Oxford University Press.

Malerba, F. 2002. Sectoral systems of innovation and production. *Research Policy* 31: 247–264.

Malerba, F. (ed.). 2004. *Sectoral Systems of Innovations: Concepts, Issues and Analyses of Six Major Sectors in Europe*. Cambridge: Cambridge University Press.

Marshall, A. 1920. *Principles of Economics*, 8th ed. London: Macmillan.

Miozzo, M., P. Desyllas, H.-F. Lee and I. Miles. 2015. Combining appropriability mechanisms for innovation collaboration by knowledge-intensive firms. Paper presented at DRUID 15, Rome, 15–17 June.

Munari, F., M. Sobrero and A. Malipiero. 2011. Absorptive capacity and localized spillovers: focal firms as technological gatekeepers in industrial districts. *Industrial and Corporate Change* 21: 429–462.

Musson, A. 1976. Industrial motive power in the United Kingdom, 1800–1870. *Economic History Review* 29: 415–439.

Nelson, R. (ed.). 1993. *National Innovation Systems: A Comparative Analysis*. New York: Oxford University Press.

Nidumolu, S., M. Subramani and A. Aldrich. 2001. Situated learning and the situated knowledge web: exploring the ground beneath knowledge management. *Journal of Management Information Systems* 18: 115–150.

Paci, M. 1991. The social bases of diffuse industrialization. *International Studies of Management and Organization* 21: 21–37.

Pavitt, K. 1984. Sectoral patterns of technical change: towards a taxonomy and a theory. *Research Policy* 13: 343–373.

Pollard, S. and P. Robertson. 1979. *The British Shipbuilding Industry 1870–1914*. Cambridge, MA: Harvard University Press.

Porter, M. 1990. *The Competitive Advantage of Nations*. New York: Free Press.

Porter, M. and C. Ketels. 2009. Clusters and industrial districts: common roots, different perspectives.

In G. Becattini, M. Bellandi and L. De Propris (eds), *A Handbook of Industrial Districts.* pp. 172–183. Cheltenham, UK and Northampton, MA, USA: Edward Elgar Publishing.

Randhawa, K. and M. Scerri. 2015. Service innovation: a review of the literature. In R. Agarwal, W. Selen, G. Roos and R. Green (eds), *The Handbook of Service Innovation.* pp. 27–52. London: Springer.

Robertson, P. 1998. Information, similar and complementary assets, and innovation policy. In N. Foss and B. Loasby (eds), *Economic Organization, Capabilities and Co-ordination: Essays in Honour of G. B. Richardson.* pp. 261–288. London: Routledge.

Robertson, P. and P. Patel. 2007. New wine in old bottles: technological diffusion in developed economies. *Research Policy* 36: 708–721.

Robertson, P., E. Pol and P. Carroll. 2003. Receptive capacity of established industries as a limiting factor in the economy's rate of innovation. *Industry and Innovation* 10: 457–474.

Robertson, P., G. Casali and D. Jacobson. 2012. Managing open incremental process innovation: absorptive capacity and distributed learning. *Research Policy* 41: 822–832.

Rogers, E. 2003. *Diffusion of Innovations.* New York: Free Press.

Rostow, W. 1960. *The Stages of Economic Growth.* Cambridge: Cambridge University Press.

Russo, M. 1985. Technical change and the industrial district: the role of interfirm relations in the growth and transformation of ceramic tile production in Italy. *Research Policy* 14: 329–343.

Ruuska, I. and M. Vartiainen. 2005. Characteristics of knowledge sharing communities in project organizations. *International Journal of Project Management* 23: 374–379.

Seck, A. 2012. International technology diffusion and economic growth: explaining the spillover benefits to developing countries. *Structural Change and Economic Dynamics* 23: 437–451.

Smith, A. 1937[1776]. *An Inquiry into the Nature and Causes of the Wealth of Nations.* New York: Modern Library.

Wenger, E. 1998. *Communities of Practice: Learning, Meaning, and Identity.* Cambridge: Cambridge University Press.

Wenger, E., R. McDermott and W. Snyder. 2002. *Cultivating Communities of Practice: A Guide to Managing Knowledge.* Boston, MA: Harvard Business School Press.

Whitaker, J. (ed.). 1975. *The Early Economic Writings of Alfred Marshall 1867–1890.* London: Macmillan.

Winter, S. 2003. Understanding dynamic capabilities. *Strategic Management Journal* 24: 991–995.

Wu, L.-W., Y.-S. Lii and C.-Y. Wang. 2015. Managing innovation through co-prodution in interfirm partnering. *Journal of Business Research* 68: 2248–2253.

3. Beyond the binaries: geographies of gender–technology relations
Jessica McLean, Sophia Maalsen and Alana Grech

Geographies of gender and technology relations are diverse, complex and intertwine public and private, social and cultural dimensions. This chapter provides an overview of some aspects of gendered relations in technology with a spatial lens. As such, it is a partial narrative of particular facets of gender and technology that geographers have been paying attention to, while drawing on work by academics from other disciplines who have developed feminist readings of gender–technology relations. The chapter begins with an overview of histories of gender–technology relations before examining the current conceptual debates in the field. Political concerns are inherent in any examination of gender and technology relations, especially given the influence of feminist critiques within this area. To close the chapter, a case study of Destroy the Joint, a feminist online group, captures an example of how geographers have engaged with social movements in online spaces in recent research, and what a geographic perspective on gender–technology relations can offer.

In her manifesto *For Space*, Massey (2005) introduced her reconceptualisation of geographic processes with the explanation and assertion that 'This is a space of loose ends and missing links' (p. 11) and we lean on her words here. The space we are looking at in this chapter – gendered dimensions in digital spaces – comprises loose ends and missing links, and appropriately hyperlinks, constant becomings and renegotiations. Yet as a chapter in and of itself, this space includes unavoidable gaps as important themes and issues are only touched upon that deserve significantly more consideration. Issues pertaining to intersectionality and transecologies, for instance, are not engaged with in substantive detail in this chapter. With this in mind, please read on and consider the power of partial perspective (Haraway 1988) that runs through our discussion of geographies of gender–technology relations.

SITUATING GENDER–TECHNOLOGY RELATIONS: A HISTORICAL OVERVIEW

Feminist theories of technology are important in grounding a spatial perspective on gender–technology relations, not least because they have brought a critical view to a field that demands such. Wajcman (2010, 143) provides an astute summary of feminist theories of technology and the development of these theories is important to situate feminist approaches to technology. Feminism is characterised by a diversity and overlapping of ideas and approaches and this multiplicity is something that enables a greater understanding of the geographies of gender–technology relations.

Although there are shifts in emphasis from early to current feminist theories of technology, the multiple feminist debates surrounding technology are united by a common goal. Wajcman notes a shared purpose 'to interrogate the gender power relations of the material world' (Wajcman 2010, 143). In other words, feminist enquiry seeks to understand the opportunities that are both denied and offered through technology dependent on gender and how this relationship between the social and technological affects lived experience. Additionally, there is room to radicalise the potential of technology as a progressive political project that can afford opportunities to redress gender imbalances and offer new forms of opportunities and liberations (Millar, 1998).

Firstly though, it is necessary to define how technology is viewed both in this paper and in broader feminist theory. We take the definition, drawing from anthropological, science and technology studies, and material culture approaches, that technology is more than the artefact, *it is the socio-material and socio-technical relations inherent to it.* This aligns with Wajcman's observation that feminist perspectives 'shift our understanding of what technology is, broadening the concept to include not only artefacts but also the cultures and practices associated with technologies' (Wajcman 2010, 143). Technology is therefore not just an object, but representative of multiple networks including all the mechanisms that went into its production, the social and cultural contexts in which it was produced, the power asymmetries associated with its production and produced through its use, and its various potential interpretations and purposes. It is an incredibly complex socio-technical product.

Earlier feminist debates around technology were technically deterministic in nature. Here, the emphasis centred on understanding how technology reproduced patriarchy (Wajcman 2010, 144). Feminists challenged the assumption that the association of technology with masculinity was a result not of biological sex difference but of cultural constructions and practices of technology and gender (Wajcman 2010, 144). Both Wajcman and Oldenziel (1999) observe the production of 'male machines' (Oldenziel 1999), through the formalization of technologically oriented careers such as engineering, managerial positions, ideals of manliness and a concurrent devaluation of female and domestic technologies, that positioned femininity as 'incompatible with technological pursuits' (Wajcman 2010, 144).

In response to this, feminists argued for greater engagement with, and opportunities for, women in technological careers. This meant reshaping institutions to accommodate women and also efforts to change the cultures of masculinity that impacted on the retention of women in these fields. The broad aim was to enable women to successfully participate in technological fields in ways that were not on masculinist terms.

The 1980s saw feminist debate shift from technological determinism and access to technology to understanding the way technology was gendered. That technology could be gendered referred to the ways in which the artefacts are informed by social factors and, in this way, the dominant social norms are embedded in technology. The dominance of masculinity within both Western science and technology was perceived as problematic owing to the roles they played in the domination of women and nature (Wajcman 2010). Approaching technology from this perspective recognised its political qualities and the radical feminist advocacy for technology based on feminine values. Aligning technology with women's values emphasised the difference between men and women and such an approach was criticised for reinforcing the essentialism of sex differences and of positioning 'women as victims of patriarchal technoscience' (Wajcman 2010, 146).

Socialist feminism, immersed in Marxian labour processes, sought to interrogate relationships between women, work and technology. Unlike technological determinism, socialist feminist theory situated technology within class relations of production and this gave further strength to the idea of technology as gendered. The division of labour was seen as a sexual hierarchy that reinforced male dominance of technology (Wajcman 2010). The idea of artefacts being shaped by and reproducing socio-cultural gender relations was reiterated. Therefore, technology was not neutral but implicated, through design and associated meanings, to the exclusion of women within a socialist feminist perspective. While the focus on the agentive capacities of technology revealed a tendency to emphasise cultures of masculinity, the lack of recognition that women themselves could be agents and reappropriate technology was also critiqued. By questioning the embedded gender hierarchies of technologies, feminist theorists highlighted the political potential and opportunities for change enabled by technology but did not give enough kudos to the potential for women to enact significant change (Wajcman 2010).

Concurrently, second-wave feminism was responding to black feminist critique, queer theory, post-colonial theory and post-modernism to develop into third-wave feminism, sometimes referred to as 'post-feminism' (Wajcman 2010, 147). Feminism in this guise recognised feminism's multiplicity, plurality and dynamic process, and sought ongoing reflections and reflexive practice. Feminism could not speak for all women owing to the differences in the situated experience of race, class, gender and sexuality, among others, and multiple approaches to feminism, which encompassed this diversity, was seen as a positive move to accommodate the variety of positionings.

A feminism that embraces its plurality also shifted from the pessimistic approaches to technology characterised by first- and second-wave feminism to being aware of the opportunities that technology of the digital age could afford (Wajcman 2010). Feminists distinguished digital technology from industrial technology, the former based on brains and networks, the latter on hierarchy and physicality, and thus digital technologies were optimistically positioned as presenting a new opportunity for women and technology.

Digital spaces, it seemed, created a space where sex differences were no longer relevant. The online represented a blurring of boundaries between body and machines, referencing the potential for disrupting gender and bodily boundaries and the associated power hierarchies, a post-human position characterised by Haraway's (1985) cyborgs.

This optimistic approach was propelled by Haraway's work. Haraway notes how Grossman's (1980) term, 'the social relations of science and technology', challenges technological determinism to suggest instead an 'historical system depending upon structured relations among people' (Haraway 1985, 165). Haraway extends the importance of this phrase, suggesting that it also denotes that 'science and technology provide fresh sources of power, that we need fresh sources of analysis and political action' (Haraway 1985; Latour 1987). Online spaces are one domain that can provide a site for the analysis advocated by Haraway. Digital spaces have evolved to offer engagement in public domains but also give users the flexibility to choose the identity of their online self and act in ways not enabled by their non-digital positioning. Users are therefore offered an opportunity to subvert dominant power structures and engage in political action, while not restricted by the gender, sex and physical confines of their selves.

As Wajcman (2010) notes, however, Haraway's writings on technology have at times been misinterpreted as an endorsement of all things digital, ignoring the continuities between technological stages and the gendered hierarchies that still persist. So while engaging with the digital world through acts including online activist groups and communities, gaming and social media, enables possibilities for hybrid, transgendered and alternative identities, there are limitations. For example, the liberating qualities of the digital are still affected by the gendered relations of the social-cultural world (Wajcman 2010; Plant 1998). The GamerGate incidents of 2015 (Todd 2015) are an illustrative example of the gender discourses of everyday life influencing behaviour and discriminatory practices in the more-than-real (McLean 2016). Gaming is diversifying on a global scale, with masculinist white stereotypes of who a gamer is now becoming irrelevant. However, many gamers resist such diversification and this was manifest in #GamerGate, where 'rape and death threats . . . were being directed primarily at scholars and women involved in the gaming industry' (Todd 2015, 64). In particular, Zoe Quinn was attacked online by her ex-boyfriend Erin Gjoni and had her personal details, including address and phone number, circulated widely online. This led to innumerable rape and death threats and continues to affect Quinn's life today. From this example, we can agree with Wajcman (2010) that to progress feminist politics of technology, technology needs to be understood as 'neither inherently patriarchal nor ambiguously liberating' (Wajcman 2010, 148).

More recent feminist approaches to technology, influenced by science and technology studies, approach technology and gender as mutually shaped by each other and thus co-produced. Technology is a socio-technical product. By recognising the relationship between gender and technology as co-produced then it follows that 'gender relations can be thought of as materialised in technology, and masculinity and femininity in turn acquire their meaning and character through their enrolment and embeddedness in working machines' (Wajcman 2010, 149).

Understanding gender and technology as mutually shaped demonstrates that the machines themselves are part of ourselves. The borders between human and object have been transgressed, most often illustrated by medical technologies that are now being incorporated into the human form through nano technologies (Milburn 2002). The broader potential of the conceptualisation of a co-produced gender and technology hybrid is the opportunity it provides for agency. We are not mere recipients of technological acts, instead we can act with and through technology as 'us' and this has relevance for a progressive politics of gender, returning to Haraway:

> The machine is not an *it* to be animated, worshipped and dominated. The machine is us, our processes, an aspect of our embodiment. We can be responsible for machines; *they* do not dominate or threaten us. (Haraway 1985, 36)

Therefore gender and technology are processes of doing, of performance, and can hold fluid meanings and be enacted for multiple purposes depending on context. Acknowledging this means acknowledging the historical and hierarchical structures embedded in both gender and technology but also recognising that neither is mute; both are agentive and work together to produce material realities and lived experience. Understanding that we have agentive possibilities through, upon and with our relationships with technology means that the technological is a politically significant realm through which to continue to do gender work. This forwarding of agency in gender–technology relations opens up potentials for technology to redress gender power asymmetries, and we expand on its geographical and spatial implications in the next section of this chapter.

GENDER AND TECHNOLOGY: CONTRIBUTIONS TO GEOGRAPHICAL THOUGHT

Our overview on the development of feminist theoretical approaches identified the relational perspective of more recent feminist work as an effective way to address the socio-technical relationships without privileging one or the other, negating both social and technological determination. The influence of feminist and post-human theoretical thought has leveraged certain openings in geographic enquiry to differently understand human–technology relationships.

Clearly, there are geographical implications to the relationship between gender and technology for, as Massey (2005, 9) argues, we must recognise 'space as the product of interrelations'. For instance, an established tradition of critical Geographical Information Science (GIS) studies since the 1990s which critiqued GIS was heavily influenced by feminist theory in its emphasis on representation (Leszczynski and Elwood 2015). In this sense, geographers were using GIS technology to highlight the discrepancies in representation and the technology became a tool to rebalance gendered, and also racial and socio-economic, relations, rather than perpetuating the absence of these inequalities from GIS work. The increasing body of research into Smart Cities is another area in which geographers are turning their attention to the spatial implications of cities run by code. However, the rapid increase in influence of digital technology means that there are many areas not yet adequately addressed by geographers. As observed by Rose (2015), very few geographers have attended to interpreting digital technologies and digital objects despite cultural production having shifted significantly since the 'new cultural geography' turn of the 1980s. Acknowledging geography's tradition of understanding the spatial implications of technology, it is well placed as a discipline to take heed of Rose's observations, and to continue this work in the digital world.

Relational approaches to the socio-technical nexus are characterised by the post-human theories exemplified by Haraway, as described above, and the possibilities for destabilising gender hierarchies and even a post-gender world. Haraway's work therefore has great application to understanding the relationship between technology and geography through a feminist or post-human framework. Most notably associated with Haraway (1988, 1991), the idea of the cyborg presents a subject that is the combination of the human and the nonhuman. This shift is anchored in an ethical stance, as Baker (2000, 102) notes, 'Cyborg-status was more than a simple and joyful release from anthropocentric values'. Instead, cyborg entities introduced fertile scope for contestation of anthropocentrism, including discussions of co-optation and domination. The cyborg idea provides space for debate, including how a body/technology hybrid 'enables us to expose, examine, and critique the ways in which the body is implicated and bound up in our understandings of art, technology and identity' (Garoian and Gaudelius 2001, 333).

Humans have the capacity and tools to envisage a hybrid human and this is not dissimilar to Deleuze and Guattari's notion of becoming animal, although Haraway criticises their 'disdain for the daily, the ordinary, the affectional rather than the sublime' (Haraway

2008, 29) with their focus on pack rather than domesticated animals. However, Deleuze and Guattari's theory of becoming offers an opportunity to envision humans as a possible becoming of other 'Becoming-Intense, Becoming-Animal, Becoming Imperceptible' (Deleuze and Guattari 1987, 256) in a way that challenges the object/subject binary. As such, notions of becoming provide an opportunity for geographers to understand the co-production of object and subject, in other words, technology and human, in a way that provides more insight into cyborg/hybrid human ways of being. The affect of challenging object/subject binary *through becoming* is explored by Lawlor (2008, 176):

> In the experience of becoming, when one is fascinated by something before oneself, when one contemplates something before oneself, one is *among* it, *within* it, together in a zone of proximity.

Both the cyborg and the notion of becoming are analogous in the way that they resist categorical thinking and insist on recognition of commingling and co-production. Brown notes that it is 'Through becoming, we join with the other animal in a zone of proximity that dissolves our identities and the boundaries that we set up between us' (Brown 2007, 262). Here, we refocus becoming to Becoming-Machine, both paralleling and reinforcing the idea of cyborg. Crucially for our purposes, Brown sees Deleuze and Guattari's becoming as decentring the position of the human agent:

> In the process, the human being moves out of a position of dominance. She slips out of the position of centrality that enabled her to establish the binary of human-animal to begin with. (Brown 2007, 262)

The cyborg offers a framework through which it is possible to escape the separation of body and technology and forge the way for an understanding that incorporates both as part of a heteroglossia, to use Haraway's term. In attempting to describe the cyborg, Hables Gray (2002, 19) claims that

> *Cyborg* is as specific, as general, as powerful, and as useless a term as *tool* or *machine*. And it is just as important. Cyborgs are proliferating throughout contemporary culture, and as they do they are redefining many of the most basic political concepts of human existence.

The relevance of the post-human theoretical approaches to geography characterised by Haraway and Deleuze and Guattari is in their boundary-challenging potentials. Here, borders are messy, unstable, reterritorialised and deterritorialised. Sarah Whatmore has applied this approach in her notion of 'hybrid geographies', emphasising the hybridity of the relationship between the human and nonhuman to rethink the nature/culture binary (Whatmore 2002). More recently, McLean and Maalsen (2013) have applied relational perspectives to online spaces of resistance, to ask how the socio-technical relationships of online activist groups can produce material changes in the physical world. Boundaries between human and technology are mutable and, aligning with third-wave feminist theories, such porous boundaries allow the mutual shaping of both technology and gender.

Focusing on the geographical conceptualisations of space, place and information technology in the era of cyberspace, Graham (1998) describes three perspectives on the implications of socio-technical relations and the production of space and place: 'substitution and transcendence', 'co-evolution' and 'recombination' (Graham 1998, 165).

Recombination is particularly relevant to the understanding of human and technological agents as mutually shaping. Graham identifies 'recombination' as a perspective that draws on actor-network theory (Callon 1986, 1991; Latour 1993) and Haraway's (1991) cyborgs. Both of these approaches argue for a relational approach to understanding technology and the social world (Graham 1998, 176–177). Graham illustrates the heteroglossia of human–technological relations using telecommunications as an example. Here, developments in communications facilitate action at a distance but do so through heterogeneous networks of people, technology, infrastructure and institutions (Graham 1998). Neither is negated by the other, it is a mutual shaping which affects how technology is developed and how that technology orders lived space. Additionally telecommunication infrastructures support other systems such as automatic teller machines, credit cards and telebanking, financial flows and data flows, all supported by a multitude of technologies such as broadband and satellite (Graham 1998). In all of these technologically facilitated acts and transactions, human agency is still current, avoiding technological determinism. The plurality of such socio-technical systems moves Graham to characterise cyberspace as 'a fragmented, divided and contested multiplicity of heterogeneous infrastructures and actor-networks' (Graham 1998, 178).

This is particularly relevant for online digital worlds of social media such as those discussed in this paper. Robyn Longhurst (2013) has paid attention to gender–technology relations with a focus on social media and the politics of birth (Longhurst 2009) and online tools for mothering. As a feminist geographer based in New Zealand, Longhurst (2009) writes that we should conceptualise digital spaces as 'cyber/space' to focus attention on the mutually constituted interrelations of online and offline spaces. Further, she argues that, while YouTube offers potential to refigure the way birth is understood, it also has served to concretise particular views on birth and censor that which is deemed inappropriate for general viewing, such as vaginal births. This speaks to Wajcman's theorising that the digital is not inherently emancipatory or oppressive (see Millar 1998) but rather holds potential for both.

Another way that geographers have engaged with technology is in the Smart City arena, although the relative recent emergence of this field means that much work has been focused on the role of software in the production of urban space, rather than the actual relationship between Smart City technology and the experience of difference, such as gender, in the coded urban spaces. Nevertheless it is pertinent to look at some of the key theoretical approaches in this area, to see how they are addressing geography and technology.

CASE STUDY: GEOGRAPHIES OF ONLINE FEMINISM

Geographies of gender–technology relations continue the important theoretical and empirical work of geographers who have examined digital spaces as changing social and institutional processes before the turn of the century (Kitchin 1997) and queried the discourses that produce artificial dualities between online and offline selves (Kinsley 2014; McLean 2016).

Relational approaches to analysis of gender–technology relations in digital spaces have provided generative opportunities to increase our understanding of online geographies. Leszczynski and Elwood (2015) have shown how technologies are shaped by ambivalence by examining new spatial media in online spaces. These technologies can 'encode and generate social exclusions along multiple axes of difference' (Leszczynski and Elwood 2015, 13) and are no panacea.

The role of the cyborg in geography is analysed by Schuurman (2002) with respect to gender and GIS practices. While noting the dominance of men employed in GIS fields, Schuurman (2002) does not agree with the assumption that GIS is inherently masculinist and positivist. Rather, and following Haraway's thinking on the ambivalence of technologies, Schuurman (2002, 261) states that 'it is as masculinist as we allow it to be'. An exemplary feminist scholar who links gender issues to spatial representations is Kwan (1999). In that landmark research, Kwan (1999) shows that women have less everyday mobility than men by demonstrating their spatial and temporal movements in terms of work and home activities.

The possibilities for productive forces of change to emerge in online spaces have also been observed. In 2013, the authors of this chapter delved into the creation of 'Destroy the Joint' (DTJ) and the feminist revitalisation that was emerging in online and offline spaces, especially in Australia. We wrote then about how DTJ was reliant on co-productive forces of change in material and discursive spaces, and was partially made possible by the social media tools Facebook and Twitter (McLean and Maalsen 2013). The ordinary affect (Stewart 2007) that drives much action in social media spaces was one of the important dimensions making the formation of DTJ possible.

To give some context to the online feminist group and its name, it is worthwhile looking at its origins. Conservative radio commentator Alan Jones was discussing the state of Australian politics with right-wing politician Barnaby Joyce and the subject of how women are performing as political leaders was raised:

> She (the Prime minister) said that we know societies only reach their full potential if women are politically participating . . . Women are destroying the joint – Christine Nixon in Melbourne, Clover Moore here. Honestly. (Quote from Farr 2012)

Destroy the Joint sprung from a hashtag initiated in a conversation between two women who were critiquing Jones's sexism. Jane Caro, a public education advocate and commentator, wrote

> Got time on my hands tonight so thought I'd spend it coming up with new ways of 'destroying the joint' being a woman & all. Ideas welcome.

Jill Tomlinson, a surgeon, responded with

> Bored by Alan Jones' comments on women destroying Australia? Join with @JaneCaro & suggest ways that women #destroythejoint.

After this exchange, thousands of tweets made humorous offerings for how women are destroying the joint, from even appearing in the workplace to taking care of children and volunteering in various contexts. The ironic hashtag attracted a substantial following. The next week, a Facebook page (www.facebook.com/destroythejoint) and Twitter (www.twitter.com/jointdestroyer) account were established and today there are 75,000 Facebook 'likers' and 16,000 Twitter followers who contribute to DTJ's ongoing efforts to stop sexism and misogyny.

Substantial achievements for DTJ, on both the micro and macro scales, have come as a result of the campaigning within these social media spaces. For instance, in the first few weeks of its creation, an online petition to withdraw advertising support for Alan Jones's radio show was successful and for a week the radio host was not on air. Throughout 2015 DTJ drew public awareness to gender-based violence, counting how many women have been killed in their domestic environments and highlighting this form of misogyny as unacceptable. This count is continuing in 2016.

In early 2014, the authors of this chapter approached DTJ to see if they would like to collaborate on a participatory mapping and survey activity, to glean greater understanding of the spatial distribution of DTJ followers, and what their priorities in addressing sexism were at that stage. This collaboration was welcomed by the DTJ group and its participants contributed spatial and qualitative data for the mapping and survey. Nearly 900 people filled in the survey and many of these participants were able to complete the mapping activity too, although there were some technical problems with the software accessibility as it relied on having a Google account to complete it. We had designed the research with this possibility in mind and so had included a question on the survey that asked for participants' postcodes. To read more about this research, go to www.geofeminisms.org or see McLean et al. (2016).

One remarkable aspect of this research was how participants viewed their interest in feminism as a globally linked project, and that different sorts of campaigns were important, from the seemingly superficial to the more substantive. A global feminist revitalisation was perceived by a majority of survey participants (Figure 3.1, bottom corner) and the diversity of this revitalisation is evident throughout the responses. Respondents stated that feminism in social media spaces can bring about substantive change but that this must be intertwined with other actions. Table 3.1 summarises the range of views held by DTJ participants on how DTJ was addressing gender inequality.

DTJ works as an umbrella organisation for multiple micro-campaigns, accommodating different priorities of participants and organisers at different times. However, some conflicts have emerged within DTJ when silencing of difference has left groups feeling marginalised. For example, in November 2015, DTJ was criticised for banning users and deleting posts on their Facebook page which had responded to DTJ's call to submissions under #beingawoman (Ellis 2015). DTJ's excuse was that the posts were repetitive and off topic, and after criticism to their response, DTJ issued an apology acknowledging that they could be more inclusive and offered to unban those who had been banned (Ellis 2015). This incident reflects the tensions inherent in a feminism characterised by multiplicity, and the contested representations of womanhood, personhood, ability and disability. Importantly, this incident also illustrates that the prejudices and power relations of cultural and social relations at a broader level influence the potential for online spaces to afford a progressive politics of resistance. Just as elsewhere, power hierarchies sometimes need to be renegotiated to be inclusive of difference.

Within geographic academic practice, gender-focused research continues to grow (see the journal *Gender, Place and Culture*) but there has been some concern about the sidelining of research into sexism as a form of discrimination (Valentine et al. 2014). Along with other scholars who are looking at sexism in online spatial media (Leszczynski and Elwood 2015) and the power of geovisualisation to reveal gender-based differences in

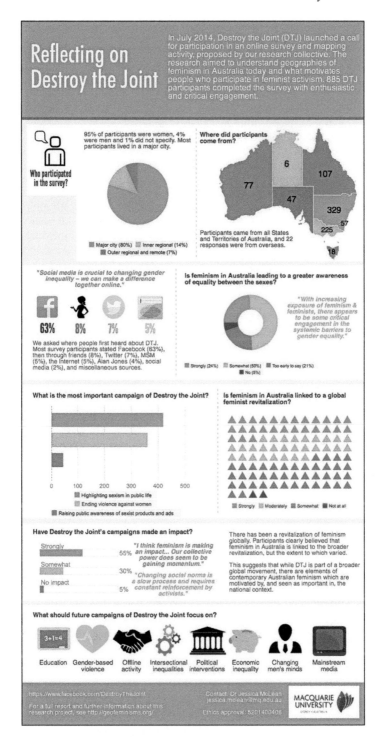

Figure 3.1   Infographic produced to summarise the data

Table 3.1   Where sexism needs to be addressed and how this is happening

Area of change	Example of participant's perception
Media	Media literacy is improving and more people are aware of the link between misogynistic language/imagery and actual violence against women
	Specific Groups like Destroy the Joint and people such as Catherine Deveny on social media. Women don't necessarily have access to other forms of media such as TV and newspapers.
Personal relationships	It raises issues many men have never had to consider such as combining children and work, and it questions behaviour men and women take for granted.
Political awareness	It's certainly exposing our enemies. It has confirmed for me that my natural political sphere, i.e. the Left, is no more supportive of women's struggles than right-wing groups. Women have to do it themselves, our allies are few.
	There is a resurgence of feminist debate which was missing for many years. This in turn has highlighted specific incidents of inequality which have been publically rectified. Even though small, the way in which these rectifications have been reported on and discussed shows awareness is improving.
Younger women	The reinvigorated interest in feminism, particularly among younger women, is providing a sense of focus for a number of discontents regarding the treatment of women in Australian society.
	I think younger women are being encouraged to examine feminism and see that it is not anti-male, but rather empowering all people to be themselves in a healthy and authentic way.
Voice	The recent wave of feminist backlash which assumes that the fight against gender inequality has been won (e.g. pay gap – despite being proved false) has allowed feminism to highlight areas in which women are being dissuaded from participating the public sphere, exploited and marginalised, as well as initiating discussions about women who don't often get a voice in everyday Australian discourse.
	It is in the public discourse. People think twice about their language (social media is watching).
Gender-based inequality	Grassroots activism is heightening social awareness of gender disparity.
	It serves to deconstruct traditional social norms which were rooted in power structures and fear.
	Feminism looks at issues like gender-based violence and workplace inequality as situations that represent a (constructed) patriarchal society. It looks at gendered issues as ones that can be changed rather than as coincidences or unchangeable facts.
	With increasing exposure of feminism and feminists, there appears to be some critical engagement in the systemic barriers to gender equality.
Social media	Awareness . . . on the Internet. Outside that world, who knows?
	Social media is crucial to changing gender inequality – we can make a difference together online.

commuting patterns (Kwan and Kotsev 2015), our research on spatialising feminist activism has contributed to (re)building focus on this axis of difference.

In this section, we have illustrated how the relationship between human agents and digital technology mutually shapes each other and co-produces space and material effects in the non-digital realm. Through DTJ and the heterogeneous networks of technology, infrastructure and people that enable it, a spatially diverse member base is able to communicate, organise and effect action. In this sense, agency is distributed and facilitated by technology, but such agentive effects could not exist without a collective of people, advocating and effecting change, united by a common cause and the reaches of the online world. It is, as quoted by Haraway earlier in this chapter, an example of us not needing to be dominated or threatened by machines, but an example of extending ourselves and enacting our agency through machines, 'The machine is us, our processes, an aspect of our embodiment' (Haraway 1985, 36).

The spatial implications of relationships between gender and technology characterised by complexity and hybridity thus are an increasingly important area for geographers to attend to.

CONCLUSION

Geographies of gender–technology relations are characterised in this chapter as diverse, complex and in constant (re)negotiation. We have focused on feminist practices and geographies of feminism herein and there are important aspects of geographies of gender–technology relations that we have not had room to consider, including transecologies and queering space. We recognise that this is a partial narrative from particular perspectives and do not claim to have achieved an all-encompassing position. What this chapter has done is provide an account of histories of gender and technology scholarly debates and how these have intersected with feminist thought and practices. We have also presented an overview of some recent work in feminist geographies of online spaces that is repositioning debates in gender–technology relations.

Revisiting the cyborg notion and tracing its continuing relevance within geographic scholarship indicates the lasting contribution of Haraway's conceptualisation. The hybrid qualities of human and technology relations are growing as we increasingly blur distinctions between online and offline interactions, and circumvent traditional distinctions between public and private spaces. Dualities in and of themselves are being refigured and abandoned with binary understandings of gender, technology and geographic space consistently receiving critical academic attention.

The emancipatory potential of new technologies is a familiar trope that has emerged in the expansion of digital spaces. However, tempering naïve optimism with realistic appreciation of the negative and positive potentialities of spaces such as social media has led to an increased understanding of how gender and technology are co-productive (cf. Wajcman 2010; McLean and Maalsen 2013).

Within geography, previous assumptions with regard to certain sub-disciplines being masculinist domains – such as GIS – are increasingly being challenged through conceptual reworkings of what GIS can and should do (e.g. Schuurman 2002), and applications of GIS technologies to answering feminist questions (Kwan 1999, 2002). Our recent research on geographies of feminism in online spaces (McLean et al. 2016; McLean and Maalsen 2013) shows the powerful accumulation of ordinary affect to produce substantive interventions. However, it also emphasises the commingling of online and offline engagements and practices by interrogating the values and possibilities of forces of change through collaborations with feminist activists.

Despite this rich work within geography, there are many opportunities for geographers to continue research into the relationship between gender and technology in a multiplicity of spaces. As mentioned previously, Smart Cities are receiving significant attention from geographers; however, there has been comparatively little research undertaken on the gendered experience of difference in this area. We also acknowledge and support Rose's (2015) advocation for cultural geographers to interpret digital technologies and digital objects, so as to keep pace with the changes in cultural production. We add to Rose's suggestion that gender be an area of analysis included in geographical research into digital technologies and objects. It is our hope that the readers of this chapter consider these possibilities in their future research.

REFERENCES

Baker, S. 2000. *The Postmodern Animal.* London: Reaktion.

Brown, L. 2007. Becoming-animal in the flesh: expanding the ethical reach of Deleuze and Guattari's tenth plateau. *PhaenEx* 2(2): 260–278.

Callon, M. 1986. Some elements of a sociology of translation: domestication of the scallops and the fisherman of St Brieuc Bay. In J. Law (ed.), *Power, Action and Belief: A New Sociology of Knowledge.* pp. 196–232. London: Routledge.

Callon, M. 1991. Techno-economic networks and irreversibility. In J. Law (ed.), *A Sociology of Monsters: Essays on Power, Technology and Domination.* pp. 196–233. London: Routledge.

Deleuze, G. and F. Guattari. 1987. *A Thousand Plateaus: Capitalism and Schizophrenia.* London: Continuum.

Ellis, K. 2015. Destroy the Joint, sure, but feminism must include disability politics. *The Conversation,* accessed 25 January 2016 from http://theconversation.com/destroy-the-joint-sure-but-feminism-must-include-disability-politics-51119.

Farr, M. 2012. Alan Jones: women are destroying the joint. news.com.au, http://www.news.com.au/national/alan-jones-women-are-destroying-the-joint/story-fndo4eg9-1226462326339.

Garoian, C. and Y. Gaudelius. 2001. Cyborg pedagogy: performing resistance in the digital age. *Studies in Art Education* 42(4): 333–347.

Graham, S. 1998. The end of geography or the explosion of place? Conceptualizing space, place and information technology. *Progress in Human Geography* 22(2): 165–185.

Grossman, R. 1980. Women's place in the integrated circuit. *Radical America* 14(1): 29–50.

Hables Gray, C. 2002. *Cyborg Citizen: Politics in the Posthuman Age.* London: Routledge.

Haraway, D. 1985. A manifesto for cyborgs: science, technology, and socialist feminism in the 1980s. *Socialist Review* 80: 65–108.

Haraway, D. 1988. Situated knowledges: the science question in feminism and the privilege of partial perspective. *Feminist Studies* 14(3): 575–599.

Haraway, D. 1991. *Simians, Cyborgs, and Women: The Reinvention of Nature.* London: Free Association Press.

Haraway, D. 2008. *When Species Meet.* Minneapolis, MN: University of Minnesota Press.

Kinsley, S. 2014. The matter of 'virtual' geographies. *Progress in Human Geography* 38(3): 364–384.

Kitchin, R. 1997. *Cyberspace: The World in the Wires.* Chichester: John Wiley.

Kwan, M. 1999. Gender and individual access to urban opportunities: a study using space–time measures. *Professional Geographer* 51(2): 210–227.

Kwan, M. 2002. Feminist visualization: re-envisioning GIS as a method in feminist geographic research. *Annals of the Association of American Geographers* 92(4): 645–661.

Kwan, M. and A. Kotsev. 2015. Gender differences in commute time and accessibility in Sofia, Bulgaria: a study using 3D geovisualisation. *Geographical Journal* 181(1): 83–96.

Latour, B. 1987. *Science in Action: How to Follow Scientists and Engineers through Society.* Oxford: Oxford University Press.

Latour, B. 1993. *We have Never been Modern.* London: Harvester Wheatsheaf.

Lawlor, L. 2008. Following the rats: becoming-animal in Deleuze and Guattari. *SubStance* 37(3): 169–187.

Leszczynski, A. and S. Elwood. 2015. Feminist geographies of new spatial media. *Canadian Geographer* 59(1): 12–28.

Longhurst, R. 2009. YouTube: a new space for birth? *Feminist Review* 93: 46–63.

Longhurst, R. 2013. Using Skype to mother: bodies, emotions, visuality, and screens. *Environment and Planning D: Society and Space* 31(4): 664–679.

Massey, D. 2005. *For Space*. London: Sage.

McLean, J. 2016. The contingency of change in the Anthropocene: more-than-real renegotiation of power relations and climate change activism in Australia. *Environment and Planning D: Society and Space* 34(3): 508–527.

McLean, J. and S. Maalsen. 2013. Destroying the joint and dying of shame? A geography of feminist revitalisation in social media and beyond. *Geographical Research* 51(3): 243–256.

McLean, J., S. Maalsen and A. Grech. 2016. Learning about feminism in digital spaces: online methodologies and participatory mapping. *Australian Geographer* 47(2): 157–177.

Milburn, C. 2002. Nanotechnology in an age of posthuman engineering: science fiction as science. *Configurations* 10(2): 261–295.

Millar, M. 1998. *Cracking the Gender Code: Who Rules the Wired World?* Toronto: Second Story Press.

Oldenziel, R. 1999. *Making Technology Masculine: Men, Women, and Modern Machines in America*. Amsterdam: Amsterdam University Press.

Plant, S. 1998. *Zeros and Ones: Digital Women 1. The New Technoculture*. London: Fourth Estate.

Rose, G. 2015. Rethinking the geographies of cultural 'objects' through digital technologies: interface, network and friction. *Progress in Human Geography* doi: 10.1177/0309132515580493.

Schuurman, N. 2002. Women and technology in geography: a cyborg manifesto for GIS. *Canadian Geographer* 46(3): 258–265.

Stewart, K. 2007. *Ordinary Affects*. London: Duke University Press.

Todd, C. 2015. GamerGate and resistance to the diversification of gaming culture. *Women's Studies Journal* 29(1): 64–67.

Valentine, G., L. Jackson and L. Mayblin. 2014. Ways of seeing: sexism the forgotten prejudice? *Gender, Place & Culture* 21(4): 401–414.

Wajcman, J. 2010. Feminist theories of technology. *Cambridge Journal of Economics* 34(1): 143–152.

Whatmore, S. 2002. *Hybrid Geographies: Natures Cultures Spaces*. London: Sage.

4. Space for STS: an overview of Science and Technology Studies
Jordan P. Howell

Science and Technology Studies – or alternatively, Science, Technology and Society (in either instance, abbreviated as STS) – is a vibrant field of academic inquiry that has emerged primarily in the US and Western Europe since the end of World War II. STS scholars and practitioners seek to illuminate both ideological and practical (in the sense of 'practice', not 'utility') dimensions of the creation and dissemination of scientific knowledge. At first glance, this would seem to be a rather sharply circumscribed area of inquiry; however, as the past 60 plus years of scholarship illustrate, the topics, theoretical approaches, research methods and modes of dissemination and implementation of STS scholarship transcend simple analyses of laboratories, inventions, and the competing ideas of 'great men'.

Rather, STS scholars have asked incisive questions challenging the notion that scientific and technological knowledge is 'revealed' or 'unveiled' in a linear progression, and highlighted instead the ways it is constructed; explored how understandings of scientific issues are translated along the path from laboratories and publications to policymakers and public understanding; and critically examined the ways in which identity categories like ethnicity, nationality and gender become enmeshed within practices and processes of scientific knowledge-making and technological development. Along the way, STS has become instituted as a field of study replete with specialist journals, academic departments and the other trappings of intellectual validity. Likewise, STS has experienced episodes of intense acrimony not only among its own practitioners disagreeing over the scope of the field and its methods but also between the discipline as a whole and the institutions at the epicenter of critique of STS: the scientists, engineers, policymakers and corporations forming the corps of the global science and technology establishment that have at different points in time had much to say in response to the findings of STS scholarship. Many readers active in geography may feel a sense of *déjà vu* from the previous paragraphs, moving as geography has from a 'science' practiced uncritically towards various state and private ends to one inspired and obligated to encompass a diversity of perspectives, methods and, ultimately, 'meanings' for society writ large (e.g. Livingstone 1992).

This chapter focuses on the origins, trends and intellectual contributions of STS, which are presented in light of the fruitful ways that the field might be combined with geography. Clearly, many traditions of scholarship within geography ask questions about the nature and construction of both scientific knowledge and technological systems. Whereas STS by definition has focused on physical- and life-scientific knowledge and technological development, historical(ly oriented) geographers and political geographers have studied the changing constructions of spatial knowledge and especially scale; cultural and human geographers have examined various constructed knowledges about populations and the places people live; political ecologists of all stripes have considered the ways in which ideologies of expertise and positivism shape modification of the physical environment via infrastructure or other human-induced change; and experts in remote sensing, GIS and cartography have conducted many critical analyses of the technologies *of* geography in the sub-field of critical GIS. This chapter aims to encourage geographers to build on these frequently *implicit* mobilizations of STS's ethos and to more *explicitly* apply the varied body of STS literature in their work. Proceeding from an historical overview of the development of the field of STS, this chapter offers rough sketches of major conceptual issues and debates within STS before highlighting some areas of overlap, connection and mutual reinforcement between geography and STS.

THE HISTORICAL DEVELOPMENT OF STS AS A FIELD

There are a number of chapter- and article-length overviews of the history of STS as a field, recounting origins, key moments and intellectual trends (e.g. Edge 1995; Bucchi 2004; Sismondo 2004, 2008; Yearley 2005). These have been synthesized here to sketch a more compressed outline of the field, and also one which may more directly highlight some of the overlaps with 20th- and 21st-century geography.

The field of STS has its most commonly agreed-upon origins in the 1960s. Although there was (and remains) a clear tradition of historical scholarship dealing with science, medicine, inventions and the like, and an even deeper philosophical tradition concerned with the nature of science (cf. Turner 2008), it was not until the more general upheavals associated with the 1960s that a particular emphasis on the roles or meanings of science in society emerged. This is perhaps understandable given the previously unforeseen levels of 'high' science and technology that erupted during World War II, ushering global society into the atomic age. Edge (1995) argued that a concern (perhaps most specifically in policymaking circles) with the economic costs of science and ways to ensure a 'return' on the investment of funds into scientific research agendas – in other words a 'science of science' (Edge 1995, 7) – can be isolated as a root of the STS tradition during this era. However, of greater significance to the growth of the field were advances in the area of the 'sociology of scientific knowledge' (sometimes abbreviated 'SSK').

The form and content of SSK before World War II and through the early 1960s is conventionally caricatured by pointing to the work of Robert Merton (1973). For example, Edge (1995) categorizes Mertonian understandings of science as linear, 'progress'-oriented (but with no ultimate endpoint), and constituting a self-contained and self-evident social system invincible to external societal forces. In contrast, the post-War era produced a new and more relativistic approach to SSK drawing on a range of social sciences and humanities to illustrate instead the intensely social – meaning, constructed – nature of scientific knowledge. This more radical approach to SSK is exemplified most strongly by the work of Thomas Kuhn (1962), whose humanistic ethos towards understanding science, scientists and scientific practice has resonated through STS to the present day.

In his overview of the STS field, Sismondo (2004) argues that the new Kuhnian paradigm opened a universe of possibility for STS as a field that continues to unfold today. Subsequently Sismondo (2008) highlights no fewer than six distinct STS traditions that he links to Kuhn's work, several of which – for example Actor-Network Theory and the idea of 'social construction' of science and technology – are frequently among the first introductions that geographers make with STS. He argues that these emerged somewhat sequentially (although not necessarily in a linear fashion) during the 1970s, 1980s and 1990s. Alongside these first Kuhnian traditions, both Edge and Sismondo highlight a parallel track of STS scholarship, concerned less with theorizations of science and critical analyses of the ways in which scientists do their work than the interface of science and society both in the public sphere and in the specific instances of science education and scientist training.

The major contours of each of these (and other) trends are the focus of the next section of this chapter; however, at this point it is worth noting that the growth of STS along these lines did not proceed unimpeded. STS, like many humanities and social science disciplines, was affected by the postmodern and deconstructionist 'turn' in scholarly research. However, the conflict over deconstructing science spilled into the public sphere in a way that many other scholarly fields did not experience. The so-called 'Science Wars' of the mid-1990s centered on at times vitriolic exchanges between STS scholars and experts in the life and physical sciences (several volumes, collections and reviews capture the nature and tone of these confrontations, e.g. Gross and Levitt 1994; Ross 1996; Sokal and Bricmont 1998; Callon 1999b; Sokal 2000; Parsons 2003). The confrontations seemed to center on one group's caricatures of the other. Experimental scientists, for example, derided some within STS for apparently arguing that the physical worlds we inhabit are in fact themselves constructed; in other words, that there is no objective material reality (and by extension, no possibility for *science* or an organized 'knowing' of that reality or any type of uniform 'law', like gravity, that would govern all physical phenomena). Likewise, some in STS excoriated scientists' claims to absolute knowledge, ridiculing practitioners' adherence to 'revealed truths' and 'laws of nature' that clearly had been stamped by the forces of capitalism or Western notions of male domination. The 'science wars' reached a nadir surrounding physicist Alan Sokal's publication of a parody STS-deconstructionist research article in the (then, nonpeer-reviewed) cultural studies journal *Social Text*, which Sokal later revealed to be a hoax and intended to illuminate the low scholarly standards and poor quantitative reasoning skills of many cultural studies and STS practitioners. The 'Sokal Affair' was regarded by many, and not only in STS, to have been a highly unsavory episode for the simultaneous abuse of 'good faith' in scholarly publishing and lax editorial practices, but also more significantly because it eradicated an opportunity to clarify in a useful way what the sciences and scholarship about the sciences (like STS) could learn from one another.

Clearly, however, the Sokal Affair was the death knell for neither STS nor experimental science. STS-specific journals, publication series and professional societies are well established and remain healthy. Funding sources for STS research can be identified at major national agencies and foundations, such as the National Science Foundation in the US. Likewise, STS departments and degree programs can be found at colleges and universities around the world, and in various institutional configurations (e.g. some housed amongst the humanities, others amongst the social sciences, and still others as components of larger business, engineering, or experimental science endeavors). Tables 4.1 and 4.2 illustrate various institutional components of the STS field around the world.

Table 4.1   *Listing of peer-reviewed STS journals and the year in which they started publication (after Agar 2015) and listing of major international STS scholarly societies*

Peer-reviewed STS[a] journals; year founded	Major STS scholarly societies; year founded
Minerva; 1962	*STS-Specific Societies*
Perspectives on Science; 1993	Asia Pacific STS Network; 2008
Physics in Perspective; 1999	European Association for the Study of Science
Public Understanding of Science; 1992	and Technology (EASST); 1981
Research Policy; 1971	Japanese Society for STS; 2001
Science & Education; 1992	Society for Social Studies of Science; 1975
Science and Public Policy; 1974	Society for the History of Technology; 1958
Science as Culture; 1987	
Science in Context; 1987	
Science, Technology & Human Values; 1976	
Social Epistemology; 1987	
Social Studies of Science; 1971	
Technology and Culture; 1959	
Valuation Studies; 2013	

Note: [a]This list specifically excludes journals in the history and philosophy of science, technology, and medicine. A listing of these additional types of journals has been compiled by Agar (2015).

MAJOR CONCEPTUAL ISSUES AND DEBATES IN STS

The journals, societies, and departments detailed above thrive on the intellectual richness of STS scholarship. This section introduces some of the major issues and theoretical and methodological approaches that can be identified in STS. Since the 1970s, STS scholarly organizations (and in particular the Society for Social Studies of Science, or 4S) have published introductory handbooks offering an overview of the field and intellectual traditions of STS, as well as highlighting areas of future growth (e.g. Jasanoff et al. 1995; Hackett et al. 2008). The list that follows is derived from examples of this type of work crafted by Sismondo (2004, 2008).

Kuhn and 'Paradigms'

Much of pre-Kuhnian history of science centered on, at least implicitly, the idea that eternal truths and laws regarding the physical world already existed and were waiting to be uncovered through the activities of great scientists. In marked contrast, Kuhn, in his *The Structure of Scientific Revolutions* (1962), argued instead that scientific knowledge is in fact constructed from somewhat arbitrary ways of seeing the world. These in turn become incorporated into a rigid and rigorous knowledge of the world, 'proven' by particular approaches to the practice of science subsequently transmitted to others via science education, laboratory apprenticeships, and so on. When anomalies to the established science cannot be explained away within the existing paradigm, a revolution

Table 4.2   STS educational programs by country

Country	Number of STS educational programs[a]
Australia	4
Austria	2
Belgium	1
Brazil	2
Canada	4
China	3
Denmark	1
Finland	1
France	10
Germany	3
India	4
Israel	1
Japan	2
Malaysia	1
Netherlands	7
Norway	2
Singapore	1
South Korea	3
Sweden	6
Taiwan	3
UK	13
US	47

Notes: [a]This table includes programs that emphasize the history and philosophy of science, technology, and/or medicine as well as STS as described in this chapter. However, it has tried to minimize inclusion of nondegree programs, such as intra-university consortia and committees.

Source: STS Next 20 (2011); STS Wiki (2015); LIS522LE Webliography STS (n.d.).

occurs that prompts a re-evaluation of accepted facts and theories. In the Kuhnian model, advances in scientific knowledge are not the result of gradual accumulation of data towards a pre-determined ending point, but rather the result of upheaval and competition between different segments of a given scientific community regarding how to handle outlying information that does not fit within an established set of explanations and practices.

Kuhn's thesis does not directly question the existence or functioning of physical forces and laws in any poststructural type of way. However, Kuhnian SSK confronts the idea that the physical world is governed by immutable rules, since the book strives to illustrate the ways in which describing and documenting those rules is a highly subjective process. As such, a great deal of subsequent work in STS has at least part, if not a great deal, of its philosophical origin in the constructivist ideas raised in *The Structure of Scientific Revolutions*, as is illustrated in the sections below.

The 'Strong Program' of STS and the Empirical Program of Relativism

The 'strong program' of STS emerged alongside the work of scholars like Bloor (1976) and MacKenzie (1981), among others. Strong program scholars adapted the Kuhnian ethos (implicitly or explicitly) to illustrate the social dimensions of constructing scientific knowledge, as opposed to simply studying scientific error and failed attempts at theory building (Sismondo 2008). Like Kuhn, scholarship in the strong program sought to move away from histories of science, medicine and technology that suggested an inevitability that scientists would uncover the 'truth' about the physical phenomena that they were observing or seeking to understand. Bloor (1976) in particular articulated the importance of naturalist explanations and weighing equally claims to truth and falsehood, urging scholars to focus on the reasons why particular claims achieve stability and acceptance within the scientific community rather than proving why one scientific explanation was right or wrong.

In a similar vein, the Empirical Program of Relativism (EPOR) examines specifically the building of knowledge claims among the scientific community. Collins (1985) argued that as the basis of much scientific consensus-building depends on the replicability of laboratory experiments and data analysis techniques, a truth claim and the work to support the claim become mutually reinforcing. That is to say, as scientists exist as a community of practitioners, claims to truth gain strength and are deemed to be the correct interpretation of natural phenomena as the work done to illustrate those claims is considered also to have been conducted in the correct fashion. Thus in EPOR, as Sismondo (2008, 15) argues, 'the constitution of scientific knowledge contains an ineliminable reference to particular social configurations'.

Laboratory Studies and Ethnographies of Science and Technology

If science is produced in places of experimentation, then it is clear why STS scholars would endeavor to understand in great detail the spaces where knowledge claims are made. Growing out of the logic of EPOR was the category of studies first produced in the late 1970s and 1980s seeking to illuminate laboratory practices. One of the most prominent, published by Latour and Woolgar (1979), adopts an avowedly anthropological approach to understanding the ways in which practitioners of science use experiments, laboratory equipment and the processes of scholarly attribution and publication to order and describe the natural world. Critically, and diverging from the 'strong program', the arguments in many laboratory studies-type projects center on the construction of *observable phenomena* themselves, as opposed to the ways in which *claims to credibility* are constructed; an even greater emphasis is placed on the working cultures, social arrangements and modes of training laboratory workers in this area of STS work than elsewhere. As a result, these types of studies (Knorr-Cetina 1981; Knorr-Cetina and Mulkay 1983; Collins 1985; Lynch 1985, 1993; Traweek 1988) seek to demonstrate that 'laboratory phenomena . . . are not in themselves natural but are made to stand in for nature; in their purity and artificiality they are typically seen as more fundamental and revealing of nature than the natural world itself', and can perhaps be traced to some of the underlying factors shaping the 'science wars' outlined above (Sismondo 2008, 15).

Despite being controversial, the ideas motivating laboratory studies and an ethnographic approach to understanding the production of science themselves contributed to three new, major tracks for research in STS. The first was a revitalized look at the history of science (e.g. Shapin and Schaffer 1989), where the approaches of laboratory studies brought fresh insights into the scientists and inventors of the past. Second, and more abstractly, was a new emphasis on technology, invention and the meaning of various 'artifacts' in society. Pioneered explicitly by Bijker, Pinch, Hughes and Law (Bijker et al. 1987; Bijker and Law 1992), but drawing from the lineage of Mumford (1963), Winner (1980), and an historical tradition of technology studies, the new Social Construction of Technology examined the ways in which the meaning of technologies as diverse as bicycles and light bulbs (Bijker 1995) to nuclear power facilities (Hecht 1998) was constructed and interpreted differently by different entities during design and use.

Actor-Network Theory

The third, and perhaps farthest-reaching new track within STS that emerged from a connection to laboratory studies is the theory-method known as Actor-Network Theory (ANT). Pioneered by Serres, Latour, Callon and Law (in association with others, and in various configurations, e.g. Law 1986, 1991, 1992, 1994; Latour 1987, 1993, 1996; Callon and Law 1995; Callon 1999a, b; Law and Hassard 1999), ANT proposes to 'flatten out' relationships between human and nonhuman entities; that is to say, as a methodological approach and theoretical orientation, ANT ascribes equal agency to the living and non-living, the human and everything else. In relationship to intellectual trends in STS, ANT builds on the notion that scientific knowledge is actually a collaboration, or in the terms of STS, an 'heterogeneous network' between scientists (and the institutional contexts in which they work), their equipment (and the various metrics and outputs of scientific equipment) and the properties (controllable or otherwise) of the materials used in experiments, prototypes and models (e.g. Latour 1988). Yet while focusing on the constructed nature of many scientific findings, ANT more heavily emphasizes the formation and maintenance of the networks themselves. As a result, a rigorous interpretation of ANT and its philosophical commitments has interesting implications for key geographic concepts like space and scale.

ANT theorists and practitioners, perhaps most notably Bruno Latour, have also applied the concepts of heterogeneity, flattening and an eradication of divisions between nature and society in public and political arenas. Latour, for example, has authored works applying ANT to public transit projects (Latour 1996), environmental politics (Latour 2004) and the legislative process itself (Latour and Weibel 2005; Latour 2009). Related to this concern for 'the political' is a strand of STS scholarship concerned with co-construction (Jasanoff 2004), or the study of 'the process of adjusting pieces of technoscience and their environments to each other, or of simultaneously creating both knowledge and institutions' (Sismondo 2008, 17). In other words, studies emphasizing co-construction look for ways in which society (or some segment of society) is made ready for a particular piece of knowledge, medical treatment or technological advance.

OTHER AREAS OF CONCERN IN STS: THE 'SCIENCE OF SCIENCE'

The sociology of scientific knowledge ushered in by Kuhn led the way for various studies of the ways in which science and technology are constructions and deal in the construction of new knowledge. However another major emphasis of work in STS has persisted alongside constructionism since the end of World War II. The 'science of science' (Edge 1995), or perhaps the 'Science, Technology, and Society' variant of STS, has investigated what might be described as more practical questions about funding mechanisms and especially the relationship between scientific innovation and state, military and capitalist goals (cf. Amsterdamska 2008), as well as a pool of related topics considering the relationship between scientific knowledge and 'the public', conceived in terms of the public's understanding of science, science education and training at various levels, and public participation in science governance issues (after Hackett 2008).

Science, Money, the State and Industry

Although a caricature of scientific inquiry might depict scholars in lab coats investigating solely that which piques their curiosity or improves humanity's understanding of the Earth and universe, the reality of scientific work is that it is tightly intertwined with profit-seeking companies, political agendas and especially in relation to the development of new technologies, military activity. One major focus has been the questions of commercialization and privatization, and how these make an impact on scientific practice. Mirowski and Sent (2002) argue that the ways in which science has (or, has not) been commercialized have directly shaped the practices that happen in the laboratory, as the construction of both scientific knowledge and new technology proceeds along different paths in academic, government and private contexts given the different aims each type of laboratory might have. While this claim has also been asserted by scholars in the ANT tradition (e.g. Latour 1996) – that economics are just another variable in the 'heterogeneously engineered' network of science – Mirowski and Sent (2008, 639) argue more forcefully for 'stressing the essential historical instability of the commercial/communal binary as instantiated in actual concrete practice'. They and others working on this topic suggest that, for example, the early 20th-century context for science, centered as it was on large private firms and captains of industry, imbued the practices of science with a completely different ethos and suite of productive practices than the emerging neoliberal, internationalized scientific regime; these changes can be perhaps most readily witnessed in the shifting institutional contexts in which scientists work vis-à-vis academic employment (and tenure) but also employment trends in other government, private and military contexts (cf. Gibbons 1994; Croissant and Restivo 2001; Croissant and Smith-Doerr 2008).

Of special note is the longstanding concern within STS for the relationships between science and military force, from ethical, constructivist and applied perspectives. Focusing on the US context, and looking back to the first half of the 20th century, Smit (1995, 598) argues that a 'highly organized and concentrated effort during World War II [saw] a great number of scientists, mainly under the auspices of the newly established Office of Scientific Research and Development, contribute to the development of a variety of new technologies (atomic bomb, radar, proximity fuse, but also penicillin) . . . [implying] a dramatic turn in the role for science and technology in future military affairs'. Understanding of the roles of scientists and inventors in developing more efficient ways to both kill and heal has been complemented by organizational studies of the linkages between the military, universities (and especially university funding) and the ethics of scholarly practices such as publication embargoes and confidentiality agreements (cf. Rappert et al. 2008).

Public Understanding of Science, Public Participation in Science Governance and Science Education

Debates over the interconnections between science, money and the state (whether the military or otherwise) have important ethical dimensions, and from a governance per- spective, are certainly important in and of themselves. On the other hand, they matter only so far as various publics understand what is at stake and how the science underlying any debate over scientific funding and practice actually works. Yet publics are more than just passive consumers of scientific knowledge or indiscriminate users of a given technol- ogy. Bucchi and Neresini (2008, 449) define public participation in science as 'organized and structured' activities wherein 'nonexperts become involved, and provide their own input to, agenda setting, decision-making, policy forming, and knowledge produc- tion processes regarding science'. Public participation in science may take many forms, ranging from protests over military research funding to patients' rights groups (e.g. Hess et al. 2008) to citizen science organizations (e.g. Irwin 1995). Regardless of the specific topic, STS work considering the role of the public centers on the fundamental question of expertise: who is considered an expert on a particular issue, and why? The (percep- tion of) the public's understanding of science, or more accurately, a given scientific paradigm, conventionally focused on the so-called 'deficit model' (Bucchi and Neresini 2008), wherein it was assumed that 'people' were essentially scientifically illiterate, and that if their technical knowledge could be improved, they would more readily accept the importance of scientific findings. In reality, this strand of STS work has clearly illustrated that, many times, expert and nonexpert narratives are not factually 'wrong' or 'right' – although they certainly can be – but rather alternative interpretations of both scientific data and the ability of parties to establish credibility. As such, public participation in science, whether through the construction of scientific knowledge, participation in the governance process or most powerfully by agitating for the recognition of environmental and human health issues minimized by the scientific establishment (e.g. early demands for researchers to study HIV/AIDS, Epstein 1996), plays a crucial counter-balancing role to expert narratives.

SPATIAL AND GEOGRAPHICAL CONNECTIONS AND IMPLICATIONS

The rough sketches made in the preceding section highlight some of the major concepts and debates in STS. These sketches suggest not only the diversity of thought within STS, but also opportunities for geographers to address research questions and problems with a fresh perspective. Every category outlined above already has examples of geographers employing STS theory and method, although perhaps not always explicitly. To highlight only a handful of examples familiar to the author: scholars in both disciplines have endeavored to study spatial dimensions of science and the role of place in the construction of scientific knowledge (Livingstone 2003; Henke and Gieryn 2008); nonrepresentational and other 'new' more-than-human geographies emphasizing relationality have very close theoretical origins to ANT (Thrift 1996, 2008; Murdoch 1997a, b, 1998; Whatmore 2002); narratives of expertise are regularly deconstructed (Whatmore 2009; Howell 2012; Lave 2012, 2015); and the question of public participation in environmental infrastructure and landscape or wildlife conservation issues is a common topic of examination (Castree and Braun 2001; Furlong 2011; Howell 2015). Programmatically, one could hardly choose a more contentious history of an academic discipline than geography to illustrate the arguments about paradigm that Kuhn made so many decades ago, or to consider the 'real-world' ethical dimensions of applying scientific knowledge to military endeavors than the checkered past of geopolitical thought.

Readers can undoubtedly think of additional examples of areas where geography and STS can be brought in to fruitful combination; if not, plenty of examples can be found in this volume. Bringing STS and geography into closer conversation will certainly augment scholarly efforts to address the increasingly technical and science-ized problems facing the world today.

REFERENCES

Agar, J. 2015. *A List of STS Journals*. UCL STS Observatory, accessed 4 January 2016 from https://blogs.ucl. ac.uk/sts-observatory/2015/02/24/a-list-of-sts-journals/.

Amsterdamska, O. 2008. Institutions and economics. In E. Hackett, O. Amsterdamska, M. Lynch and J. Wajcman (eds.), *The Handbook of Science and Technology Studies*. pp. 631–634. Cambridge, MA: MIT Press.

Bijker, W.E. 1995. *Of Bicycles, Bakelites, and Bulbs: Toward a Theory of Sociotechnical Change*. Cambridge, MA: MIT Press.

Bijker, W. and J. Law. 1992. *Shaping Technology/Building Society: Studies in Sociotechnical Change*. Cambridge, MA: MIT Press.

Bijker, W., T. Hughes and T. Pinch. 1987. *The Social Construction of Technological Systems: New Directions in the Sociology and History of Technology*. Cambridge, MA: MIT Press.

Bloor, D. 1976. *Knowledge and Social Imagery*. London: Routledge & Kegan Paul.

Bucchi, M. 2004. *Science in Society: An Introduction to the Sociology of Science*. New York: Routledge.

Bucchi, M. and F. Neresini. 2008. Science and public participation. In E. Hackett, O. Amsterdamska, M. Lynch and J. Wajcman (eds.), *The Handbook of Science and Technology Studies*. Cambridge, MA: MIT Press.

Callon, M. 1999a. Some elements of a sociology of translation: domestication of the scallops and the fishermen of St. Brieuc Bay. In M. Biagioli (ed.), *Science Studies Reader*. pp. 67–83. New York: Routledge.

Callon, M. 1999b. Whose imposture? Physicists at war with the third person. *Social Studies of Science* 29(2): 261–286.

Callon, M. and J. Law. 1995. Agency and the hybrid collectif. *South Atlantic Quarterly* 94(2): 481–507.

Castree, N. and B. Braun. 2001. *Social Nature: Theory, Practice, and Politics*. Malden, MA: Blackwell.

Collins, H.M. 1985. *Changing Order: Replication and Induction in Scientific Practice*. London: Sage.

Croissant, J. and S.P. Restivo. 2001. *Degrees of Compromise: Industrial Interests and Academic Values*. Albany, NY: State University of New York Press.

Croissant, J. and L. Smith-Doerr. 2008. Organizational contexts of science: boundaries and relationships between university and industry. In E. Hackett, O. Amsterdamska, M. Lynch and J. Wajcman (eds.), *The Handbook of Science and Technology Studies*. Cambridge, MA: MIT Press.

Edge, D. 1995. Reinventing the wheel. In S. Jasanoff, G. Markle, J.C. Petersen and T. Pinch (eds.), *Handbook of Science and Technology Studies*. pp. 3–23. Thousand Oaks, CA: Sage.

Epstein, S. 1996. *Impure Science: AIDS, Activism, and the Politics of Knowledge*. Berkeley, CA: University of California Press.

Furlong, K. 2011. Small technologies, big change: rethinking infrastructure through STS and geography. *Progress in Human Geography* 35(4): 460–482.

Gibbons, M. 1994. *The New Production of Knowledge: The Dynamics of Science and Research in Contemporary Societies*. London: Sage.

Gross, P. and N. Levitt. 1994. *Higher Superstition: The Academic Left and its Quarrels with Science*. Baltimore, MD: Johns Hopkins University Press.

Hackett, E. 2008. Politics and publics. In E. Hackett, O. Amsterdamska, M. Lynch and J. Wajcman (eds.), *The Handbook of Science and Technology Studies*. pp. 429–432. Cambridge, MA: MIT Press.

Hackett, E., O. Amsterdamska, M. Lynch and J. Wajcman (eds.), 2008. *The Handbook of Science and Technology Studies*, 3rd ed. Cambridge, MA: MIT Press.

Hecht, G. 1998. *The Radiance of France: Nuclear Power and National Identity after World War II*. Cambridge, MA: MIT Press.

Henke, C. and T. Gieryn. 2008. Sites of scientific practice: the enduring importance of place. In E. Hackett, O. Amsterdamska, M. Lynch and J. Wajcman (eds.), *The Handbook of Science and Technology Studies*. Cambridge, MA: MIT Press.

Hess, D., S. Breyman, N. Campbell and B. Martin. 2008. Science, technology, and social movements. In E. Hackett, O. Amsterdamska, M. Lynch and J. Wajcman (eds.), *The Handbook of Science and Technology Studies*. Cambridge, MA: MIT Press.

Howell, J. 2012. Risk society without reflexive modernization? The case from northwestern Michigan. *Technology in Society* 34: 185–195.

Howell, J. 2015. 'Modes of governing' and solid waste management in Maui, Hawaii, USA. *Environment and Planning A* 47(10): 2153–2169.

Irwin, A. 1995. *Citizen Science: A Study of People, Expertise and Sustainable Development*. London: Routledge.

Jasanoff, S. 2004. *States of Knowledge: The Co-Production of Science and Social Order*. London: Routledge.

Jasanoff, S., G. Markle, J. Petersen and T. Pinch (eds.). 1995. *Handbook of Science and Technology Studies*. Thousand Oaks, CA: Sage.

Knorr-Cetina, K. 1981. *The Manufacture of Knowledge: An Essay on the Constructivist and Contextual Nature of Science*. Oxford: Pergamon Press.

Knorr-Cetina, K. and M. Mulkay. 1983. *Science Observed: Perspectives on the Social Study of Science*. London: Sage.

Kuhn, T. 1962. *The Structure of Scientific Revolutions*. Chicago, IL: University of Chicago Press.

Latour, B. 1987. *Science in Action: How to Follow Scientists and Engineers Through Society*. Cambridge, MA: Harvard University Press.

Latour, B. 1988. *The Pasteurization of France*. Cambridge, MA: Harvard University Press.

Latour, B. 1993. *We Have Never Been Modern*. Cambridge, MA: Harvard University Press.

Latour, B. 1996. *Aramis, or, The Love of Technology*. Cambridge, MA: Harvard University Press.

Latour, B. 2004. *Politics of Nature: How to Bring the Sciences Into Democracy*. Cambridge, MA: Harvard University Press.

Latour, B. 2009. *The Making of Law: An Ethnography of the Conseil d'Etat*. Cambridge: Polity.

Latour, B. and P. Weibel. 2005. *Making Things Public: Atmospheres of Democracy*. Cambridge, MA and Karlsruhe, Germany: MIT Press and ZKM/Center for Art and Media.

Latour, B. and S. Woolgar. 1979. *Laboratory Life: The Social Construction of Scientific Facts*. Beverly Hills, CA: Sage.

Lave, R. 2012. Bridging political ecology and STS: a field analysis of the Rosgen Wars. *Annals of the Association of American Geographers* 102(2): 366–382.

Lave, R. 2015. The future of environmental expertise. *Annals of the Association of American Geographers* 105(2): 244–252.

Law, J. 1986. *Power, Action, and Belief: A New Sociology of Knowledge?* London: Routledge.

Law, J. 1991. *A Sociology of Monsters: Essays on Power, Technology, and Domination*. London: Routledge.

Law, J. 1992. Notes on the theory of the actor-networks: ordering, strategy, and heterogeneity. *Systems Practice* 5(4): 379–393.

Law, J. 1994. *Organizing Modernity*. Cambridge, MA: Blackwell.

Law, J. and J. Hassard. 1999. *Actor Network Theory and After*. Malden, MA: Blackwell.

LIS522LE Webliography STS. n.d. *STS Academic Programs*, accessed 10 January 2016 from https://sites.google.com/site/lis522lewebliographysts/Home/sts-academic-programmes.

Livingstone, D. 1992. *The Geographical Tradition: Episodes in the History of a Contested Enterprise*. Oxford: Blackwell.

Livingstone, D. 2003. *Putting Science in its Place: Geographies of Scientific Knowledge*. Chicago, IL: University of Chicago Press.

Lynch, M. 1985. *Art and Artifact in Laboratory Science: A Study of Shop Work and Shop Talk in a Research Laboratory*. London: Routledge & Kegan Paul.

Lynch, M. 1993. *Scientific Practice and Ordinary Action: Ethnomethodology and Social Studies of Science.* Cambridge: Cambridge University Press.

MacKenzie, D. 1981. *Statistics in Britain, 1865–1930: The Social Construction of Scientific Knowledge.* Edinburgh: Edinburgh University Press.

Merton, R. 1973. *The Sociology of Science: Theoretical and Empirical Investigations.* Chicago, IL: University of Chicago Press.

Mirowski, P. and E.-M. Sent. 2002. *Science Bought and Sold: Essays in the Economics of Science.* Chicago, IL: University of Chicago Press.

Mirowski, P. and E.-M. Sent. 2008. The commercialization of science and the response of STS. In E.J. Hackett, O. Amsterdamska, M. Lynch and J. Wajcman (eds.), *The Handbook of Science and Technology Studies.* Cambridge, MA: MIT Press.

Mumford, L. 1963. *Technics and Civilization.* New York: Harcourt.

Murdoch, J. 1997a. Inhuman/nonhuman/human: actor-network theory and the prospects for a nondualistic and symmetrical perspective on nature and society. *Environment and Planning D: Society and Space* 15(6): 731–756.

Murdoch, J. 1997b. Towards a geography of heterogeneous associations. *Progress in Human Geography* 21(3): 321–337.

Murdoch, J. 1998. The spaces of actor-network theory. *Geoforum* 29(4): 357–374.

Parsons, K. 2003. *The Science Wars: Debating Scientific Knowledge and Technology.* Amherst, NY: Prometheus Books.

Rappert, B., B. Balmer and J. Stone. 2008. Science, technology, and the military: priorities, preoccupations, and possibilities. In E. Hackett, O. Amsterdamska, M. Lynch and J. Wajcman (eds.), *The Handbook of Science and Technology Studies.* Cambridge, MA: MIT Press.

Ross, A. 1996. *Science Wars.* Durham, NC: Duke University Press.

Shapin, S. and S. Schaffer. 1989. *Leviathan and the Air-Pump: Hobbes, Boyle, and the Experimental Life.* Princeton, NJ: Princeton University Press.

Sismondo, S. 2004. *An Introduction to Science and Technology Studies.* Malden, MA: Blackwell.

Sismondo, S. 2008. Science and technology studies and an engaged program. In E. Hackett, O. Amsterdamska, M. Lynch and J. Wajcman (eds.), *The Handbook of Science and Technology Studies.* pp. 13–31. Cambridge, MA: MIT Press.

Smit, W. A. 1995. Science, technology, and the military. In S. Jasanoff, G. Markle, J.C. Petersen and T. Pinch (eds.), *Handbook of Science and Technology Studies.* pp. 598–626. Thousand Oaks, CA: Sage.

Sokal, A. 2000. *The Sokal Hoax: The Sham that Shook the Academy.* Lincoln, NE: University of Nebraska Press.

Sokal, A. and J. Bricmont. 1998. *Fashionable Nonsense: Postmodern Intellectuals' Abuse of Science.* New York: Picador USA.

STS Next 20. 2011. *STS Program List,* accessed 10 January 2016 from http://stsnext20.org/stsworld/sts-programs/.

STS Wiki. 2015. *Worldwide Directory of STS Programs,* accessed 10 January 2016 from http://www.stswiki.org/index.php?title=Worldwide_directory_of_STS_programs.

Thrift, N. 1996. Inhuman geographies: landscapes of speed, light, and power. In N. Thrift (ed.), *Spatial Formations.* pp. 256–310. London: Sage.

Thrift, N. 2008. *Non-representational Theory: Space, Politics, Affect.* New York: Routledge.

Traweek, S. 1988. *Beamtimes and Lifetimes: The World of High Energy Physicists.* Cambridge, MA: Harvard University Press.

Turner, S. 2008. The social study of science before Kuhn. In E. Hackett, O. Amsterdamska, M. Lynch and J. Wajcman (eds.), *The Handbook of Science and Technology Studies.* pp. 34–62. Cambridge, MA: MIT Press.

Whatmore, S. 2002. *Hybrid Geographies: Natures, Cultures, Spaces.* London: Sage.

Whatmore, S. 2009. Mapping knowledge controversies: science, democracy and the redistribution of expertise. *Progress in Human Geography* 33(5): 587–598.

Winner, L. 1980. Do artifacts have politics? *Daedalus* 109(1): 121–136.

Yearley, S. 2005. *Making Sense of Science: Understanding the Social Study of Science.* Thousand Oaks, CA: Sage.

PART II

COMPUTATIONAL TECHNOLOGIES

5. Code/space and the challenge of software algorithms[1]

Martin Dodge

Software has replaced a diverse array of physical, mechanical, and electronic technologies used before the 21st century to create, store, distribute and interact with cultural artifacts. It has become our interface to the world, to others, to our memory and our imagination – a universal language through which the world speaks, and a universal engine on which the world runs.

Manovich 2013

How many waking hours in a day are you focused on your mobile phone, a laptop screen, the smart television on the wall, a connected watch on your wrist or the GPS display and digital dashboard in your car? Lives are lived through computer screens, which are really windows into a space of code. The spaces are brought into being through software algorithms working on data in ways and at speeds that are hard, if not impossible, for people to comprehend. It's not possible to really see code as code and even if we did few would have the capacity to meaningfully decipher it. Yet code is an active and important social actor; it is indispensible and unavoidable in the conduct of daily life. Code changes how space is produced and poses a significant challenge to unpack and explain.

Myriad systems of software are now essential to the functioning of many societies. Software makes a difference to how social, spatial and economic life takes place. It is a vital element in the operation and governance of transport and utilities infrastructures, the planning and maintenance of city spaces, consumption practices and work processes in virtually all sectors of the economy, as well as personal life in domestic settings. To date, most analysis by geographers and allied social scientists has been focused on the technologies that software enables, rather than the underlying code (data inputs, information stores and decision-making algorithms) that does work in the world and is capable of bringing particular socio-spatial formations into being (Figure 5.1).

An indicator of the contemporary significance of code is the huge economic value accruing to most successful companies that develop software and provide digital services. Software businesses have grown rapidly and come to rival and even eclipse established corporate giants in petroleum production, retail and automobile manufacture in terms of stock market capitalization. Corporations like Microsoft, Alphabet (parent company of Google), Facebook, SAP, Oracle and Apple are worth tens of billions of dollars and are often highly profitable. For two decades Microsoft was the largest software firm, by a significant margin, and it still had revenues in 2015 of $93.6 billion and had a market value of $416 billion (March 2016). In terms of economic scale (and media profile and cultural cache) Microsoft has been usurped by the likes of Google in recent years. The new parent company Alphabet has only 61,000 employees but its algorithms and vast databases are depended on by hundreds of millions of people every day to solve all manner of tasks and consequently investors value the business massively (capitalization was $516 billion in March 2016).

Figure 5.1   *Control rooms with banks of screens to monitor myriad processes and specialized server computers in data centers. Such spaces are essential to contemporary society but often hidden away from sight; with restricted access they are designed specifically for the work of software and data computation, their functional value depending on code. They are code/space*

Beyond the influence of individual market-leading tech corporations, software is elemental in new rounds of capitalism, seen most overtly in the dot-com boom in the late 1990s, the rise of social networks more recently and the current rapid growth in the so-called 'sharing economy', which is dramatically reshaping established markets. For example in the fields of urban mobility and short-term rental space the success of new economic models from companies like Uber and Airbnb rests solely on their innovative code with dynamic databases of available resources and clever matching algorithms often exploiting geolocation information from smartphone apps, along with seamless online payment systems and means of creating trust in virtualized exchange through reviews and recommendations.

There is a need to be alert to the organization cultures, political economy and geographical locations in the development of the software underpinning social networking, smartphones and the sharing economy. The code is conceived and written by people, usually employed by corporations, and working in particular places. Consequently software is not a neutral agent; it embodies the values and biases of those who write it. Work by economic geographers and regional scientists shows how much the creation, maintenance and marketing of software products is heavily concentrated in relatively few places and conducted by a highly specialized workforce, despite the rhetoric that often surrounds tech companies (e.g. Andreosso-O'Callaghan et al. 2015; Arora and Gambardella 2006; Cook et al. 2013). Distinct but closely interconnected to power of software there has been massive growth in the collection of near-continuous streams of personal data on millions of people. Whenever a person interacts with a software system – every login, swipe of a card, social network post, online payment, opening of an app, clicking on a hyperlink – is logged and often tied to identifiers and geolocation details. According to the influential security consultant Bruce Schneier (2016, no pagination), '[d]ata brokers save everything about us they can get their hands on. This data is saved and analyzed, bought and sold, and used for marketing and other persuasive purposes. And because the cost of saving all this data is so cheap, there's no reason not to save as much as possible, and save it all forever'. As such what has been termed a 'data shadow' follows people and informs how they are treated in future interactions with businesses, government agencies and other institutions that have access to various profile and risk scores. There is scope for more research by human geography, examining how the labor of software writing and the collection of personal data are both placed and scaled, shaped by cultural values and other contextual factors.

It is clear that software-enabled technologies, digital services and social networks, and new kinds of ecommerce predicated on vast silos of personal information have brought many benefits, although these are not shared equally or available to all. However as lives become more dependent on correct operation of software and automated evaluation records in data shadows there are growing risks and a scope for new kinds of malfeasance and criminality. One way that these risks arising from increasing dependency on software have been highlighted is through the impacts of thefts of personal information from network databases, often compromising the privacy of millions of people and putting them at danger of identity and financial frauds. Some high-profile recent cases include the hacking of systems of online dating website Ashley Madison and the public release of confidential details on the members (cf. Hern 2015), repeated break-ins to the customer database of the telecommunications company TalkTalk (cf. BBC News 2015) and the politically damaging theft of many millions of records from the US Office of Personnel Management with consequences for personal safety and national security (cf. Davis 2015). Beyond thefts of data, many analysts point to escalating risks from deliberate sabotage of software, with attacks conducted virtually but causing real-world distribution to crucial systems and infrastructure. A potent recent example is the sophisticated cyber-attack on the software controlling the electricity grid that blacked out large parts of Ukraine on 23 December 2015 (Zetter 2016). The motives of the attackers in this case are unclear but, given the geopolitics of this region, suspicion has focused on the role of the Russian state in sponsoring or directing the sabotage. It seems that cyberwarfare fought through computer networks using code, but with impacts on cities and essential services, which was once the realm of science fiction and fear-mongering speculation, is fast becoming a reality.

DISCOURSE SUPPORTING SOFTWARE

It is important to realize that there are distinct sets of discourses that are at work promoting ever deeper algorithmic automation of society and counteracting criticism of code and the risks it brings. Computer technology more generally but software specifically is subtly, and not so subtly, advocated as the *only* possible way to solve contemporary social problems by many actors (e.g. security from terrorism can only be tackled by more software surveillance and monitoring of online activity). Software is promoted endlessly as an *essential* means to keep economies competitive in the 'global race'. It is important to interrogate the discourses that are constructed by companies with direct vested interests, governments and other institutions (including universities) to create a discursive regime that supports and normalizes the development and roll-out of software-dependent infrastructures and processes.

Above all, the discursive regime of code seeks to present a commonsensical façade – it simply 'makes sense' to invest in software technologies. Using lobbying, media campaigns, direct advertising, inducing early adopters and trendsetters, celebrity endorsements, employee training and educating children, the argument for entrusting more and more of everyday life and its governance to software systems is made with reference to incontrovertible issues of efficiency, making money, time-savings, security and safety and personal empowerment. (Such discursive regimes are produced by vested interests with respect to notions of 'smart homes' and 'monitoring of bodily performance', as discussed below.)

The power of discourse, as Michel Foucault and others have pointed out, lies in persuading people to its logic – to believe and act in relation to this logic – without thinking there might be alternative ways to proceed. With social conventions and cultural norms of state capitalism it can be hard to resist such logic that technology can make society more secure and productive. These discourses are often promoted by key actors in government in tandem with business elites and consultants, driven by the interests of corporate capitalism and, increasingly, the agenda of neoliberalism focused on exploiting software as far as possible to cut costs, outsource, de-skill and reduce staffing, and reconfigure public services for profit within a target-driven culture. People who cannot or choose not to participate are easily marginalized (e.g. requirements to register for welfare services through software systems, or being excluded from cheaper prices from online-only services channels).

Importantly the discursive regime for more deployment of software does not operate solely from the top downward, but also through diffused microcircuits of power, the outcome of processes of local regulation, self-regulation and small notes of resistance. As such, people are not simply passive subjects drawn along in unproblematic ways by a desire for ever more software. Notes of contestation are made by civil rights activists and privacy groups, skeptical journalists and critical academic voices; more broadly there is a kind of passive reluctance and apathy in regards to exaggerated claims for software (e.g. around ease of use or productivity gains). Although it has been argued, given the increasing power and role of software, that resistance to digital technologies has been remarkably mute, as Thrift and French (2002, 313) note, '[e]ven though software has infused into the very fabric of everyday life – just like the automobile – it brings no such level of questioning in its wake'.

THE NATURE OF CODE AND NEED FOR 'SOFTWARE STUDIES'

Software consists of lines of code – specific decision-making instructions and mathematical algorithms that, when combined and supplied with appropriate input data, produce routines and programs capable of complex digital functions and crucially performing appropriate ['correct'] actions. Put simply, the action of software can instruct computer hardware (the physical, digital circuitry) in an unambiguous way about what to do (which in turn can engender action in other machinery, such as switching on electrical power).

Although code in general is hidden, invisible inside the machine, it can produce visible and material effects in the world.

From humble beginnings in the 1940s, software can now be hugely complex and in many, varied forms, from abstract machine code and assembly language ('low level') to libraries and application programming interfaces and up to more formal programming languages, applications, user-created macros and simple scripts ('high level'; Figure 5.2). Often these forms are nested together or arranged into hierarchically connected libraries of code to produce effective entities. These might be hard-coded applications embedded on chips, specialized applications (banking software, traffic management systems), generic user applications (word processors, Web browsers, video games) and large operating systems (Windows, Mac OS, Linux) that run on a variety of hardware platforms (smartphone, dedicated units, PCs, servers).

The phenomenal growth in software development and deployment stems from the fact that digital code (unlike analog forms of instructions) has executable properties, that is, how it codifies the world into rules, routines, algorithms, and databases and then uses these to do work in the world. This means that it can run by itself, and although software is not sentient or self-conscious, it does exhibit some of the characteristics of being alive. Thrift and French (2002, 310) describe the self-operative nature of software as being 'somewhere between the artificial and a new kind of natural, the dead and a new kind of living' and having 'presence as "local intelligence"'. This property of being 'alive' (executable and self-active) is significant because it means that code can make things do work in the world in an autonomous fashion – that is, it can receive inputs and process data, evaluate diverse situations, reach complex decisions and, most significantly, act without human oversight or authorization. As an illustration of this, at any moment when you are interacting with a laptop or smartphone there are likely to be hundreds of other pieces of code and larger programs performing their own work beyond human awareness (Figure 5.3). Moreover code has pervaded non-computer-like objects and everyday environments in often subtle and opaque ways, so software is forming a 'technological unconscious' that is noticed only when it performs incorrectly or fails (Graham and Thrift 2007). Another important, related development is that of machine-to-machine communications with smart devices and tagged objects connected to through the so-called 'Internet of things', which is creating a parallel digital ecology that exists without human authorization or even awareness of what is happening. A consequence of the evolving power of software and spatial diffusion is that the things people interact with appear to be 'automagical' in nature in that they work in ways that are not visible to most people, and this produces complex outcomes that are not easily accounted for by everyday experience of the analog and non-digital.

Executable code with 'automagical' algorithms for reaching automated decisions that must be correct and unchallengeable (within its operating domain and input parameters) has clear social and political implications. Yet the internal workings of software have been poorly theorized and empirically little studied from a social sciences perspective beyond a handful of formative texts by cultural and new media theorists (including Manovich (2013), Fuller (2008), Galloway (2004) and Mackenzie (2010)). Instead, software has been understood from a technical, instrumental perspective that treats it as largely an immaterial, stable, neutral product, rather than as a complex, multifaceted, mutable set of relations created through diverse sets of discursive, economic and material practices.

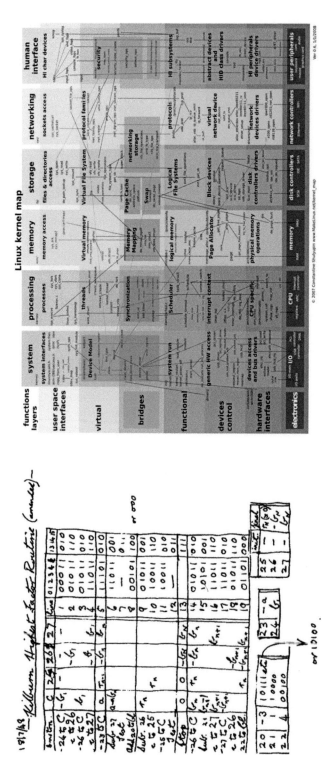

Source: Left image courtesy of G. Tootill. Right image courtesy of Constantine Shulyupin, www.MakeLinux.net/kernel_map.

Figure 5.2   High-level representations of code and the juxtaposition have been chosen to illustrate the scale of change. On the left is a handwritten notion of what is claimed to be one of first computer programs, written by Tom Kilburn in 1948 as part of the pioneering computer research at the University of Manchester. On the right is the 'Linux Kernel Map', created by Constantine Shulyupin, displaying the hierarchy of software components at the core of this operating system and how they work together

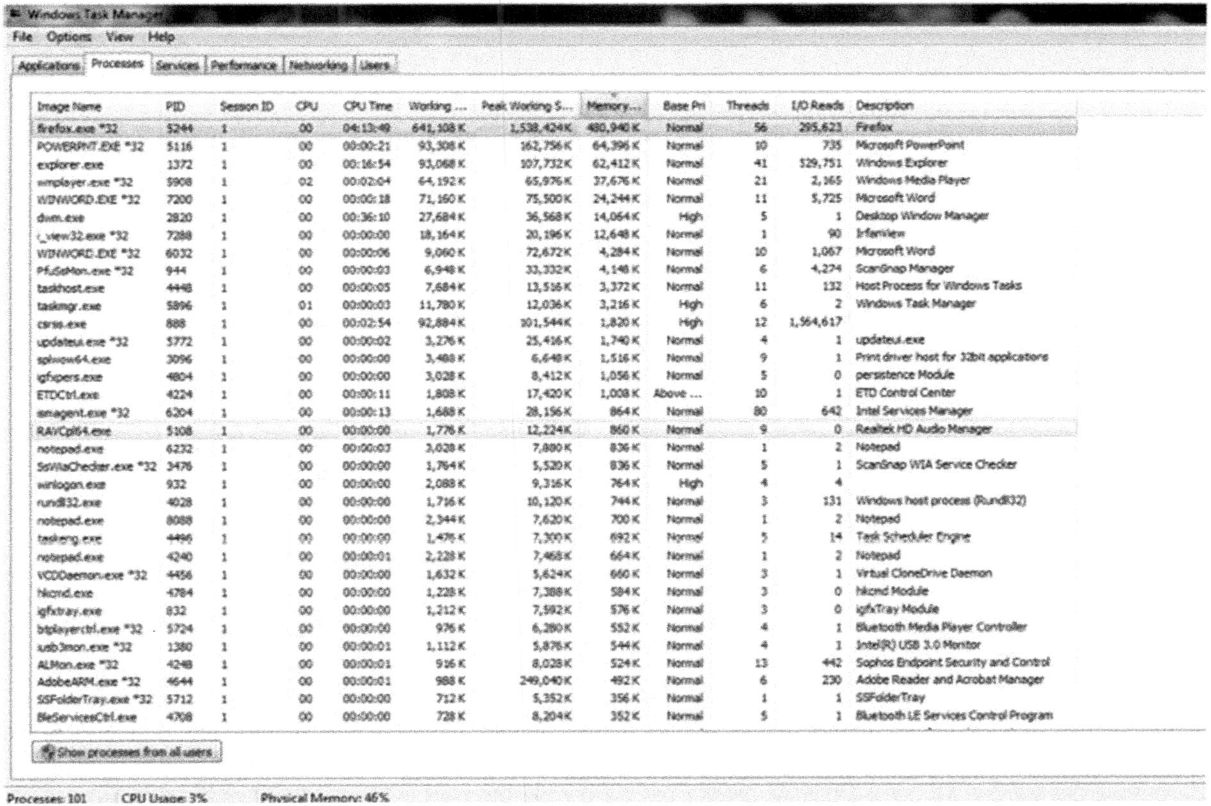

Windows Task Manager

File Options View Help

Applications | Processes | Services | Performance | Networking | Users

Image Name	PID	Session ID	CPU	CPU Time	Working ...	Peak Working S...	Memory ...	Base Pri	Threads	I/O Reads	Description
firefox.exe *32	5244	1	00	04:13:49	93,508 K	1,538,424K	480,940 K	Normal	56	295,621	Firefox
POWERPNT.EXE *32	5116	1	00	00:00:21	93,068 K	162,798K	64,396 K	Normal	10	735	Microsoft PowerPoint
explorer.exe	1372	1	00	00:16:54	64,152 K	107,732K	62,412 K	Normal	41	129,751	Windows Explorer
wmplayer.exe *32	5908	1	02	00:02:04	64,152 K	65,976K	37,676 K	Normal	21	2,165	Windows Media Player
WINWORD.EXE *32	7200	1	00	00:06:18	71,160 K	75,500K	24,344 K	Normal	11	5,725	Microsoft Word
dwm.exe	2820	1	00	00:36:10	27,684 K	36,568K	14,064 K	High	5	1	Desktop Window Manager
i_view32.exe *32	7288	1	00	00:00:06	18,164 K	20,196K	12,648 K	Normal	1	90	IrfanView
WINWORD.EXE *32	6032	1	00	00:00:06	9,060 K	72,672K	4,284 K	Normal	10	1,067	Microsoft Word
P4.sdxlsrvr.exe *32	944	1	00	00:00:03	6,948 K	33,332K	4,148 K	Normal	6	4,224	ScanSnap Manager
taskhost.exe	4448	1	00	00:00:05	7,684 K	13,536K	3,372 K	Normal	11	132	Host Process for Windows Tasks
taskmgr.exe	5896	1	01	00:00:03	11,780 K	12,036K	3,216 K	High	6	2	Windows Task Manager
csrss.exe	888	1	00	00:02:54	92,684 K	101,544K	1,820 K	High	12	1,964,617	
updates.exe *32	5772	1	00	00:00:02	3,276 K	25,456K	1,740 K	Normal	4	1	updates.exe
spoolsv64.exe	3096	1	00	00:00:00	3,488 K	6,648K	1,516 K	Normal	9	1	Print driver host for 32bit applications
gfxsrvc.exe	4804	1	00	00:00:00	3,028 K	8,412K	1,056 K	Normal	5	0	persistence Module
ETDCtrl.exe	4224	1	00	00:00:11	1,808 K	17,420K	1,008 K	Normal	10	1	ETD Control Center
seagent.exe *32	6304	1	00	00:00:13	1,688 K	28,156K	864 K	Normal	80	642	Intel Services Manager
RAVCpl64.exe	5108	1	00	00:00:00	1,778 K	12,224K	860 K	Normal	9	0	Realtek HD Audio Manager
notepad.exe	6232	1	00	00:00:03	3,028 K	7,880K	836 K	Normal	1	2	Notepad
SnVisChecker.exe *32	3476	1	00	00:00:00	1,764 K	5,529K	836 K	Normal	5	1	ScanSnap WIA Service Checker
winlogon.exe	932	1	00	00:00:00	2,088 K	9,316K	764 K	High	4	4	
rundll32.exe	4528	1	00	00:00:00	1,716 K	10,120K	744 K	Normal	3	131	Windows host process (Rundll32)
notepad.exe	8588	1	00	00:00:00	2,344 K	7,620K	709 K	Normal	1	2	Notepad
taskeng.exe	4444	1	00	00:00:00	1,476 K	7,305K	692 K	Normal	5	14	Task Scheduler Engine
notepad.exe	4240	1	00	00:00:01	2,228 K	7,465K	664 K	Normal	1	2	Notepad
VCDDaemon.exe *32	4456	1	00	00:00:00	1,632 K	5,624K	660 K	Normal	3	1	Virtual CloneDrive Daemon
hkcmd.exe	4384	1	00	00:00:00	1,228 K	7,388K	584 K	Normal	3	0	hkcmd Module
gfxtray.exe	832	1	00	00:00:00	1,212 K	7,592K	578 K	Normal	3	1	igfxTray Module
btplayerctrl.exe *32	5724	1	00	00:00:00	976 K	6,280K	552 K	Normal	4	1	Bluetooth Media Player Controller
usb3mon.exe *32	1380	1	00	00:00:01	1,112 K	5,876K	544 K	Normal	4	1	Intel(R) USB 3.0 Monitor
ALMon.exe *32	4248	1	00	00:00:01	916 K	8,023K	524 K	Normal	13	442	Sophos Endpoint Security and Control
AdobeARM.exe *32	4644	1	00	00:00:01	988 K	249,004K	452 K	Normal	6	230	Adobe Reader and Acrobat Manager
SSFolderTray.exe *32	5712	1	00	00:00:00	712 K	5,352K	356 K	Normal	1	1	SSFolder Tray
BtServicesCtrl.exe	4708	1	00	00:00:00	728 K	8,204K	352 K	Normal	5	1	Bluetooth LE Services Control Program

Show processes from all users

Processes: 101   CPU Usage: 3%   Physical Memory: 46%

Source: Author image.

Figure 5.3   *A display of the numerous separate pieces of software running on a typical laptop (controlled by Windows7 operating system) and various installed applications and utilities. Only a few are directly initiated by the person using the laptop and much of what is running is not meant to interact with people at all*

However, in the last few years efforts have been made to develop 'software studies' as a field of scholarly enquiry that seeks to create an expanded understanding of code that extends significantly beyond the technical and examine the culture of digital computation. It largely uses culturally informed epistemologies and theoretical critiques to ask how the social world itself is captured within code in terms of algorithmic potential and formal data descriptions. Software studies focuses attention squarely on the nature of code and not simply its effects, and it conceptualizes software as both a product of the world and a producer of the world in a way that recognizes that its production and work are not deterministic and universal in form. According to Lev Manovich (2013, 10), an influential theorist in the field, 'I think that Software Studies has to investigate both the role of software in forming contemporary culture, and cultural, social, and economic forces that are shaping development of software itself'. From a geographical perspective, it also means recognizing that the potential work of software is contingent on the spatial context and is the product of people in time and particular places; furthermore its use can transform these times and places, and as such the same code may work in different ways in different grounded situations.

Matthew Fuller (2008) argues that software studies proposes that code can be seen as an object of study and an area of practice for kinds of thinking and areas of work that have not historically authored or owned software, or indeed often had much to say about it. In this regard there is much that needs to be said by geographers who have traditionally not had much to say about the spatiality of software. One way to begin this is to analyze the way in which code can, quite literally, bring unique kinds of spaces into being. In some circumstances software is enrolled so completely into socio-spatial relations that they are *dependent* on the effective operation of code; these are what Dodge and Kitchin (2005) have called code/spaces.

THE NOTION OF CODE/SPACES

Here we shall consider the conceptual model of code developed by the author working with Rob Kitchin. As a kicking off point Kitchin and Dodge (2011) made the case that, to understand this large-scale structural change flowing out of the technicity of software, one needs to develop rich historico-geographical accounts of the contexts in which code is embedded into workplaces and labor practices over time and the new kinds of future trajectories this enables. Otherwise, code is simply seen as an abstract, exogenous factor rather than a socially embedded variable.

To develop their account of code, Dodge and Kitchin (2005) advanced the idea that software can be embedded in everyday life at series of scales/levels of activity. They proceeded firstly by defining the range of forms of software into a four-level hierarchy: (i) individual coded objects, which can be groups together or linked to form (ii) coded infrastructures, which in turn are controlled by and also carry (iii) coded processes. Coded objects, infrastructures and processes, in turn, can combine together to form larger (iv) coded assemblages. This hierarchy enables the software, through its varying degrees of technicity (power and productive capacity for work) to transduce space, that is it brings new spatial formations into existence to solve a problem or perform a task. Furthermore these spatial transductions through the enrollment of software we theorized at three distinct levels of intensity: (i) code/space; (ii) coded space; and (iii) background coded space.

In the most intensive level of transduction the technicity of software is so significant that the space brought into being *depends* on the operation of the code. There is a dyadic relationship between the existence of the space and the execution of the code – hence the co-joint term 'code/space' – and if the software fails to operate then the space is not produced. Code/space occurs when software and the spatiality of everyday life become mutually constituted, that is, produced through one another. Here, spatiality is the product of code, and the code exists primarily in order to produce a particular spatiality. People regularly coproduce code/spaces, even if they are not always aware they are doing so, and as we consider in the next two empirical sections of this chapter, they are increasingly common in a range of everyday contexts and domestic situations (cf. Dodge and Kitchin 2009). Any space that is *dependent* on software-driven technologies to function as intended constitutes a code/space: workplaces dependent on office applications such as spreadsheets, shared calendars, information systems, networked printers, e-mail and intranets; aspects of the urban environment reliant on building and infrastructural management systems; many forms of transport, including nearly all aspects of air travel and substantial portions of automobility and rail transport (Figure 5.4); and a majority of the components of the telecommunications, media, finance and entertainment industries. Such is the reliance by governments and businesses on a raft of office applications and larger software systems that it is now unthinkable to backtrack to pre-digital paper-based processes: the nature of tasks has changed, staff levels have been reduced and deskilled in many cases, and operational networks and transactions have become much more complex and interdependent.

In the second level of intensity, what we term 'coded space', the transduction is mediated by code but the relation is not dyadic so if the software were to fail to operate for whatever reason the space would still be produced as intended to solve a problem or perform a task. However, the nature of the spatial transduction without software is potentially a less efficient solution to the problem or a more costly way to perform a task (e.g. failure of computer system forces workers to use a 'manual' backup procedure that is much more labor intensive, slower and more costly, or perhaps occurs with reduced safety). Here, the role of software is often one of augmentation, facilitation, monitoring and so on rather than control and regulation.

The lowest intensity might be thought of as a transduction-in-waiting, what Dodge and Kitchin (2005) termed 'background coded space', a situation when software is present in the environment and has the potential to mediate a solution if activated. Code could make a difference but is inert unless consciously evoked. Much of people's ordinary living in Western cities occurs in 'background coded space' where people are surrounded by coded objects, coded infrastructures and coded processes that can be called upon in myriad ways to solve a problem or perform a task. For many people much of their deliberative daily activities proceed in coded space – that is, places animated by software and where their practices are routinely augmented by algorithms and software without human intervention. It can plausibly be argued that an increasing amount of time, for a larger range of individuals, is spent within bubbles of code/space where the spatial transduction depends wholly on software to achieve the tasks and solve problems they face.

Source: Guillaume Grandin, www.guillaumegrandin.eu.

*Figure 5.4   Many of the places that people live in; the offices, shops and factories they
work in; and the vehicles they travel in are code/spaces, at least part of the
time. A particularly high-profile example would be the jetliner which is code/
space with flight dependent on multiple software systems and distinct coded
objects. The display screens in the cockpit are small portals into the larger
realm of code that monitors and controls much of the aircraft's performance,
and software is primary determinant on how the pilots work*

DELINEATING CODE: (1) DOMESTIC AUTOMATION AND EVERYDAY LIVING

The significance of code might be more apparent in 'high-tech' public spaces, at airports,
in contemporary offices with a computer display on every desk or the robotic production lines in highly automated manufacturing plants, but software is being enrolled ever
more deeply to transform the spatiality of people's homes and the daily self-reproduction
of their bodies. The domestic realm has become coded space for most people living in
countries in the Global North and code/space for some.

Throughout the modern period the home has been a site of technologies to augment
daily activities and, supposedly, to reduce the burden of domestic labor through automation. Like earlier rounds of electromechanical appliances and electronic gadgets that
were initially a novelty and luxury goods, but became taken-for-granted and sunk into

the background of most domestic spaces, so software-enabled technologies have been transforming the performance of home life over the last 20 years or so, helping people solve everyday tasks (such as cleaning, cooking, recreation, shopping, saving and sharing family memories, maintaining personal hygiene and well-being). Most homes today are populated with lots of distinct coded objects, a good number of which are becoming connected together (by wifi and other radio communication) and able to interact with each other and outside companies and institutions through coded processes. Most particularly for those with money and the inclination, many of the 'smart home' fantasies from the early decades are becoming a reality, although the degree to which the latest digital gadgets and software-enabled appliances are really improving people's lives and making their home a happier place is debatable. The promise of more leisure time, particularly for women who conventionally carry most of the burden of maintaining the home, through purchasing new appliances is a longstanding myth (cf. Cowen 1983); if anything adding software to domestic spaces simply adds in new layers of complexity and stresses! (Think of new burdens of software security, remembering passcodes, changing passwords, ensuring compatibility and needs to update this and that to keep it working; such work was never needed with an analog television set!)

While we need to be critical of upbeat, almost utopian, predictions of better living in software-enabled 'smart homes' that are promulgated powerfully by vested commercial interests, there is no doubt that code has come into homes in developed countries in significant ways. An overt example of the potential of code object to change domestic space into coded space is evident in the deployment of small home robots designed to perform routine tasks in a largely autonomous way using software algorithms and awareness of aspects of its environment from sensor inputs (Figure 5.5). While such robots remain rather a gimmick, a more prosaic and potent impact of code is in enhancing home entertainment technologies and the depth of consumption of digital media over the last decade; for many people, key aspects of their recreation at home are now dependent on software and coded objects. (Although as with any technological evolution, there is never a complete substitution of one form for another, and it is likely that there are a good number of homes happily using analog VHS videos and or reveling in the cult status of vinyl records!) Today, the domestic spaces of affluent households are literally overflowing with all kinds of computer devices for delivery entertainment content and keeping people 'occupied' (and consuming): videogame consoles, interactive toys, digital radios, tablets, laptops, webcams, camcorders, media players, ebook readers, and so on. Screens are in almost every room, including the bathroom, and many people take their smartphone or tablet into the toilet!

The television set, the central technological component of domestic recreation for the majority of homes since the 1960s, has become a 'smart tv' with powerful software, added features and complex screen interfaces. In the UK the last broadcast analog television ended in 2012, making this media only available digitally and dependent on coded objects to process the signal, while in the last few years more people have stopped following fixed television schedules and become consumers of on-demand streaming media, which depends wholly on software to deliver the right show to the viewer. As such, to watch television is to bring into being code/space. There are claims of more choice and interactivity but new, so-called 'smart tv' sets with two-way connection to the Internet also require 'looking after' in ways that were never needed with 'dumb' cathode ray tube

Source: Author scans of magazine advertisements, 2014.

Figure 5.5 *The possibility of software to make life easier, as seen in two magazine advertisements promoting domestic robots to potential purchasers. The claims to be able to automate specific tasks that are typically seen as onerous and thereby offering the owners prospects of more 'free time' is a longstanding technological discourse and still a powerful one*

sets plugged into analog aerial sockets. Functionally many of the latest television sets log usage patterns, and some have environment sensors to make them aware of their surroundings and how they are being used, coupled with the ability to send this data out of the home as well as receiving media. A particular risk was flagged concerning those smart tv sets with voice-activated controls because they listen to all conversation in the room to offer enhanced usability but with concomitant concerns about privacy of the domestic realm (Schneier 2015). Television sets use code to watch and record the watching household and there is an analogy to be drawn here to the 'telescreens' imagined by Orwell in his dystopian novel *1984*.

Overt coded objects like large smart televisions hanging in pride of place on the living room wall, white appliances with digital controls in the kitchen and clever little robots for cleaning are easily visible instances of software in the home but perhaps more significant will be the enrollment of software to provide a management system of the whole domestic space and environment, potentially making the whole assemblage into a code/space. More sophisticated building management systems (which have typically been associated with high-end office complexes and the like) are now being heavily promoted to the domestic market with the ability to coordinate all manner of mundane but vital tasks, heating and air conditioning, access and security monitoring, fire/flood warning through sensors, and smart meters able to adaptively tune utility usage. Such home 'hubs' promise more control, for example residents can have remote access to services at home from a smartphone app (e.g. tweaking the heating before arriving home), and also ease of use as the software will take care of things for you through self-learning of household patterns and auto-scheduling services. Promotional rhetoric rests on powerful discourses of efficiency, comfort and risk-reduction. For example, one of the current market-leading 'smart thermostat' products, NEST, asserts that its controller system 'uses state of the art machine learning algorithms to determine the thermal properties of each home individually. And once the thermostat understands how the house warms up and cools down, it allows it to keep users much more comfortable while maximizing their energy savings' (NEST 2015, 3). The understanding of the physical environment of individual homes and the daily behavior patterns of the household through software in such management systems could well lead to more efficient energy usage and some saving of money, but it comes at the cost of a 'spy' in the home, compromising the accepted private sanctity of domestic space by sending out details of consumption of water and power to utilities companies. This is coupled with a significant risk that such home management hubs, accessible through the Internet, could be hacked by those with malicious intent (Hernandez et al. 2014).

While there is scope to be critical of these home management systems and claims made by the vendors, there is a strong case to be made that safer living through software in terms of home telecare systems will become ever more significant in the next decade (although with attendant risks and great loss of privacy). A key challenge facing many affluent developed countries is coping with much larger aged populations, including many physically frail and cognitively forgetful people who desire to live independently in their own homes for as long possible. Telecare systems are promoted as a solution with software able to continuously, automatically and remotely monitor daily behavior, looking for changes in routines and ready to respond to real-time emergencies (such as the user pressing a wearable panic alarm). For example, detectors on the fridge door, kettle and oven provide data streams for software algorithms to process and give

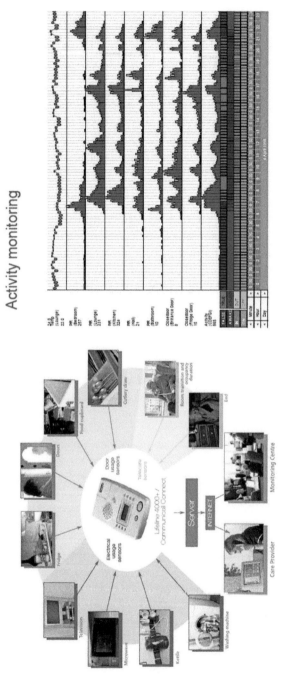

Source: Author scans of magazine advertisements, 2014.

Figure 5.6 *Two images that illustrate telecare. (left) An info-graphic from a marketing brochure of a leading domestic telecare system in the UK indicating the range of possible routine activities that are monitored by sensors linked to a local control box and onwards to software systems in distance control center. (right) Image of the kind of temporal pattern analysis that can be presented by the telecare system to monitoring staff as a tool to spot behavioral change that may indicate a problem with the person*

indications that the householder might be failing in their ability to prepare hot drinks and daily meals (Figure 5.6). As more sophisticated telecare monitoring systems become available so (affluent) pensioners and the disabled in particular can live in an ever more coded cocoon. They may feel safer and potentially are kept safer but it is less clear how well code can really care for people beyond monitoring their physical performance and material circumstances.

Software has crossed the threshold and is infusing into the large majority of homes in a significant but still partial way, and while its presence will undoubtedly grow, it will remain haphazard and incomplete. Some of it is also strongly tied to income levels and social status in buying into the latest rounds of software technologies and 'smart' gadgets. Much is coded space in which the activity could continue even if the software monitoring failed and the log was not updated, although watching your favorite tv series on a tablet through a streaming service is code/space.

DELINEATING CODE: (2) PERFORMING SOFTWARE SURVEILLANCE OF THE SELF

Of course code is not just monitoring the health and well-being of frail elderly people via expensively installed telecare systems; there is now a whole plethora of wearable gadgets and software apps for those interested in self-surveillance of their bodily performance (Figure 5.7). Wristband trackers are marketed for serious sports types, for those keen to improve their fitness or just those with a daily prurient fascination with how far they have walked in the last 24 hours, how many calories were nominally burnt and their sleep pattern over the previous week. This new breed of coded objects key into all kinds of consumerist trends around health and body and technological design is meant to appeal to the affluent and gadget obsessed, with some promoted by existing high-profile brands (e.g. Nike with its Sportband) and 'cool' companies with 'must-have' cache (like the Jawbone UP or Fitbit Flex). Evaluation of users shows that the act of tracking becomes important in and of itself, and the build-up of a log of performance becomes information to play with, contemplate in idle moments and perhaps worry over ('why has my fitness level dropped this month?'; cf. Rooksby et al. 2014; Williamson 2015). The tracker hardware communicates seamlessly with software on the owner's smartphone and the attendant app does not just present the data – its algorithmic capacity means that it acts as an virtual trainer able to encourage greater efforts and actively making suggestions on how to hone bodily performance in the future. Another aspect, made possible by software, is the sharing of bodily performance with family, friends or peer groups online; this can engender pride, sense of self-worth or just be about ego and competition.

Wristband has been a rapidly developing market segment in recent years as more people see the appeal of so-called 'self-quantification', but in many ways it is merely a parallel to much more significant and more medically determined monitoring of the body through software (e.g. coded objects and coded processes enrolled in management of chronic morbidity in the home settings such as heart conditions with blood pressure testing and diabetes with blood sugar testing pens; cf. Lupton 2013). Of course intimate body monitoring, for example using scales to watch weight changes, is a longstanding practice for many, but it is now enhanced with code as digital data that can easily be

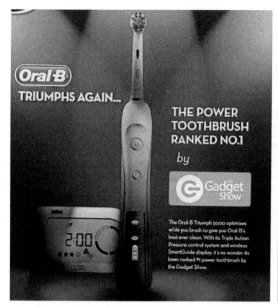 

Source: Author photograph of product packaging, 2015.

Figure 5.7   Two examples of discourses reflecting code/space. (left) Product manufacturers seek new rounds of consumption by linking into the self-surveillance of the body, with electric toothbrush morphing into a code object with the promise of better oral hygiene with software monitoring and regulating the person's performance. (right) An example of the promotional rhetoric for Fitbit, currently one of the leading wearable personal performance trackers

stored and graphed for trends on a smartphone app; for many this kind of algorithmic analysis of their body will be helpful but for others it is more likely to be anxiety inducing, with data displays keying into all kinds of latent worries around diet, (un)healthy eating and the specter of obesity. Other examples include personal air pollution sensors aimed at the growing number of asthma suffers, and for those seeking to get pregnant the use of software to better monitor menstruation and ovulation cycles (perhaps a positive result being regarded as coded conception!). More generally the algorithms in smartphone apps have enabled rapid development of the home medical market and allowed for an intensive degree of self-monitoring and wider collection of health status data. In 2012, 69% of smartphone owners in America reported tracking at least one health indicator such as weight, diet or exercise (Fox and Duggan 2013). The resultant data can be exploited to trigger highly personalized interventions and can be stored in large databases with the potential to improve healthcare research and plan better public health strategies. There is also huge commercial interest in such individual and detailed data.

EPILOGUE: FURTHER CHALLENGES IN COMPREHENDING CODE/SPACE

Further analysis by human geographers and other social scientists to understand more fully the nature of software and where coded space occurs and of how code/spaces come into being will require investigations in specific places and developing means to interrogate how algorithms are working in everyday contexts. This will be a challenge for several reasons, firstly because the software sector is so dynamic and exhibits such fierce competition between corporations (for example in smartphone apps); this means that there is a fast turnover in applied code and updates in algorithms. Considered scholarship struggles when the target of analysis never stands still. Moreover, much of this executable code is proprietary and owned by corporations that, understandably, want to keep the operations of their key algorithms as trade secrets for fear of losing competitive advantages or being exposed for having questionable products (e.g. code that makes judgments that unwittingly discriminate against some users, or are deliberately biased in generating results for commercial gain or erroneous in particular circumstances, or simply with unfixed security flaws; cf. Kirchner 2015; Ziewitz 2016; Zook and Graham 2007). The temptation will be to analyze publicly available forms of code (such as online media and open source software development) rather than investigating the harder to access code that matters most to daily life and the ongoing production of the society (e.g. control systems of the utility companies, the scheduling software of transport systems, the tasking of security personnel, the calculation of insurance rates and mortgage risks, etc.).

Beside speed of change and the issue of corporate secrecy, the core code at the center of major software products comprises archetypical 'Black Box' operations with data going in and results coming out but no detail on the interior workings. Even if researchers were able to break open the 'black boxes' of commercial software it is not clear whether they could really interpret what they find inside. While there might be valid political reasons to seek such algorithmic transparency, there are likely to be practical issues of how to read and understand what is there. Without significant experience in mathematics, coding architectures and arcane syntax, even for very diligent researchers the algorithms will remain unfathomable. (As a simple illustration of the scale of the interpretative challenge, just look at the scripting behind a typical webpage and see if you can work out what it is doing and how this relates to what you see in the browser; Figure 5.8.) Assuming social science researchers could develop the necessity vocabulary and grammar to begin to decode the obtuse workings of algorithms at the heart of software systems, like a smartphone, there is no guarantee that this will produce meaningful knowledge in a social science sense. Will identifying individual decision points and operations in an algorithm provide recognizable attributes with social context or relatable properties to use as handholds for critical geographical analysis? As Mackenzie (2009, 1295) showed in his analysis of digital signals processing performed in mobile telephones, 'the algorithmic processes . . . offer a strong challenge for research . . . in their somewhat stunning complexity, they seem to bear only a tangential relation to the powerful dynamics of belonging, participation, separation, and exclusion typical of contemporary network cultures'.

Besides trying to examine source code directly, Kitchin (2014, 17–22) suggests some other ways of researching algorithms, and their wider significance, including researchers reflexively producing their own code, reverse engineering, interviewing designers

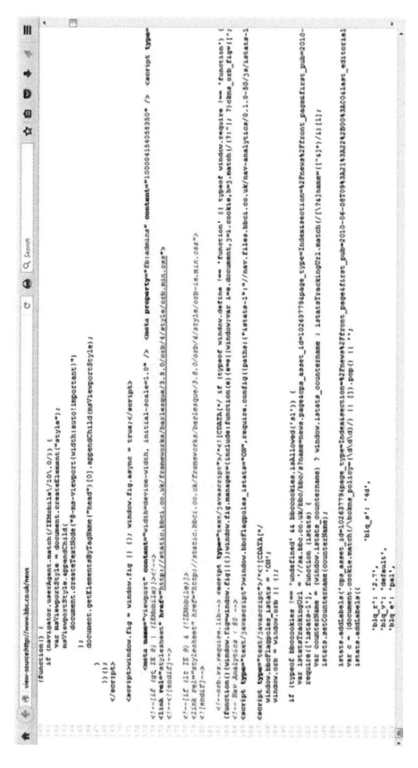

Source: Author image.

Figure 5.8 *The view of the source code for a typical webpage. Shown here is a small part of the complex scripting and opaque relation to how the actual information content appears to the user*

and conducting ethnographies of coding teams and examining how algorithms work in specific places. Even if such research becomes possible, a further problem is how to communicate the workings of an algorithm to a wider audience without resorting to mathematics or repeating the logical operations in a slightly different annotated form. An effort in this direction by Cormen (2013) is worthwhile but demonstrates how hard it is in practice to represent and explain algorithms to the layperson. There may also be scope for applying spatial visualization to map out the flows of data logic of decisions made in a given algorithm.

Ultimately, the challenge of understanding software is hard going for human geographers because the overwhelming majority cannot code! While academic geographers are very much software-embedded workers, like many other occupations and professions, they generally do not need to write code to tackle their daily scholarly tasks. This inability to program software (held generally across humanities and social sciences) is a problem as there is a real danger of becoming thoroughly alienated from a key source of creative power in working practices going forward. As van Kranenburg (2008, 23) put it, in an analogous context:

> If as a citizen you can no longer fix your own car – which is a quite recent phenomenon – because it is software driven, you have lost more than your ability to fix your own car, you have lost the very belief in a situation in which there are no professional garages, no just in time logistics, no independent mechanics, no small initiative. . . . [Citizens] become helpless very soon, as they have no clue how to operate what is running in the background, let alone fix things if they go wrong. As such, 'ambient intelligence' presumes a totalising, anti-democratic logic.

So while geographers are making some progress in describing the consequences of code for everyday activities, before proceeding much further they will need to take some courses in computer science to really begin to expose algorithmic pinch-points and explain in precise ways how a piece of software, working in a particular place, can bring code/space into being.

NOTE

1. Some of the material discussed in this chapter is based on collaborative research with Rob Kitchin and published in our 2011 book *Code/Space* (MIT Press) and other papers. Every effort has been made to trace all the copyright holders but if any have been inadvertently overlooked the publishers will be pleased to make the necessary arrangement at the first opportunity.

REFERENCES

Andreosso-O'Callaghan, B., H. Lenihan and P. Reidy. 2015. The development and growth of the software industry in Ireland: an institutionalized relationship approach. *European Planning Studies* 23(5): 922–943.

Arora, A. and A. Gambardella. 2006. *From Underdogs to Tigers: The Rise and Growth of the Software Industry in Brazil, China, India, Ireland, and Israel*. Oxford: Oxford University Press.

BBC News. 2015. TalkTalk cyber-attack: website hit by 'significant' breach. *BBC News Online*, 23 October, accessed from www.bbc.co.uk/news/uk-34611857.

Cook, P., G. Searle and K. O'Connor. 2013. *The Economic Geography of the IT Industry in the Asia Pacific Region*. Abingdon: Routledge.

Cormen, T.H. 2013. *Algorithms Unlocked*. Cambridge, MA: MIT Press.

Cowen, R.S. 1983. *More Work for Mother: The Ironies of Household Technology from the Open Hearth to the Microwave*. New York: Basic Books.

Davis, J.H. 2015. Hacking of Government computers exposed 21.5 million people. *New York Times*, 9 July, accessed from www.nytimes.com/2015/07/10/us/office-of-personnel-management-hackers-got-data-of-millions.html.

Dodge, M. and R. Kitchin. 2005. Code and the transduction of space. *Annals of the Association of American Geographers* 95: 162–180.

Dodge, M. and R. Kitchin. 2009. Software, objects, and home space. *Environment and Planning A* 41(6): 1344–1365.

Fox, S. and M. Duggan. 2013. Tracking for health. Pew Research Center, 28 January, accessed from www.pewinternet.org/2013/01/28/tracking-for-health/.

Fuller, M. 2008. *Software Studies: A Lexicon*. Cambridge, MA: MIT Press.

Galloway, A.R. 2004. *Protocol: How Control Exists After Decentralization*. Cambridge, MA: MIT Press.

Graham, S. and N. Thrift. 2007. Out of order understanding repair and maintenance. *Theory, Culture & Society* 24(3): 1–25.

Hern, A. 2015. Infidelity site Ashley Madison hacked as attackers demand total shutdown. *Guardian*, 20 July, accessed from www.theguardian.com/technology/2015/jul/20/ashley-madison-hacked-cheating-site-total-shutdown.

Hernandez, G., O. Arias, D. Buentello and Y. Jin. 2014. Smart Nest Thermostat: a smart spy in your home. Blackhat Conference, 2–7 August, accessed from http://www.blackhat.com/docs/us-14/materials/us-14-Jin-Smart-Nest-Thermostat-A-Smart-Spy-In-Your-Home-WP.pdf.

Kirchner, L. 2015. What we know about the computer formulas making decisions in your life. *ProPublica*, 30 October, accessed from https://www.propublica.org/article/what-we-know-about-the-computer-formulas-making-decisions-in-your-life.

Kitchin, R. 2014. Thinking critically about and researching algorithms. The Programmable City Working Paper, Maynooth University, Ireland, accessed from http://papers.ssrn.com/abstract=2515786.

Kitchin, R. and M. Dodge. 2011. *Code/Space: Software and Everyday Life*. Cambridge, MA: MIT Press.

Lupton, D. 2013. Quantifying the body: monitoring and measuring health in the age of mHealth technologies. *Critical Public Health* 23(4): 393–403.

Mackenzie, A. 2009. Intensive movement in wireless digital signal processing: from calculation to envelopment. *Environment and Planning A* 41(6): 1294–1308.

Mackenzie, A. 2010. *Wirelessness: Radical Empiricism in Network Cultures*. Cambridge, MA: MIT Press.

Manovich, L. 2013. *Software Takes Command*. London: Bloomsbury Academic.

NEST. 2015. Thermal model and HVAC control. White Paper, accessed from https://nest.com/downloads/press/documents/thermal-model-hvac-white-paper.pdf.

Rooksby, J., M. Rost, A. Morrison and M. Chalmers. 2014. Personal tracking as lived informatics. Paper presented at the CHI 2014 Conference, 26 April to 1 May, Toronto, accessed from http://dx.doi.org/10.1145/2556288.2557039.

Schneier, B. 2015. Your TV may be watching you. CNN, 12 February, accessed from http://edition.cnn.com/2015/02/11/opinion/schneier-samsung-tv-listening/.

Schneier, B. 2016. Data is a toxic asset. CNN, 1 March, accessed from http://edition.cnn.com/2016/03/01/opinions/data-is-a-toxic-asset-opinion-schneier/.

Thrift, N. and S. French. 2002. The automatic production of space. *Transactions of the Institute of British Geographers* 27(3): 309–335.

van Kranenburg, R. 2008. The internet of things: a critique of ambient technology and the all-seeing network of RFID network notebooks, accessed from http://networkcultures.org/wpmu/portal/publications/network-notebooks/the-internet-of-things/.

Williamson, B. 2015. Algorithmic skin: health-tracking technologies, personal analytics and the biopedagogies of digitized health. *Sport, Education and Society* 20(1): 133–151.

Zetter, K. 2016. Inside the cunning, unprecedented hack of Ukraine's power grid. *Wired News*, 3 March, accessed from www.wired.com/2016/03/inside-cunning-unprecedented-hack-ukraines-power-grid.

Ziewitz, M. 2016. Governing algorithms: myth, mess, and methods. *Science, Technology, & Human Values* 41(1): 3–16.

Zook, M. and M. Graham. 2007. The creative reconstruction of the internet: Google and the privatization of cyberspace and digiplace. *Geoforum* 38: 1322–1343.

6. Understanding locational-based services: core technologies, key applications and major concerns
Daniel Sui

INTRODUCTION: LBS AND THE GEOSPATIAL REVOLUTION

As an integral part of the on-going geospatial revolution in society at large (http://geospatialrevolution.psu.edu), location-based services (LBS) generally refers to all the information services that exploit the ability of technology to know where objects or people are located, and to modify the information it presents accordingly. Different from other services developed so far, LBS is inherently distributed, mobile and potentially ubiquitous (Heinemann and Gaiser 2014). LBS can be offered either in the push (business to consumers) or pull (consumers to business) mode. The demands for various kinds of LBS from individuals, businesses and government agencies have been growing steadily. It is estimated that the global LBS market value will grow from $12 billion in 2015 to $55 billion by 2020 (http://www.marketsandmarkets.com/Market-Reports/location-based-service-market-96994431.html). The rapid growth of LBS during the past decade is a result of the accelerated convergence of new information and communication technologies such as mobile telecommunication systems, broadband Internet, Wi-Fi, location-aware technologies for both indoors and outdoors, and advances in GIS and spatial big data analytics. The convergence of these technologies has enabled us for the first time in human history to track down pretty much anybody, anything, anytime and anywhere on/near the surface of planet Earth. Consequently, this new capability has profound implications for industry/business, government operations, academic research and people's daily lives. To understand the full implications of the geospatial revolution, an appreciation of the different aspects of LBS will be essential. This chapter reviews LBS's core technologies, key applications and major social impacts/concerns.

LBS: CORE TECHNOLOGIES

LBS is made possible through the convergence of location, communication and information technologies. Essentially, four basic components are required to provide the basic location-based services (Figure 6.1): the end user's mobile device with a positioning component; the service provider's application software; a mobile network to transmit data and requests for service; and a content provider to supply the end user with geo-specific information (Ahson and Ilyas 2010). In most countries, location-based services are permission-based as required by law. What this means is that the end user must opt in to share their locational information and give permission to the service to use that information. In most cases, this entails installing some kind of LBS application software on the end user's device and accepting a request to allow the service to obtain its location.

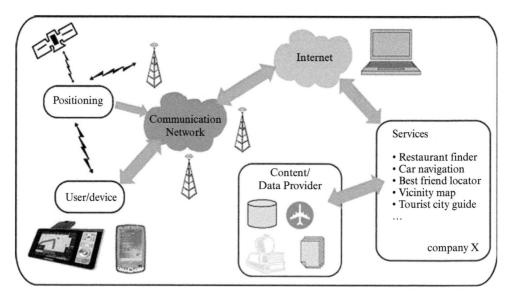

Source: https://rightdealrightnow.files.wordpress.com/2010/08/clip_image008.gif.

Figure 6.1   LBS core technologies

The core technology for LBS is based upon the advances of both indoor and outdoor locating technologies and their full integration with various mobile devices. Until recently, geospatial technologies have been described as a '15% technology' as far as human activities are concerned. This is due to the fact that most humans spend on average 85% of their lives indoors rather than outdoors, but until recently we did not have very effective locating technology to track people or objects inside buildings. In recent years, indoor LBS has also grown considerably owing to advances in indoor locating technologies, making geospatial technologies reach out beyond the 15% of outdoor human activities (Werner 2014). Both indoor and outdoor navigation relies heavily on the concept of triangulation, but specific locating techniques may differ. An indoor positioning system (IPS) to locate objects or people inside a building typically relies on radio waves, magnetic fields, acoustic signals or other sensory information collected by mobile devices. Although there are several commercial systems available in the market, it is still a rapidly evolving field and thus there is no standard for an IPS system. In general, IPS systems can use a combination of the following: (1) distance measurement to nearby anchor nodes (nodes with known positions, e.g. Wi-Fi access points); (2) magnetic or light-based positioning; and (3) dead reckoning, a set technique to determine location based upon one's previous locations. These technologies either actively locate mobile devices and tags or provide ambient location or environmental context for devices to get sensed. The localized nature of an IPS has resulted in design fragmentation, with systems making use of various optical, radio or even acoustic technologies.

Although the specific technologies used in determining outdoor locations are quite different from those for indoor locations, the principles for locating an object in space are more or less the same. Generally speaking, geotracking – to obtain the locational coordinates of the target being tracked – has mostly relied on three primary techniques for determining location (Bartlett 2013): (1) triangulation, which can be done via either lateration (using multiple distance measurements between known points) or angulation (using measured angles or bearings relative to points with known separation); (2) proximity, which measures nearness to a known set of points; and (3) scene analysis, which examines a view from a particular vantage point using high resolution aerial or satellite photographs. Global positioning systems (GPS) use radio time-of-flight lateration. Radio Frequency Identification (RFID)-based active badges, or smart floors, use infrared cellular proximity. Motion Star Radar uses a combination of scene analysis and lateration. These different technologies have different coverage, spatial–temporal resolution, and accuracy, and have increasingly been embedded into a variety of consumer products such as car navigation systems, smart cellular phones and even printers, underwear and cosmetics. Custom-designed tracking devices are also available in the market, such as Digital Angel, Wherify child locator and various other types of WhereWare (Pfeiffer 2003). Triangulation-based GPS is the most commonly used location technology for determining outdoor locations of people and objects. GPS chips are increasingly an integral part of mobile handsets, tablets and laptops with typical location-sensing accuracy of around a couple of meters. GPS can be used to calculate location information independently without relying on other technologies in places where the required line-of-sight to the sky is not obscured. The only drawback is that the location information from GPS is often in the form of raw coordinates (e.g. latitude and longitude). For most mobile applications, simple latitude and longitude information are insufficient, and they have to be transmitted to a third party in order to obtain maps or other contextual information based on the device's location.

Among these different positioning technologies, perhaps the most significant one is the global navigation satellite system (GNSS), which has been instrumental in promoting the development of location-based industry. As of 2016, the US NAVSTAR GPS is the only fully operational GNSS. However, the global dominance of positioning systems by the US is eroding with the rapid development of GNSS in the EU, Russia, China, Japan, France and India. The European Union's Galileo positioning system is a GNSS in the initial deployment phase, having been operational since 2013. The Russian GLONASS is a GNSS in the process of being restored to full operation. China has indicated that it may expand its regional Beidou navigation system into a global system. The current Beidou-1 system (made up of four satellites) is experimental and has limited coverage and application. However, China has planned to develop a complete global satellite navigation system consisting of 35 satellites (known as Compass or Beidou-2). The Quasi-Zenith Satellite System is a proposed three-satellite regional time-transfer system and enhancement for GPS covering Japan. The first satellite was launched in 2010 and the whole system is scheduled to be completed in 2018. Doppler Orbitography and Radio-positioning Integrated by Satellite is a French precision navigation system. Obviously, more and more countries have recognized the growing importance of GNSS to their military operations as well as various civilian applications, which explains their growing investment in the global navigation infrastructure to develop their own independent systems.

Besides GPS, a cellular base station is also commonly used to determine location. Cellular handsets must constantly register their presence with the nearest base station in order to establish service even in standby mode. Because the cell phone service provider has the information on the exact location of each base station, the location of the handset can be estimated within the coverage area. The radius covered can vary greatly, from several miles down to a city block or even an individual business or residential district, depending on the cell density and network architecture (FCC 2012). Accuracy of locational information can be improved significantly by triangulating between over-lapping cell sectors and is often used by providers to improve the E-911 response and to monitor coverage. TruePosition's LBS is based upon cellular base. Currently, GPS signals are lost inside most buildings, but triangulation-based technologies (similar to GPS) have been developed for indoor navigation. For example, relying on the same principles as GPS, Locata, an Australian company, offers beacon signals that cover large areas with the capability of penetrating walls. The In-Location Alliance, a collaboration with 23 companies specialized in indoor navigation, developed a Bluetooth-based (sent via beacons) indoor locating solution (Schutzberg 2013). Google, Apple and Microsoft have developed indoor locating technologies based upon Wi-Fi signals. The Wi-Fi technologies in handheld devices can be used to scan their surroundings for known or open networks. LBS using Wi-Fi relies on active surveys of an area to note the unique identifier and location of each Wi-Fi base station (Sladden and Girimaji 2016). These may include everything from hotspots in coffee shops and hotels to residential and business networks. When a Wi-Fi-enabled device accesses a location service, the browser or application may send to the service the coordinates of Wi-Fi networks it currently 'sees', enabling the current location to be triangulated.

RFID technology relies on storing and remotely retrieving data using radio waves. An RFID tag can be attached or embedded to a product, animal or person for identification or tracking. An RFID chip typically is composed of a tiny antenna attached to a tiny ID chip, and it can be passive (without an internal power source) or active (with a built-in power source). As GPS signals are usually too weak to be used for tracking purposes inside a building, the development of RFID technologies, especially its full integration with GIS, enables the potential tracking and mapping of everything on the surface of the Earth. EPCglobal, a system of tags to create globally unique serial numbers for all products using RFID technology, may be a concrete step forward to stitch together a new electronic skin for the Earth. Other indoor locating technologies currently being developed include ByteLight (based upon light) and IndoorAtlas (based upon magnetic fields). As of today, most indoor positioning technologies can detect the location of a person or object, but not always its orientation and directions. Vertical information like high-rise buildings still needs significant improvement. Also, the true indoor version of GPS has yet to be developed, thus indoor navigation usually relies on a hybrid of two or more technologies. For example, AISEL411 taps both Wi-Fi and MEMS (for the retail market). Last, but certainly not least, self-reporting through verbal description and geocoding via social media, especially location-based social media, can also enable service providers and other users to actually access an individual's locational information. This new form of self-reporting is also known in the literature as sousveillance (Mann et al. 2003) or participatory sensing. LBS has apparently accelerated to the transition from surveillance to sousveillance and participatory sensing with explicit geographic information. More people these days can broadcast their whereabouts via social media, which has increasingly become an important source of locational information.

KEY APPLICATIONS

Locational information plays an increasingly important role in people's daily lives, a variety of business operations and a growing number of government services and programs (FCC 2012). Although quite a few LBS applications overlap and often defy a single label, key LBS application areas can be classified according to three primary service goals – facilitating individual lives, business operations and government services.

LBS Applications at the Individual/Personal Level

LBS can be offered to individual consumers at different levels. At the most primitive level LBS can provide users with raw coordinates on location, which can be used for navigation and trip planning. A GPS or a cellular radio system can supply the coordinate information. From the raw coordinates it is possible to provide individuals with a grid reference that would enable locating themselves with a distance and direction for navigation. Once the locational coordinates of a person or an object are known, we can use this locational information to navigate, to plan one's own physical activity, track children, the elderly (especially those with Alzheimer's disease) and lost/stolen property (FindMyiPhone, MyChevrolet). Consumers can use the basic LBS, such as Lookout, OnStar, MyLink or Zillow, to search for various services, products, people or housing.

At a more sophisticated level, raw coordinates can be integrated or combined with other web-based information or information from a local GIS for locations of interest to indicate user locations on a street map and provide more detailed information about various local places. Dynamic information could be converted into services by matching the changing environment and historic preferences of the user. Location-based social networking (such as Foursquare, Loopt, FamilyLocation, Adrient, Tagg, FindFriends, Gowalla, Facebook Places and Yelp), location-based games (Scabble, Tourality, Geocaching, Pac Manhattan, Tourality, CanYouSeeMeNow.co.uk) and location-based assistive health-related services (Boulos 2002) are all commonly available LBS applications at the individual level. With the accelerated development of autonomous (driverless) vehicles and civilian drones, it can be expected that LBS will be more adapted to diverse mobile platforms. As part of the growing sousveillance, various wearable devices (such wrist bands, ear plugs and ankle bands) are becoming popular devices to track one's own physical activities or daily routines in space and time. In addition, moving beyond utilitarian applications, recent years have also witnessed the growing application of LBS for artistic purposes, such as locative art (gpsdrawing.org) and location-based story telling (mapstory.org). Indoor LBS has not only been used for navigation in public spaces (such as hospitals, shopping malls, and airports), but it is also an integral part of location-based advertising, providing just-in-time information via audio for tours, video and augmented reality experiences for connecting people of interest in proximity to one another.

LBS Applications for Business Operations

LBS is not only changing people's lives profoundly at the individual level, but is also becoming one of the major forces shaping and reshaping business practices across different economic sectors because of the growing availability of the locational information on consumers, service providers, vehicles and moving objects in real time. For example, LBS is one of the driving technologies responsible for the emerging sharing economy based upon mobile commerce and collaborative consumption. Although we do not have a commonly accepted definition of what precisely constitutes the 'sharing economy', also known in the literature as the 'collaborative economy', 'collaborative consumption' or 'peer economy', it generally describes the emergence of a new type of business built on the sharing of resources, allowing customers to access goods or services when needed (e.g. Uber, Airbnb and Car2Go; Badger 2013; Botsman 2013). According to a recent *Time* magazine cover story (Stein 2015), the explosive growth of the sharing economy will have profound, far-reaching impacts on society along multiple fronts. While sharing goods has always been a common practice among friends, family and neighbors, in recent years, the concept of sharing has moved from a community practice to a viable business model. Knowing where people or goods are located is obviously a prerequisite for the sharing economy. Without LBS, services like Uber and its derived services (sharing a taxi with a stranger along the way) are impossible to implement. Indeed, meaningful sharing will not take place unless we know where those who can provide a service (e.g. a ride) and those who need it are located in real time. Besides the sharing economy, LBS has also been increasingly integrated into growing business sectors that potentially can play a disruptive role. In the advertising industry, location has been a key to customizing marketing and advertising messages to potential customers at specific places or locations. The insurance industry has been exploring the possibility of location-based or usage-based premium policies, part of the location-based pricing/tariff for business operations. Most important, LBS also promotes peer-to-peer transactions as a part of growing disintermediation – the elimination of the middleman – a new reality every business must face in the sharing economy based upon LBS. For LBS applications in business, a commonly used method/technique is the so-called geofencing. A geofence is a virtual barrier or boundary that is created using LBS. Once a geofence is established, it allows an administrator to set up the system so that, when a device enters (or exits) the boundaries defined by the geofence, a text message or email alert is sent to geofence administrators. Web-based maps or digital globes such as Google Earth or Microsoft's Virtual Earth are often used by many geofencing applications to allow administrators to define boundaries on top of a satellite or map view of a specific geographical area. Many business LBS applications are based upon geofencing (Table 6.1).

LBS Applications for Government Services

Governments play a decisive role in providing LBS for both customers and businesses. Governments in various countries not only maintain control of location/positioning satellites in orbits, but have also increasingly relied on LBS to improve their services as well, especially in areas related to public safety and emergency management. Enhanced 911 (E- 911) and Amber Alert in the US from local to federal levels are potentially game changers for improving emergency management. E-911 relies primarily on the caller's location to determine the deployment of resources and shortest possible route to the site that needs assistance. Since 1 January 2013, AMBER Alerts™ has been automatically sending alerts (via its location-based Wireless Emergency Alerts (WEA) program) to millions of cell phone users based upon their geographic location. To receive these messages, users must

Table 6.1   LBS applications based upon geofencing

Use	Example
Fleet management	When a truck driver breaks from his route, the dispatcher receives an alert.
Human resource management	An employee smart card will send an alert to security if an employee attempts to enter an unauthorized area.
Compliance management	Network logs record geofence crossings to document the proper use of devices and their compliance with established rules.
Marketing	A restaurant can trigger a text message with the day's specials to an opt-in customer when the customer enters a defined geographical area.
Asset management	An RFID tag on a pallet can send an alert if the pallet is removed from the warehouse without authorization.
Law enforcement	An ankle bracelet can alert authorities if an individual under house arrest leaves the premises.

Source: http://whatis.techtarget.com/definition/geofencing.

have a WEA-enabled phone. By default, users are automatically enrolled for the three different levels of alerts: President, Imminent Threat and AMBER alerts (www.amberalert.gov). The addition of AMBER alerts to this notification system is a result of a partnership between CTIA and the wireless industry, the Federal Communications Commission (FCC) and the Federal Emergency Management Agency. Moving forward, AMBER alerts may have profound implications for law enforcement at local, state and federal levels. Under the new paradigm of citizens as sensors (Goodchild 2007), individuals and informal groups of individuals popularly known as smart mobs (Rheingold 2002), can record, describe and disseminate information on activities at an either individual or collective level from the perspective of the participants on a voluntary basis through crowd sourcing. These activities have contributed to the phenomenal growth of volunteered geographic information (VGI), and government agencies from local to international levels have been motivated and mobilized to use VGI to improve various kinds of services (Haklay et al. 2014). In recent years, the FCC has also aimed to use indoor navigation to provide timelier and more effective emergency services, but to make indoor LBS work better, we need other ancillary/supporting information through indoor maps. During the early days of the Obama administration, top officials called for making government operations and public policy decision-making more sensitive to local conditions and circumstances (Douglas 2010). This place-based decision-making will continue to bring about more demand for locational information of various kinds and LBS can be expected to play even more important roles for providing government services.

MAJOR CONCERNS

Similar to many other technological innovations in history, the widespread applications of LBS in society have also caused growing concerns in recent years. These concerns can be broadly grouped into three categories: eroding personal privacy; exacerbating social inequity; and impairing environmental sustainability.

LBS and Personal Privacy

The increasing applications of LBS have intensified society's concerns over privacy as potentially troubling apps such as *Girls Around Me* (http://girlsaround.me) that can be downloaded for free from iTunes,[1] or condom use can be mapped using precise latitude and longitude coordinates (http://wheredidyouwearit.com). The widespread applications of LBS and location-based social media have contributed to high-tech-based stalking of both men and women in recent years. As a worst-case scenario, some scholars have also contemplated the possibility of an extreme form of geofencing – geoslavery in the age of LBS (Dobson and Fisher 2007). The growing applications of geospatial technologies, especially through the development of human tracking devices and various kinds of LBS, have introduced a new potential for real-time locational control and monitoring that extend far beyond privacy and surveillance *per se*. Scholars have warned that society must address a new form of slavery characterized by location control – geoslavery, now looming as a real, immediate and global threat. According to Dobson and Fisher, geoslavery refers to 'a practice in which one entity (the master) coercively or surreptitiously monitors and exerts control over the physical location of another individual (the slave)' (Dobson and Fisher 2007, 47). With the current geospatial technologies readily available and affordable (US$300–500), an individual can routinely control the time, location, speed and direction of another individual for each and every movement, even without the permission of the individual being monitored. The fact that child slavery, sex slavery and other abusive behaviors (such as stalking) are still rampant worldwide (e.g. http://www.iabolish.com/slavery_today.htm) begs one to ponder where the increasing use of LBS will lead us in the near future. The potential for geoslavery has raised the issue of whether locational privacy should be universally protected as a fundamental human right (Monmonier 2002; Dobson and Fisher 2007).

LBS and Social Inequity

Besides privacy issues at the individual level, there have been growing concerns about the social impacts of the sharing economy relying on various kinds of LBS. The major concern in this regard is that the sharing economy based upon LBS has become a disruptive force to transform the regular labor force into an army of part-time contractors (Schor 2014). True, Uber drivers have flexible work hours and they can choose whether to work or not, but they essentially provide taxi services without the benefits or protection that traditional taxi drivers or workers used to have. Critics of the sharing economy argue that the new sharing economy can be very exploitative. While owners of the platform become millionaires or even billionaires in a few years, the actual workers on the ground just barely make the minimum wage. Recent income disparities in the US seem to confirm this unintended side effect of this economic transformation. Apparently most workers in the sharing economy are really getting a raw deal (Hill 2015). An interdisciplinary group of scholars has recently called for a more critical and comprehensive examination of the exploitative nature of the open/sharing economy for accumulation of profits in the age of hyper-capitalism (Ettlinger 2015; Reich 2015).

The growing use of LBS and its concomitant generation of spatial big data have also contributed to geographical profiling at both individual and community levels by corporations. The geographical profiling of people and places has improved the efficiency of marketing and advertising, even for some disadvantaged communities (Dillahunt and Malone 2015), but at the same time, it has also contributed to a new form of redlining that may inadvertently deteriorate or even deny services to certain areas or people, thus exacerbating the digital divide that will further perpetuate social inequities (Edelman and Luca 2014; Thebault-Spieker et al. 2015).

LBS and Environmental Sustainability

Besides privacy and social equity, environmental implications of the sharing economy and LBS also require further study (*The Economist* 2015; Knowledge@wharton 2015). At one level, the new sharing economy seems to promote a more efficient use of materials and energy by knowing where people and objects are located in real time via LBS, but it is premature to argue that the LBS-supported sharing economy is green. Further empirical studies are needed on the environmental impacts of the sharing economy, especially when the full picture of the entire life cycle of products and services is considered. So far the literature provides only partial and conflicting evidence. For example, Uber or Lyft have improved urban residents' mobility, but their full impacts on urban traffic and energy consumption are unknown. Further studies are needed in order to develop a more sensible policy to guide the development of the sharing economy and LBS in the coming years with the goal of economic efficiency, social equity and environmental sustainability.

SUMMARY AND CONCLUSIONS

LBS today represents only the beginning of a series of technological innovations that can potentially impact society in many ways at various scales, ranging from surveillance and the invasion of personal privacy to technologically induced changes in human spatial behavior. For the businesses community, LBS may engender drastic spatial structuring of business operations from retail trade to the insurance industry. LBS has also exhibited ample potential to improve government services and public policy decision-making. Perhaps most profoundly of all, further penetration of LBS in people's daily lives will mark the beginning of what Bill Gates (1995) envisioned as a 'documented life'. With more and more people's daily routines recorded at very fine spatial and temporal resolutions, massive amounts of data at the individual level will be accumulated with the growing popularity of LBS. Thus, LBS has the potential to provide novel sources of data for researchers in a variety of disciplines in both physical and social sciences, further promoting the spatial turn across the disciplines. Contrary to the early (in retrospect, premature) announcement of the death of distance (and thus the irrelevance of geography), the on-going geospatial revolution in society in general and the rapid growth of LBS during the past decade in particular have once again proved that the location of anything is everything in the age of big data. This revenge of geography is manifested in industry, government operations, academic research, and most importantly, in people's daily lives. The convergence of multiple technologies has contributed to the rapid growth of LBS during the last decade. Just like all other major technological innovations throughout human history, LBS has its own hidden and unintended consequences. Instead of indulging ourselves in a blind technological somnambulism, this chapter also highlights the potential threat of LBS to personal privacy, social equity and environmental sustainability. Moving forward, it is an imperative to optimize the benefits of LBS for society and at the same time develop effective/efficient means to address potential problems that powerful LBS can bring to society and individuals.

NOTE

1. The app *'Girls Around Me'* was removed from Apple's app store owing to mounting public pressure in the US, especially from women's groups, but similar programs like MoMo are available on the web to download in many other countries.

REFERENCES

Ahson, S. and M. Ilyas. 2010. *Location-Based Services Handbook: Applications, Technologies, and Security*. New York: Wiley.
Badger, E. 2013. Share everything: why the way we consume has changed forever. The Atlantic City Lab, 4 March, accessed 1 November 2013 from http://www.theatlanticcities.com/jobs-and-economy/2013/03/share-everything-why-way-we-consume-has-changed-forever/4815/.
Bartlett, D. 2013. *Essentials of Positioning and Location Technology*. Cambridge: Cambridge University Press.
Botsman, R. 2013. The sharing economy lacks a shared definition. *Fast Company*, 21 November, accessed 6 March 2015 from http://www.fastcoexist.com/3022028/the-sharing-economy-lacks-a-shared-definition#9.
Boulos, M. 2002. Location-based health information services: a new paradigm in personalised information delivery. *International Journal of Health Geographics* 3(3), accessed from http://ij-healthgeographics.biomedcentral.com/articles/10.1186/1476-072X-2-2.
Dillahunt, T. and A. Malone. 2015. The promise of the sharing economy among disadvantaged communities. In *Proceedings of the ACM 32nd International Conference on Human Factors in Computing Systems (CHI)*, accessed 7 March 2015 from http://www.tawannadillahunt.com/wp-content/uploads/2012/12/pn0389-dillahuntv2.pdf.
Dobson, J. and P. Fisher. 2007. The panopticon's changing geography. *Geographical Review* 97(3): 307–323.
Douglas, D. 2010. Place-based investments, accessed 10 April 2016 from https://www.whitehouse.gov/blog/2010/06/30/place-based-investments.
Edelman, B. and M. Luca. 2014. Digital discrimination: the case of Airbnb.com. Harvard Business School NOM Unit Working Paper 14-054, accessed 7 March 2015 from http://www.hbs.edu/faculty/Publication%20Files/14-054_e3c04a43-c0cf-4ed8-91bf-cb0ea4ba59c6.pdf.
Ettlinger, N. 2015. Accumulation by desperation: open innovation and networks and the predicament of labor. *New Left Review* 89: 89–100.
FCC. 2012. Location-based services report. Federal Communication Commission, accessed 10 April 2016 from https://www.fcc.gov/document/location-based-services-report.
Gates, W. 1995. *The Road Ahead*. New York: Viking.
Goodchild, M. 2007. Citizens as sensors: the world of volunteered geography. *Geojournal* 69: 211–221.
Haklay, M.E., V. Antoniou, S. Basiouka, R. Soden and P. Mooney. 2014. Crowdsourced Geographic information use in government. Global Facility for Disaster Reduction and Recovery (GFDRR). London: World Bank, accessed 21 December 2015 from http://discovery.ucl.ac.uk/143316.
Heinemann, G. and C.W. Gaiser. 2014. *Social – Local – Mobile: The Future of Location-based Services*. Berlin: Springer.
Hill, S. 2015. *Raw Deal: How the 'Uber Economy' and Runaway Capitalism Are Screwing American Workers*. New York: St Martin's Press.
Hill, S. 2015. New economy, new social contract. Washington, DC: New America Foundation. ttps://static.newamerica.org/attachments/4395-new-economy-new-social-contract/New%20Economy,%20Social%20Contract_Final.e38da6a16d714e58bbbd7462e83852d5.pdf.
Knowledge@wharton. 2015. How green is the sharing economy, accessed 16 March 2016 from http://knowledge.wharton.upenn.edu/article/how-green-is-the-sharing-economy.
Mann, S., J. Nolan and B. Wellman. 2003. Sousveillance: inventing and using wearable computing devices for data collection in surveillance environments. *Surveillance & Society* 1(3): 331–355.

Monmonier, M. 2002. *Spying with Maps*. Chicago, IL: University of Chicago Press.

Pfeiffer, E. 2003. WhereWare. MIT Technology Review, accessed from https://www.technologyreview.com/s/402018/whereware.

Reich, R. 2015. Why the sharing economy is hurting workers – and what must be done, accessed 10 April 2016 from http://robertreich.org/post/134080559175.

Rheingold, H. 2002. *Smart Mobs: The Next Social Revolution*. Cambridge, MA: Perseus.

Schor, J. 2014. Debating the sharing economy, accessed from http://www.greattransition.org/publication/debating-the-sharing-economy.

Schutzberg, A. 2013. Ten things you need to know about indoor positioning, accessed 2 April 2016 from http://www.directionsmag.com/entry/10-things-you-need-to-know-about-indoor-positioning/324602.

Sladden, D. and J. Girimaji. 2016. *Connected Mobile Experiences and Location Based Services: Understanding Indoor and Outdoor Location Technologies using Wifi, BLE, iBeacon and Other Sensors*. Indianapolis, IN: Cisco Press.

Stein, J. 2015. Strangers crashed my car, ate my food and wore my pants. Tales from the sharing economy (cover story). *Time*, 9 February, accessed 1 March 2015 from http://time.com/3686877/uber-lyftsharing-economy/.

Thebault-Spieker, J., L. Terveen and B. Hecht. 2015. Avoiding the south side and the suburbs: the geography of mobile crowdsourcing markets' urban environments. CSCW 2015, 14–18 March 2015, Vancouver, accessed 16 March 2016 from http://www-users.cs.umn.edu/~bhecht/publications/MobileCrowdsourcingSES_CSCW2015.pdf.

The Economist. 2015. Is the 'sharing economy' sustainable?, accessed 16 March 2016 from http://www.economistinsights.com/sustainability-resources/opinion/sharing-economy-sustainable.

Werner, M. 2014. *Indoor Location-based Services: Prerequisites and Foundations*. Berlin: Springer.

7. Virtual realities, analogies and technologies in geography
Michael Batty, Hui Lin and Min Chen

From one perspective, all digital representation and simulation imply a virtual reality (VR) in which the world is abstracted and then manipulated using various kinds of computation. Indeed one could argue that any kind of abstraction, hence our ability to stand back from the real world and make sense of it, involves constructing a virtual reality, although in terms of digital representation, the way we manipulate such a world is very different from anything prior to the introduction of computers in the middle of the last century. Digital computers are universal machines and hence the worlds that can be represented contain a generic mode of manipulation that involves producing changes to the virtual world that are impermanent and disappear (notwithstanding that such simulations can be captured digitally) once the devices are switched off. Strict virtual reality systems, however, are somewhat narrower than digital representation, and simulation in general for the virtual worlds we will introduce here is characterised by an intrinsic link between the computable world itself and the way we as users of this world interact directly with it. Unlike the wider spectrum of computer models, for example, virtual reality systems engage the user interactively in that the user changes the world with the world also changing the behaviour of the user. Virtual reality systems always involve this two-way traffic; they are thus said to be 'immersive', meaning that users interact directly with the world. Yet as we will see, there are many different kinds of immersion across a continuum from relatively passive interaction to complete immersion where there is little obvious difference between the user and the world itself.

Our implicit definition of VR is clearly illustrated in its short history. In the mid-1980s, Jaron Lanier, who is the accredited populariser of the term, devised various artefacts that enabled users to directly interact with computable objects in which they could manipulate their own actions using devices such as the data glove and 3D goggles (Lewis 1994). Right from the beginning, this kind of interaction came to characterise VR with the user seeing, pointing at and touching the digital media in diverse ways, hence changing its form through literally 'hands-on' computation. Just as games were integral to driving the PC revolution forward, they were also instrumental in pushing VR. By the mid-1990s, gloves and related devices were followed by entire environments where the user could be immersed in a physical space such as a theatre or a room (sometimes called a CAVE) whose walls were controlled by computable media and in which users could interact with this media in 3D through various modes of touch.

Most VR systems until then were based on physically connecting users to machines, but by the late 1990s, once computers had more or less converged with communications through the Internet, VR came to be extended to networked environments. Online games emerged while virtual worlds were constructed using network software producing environments that, although local, were controlled by users at remote locations. In particular, the emergence of virtual worlds represents a synthesis of networked communications, 3D virtual environments and interactive gaming platforms where users might appear alongside each other as avatars. These avatars can interact with other computable objects, some controlled by other users acting autonomously with respect to other objects in the scene but with some objects simply acting as programmed by developers outside the immediate VR scene. Such worlds enable many different kinds of feedback between humans and computable objects which generate realities that are very different from the real world, that enrich the real world through such interactions as well as providing environments to explore future worlds as yet unrealised.

Currently we can distinguish between several types of VR environment that link users to computable representations and simulations in two-way fashion. First we have *standalone personal environments* which represent the lineage from Lanier's data glove to contemporary headsets such as Samsung's Galaxy Gear VR which are primarily used for gaming but which are now completely affordable and represent the wave of the future in personal VR. High-end versions such as those produced by Oculus VR might also be networked and it now goes without saying that all of the systems available can exist in networked desktop form. Second, there are *purpose-built environments* such as VR theatres and CAVEs which enable large numbers of users to participate in a virtual scene, where human–human as well as human–computer interaction is important. These systems are increasingly used for professional and scientific purposes, although they can be used for gaming. Third, there are virtual worlds that exist on the desktop of networked environments in which human users can appear as avatars alongside computable objects with whom they interact. The Unity platform provides such a virtual world with the focus on many uses from gaming to professional digital design. Fourth, there are mixed environments that consist of a mixture of human interaction, global network access, gaming environments and general purpose access to the digital world, often using various human–computer interactive tools such as those based on point and click, even sensory inputs. These are hard to classify but they do represent the most widely used of all VR environments, largely because they consist of stitching very different forms of human–computer interaction together. Last but not least, there are hybrid analogue environments where digital representation is projected back onto the real world and where human interactions are both with analogue and digital representations of the same phenomena. Sand tables, physical data tables, even touch tables and all kinds of surface access represent the contemporary realisation of these developments.

In our survey of VR technologies in geography, we cannot provide examples of all of the many different types of system. We will thus avoid gaming software although a considerable body of such software does exist for both educational and professional purposes and this is relevant to our wider quest of grounding the role of VR in other technologies in geography. Our focus here will be on how VR technologies have developed in relation to other software technologies such as geographic information systems (GIS), which in turn embody spatial analytics, simulation modelling and remote sensing amongst other developments in digital geography. We will also focus here on virtual environments that are integrated in some sense with GIS and computer-aided design methods and we will follow a line of development that embraces all of these technologies under the banner of virtual geographic environments or VGEs (Lin and Batty 2011). We define these environments as 'computer-based digital spaces that we can observe, participate in, and experience in person' (Lin et al. 2015, 493). In particular, VGEs are something more than virtual realities *per se* for we argue that they deal with functions of the environment that are process-based. To an extent, whenever we define complex topics such as VR, we encounter many new definitions that we must be specific about, and by process-based, we mean that such VGEs embody geographic processes that evolve to forms of spatial organisation in space and time; in short, they embody geographic models of various kinds and this will be our criterion for the selection of examples in the rest of this chapter.

We will begin with traditional standalone desktop environments which traditionally have been associated with single users but are rapidly being ported to the networked world. Our focus will be on 3D representations of cities which extend the idea of the digital map and immediately embrace the key idea in VR that the user moves through the scene. Motion is central to VR, usually through the way users interact with the media but also in VGEs through the processes that are embodied in the scene. We will then move to networked environments, which give equal importance to the user and the scene within virtual worlds, either on the desktop but increasingly across web-based portals that define the ways in which such worlds are configured. We will not spend very much time dealing with purpose-built hardware–software environments such as VR theatres for these have become commonplace. In many professional and now in retail environments, large-screen displays with some multimedia are widely available and even TV technologies are embracing VR in the home. The cutting edge is now much more focused on usage and the processes of engagement with the media than on the technology and in the examples we show we will focus on these features. We will then discuss mixed media realities that we call VGEs and illustrate how we can integrate different kinds of media in desktop and networked form. Last, we move back from the virtual to the real world, illustrating how we might think of virtual realities as analogies of the real world where we mix VR and ordinary human engagement. As we have implied, VR is now everywhere in contemporary life, accessible from the most common of our devices, our TVs and our smart phones. What we will provide here is a snapshot of this world as it pertains to geography, particularly but not exclusively to the geography of the city. To see how far we have come, look at Fisher and Unwin's (2001) edited collection of 15 years ago entitled *Virtual Reality in Geography*. Read the progress report by Batty (2008) a little later but written 10 years ago now and then consider this chapter and speculate on how this world will continue to change over the next 15 years, to the year 2030.

3D REPRESENTATIONS: VIRTUAL CITY MODELS

When Ira Lowry (1965) wrote his seminal article 'A short course in model design', there was a widely known distinction between iconic models and symbolic models. Symbolic were mathematical models whose functions could be formalised and thus programmed, and hence implemented on digital computers for predictive purposes. They were regarded as an order more powerful and influential than iconic models which were essentially more superficial representations of systems with little or no predictive power. In the urban domain, such iconic models were essentially architectural constructions, often of the building volumes, at a much smaller scale than the real thing that could only be viewed from the outside. Interiors required separate iconic model representations. In the 1960s, there was barely any recognition that such icons could be made digital and it was only the rise of computer graphics where the graphic content was stored in the frame buffer of a computer – often as an add-on to computer memory – that led to digital representations. It was not until the 1980s when personal computers accelerated the rise of graphics and graphical user interfaces that such digital forms could be constructed routinely (Batty 1987) and even then it would require the much more powerful miniaturisation that marked chip production in the 1990s to really reach the point where 3D iconic representations became truly digital.

The first 3D urban models were developed by large architectural firms such as Owings Skidmore and Merrill in the early 1980s. These were wire frame affairs where the buildings lacked solidity but right from the beginning the way to view these models was to interact within them: to fly through the scene and to navigate within the media. Thus 3D representation required more than simply passive map viewing, which of course marked the development of computer cartography and GIS. The complexity of 3D was such that the models required direct interaction with the user and this still largely consists of navigation, although more recently these models have been populated with many attributes that can be picked up and interrogated analytically as part of the navigation process. In fact by the early 2000s, another development had taken place. From the 1980s, first mini-computers, then workstations and finally PCs embraced graphics algorithms that were hardwired into the machines themselves with 3D rendering, hidden-line elimination and all the transformations required to scale and rotate objects around being embodied within the geometry engines, which became integral to the machines themselves. This enabled almost instantaneous rendering on-the-fly, which is an obvious prerequisite to acceptable fly-through and navigation.

By the early 2000s, the relevant hardware and software made the construction of these 3D environments routine with systems ranging from ERSI's ArcGIS to Autodesk's 3ds Max. Quite large-scale 3D city models with thousands of blocks can be constructed in a matter of hours if terrain, building polygon and height data from LIDAR are readily available. We built our first models for London only a year or so before Google Earth released its 3D globe in 2004 and began populating this with 3D building blocks (Batty and Hudson-Smith 2005). We illustrate such a 3D block model for the centre of London in Figure 7.1(a), where it is immediately obvious that there is little detailed rendering. Out of about 50,000 building blocks in this model, only some 1% are rendered in high detail for the purpose of this kind of virtual reality is to provide a 3D version of the 2D map – to populate the 3D content with attribute data pertaining to the building blocks while also enabling users to layer different surfaces across the scene associating the underlying geometry with other key geographic layers such as pollution, flooding and so on. In Figure 7.1(b), we show the same model in Google Earth, which enables us to use all the functionality of that reality (such as StreetView and so on) to inform the user about the scene. In Figure 7.1(a) we also show how the River Thames would begin to flood the South Bank were it to rise by 1 metre (which is what the IPCC are forecasting for the North Sea off eastern Britain by the year 2100) and in Figure 7.1(c) we show the layer of nitrogen dioxide, a particular pollutant associated with road traffic, in an area of central London adjacent to Parliament.

(a) The 2003 model with flooding

(b) The model in Google Earth

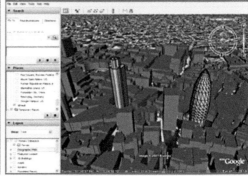

(c) The pollution NO$_2$ layer

(d) The current model system

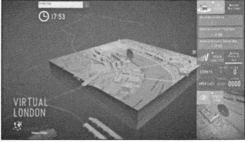

Note: The model is constructed so that a user can query all its attributes, pulling up various 3D and map layers data as and when required while populating the model with real-time movement.

Figure 7.1   Various renditions of a traditional virtual reality: the 3D London model

Figure 7.1(d) is our current model that is designed to provide considerably more flexibility to the user in that the move to using these block models as visual databases is taken to an extreme. The long-standing idea of using the model to visualise the aesthetics of the city is not within the mission of this kind of modelling for the interface is designed as a working link to the kinds of data the user can generate on-the-fly. Various analytics are accessible through this interface while the model can also be populated by importing movement streams generated from data and predictions from agent-based models of pedestrian and traffic flow. Another feature that has emerged from these kinds of 3D block model is to populate them with data that is not usually associated with 3D representations. In the development of real-time streaming of data from computers and sensors embedded in the built environment or used in a mobile context to capture media as through smart phones, such data can provide a visual animation of how geographic space and its attributes are continually changing. Rather than provide an animated map – and such maps are also part of the wider domain of virtual reality – what we

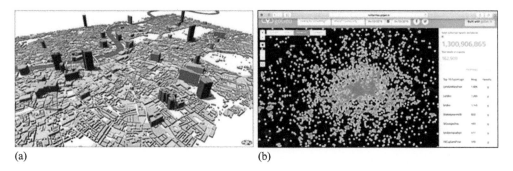

(a)   (b)

Figure 7.2   Representations of real-time tweets: (a) 3568 tweets in central London
over 15 hours and 40 minutes (1 July 2013); (b) geotagged tweets from
GoGeo's Twitter access over a 48 hour period in central London (23–25
April 2016)

demonstrate here is how we can use the 3D model to tag data that changes in real time
to the building blocks. If we then associate the numerical values of the data to the building heights – in effect, we are proposing a mapping of a real-time locational dataset
to the locations associated with the building blocks – we can vary the building heights
according the real-time data. In Figure 7.2(a), we show another virtual London model
that was built using ESRI's City Engine software, where the heights of the buildings
are proportional to the number of persons located at those buildings who are sending
tweets which are tagged to the geographical coordinates of the place. The key features of
such a visualisation is to associate the number and time of the tweets with the buildings
(http://en-topia.blogspot.hk/2013/07/tweetcity-re-populating-london.html). The snapshot in Figure 7.2 is taken from 15 hours' worth of geotagged tweets in central London
on 1 July 2013; it is clear that the dominant places where people tweet is in places of
entertainment, such as museums, places where people come together at transport hubs,
such as mainline railway stations, as well as places where people cluster for shopping.
In Figure 7.2(b) we associate the 3D visualisation of tweets in London with a real-time
map of who is tweeting taken from the GoGeo site (www.gogeo.io), which displays the
last 48 hours' worth of geotagged tweets worldwide. We show the intensity of the tweets
and it is possible to zoom in and pull up the text of individual tweets. We have not seen
anyone link this massive database of tweets to 3D visualisation – linking Figure 7.2(a) to
(b) – but this seems eminently feasible and such possibilities of extending this kind of
virtuality now appear endless.

To an extent we have already moved beyond traditional desktop VR. The current
Virtual London model is networked as is the visualisation of real-time data within it. It
is now quite straightforward to give access to many users and this changes the nature
of traditional VR quite substantially. However, to progress we now need to outline the
virtual worlds movement which has long been headed for convergence with 3D media,
and indeed requires this for its continued realisation and development.

VIRTUAL WORLDS: MIXING HUMANS, COMPUTABLE AGENTS AND GEOGRAPHIC MOTION

Our examples so far have mixed various computer technologies in diverse ways and do not show any of the purist focus of VR characterising earlier developments. In our extensions to 3D technology on the desktop, networked systems enable agents and real users to be mixed while real-time streaming of data associated with the emergence of the smart city is now providing a strong direction to such virtual worlds. Figure 7.3 indicates the kinds of developments that have taken place during the last 30 years. The picture on the left shows two well-known British planners pointing to an iconic model for the reconstruction of the town of Plymouth, which was heavily bombed during the Second World War. The picture from 1944 shows all the science that has since become computerised. Of course the iconic model has become digital and the participants can now appear as avatars in a virtual world where they engage in immediate interaction even though they might be linked in from remote sites. Other features of the traditional scenes as portrayed in the maps on the wall show flow systems and histograms which have all become computerised and whose data can be easily ported now to the virtual world. In the right panel of Figure 7.3, traditional iconic models of inner and central London are augmented by various digital technologies, mixing the real with the virtual, the direct with the augmented in different blends, and this is fast becoming the norm in cutting edge virtual reality systems.

There are many types of virtual world in which these realities can be constructed and in the one shown in the centre panel of Figure 7.3 – Second Life – gained enormous momentum during the early 2000s. However, continual innovation almost guarantees a fast turnover of software systems which make use of new media, and one of the

Notes: (left) James Paton Watson and Sir Patrick Abercrombie use an iconic model of Plymouth (1944) to point to specific developments while (centre) two avatars controlled by users from remote places engage in discussion to position the Gherkin building in the City of London (2005). Professionals (right) associated with the planning of London examine large-scale iconic models of London augmented with various information technologies that complement traditional media (2005).

Figure 7.3   From real to virtual worlds: professionals and iconic models to avatars and networked digital 3D environments

platforms that we are illustrating in this chapter is Unity 3D. In Figure 7.1(d) we show a working prototype of this platform where the media is the virtual 3D city through which a user can navigate and upload and download attribute data associated with any location which has been populated with such data. Unity is not strictly a virtual world in the sense of Second Life but this is simply a matter of taste in our view for Unity is essentially a games engine which has multi-user capabilities through the usual kinds of networked systems that enable many users to interact in a virtual environment. In fact, VR systems like Unity tend to be more restrictive than massively multi-player online worlds but much depends on the extent to which the user requires rapid motion in navigation. In the somewhat more research-based worlds that we are involved with here, 'shoot'em up' capabilities are not required and thus there is an increasingly wide choice of platforms. Moreover, more custom-built platforms are likely to characterise geographic applications such as the one that has been built for the campus of the Chinese University of Hong Kong, which we will demonstrate here.

In Figure 7.4, we show the kind of virtual environment constructed for a major

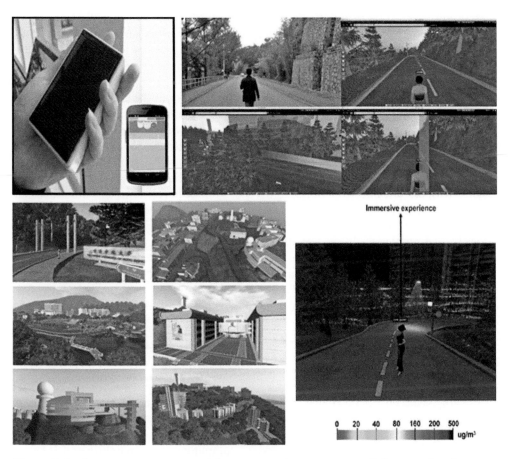

Figure 7.4   Inputting environmental data – cognitive responses to light, sound and pollution – into a virtual world through real-time sensing

university that can be used for many different purposes. Our focus here is not on naviga-tion through the campus *per se* but on how we can capture and import information about the experience of the real users and how this can be incorporated in virtual movement through the campus. Imagine a 'user' or participant walking through the campus and being equipped with sensing devices that pick up noise and pollution that is experienced as the users encounter different sensory experiences relating to what they see and hear and what they are exposed to (Hu et al. 2011). This kind of data can be captured via mobile devices which we show in Figure 7.4 – sensing devices which are portable and linked to smart phones that enable these sensory data, which is captured to be imported continually into an archive of data on the web that is thence picked up by the apps on the smart phone. This then controls the user as an avatar providing immediate feedback as to the way the environment is perceived. This is an ambitious project that seeks to encode the avatar's behaviour into a form that is influenced by the sensed data (Che et al. 2014). In short the virtual world and its avatars behave according to the real-world data that is imported into the world and users thus condition the behaviours of the virtual agents. One can imagine generating better behavioural data from this kind of experience which in turn produces much more realistic navigation in virtual worlds (Hu et al. 2014).

All we can do to give some sense of how we build such a virtual environment is to show the impressionistic collage of devices and pictures in Figure 7.4. With the emergence of low-power and inexpensive sensing and wireless communication devices, the kind of crowdsourcing and mobile sensing we show in Figure 7.4 has the ability to monitor a wide variety of environmental changes and human activity at the city scale. Using geo-sensor and mobile data-collection approaches, massive and timely environmental data can be acquired for simulation and analysis through a virtual world. To achieve this, in Figure 7.4 (at the top left) a portable kit based on a sensor, mobile application and an external network server collects environmental information by specific sensors and communicates using smart phones via Bluetooth transferring the measured data (e.g. air quality, temperature, humidity, noise) to the user's phone in real-time. The mobile sensor along with a hand-held sensor system makes up the basic environmental data collection station. As an integrated platform for data management, analysis and representation, this is our basic example of a VGE. The virtual world – a university campus composed of an interaction between a natural and social environment – is characterised by hilly terrain, complex buildings, composite road and path networks, unique geographic phenomena and social behaviours. In this example, the project is building a sharable virtual cogni-tive space that can help in enhancing our understanding of environmental protection concepts and awareness.

THE MOVE TO VGEs

VGEs are essentially digital environments that integrate diverse VR technologies. They have strong graphic image-ability as do all VR systems, but they also link input data to process simulations of many kinds (Lin et al. 2013a). They are beyond representation *per se*, although they invoke all the main features of computer cartography, GIS and 3D visu-alisation through models of geographic systems that are driven by natural and human processes. The key to VGEs is that they simulate processes in the geographic environment

Figure 7.5   *A collaborative virtual geographic environment showing regional sources of different air pollutants in map form (left) and 3D (right)*

as well as engaging multiple users in analysis. Unlike VR games, environments may be networked to involve many users; VGEs are more analytical in their usage and enable professional and research users to engage in analysis in a coordinated but active way. They thus represent a form of crowdsourcing but without the frenetic activity that characterises many games and they also require users to be expert or at least to be exposed to learning and generating greater expertise about the project and problem of their concern. To an extent, the virtual world of the university campus and the extended model of Virtual London that we illustrated in previous sections were VGEs but at a much finer scale and did not involve the kinds of analytics that characterise the examples here, which are at a much larger, coarser regional and global scale.

In Figure 7.5, we extend our modelling of air pollution from the campus-scale virtual environment to the regional where we show that environmental pollution has a wide sphere of influence, high occurrence, strong outbursts and intense spatial and temporal incidence owing to global climate change and related human activities. Its impact on regional ecological security and sustainable development is substantial and thus a collaborative system for sensing and measuring its impact is required and this lies at the basis of this collaborative VGE where we are able to carry out a distributed regional environmental simulation. These experiments not only help construct environments that are difficult to build, but also save costs and resources in developing policies to mitigate the dis-benefits that such environmental pollution generates. The system is a convenient way to package information for experts and decision-makers who are able to communicate through the VGE, and collaborate in designing and testing the impact of environmental policies. For instance, this VGE is designed for the analysis of air and water pollution management across the Pearl River Delta, which includes over 60 million people in Guangdong province, Hong Kong and Macau. The system integrates point, line and surface source data, is able to reconstruct data for coupled climate and pollutant dispersion models while imitating related problems, and can advance forward different early-warning programmes through joint-selection models. Decision-makers can engage in discussion of many policy initiatives that relate to such early-warning systems as well as coordinate environmental policy through many different agencies across a very diverse region (Lin at al. 2013b).

Moving to the national and global scales, VGEs tend to be less visually interactive but nevertheless embrace many more decision-makers of different types reflecting local elements in the global picture. Global change is both a complicated and comprehensive issue involving population, land use, the Earth system and multiple agencies reflecting many kinds of mission and politics. In this context, human activities are affecting the Earth system at an unprecedented rate. Meanwhile, urbanisation, the continuing but changing population explosion and its consequences for migration coupled with an excessive use of resources has severely hindered sustainable development. Utilising a multi-dimensional representation at the geospatial system science level, VGEs can visualise the spatiotemporal variational processes of the Earth system's mechanism that help define the impact of human activities on the global system. VGEs that encapsulate global change are being built as open platforms for collaborative simulations and evaluations focused on model integration, reuse, optimisation and collaboration. Model integration considers how scales which are mismatched can be fused, how different VGE models developed by different teams with different development styles can be integrated and reused, while flexible patterns of interaction and collaboration for different modes of collaborative modelling and optimisation can be developed (Hu et al. 2011, 2014; Che et al. 2014). These systems are rapidly being developed at the present time and promise to extend VR technologies way beyond their traditional game-like usage, merging more generally with a massive range of new technologies from real-time sensing and the streaming of big data to global decision-making which is informed by models and data that are essentially being developed in remote locations and accessible through cloud computing.

THE REAL AND THE VIRTUAL: ANALOGIES INCORPORATING MIXED, AUGMENTED AND BLENDED ENVIRONMENTS

Earlier in this chapter we illustrated traditional iconic models and how these had become digital, forming the essence of 3D virtual environments such as those that characterise the virtual city. In the rapid move towards the digitisation of society, there is now a feeling that something is being lost by separating the virtual from the real, and there are a variety of initiatives in VR which are moving these technologies back to the material world. We have found that it is extremely difficult to engage users to their full extent in a purely virtual world where they appear as avatars even if their presence is controlled by themselves from remote locations. In short, users are much more receptive to discussion of the problems that virtual worlds seek to represent and motivate if they are able to do this in the context of the real world. To this end, there are various ways in which physical projections of the virtual world onto the material might be used to structure this kind of engagement. For example, touch tables of various kinds, projecting data and simulations onto the physical medium of the screen as a table, the table as formed by sculptural surfaces such as sand or by harder physical media such as wood or plastic, are being used in ingenious ways. In Figure 7.6(a) we show how data associated with Greater London which is generated digitally can be projected back onto what we call the London Data Table, which engages people in discussion, analytical thinking and design in ways that are much richer than solely within a virtual world. In Figure 7.6(b), we show SimTable, which is a sand table developed by the Redfish Group in Santa Fe, New Mexico to simulate hazardous diffusion

Figure 7.6 *Projecting the virtual onto the material: (a) the London Data Table (wood); (b) RedFish's SimTable (sand). See https://www.simtable.com/ portfolio-items/a-defense-case-study-here/*

Figure 7.7 *Flying through virtual London: augmented reality using Kinect Xbox console (main picture) and an Oculus Rift headset fly-through (inset)*

of wild fire and associated phenomena where a cellular automata model simulates the spread while the user activates various policies to stop the spread by pointing at the displayed visual functions in the media on the surface of the sand. The same kinds of rich engagement as in the case of the London table are encapsulated by this media.

There are many other blends of the virtual with the material. In Figure 7.7, we illustrate how a person can fly through a virtual world built from Google Earth in which various other media exist, navigating by performing various actions – essentially flying through

Figure 7.8   A model of the London riots: choosing policies that minimise damage using material objects – Lego police and vehicles – to control policing and to contain rioters

the scenes as a bird might navigate. This facility is built using Microsoft's Kinect – the motion sensor for the Xbox 360 gaming console – which provides an intuitively obvious medium for navigation without any obvious controller other than the human user. This is a kind of augmented reality but in the same scene we show the more immersive form where the user wears the Oculus headset whose goggles converge on the 3D experience of flying through a virtual city. As we noted above, these technologies are now quite afford-able, such as the Samsung Gear VR, and there are now countless ways of enabling many users to coordinate and share experiences using such technologies.

Our last example shows how we can augment the reality by using material objects to activate and control a simulation of movements across a city region. We built a model of the London Riots in 2012 using agent-based simulation (Davies et al. 2013) whose media is projected on the touch table shown in Figure 7.8. To manage the riots there was a considerable police presence and we enabled this using icons of the police agents to activate certain policies and policing. The agents' positioning above the table and their activation through pressure/touch enabled the riots to be contained – the diffusion of the other agents to be manipulated – and to this was added a cost–benefit calculation associ-ated with minimising the damage that was done by the rioters. In this sense, it is easy to see how users of the system can configure policies that will optimise the role of policing.

CONCLUSIONS: WHITHER VR IN GEOGRAPHY?

We are still in the midst of a digital revolution that has moved computation from the main frame to the smart phone and has seen digital usage spread out from the laboratory to the city, the nation and the globe. Virtual realities now dominate usage to the point where the traditional usage which was largely immersive is now being augmented by all kinds of media juxtaposed in diverse and often unusual ways. What we have not explored here, which is almost bound to be significant in the next decade, is the spreading out of computers into ourselves. The spaces we inhabit are being rapidly computerised but our own bodies will be the next frontier as modern medicine continues to embrace computation. Medicine as software will continue in the treatment of illness but the notion that we will embed computers into ourselves is likely to generate all kinds of unprecedented and strange uses that will move this field on once again as we find new ways to communicate with one another.

Much of what we have discussed in this chapter involves ways in which we can now interface these technologies with one another as well as with ourselves. We have not, however, talked very much about the smart city and the way our environments are becoming digital. Most of our uses of VR have been to address traditional geographies, of nature or of human spaces that we have generated, not about how computers are entering those geographies and changing them. After all geography is about how we interpret our spatial world, and as we change it through technology so our geography will change. Perhaps the most challenging aspect of computerised technologies is the notion that in the past we have used such technologies to study material space but these same technologies are now being used to construct digital spaces. This idea of using machines to study spaces composed of the same machines presents a recursion in thinking that is both profound and challenging. What is clear is that we are but at the beginning of a long journey through the rest of this century when reality is being reconstituted in virtual terms and this promises to change the geography of technology as well as the technology of geography in ways that will continue to challenge and amaze.

ACKNOWLEDGEMENTS

For the media illustrated here, thanks are due to Andy Hudson-Smith, Steve Evans, Lyzette Zeno Cortes and Gareth Simons for the 3D London models; Steven Gray and Ollie O'Brien for the data tables; Stephan Hugel and Flora Roumpagni for the 3D tweet visualisation; Steve Guerin of Redfish, Santa Fe, NM for the SimTable simulation; and Lan You, Mingyuan Hu, Yulin Ding, Weitao Che, Chunxiao Zhang and Fan Zhang for the VGE work.

REFERENCES

Batty, M. 1987. *Microcomputer Graphics: Art, Design and Creative Modelling.* London: Chapman and Hall.
Batty, M. 2008. Virtual reality in geographic information systems. In J. Wilson and A. Fotheringham (eds.), *The Handbook of Geographic Information Science.* pp. 317–334. Oxford: Blackwell.
Batty, M. and Hudson-Smith, A. 2005. Urban simulacra: from real to virtual cities, back and beyond. *Architectural Design* 75(6): 42–47.
Che, W., Lin, H., Hu, M. and Lin, T. 2014. Reality–virtuality fusional avatar based noise measurement and visualization in online virtual geographic environments. *Annals of GIS* 20: 109–115.
Davies, T., Fry, H., Wilson, A. and Bishop, S. 2013. A mathematical model of the London riots and their policing. *Scientific Reports* 3: 1303. doi:10.1038/srep01303.
Fisher, P. and Unwin, D. (eds.). 2001. *Virtual Reality in Geography.* London: Taylor and Francis.

Hu, M., Lin, H., Chen, B., Chen, M., Che, W. and Huang, F. 2011. A virtual learning environment of the Chinese University of Hong Kong. *International Journal of Digital Earth* 4: 171–182.

Hu, M., Lin, H. Che, W., Lin, T. and Zhang, F. 2014. Combining geographical and social dynamics in dynamic 3D environments. In T. Bandrova, M. Konecny and S. Zlatanova (eds.), *Thematic Cartography for the Society*. pp. 191–208. New York: Springer International.

Lewis, P. 1994. Sound bytes; he added 'virtual' to 'reality'. *New York Times*, 25 September, accessed from http://www.nytimes.com/1994/09/25/business/sound-bytes-he-added-virtual-to-reality.html

Lin, H. and Batty, M. (eds.). 2011. *Virtual Geographic Environments*. Redland, CA: ESRI Press.

Lin, H., Chen, M. and Lu, G. 2013a. Virtual geographic environment: a workspace for computer-aided geographic experiments. *Annals of the Association of American Geographers* 103: 465–482.

Lin, H., Chen, M., Lu, G., Zhu, Q., Gong, J., You, X., Wen, Y., Xu, B. and Hu, M. 2013b. Virtual geographic environments (VGEs): a new generation of geographic analysis tools. *Earth-Science Reviews* 126: 74–84.

Lin, H., Batty, M., Jørgensen, S., Fu, B., Konecny, M., Voinov, A., Torrens, P., Lu, G., Zhu, A.X., Wilson, J., Gong, J., Kolditz, O., Bandrova, T. and Chen, M. 2015. Virtual environments begin to embrace process-based geographic analysis. *Transactions in GIS* 19(4): 493–498.

Lowry, I. 1965. A short course in model design. *Journal of the American Institute of Planners* 31: 158–165.

PART III

COMMUNICATIONS
TECHNOLOGIES

8. Fiber optics: nervous system of the global economy
Barney Warf

Among the various ways in which human beings have engineered the Earth's surface, the contemporary worldwide fiber optics network surely ranks as one of the largest, more important, and impressive for its sheer size and impact. Fiber optics lines – the seamlessly integrated network of glass wires about the size of a human hair, bundled together in cables of several thousand – form the core of the global telecommunications infrastructure. Indeed, far more than any other technology, such as copper cables, microwaves or satellites, fiber optics supply the vast bulk of data, voice and video transmission services around the world. Because of their capacity to deliver high volumes of information rapidly and securely (e.g. via broadband), fiber optics cables form the backbone of the Internet as well as private corporate lines, and are widely used in the electronic media for commercial and residential purposes (e.g. cable television). The technology is thus central to understanding contemporary economic, political and cultural transformations.

This chapter offers an overview of fiber optics as a technology, an industry and a force within the contemporary world. It begins with a brief history of how this phenomenon came to be, including the long history of scientific innovation behind it. Second, it situates and contextualizes fiber optics within the contemporary information-intensive global economy. Unfortunately, this issue has often been approached in apolitical and technocratic terms that ignore the social origins and consequences of the industry. Third, it turns briefly to the urban dimensions of this technology, the ways in which it is implicit in folding and refolding the spatiality of urban accessibility. Fourth, it maps out the global geography of fiber optics, focusing on the two major markets across the Atlantic and Pacific Oceans. Fifth, it explores three consequences of the fiber boom of the 1990s, including a wave of corporate failures, the emergence of so-called 'dark fiber', and the challenge that fiber poses to the satellite industry.

A BRIEF HISTORICAL OVERVIEW OF FIBER OPTICS

Fiber optics are long, thin, flexible, highly transparent rods of quartz glass (or less commonly, plastic) about the thickness of a human hair (Figure 8.1) that can transmit light signals through a process of internal reflection, which retains light in the core and transforms the cable into a waveguide (Agrawal 2002; Freeman 2002; Crisp and Elliot 2005; Hecht 2015). They can transmit voice, video or data traffic at the speed of light (299,792 km/s); because light oscillates much more rapidly than other electromagnetic wavelengths (200 trillion times per second in fiber cables v. two billion per second in a cellular phone), such lines can carry much more information than other types of telecommunications. Modern fiber cables contain up to 1000 fibers each and are ideal for high-capacity,

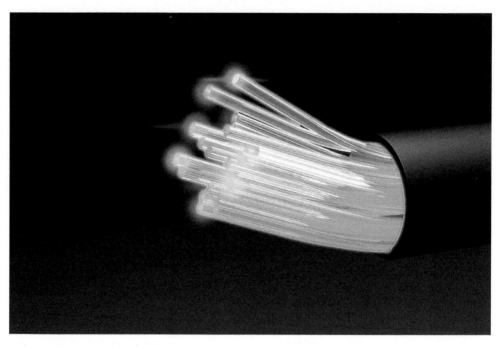

Source: https://commons.wikimedia.org/wiki/Category:Fiber_optics#/media/File:Glasfaser_-_Fibreglass. jpg.

Figure 8.1   Close up of fiber optics lines

point-to-point transmissions. Moreover, fiber cables do not corrode or conduct electricity, which renders them immune to electromagnetic disturbances such as thunderstorms.

Their development reflects a long history of experimentation and technological change. The origins of fiber optics go back to Jean-Daniel Colladon at the University of Geneva, who demonstrated light guiding in 1841. Subsequent experiments in 1870 by British physicist John Tyndall, who used moving water through curved rods to conduct light, showed that optical signals could be bent and that light therefore did not need always to travel in a straight line. In 1880, William Wheeling patented the method of 'piping light' through mirrored pipes. Alexander Graham Bell's 'photophone' in the 1880s transmitted voice signals on a beam of light; the concurrent introduction of Thomas Edison's light bulb enhanced the popularity of technologies of light. In the 1920s, Scottish television inventor John Baird and Clarence Hansell in the United States patented the idea of using transparent rods to transmit images (Hecht 1999). In the 1950s, experiments by Brian O'Brien at the American Optical Company and Narinder Kapany (who coined the term 'fiber optics') at the Imperial College of Science and Technology in London developed a fiberscope, or forerunner to contemporary fiber optics, a technology that led to laparo-scopic surgery. The introduction of a dense coat, or cladding, around the glass core, by Lawrence Curtiss of the University of Michigan, prevented the loss of light and led to near-perfect internal reflection within the core of the cable. In the 1960s, the use of laser diodes in helium–neon gas perfected this technique at Bell Labs in New Jersey. In 1956

British physicist Charles Kao showed that light attenuation was caused by impurities in the glass and suggested optimal maximum levels of glass purity for long-distance transmission. Ten years later, Robert Maurer, Donald Keck and Peter Schultz of the Corning Glass Works (later Corning Inc., now the largest provider of fiber cable in the world) developed rods of pure fused silica that greatly reduced light attenuation to the levels that Kao specified. In 1960, Theodore Maiman of the Hughes Research Laboratories in Malibu, California produced the first operational laser.

As computer equipment became rapidly more sophisticated and widespread, US military uses of fiber optics began as it deployed them for communications and tactical systems. In 1975 computers at the NORAD headquarters in Cheyenne Mountain were linked by fiber optics. The technology was also central to the development of the Internet. Indeed, much of the durability and reliability of the Internet reflects its military origins, for its original purpose was to allow communication among computers in the event of nuclear war.

Simultaneously, the microelectronics revolution initiated enormous decreases in the cost of computers and exponential increases in their power and memory, making communications the primary bottleneck to corporate productivity. As fiber optics increasingly appeared to meet rising demand in this sector, corporate applications rose steadily (Olley and Pakes 1996; Jorgenson 2001). In 1977, AT&T installed the first telephone lines to use fiber optic cables, a network 40 km in length that could carry 672 voice channels, beneath downtown Chicago; GTE followed immediately in Boston (Goff 2002). However, it was during the massive global changes in the world economy at the end of the 20th century that fiber came into its own as the dominant medium of telecommunications. Telephone companies and other providers of telecommunications services began rapidly replacing older copper wire cables with fiber optics, which many observers expect to become virtually the only telecommunications transmission technology in the future. Fiber optics facilitated the explosive growth of e-commerce, which includes both business-to-business transactions as well as those linking firms to their customers, including electronic data interchange (EDI) systems, digital advertising, online product catalogs, the sharing of sales and inventory data, submissions of purchase orders, contracts, invoices, payments, delivery schedules, product updates and labor recruitment. Indeed, fiber optics arguably transformed the Internet from a communications to a commercial system, accelerating the pace of customer orders, procurement, production and product delivery (Malecki 2002). In addition, fiber optics are used in a variety of scientific and medical equipment.

The fiber cable itself comprises a relatively small share of the total cost of an undersea cable system. Thus, improvements in fiber optics capacity and efficiency in the 1990s rested on other components of the system. Because signals inescapably attenuate during transmission, repeaters are necessary to maintain the fidelity of optical signals. The first generation of repeaters converted optical signals into electronic voltage in order to amplify them, then reconverted them to optical signals; early fiber cables required frequent repeaters, often every 5–10 km. As the purity of fiber cables improved, and as repeaters improved in power, the need for repeaters decreased accordingly. In 1991, optical amplifiers, which remove the need to convert light to electronic signals, such as the erdium doped fiber amplifier (EDFA), improved the efficiency of transmission over electronic amplifiers by a factor of 100. The TAT-12 line, installed in 1995, was the first long-haul system to use EDFA technology. Today, in long-haul cables, repeater distances are as high as 500–800 km. Similarly, dense wavelength division multiplexing, first developed in the 1970s, made it possible to transmit multiple wavelengths over a single fiber. As a result of these numerous improvements, fiber's bandwidth capacity increased more than 500-fold, from 10 mbps in the 1970s to as high as 50 terabits per second in 2005.

THEORIZING THE GROWTH OF FIBER-BASED CAPITALISM

As numerous observers have pointed out, global capitalism in the late 20th century underwent an enormous sea-change. Telecommunications constitute an integral part of this transformation. The ability to transmit vast quantities of information in real time over the planet is crucial to what Schiller (1999) calls digital capitalism. Large transnational corporations with offices and plants located in multiple national markets require intense coordination of the activities of vast numbers of employees working within highly specialized corporate divisions of labor. Information acquisition, processing and dissemination lie at the heart of many such activities. The exploding demand for high bandwidth corporate communications has thus been a major force behind the growth of the international communications infrastructure. For Castells (1996), this transformation is mirrored in the space of flows and the new geometries that accompany it, which wrap places into highly unevenly connected networks, typically benefiting the wealthy at the expense of marginalized social groups. However, the global space of flows is far from randomly distributed over the Earth's surface: rather, it reflects and reinforces existing geographies of power concentrated within specific nodes and places, such as global cities, trade centers, financial hubs and corporate headquarters. Indeed, because the implementation of fiber lines reflects the powerful vested interests of international capital, these systems may be seen as 'power-geometries' (Massey 1993) that ground the space of flows within concrete historical and spatial contexts.

Financial and producer services firms were at the forefront of the construction of fiber networks in large part because they allowed the deployment of electronic funds transfer systems, which comprise the nervous system of the international financial economy, allowing banks to move capital around at a moment's notice, arbitrage interest rate differentials, take advantage of favorable exchange rates and avoid political unrest (Langdale 1989; Warf 1995). Fiber carriers are heavily favored by large corporations for data transmissions and by financial institutions for electronic funds transfer systems, in large part because of the higher degrees of security and redundancy this medium offers. Such networks give banks an ability to move money around the globe at stupendous rates: subject to the process of digitization, information and capital become two sides of the same coin. Liberated from gold, traveling at the speed of light, as nothing but digital assemblages of zeros and ones, global money performs a syncopated electronic dance around the world's neural networks in astonishing volumes. In this context, finance capital is not simply mobile, it is *hypermobile*, i.e. it moves in a continual surge of speculative investment that never materializes in physical, tangible goods. The world's currency markets, for example, trade more than US$4 trillion every day, dwarfing the US$35 billion that changes hands daily to cover global trade in goods and services. In the securities markets, fiber optics facilitated the emergence of 24 hour/day trading, linking stock markets through computerized trading programs.

Deregulation was also a fundamental part of the growth of the global fiber optics system. This process was initiated by the United States with the breakup of AT&T in 1984, which had long enjoyed a monopoly over domestic telephony and was broken up by an antitrust suit. Deregulation opened the door for a proliferation of new fiber optics service providers such as MCI, which grew to become the second largest provider in the world. Sprint arose as the first corporate telecommunications provider entirely based on fiber optics; others such as Qwest followed shortly. In the United States, the 1996 Telecommunications Act further eliminated regulatory oversight, effectively ending the boundaries between local and long-distance traffic and opening the door to a wave of mergers and acquisitions (Warf 2003). Soon thereafter, British Telecommunications, France Telecom and Deutsche Telekom were partially or totally sold *en masse* to private investors, and in Japan, the monopoly long held by Nippon Telegraph and Telephone was broken by government fiat (although like France Telecom, it remains largely publicly owned). The World Trade Organization's Basic Telecommunications Agreement, which went into effect in 1998, also fostered competition around the world. Today, state-owned or regulated telecommunications monopolies are increasingly rare around the world. In 2005, roughly 1000 fiber optics and two dozen public and private satellite firms competed to provide international telecommunications services, the vast majority of which originated in economically developed countries. The consequences for the market structure of telecommunications have been dramatic, including new competitors, improved service and rapidly falling costs, although Graham and Marvin (1996) note that, in this climate, providers may freely engage in 'cherry picking', i.e. servicing only high-profit clients at the expense of the needy and disempowered.

Large fiber networks are generally owned and operated by consortia of firms. Until the 1990s, all commercial fiber lines were built, used and paid for by a handful of monopoly carriers such as AT&T, British Telecom and Japan's Kokusai Denshin Denwa, known informally in the industry as 'The Club'. The Club system allowed telecommunications carriers to construct and own undersea cables and to serve as their users or vendors. Typically, landing facilities are owned by carriers from the country in which the facility is located but the 'wet links' (undersea cables) are jointly owned by Club members. Under the Club system, AT&T, for example, ventured aggressively into the international fiber optics market as it globalized in the face of declining market share in the United States, often by entering strategic alliances that stretched across national borders (Warf 1998). Similarly, Sprint affiliated with France Telecom and Deutsche Telekom to form Global One in 1996, and AT&T and British Telecom acquired a 30 percent share of Japan Telecom. Table 8.1 lists the major submarine cable networks in place in 2003 for the two largest markets across the Atlantic and Pacific Oceans. Under the Club system, capacity was allocated and payments made before or during construction of the network. Members were required by national regulators to sell capacity to non-members on a non-discriminatory basis close to cost. Allegations arose that Club members discriminated against new entrants by offering disadvantageous conditions of membership, such as capacity prices. However, as deregulation encouraged new entrants into the cable markets, the Club system began to fragment. Private systems, in which carriers invite non-carrier investors such as banks, emerged as an alternative system, and recently, non-carrier systems have also appeared.

Table 8.1   Major trans-Atlantic and trans-Pacific fiber optics cables

Name	Operational capacity	Date	Landing station locations
Trans-Atlantic			
TAT-8	560 mbps	1988	United States, UK, France
PTAT-1	1.26 gbps	1989	United States, UK, Bermuda, Ireland
PTAT-2	1.26 gbps	1992	United States, UK
TAT-9	1.12 gbps	1992	United States, UK, France, Spain, Canada
TAT-10	1.12 gbps	1992	United States, Germany, Netherlands
TAT-11	1.12 gbps	1993	United States, UK, France
TAT-12	5 gbps	1995	United States, UK
TAT-13	5 gbps	1995	United States, France
Gemini	2.5 gbps	1998	United States, UK
AC-1	2.5 gbps	1999	United States, UK, Germany
Columbus 3	2.5 gbps	1999	United States, Spain, Portugal, Italy
TAT-14	10 gbps	2000	United States, UK, France, Netherlands, Germany
FLAG Atlantic	10 gbps	2001	United States, UK, France
Apollo	10 gbps	2002	United States, UK, France
Trans-Pacific			
HAW-4/TPC-3	560 mbps	1989	California, Hawaii, Guam, Japan
GPT	280 mbps	1989	Guam, Philippines, Taiwan
H-J-K	280 mbps	1990	Hong Kong, Japan, South Korea
NPC	1.26 gbps	1990	Oregon, Alaska, Japan
TASMAN-2	1.12 gbps	1991	Australia, New Zealand
TPC-4	1.12 gbps	1992	California, Canada, Japan
HAW-5	1.12 gbps	1993	California, Hawaii
PacRim East	1.12 gbps	1993	Hawaii, New Zealand
PacRim West	1.12 gbps	1994	Australia, Guam
TPC 5/6	5 gbps	1995	California, Oregon, Hawaii, Japan
KJG	1.12 gbps	1995	South Korea, Japan, Guam
TPC-5	5 gbps	1996	California, Hawaii, Guam, Japan
Southern Cross	2.5 gbps	1999	California, Hawaii, Fiji, Australia
China-US	2.5 gbps	1999	California, Hawaii, Guam, South Korea, Japan, China, Taiwan
PC-1	10 gbps	2000	Japan, United States
Japan–US	10 gbps	2000	Japan, United States
FLAG Pacific 1	10 gbps	2002	Japan, United States, Canada

Source: Smith (2003).

URBAN GEOGRAPHIES OF FIBER OPTICS

Starting in the 1980s, telecommunications firms began to build a large interurban network of fiber optics lines in the United States, whose aggregate networks exceeded 50 million km by 2001. The largest fiber optic lines (T3, OC-3, OC-4 and OC-12) connect a handful of large metropolitan areas, whose comparative advantage in producer services has benefited significantly from publicly installed telecommunications systems (e.g. see http://cybergeography.planetmirror.com/cables.html). While the largest metropolitan regions are well served (particularly New York, Chicago, Washington, DC, Atlanta, Los Angeles and Seattle), many other areas (such as the rural South) have few connections. High-capacity fiber lines are particularly important in regard to access to high-density material, e.g. graphical content on the WWW. For high-volume users (typically large service firms), for whom the copper cables used by telephone companies are hopelessly archaic, these lines are an absolute necessity. For large real estate developers, fiber capability has emerged as a critical issue in determining the price and attractiveness of corporate office space, indicating that relative space via connectivity is as important as accessibility via conventional transportation. Moreover, numerous cities have taken the initiative to establish their own municipal fiber networks as part of their economic development strategies to attract firms rather than wait for the private sector, often in the form of public–private partnerships. In such cases, fiber lines are often packaged along with the other municipal utilities such as water, electricity and natural gas. Thus, a grid of fiber lines surrounding the core of cities has become an indispensable part of urban comparative advantage. Rural areas, in contrast, often suffer a distinct disadvantage in terms of this digital divide (Gabe and Abel 2002).

Fiber optics providers prefer large metropolitan regions where dense concentrations of corporate and residential clients allow them to realize significant economies of scale and where frequency transmission congestion often plagues satellite traffic (Singhi and Long 1998). So-called 'global cities' such as New York, London and Tokyo (Sassen 1991) are prime beneficiaries, using fiber optics lines to spread their sphere of influence around the planet. For example, the Atlanta metropolitan region exhibits 644,000 km of fiber optic lines, which have been important to the revival of downtown regions and enhanced its competitive position within the national urban hierarchy (Walcott and Wheeler 2001). Within cities, fiber lines accelerate the creation of wealth by corporate elites, generating geographies of inequality in which the wired and the wireless, the haves and have-nots of the information age live in close proximity; even in the most networked of cities, there exist large disenfranchised groups who pay the costs of the digital economy but reap relatively few of the benefits. In contrast with metropolitan areas, rural areas, with relatively small populations and low market potential, hold little market appeal. This urban bias, and the social schisms it deepens, is replicated at the international scale; Graham (1999) notes that the skein of fiber cables linking the world's major cities is vital to their role and domination over the world economy. Despite the mythologized notion that fiber optics lines erase spatiality, therefore, it is evident that the geographic impacts of this technology are highly selective.

The growth of fiber optics for commercial and residential purposes, such as cable television, assumes that local lines are effectively linked to high-capacity backbone routes. However, this connection often confronts the 'last mile' problem, the gap between a facility or client and a Point of Presence, the point at which the facilities of an inter-exchange carrier are accessible. Telecommunications and cable television companies have devoted substantial resources to overcoming this problem, and as a result, broadband access has improved gradually. Some, such as Verizon, have pioneered the development of fiber-to-the-premises networks.

GEOGRAPHIES OF GLOBAL FIBER OPTICS NETWORKS

Despite exaggerated popular claims that telecommunications render distance meaningless (e.g. Cairncross 1997), the geography of fiber optic lines reflects the accumulated imprints of successive rounds of investment in space and time. The placement of terrestrial networks reflects the complex ways in which space, the global economy and technology are wrapped up in each other. Spurred by the growth of information-intensive services and predictions of unending growth in Internet traffic, telecommunications companies undertook an orgy of fiber optic cable construction in the 1980s and 1990s.

Laying transoceanic fiber cables entailed a host of technical and organizational issues, a process that extends back to 19th-century attempts to cross the oceans with telegraph lines (Hugill 1999). In addition to the costs of purchasing fiber, telecommunications companies must pay for the laying of fiber across the ocean floor and the installation of 'manholes', on-shore bunkers designed to allow access for repairs. AT&T's Submarine Systems, the world's largest supplier of undersea telecommunications systems, operates a fleet of six cable ships to service its 230,000 km of undersea cable. Submarine lines must be routed to avoid seismic activity (earthquakes and undersea avalanches), ships' anchors, deep sea currents, fishing trawlers and military activities, and must be armored against sharks, which are attracted by electromagnetic emissions. While the original lines were point-to-point, the development of submarine branching units allowed multiple points to be served simultaneously, leading to more complex network configurations. Moreover, most submarine cables today are 'self-healing', meaning that they offer redundant capacity and high resiliency, so that the loss of one link can be easily and rapidly compensated by others. Today, the world's fiber system totals more than 25 million km in length, connecting all of the world's continents except Antarctica (Figure 8.2).

The geography of global fiber networks centers primarily upon two distinct telecommunications markets crossing the Atlantic and Pacific Oceans, connecting two of the major engines of the world economy, North America and East Asia (Chaffee 2001). In 1988, in conjunction with MCI and British Telecommunications, AT&T initiated the world's first transoceanic fiber optic cable, Trans-Atlantic Telecommunications (TAT-8), which could carry 40,000 telephone calls simultaneously. The trans-Atlantic line was the first of a much broader series of globe-girdling fiber lines that AT&T erected in conjunction with a variety of local partners. Because large corporate users are the primary clients of such networks, it is no accident that the original and densest web of fiber lines connects London and New York, a pattern that extends historically to the telegraph and telephone (Hugill 1999). The next generation, TAT-9 and TAT-10, which began in 1992, could carry double the volume of traffic of TAT-8. The third generation, TAT-11 to TAT-13, was the first to use EDFA rather than older repeaters. Newer generations of cable were even more powerful.

Recent growth in fiber optics has largely shifted to Asia (Malecki and Wei 2009). Starting with the Trans-Pacific Cable (TPC-3) in 1989, connecting the New York and Tokyo stock exchanges, a growing web of trans-Pacific lines mirrored the rise of East Asian trade with North America, including the surging economies of the Newly Industrialized Countries. In 1996, the first all-fiber cable across the Pacific, TPC-5, was laid. In 2006, a consortium including Verizon and five Asian providers announced plans to lay a 17,700 km United States–China link that would support 1.28 terabits of information – 60 times the capacity

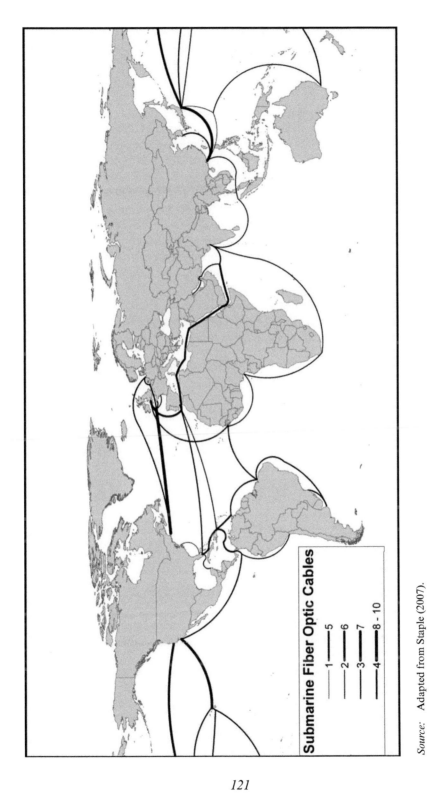

Source: Adapted from Staple (2007).

Figure 8.2   The world's major fiber optic cables

of the next largest cable – in time for the Beijing Olympics in 2008 (Shannon 2006). In 2007, Google announced the purchase of large quantities of trans-Pacific fiber cable with the aim of launching a multi-terabit Unity service in 2009.

The complex interplay of deregulation, globalization and technological change increased the international transmission capacities and traffic volumes for fiber optics carriers explosively. Between 1988 and 2003, for example, trans-Atlantic fiber optic cable capacity increased from 43,750 voice paths to 45.1 billion (103,000 percent), while across the Pacific Ocean, cable carriers' capacity rose from 1800 voice paths to 1.87 billion (an astonishing 1.6 billion percent).

In addition to the two major markets, fiber lines have extended into several newer ones. In 1997, AT&T, NYNEX and several other firms (including, for the first time, non-telecommunications firms) opened the self-healing Fiberoptic Link Around the Globe (FLAG), a system that eventually expanded to 55,000 km connecting Europe and Southeast Asia. The world's longest submarine telecommunications network (Denniston 1998), FLAG filled a void in undersea cable capacity between Europe, the Middle East and Asia. It also hooked into regional systems such as the Asia Pacific Cable Network, a 12,000 km system linking Japan, South Korea, Taiwan, Hong Kong, the Philippines, Thailand, Vietnam and Indonesia, as well as the Caribbean Fiber System (i.e. the Eastern Caribbean Fiber System, Antillas 1, Americas 1 and Columbus 2). Unlike earlier systems, FLAG allowed carriers to purchase capacity as needed, rather than compelling them to purchase fixed quantities.

Africa is surrounded by an interlinked series of fiber systems. The first of these includes a dense network that extends across the Mediterranean Sea and exits via the Suez Canal, such as the Columbus III, SeaMeWe and FLAG lines, each of which has extensions into cities in Algeria, Tunisia, Libya and Egypt. The widely publicized Africa ONE (*O*ptical *NE*twork) system, designed to surround the continent, collapsed in the dot com crash of the early 2000s. In its wake, consortia of telecommunications companies led by AT&T, Sprint, Vodacom and Verizon gradually pieced together a network on the western side of the continent. A third system, centered on East Africa, includes the privately funded, 17,500 km-long Seacom cable completed in 2008 and owned mostly by African investors, which links to 21 countries. In 2010, the Eastern Africa Submarine Cable System, or EASSy, also came online, further adding to that region's supply of telecommunications services.

THREE CONSEQUENCES OF THE FIBER OPTICS BOOM

The massive surge of supply in the global fiber optics industry generated three distinct, important, but unintended effects: oversupply and economic crisis; the growth of 'dark fiber'; and a serious challenge to the satellite industry.

As with all industries in which supply increases markedly more rapidly than demand, the explosive growth in the world's fiber optics capacity ultimately led to overcapacity and declining utilization rates (ElBoghdady 2001). As the growth of the world's fiber optics networks' transmission capacity outstripped the rise in demand, transmission prices plunged in a deflationary spiral throughout the first decade of the 21st century, often by as much as 90 percent. Telecommunications corporate stock prices plunged, forcing numerous fiber optics firms into debt and bankruptcy and most others into financial restructuring. The list of casualties from this debacle in the early 2000s included Global Crossing, Metromedia Fiber Network, Viatel, MCI/Worldcom, Williams Communications, Winstar Communications and PSINet. Some victims were purchased by buyers eager to become players on the global stage: in 2003, for example, FLAG Telecom was bought by the Indian wireless services provider Reliance. Low fiber transmission prices, in turn, helped to keep down the costs of telephone calls and other applications of the technology.

Overall capacity utilization rates fell below 50 percent, leading to large quantities of unused 'dark fiber'. With considerable amounts of dark fiber, corporate clients often lease excess backhaul capacity from former monopolies in order to connect domestic networks to the international system. In addition to system overcapacity, dark fiber reflects the high costs of planning and installing fiber lines, which leads providers to lay more than is necessary in anticipation of rising future demand. For example, a utility company may deliberately install dark fiber in the expectation of leasing it to a cable television company in the future. In addition, however, dark fiber also came to mean the leasing of unused fiber capacity from network service providers. (Indeed, some companies specialize in this market.)

A third consequence of the explosion of fiber capacity was mounting competition with the besieged satellite industry, with which fiber optics are quasi-substitutable (Pfeifenberger and Houthakker 1998). While satellites are ideal for point-to-area distribution networks common in the mass media, especially in low-density regions, fiber optic lines are preferable for point-to-point communications, especially when security is of great concern (Maclean 1995). Before the explosive growth in fiber capacity in the 1990s, satellites were traditionally more cost-effective for transmission over longer distances (e.g. more than 800 km), while fiber optic lines often provided cheaper service for shorter routes (Langdale 1989). The rise of the integrated global fiber network, however, steadily eroded satellites' share of global traffic in data and video transmission services. Despite the pitch by satellite operators that satellites could provide Internet backbone services as a way to bypass terrestrial congestion, fiber remains by far the preferred technology. Satellites simply cannot offer sufficient security or backup capacity to be economically competitive with fiber. In 2003, fiber optics carriers comprised 94.4 percent of worldwide transmission capacity (up from 16 percent in 1988), including 91.3 percent across the Pacific and 95.2 percent across the Atlantic Ocean (Warf 2006).

CONCLUDING THOUGHTS

Fiber optics are one of the great transformative innovations to emerge from the microelectronics revolution of the late 20th century. The development of this technology was the culmination of a long history of research by individuals, universities, the military and corporations, and led to a mode of telecommunications significantly more powerful, secure and rapid than competing technologies. However, because technology is a social, not simply technical, phenomenon, the dramatic expansion in fiber optics capacity and utilization reflects the historically specific circumstances of global capitalism as it emerged from the crises of the 1970s and the end of the post-Second World War economic boom. Fiber optics were ideal for the information-intensive nature of financial and producer services, particularly when security is of critical importance, and firms in this sector comprised the driving force behind the demand that propelled a vast global network of lines in the 1980s and 1990s. Whereas the two largest markets are those stretched across the Atlantic and Pacific Oceans, newer networks have increased the reach of fiber optics into Latin America and Africa.

At multiple spatial scales, from the urban to the nation to the world, therefore, fiber optics lines realigned the geographies of centrality and peripherality. Far from annihilating space, therefore, the industry reconfigured it. However, the logic that propelled the industry to such prominence also gave rise to the overcapacity and the end of the 'dot com' boom in the late 1990s and early 2000s, initiating a severe period of corporate retrenchment and restructuring. Unused capacity – dark fiber – appeared in both planned and unplanned forms. The dramatic decline in prices that accompanied this trend posed severe competitive problems for fiber optics providers, including a wave of bankruptcies, but also eroded the market share held by substitutes to fiber such as satellite services.

What does the future hold for this industry? In the short term, the substantial over-capacity in long-distance fiber generated by the boom of the 1990s will be difficult to overcome. Future market potential probably rests in the provision of services to residences (e.g. cable television and high-speed Internet), if the last mile problem may be conquered effectively. The wireless revolution may also pose a competitive challenge to fiber optics. It is evident from these remarks that fiber exhibits the dynamism and fluidity characteristic of the telecommunications sector as a whole.

REFERENCES

Agrawal, G. 2002. *Fiber-optic Communication Systems*. New York: Wiley.
Cairncross, F. 1997. *The Death of Distance*. Boston, MA: Harvard Business School Press.
Castells, M. 1996. *The Rise of the Network Society*. Oxford: Blackwell.
Chaffee, C. 2001. *Building the Global Fiber Optics Superhighway*. New York: Springer.
Crisp, J. and Elliot, B. 2005. *Introduction to Fiber Optics*, 3rd ed. London: Newnes.
Denniston, F. 1998. FLAG – Fiber-Optic Link Around the Globe. *Sea Technology*, February: 78–83.
ElBoghdady, D. 2001. Fiber-optic firms face issue of overbuilding. *Washington Post*, 28 February, 1.
Freeman, R. 2002. *Fiber Optics Systems for Telecommunications*. New York: Wiley.
Gabe, T. and Abel, J. 2002. Deployment of advanced telecommunications structure in rural America: measuring the rural divide. *American Journal of Agricultural Economics* 84: 1246–1252.
Goff, D. 2002. *Fiber Optics Reference Guide*, 3rd ed. Woburn, MA: Focal Press.
Graham, S. 1999. Global grids of glass: on global cities, telecommunications, and planetary urban networks. *Urban Studies* 36: 929–949.
Graham, S. and Marvin, S. 1996. *Telecommunications and the City: Electronic Spaces, Urban Places*. London: Routledge.
Hecht, J. 1999. *City of Light: The Story of Fiber Optics*. Oxford: Oxford University Press.
Hecht, J. 2015. *Understanding Fiber Optics*, 5th ed. Auburndale, MA: Laser Light Press.
Hugill, P. 1999. *Global Communications since 1844: Geopolitics and Technology*. Baltimore, MD: Johns Hopkins University Press.
Jorgenson, D. 2001. Information technology and the U.S. economy. *American Economic Review* 91: 1–32.
Langdale, J. 1989. The geography of international business telecommunications: the role of leased networks. *Annals of the Association of American Geographers* 79: 501–522.
Maclean, G. 1995. Will fiber optics threaten satellite communications? *Space Policy* 11: 95–99.
Malecki, E. 2002. The economic geography of the Internet's infrastructure. *Economic Geography* 78: 399–424.
Malecki, E. and Wei, H. 2009. A wired world: the evolving geography of submarine cables and the shift to Asia. *Annals of the Association of American Geographers* 99: 360–382.

Massey, D. 1993. Power-geometry and a progressive sense of place. In J. Bird, B. Curtis, T. Putnam, G. Robertson and L. Tickner (eds), *Mapping the Futures: Local Cultures, Global Change.* pp. 59–69. London: Routledge.

Olley, G. and Pakes, A. 1996. The dynamics of productivity in the telecommunications equipment industry. *Econometrica* 64: 1263–1297.

Pfeifenberger, J. and Houthakker, H. 1998. Competition to international satellite communications services. *Information Economics and Policy* 10: 403–430.

Sassen, S. 1991. *The Global City: New York, London, Tokyo.* Princeton, NJ: Princeton University Press.

Schiller, D. 1999. *Digital Capitalism: Networking the Global Market System.* Cambridge, MA: MIT Press.

Shannon, V. 2006. Group plans to build China–U.S. fiber optic link. *New York Times*, 19 December, accessed 22 January 2016 from http://www.nytimes.com/2006/12/19/technology/19cable.html?pagewanted=all.

Singhi, M. and Long, H. 1998. New undersea cable developments and satellite services: toward complementary coexistence in the 21st century. In *Proceedings of the 20th Pacific Telecommunications Conference.* pp. 566–569. Honolulu: Pacific Telecommunications Council.

Smith, D. 2003. *Digital Transmission Systems.* London: Springer.

Staple, G. 2007. *Telegeography 2006: Global Telecommunications Traffic Statistics and Commentary.* Washington, DC: Telegeography Inc.

Walcott, S., and Wheeler, J. 2001. Atlanta in the telecommunications age: the fiber-optic information network. *Urban Geography* 22: 316–339.

Warf, B. 1995. Telecommunications and the changing geographies of knowledge transmission in the late 20th century. *Urban Studies* 32: 361–378.

Warf, B. 1998. Reach out and touch someone: AT&T's global operations in the 1990s. *Professional Geographer* 50: 255–267.

Warf, B. 2003. Mergers and acquisitions in the telecommunications industry. *Growth and Change* 34: 321–344.

Warf, B. 2006. International competition between satellite and fiber optic carriers: a geographic perspective. *Professional Geographer* 58: 1–11.

9.   The Internet as geographic technology
Aharon Kellerman

The Internet by its very nature constitutes a geographic technology, and this is so in several respects: its user interface and its mode of operation by users imitate patterns and elements that typify the structure and use of physical space; it provides for an extremely fast mobility of all types of information; it has become a crucial system for people's mobility in physical space; and finally, and most recently, it enables its users to perform through its activities that have traditionally been performed in physical space. In other words, the Internet is simultaneously a system with a geographic structure; a system which is based in cyberspace, and thus facilitates the electronic movement of information; and it is further a system that supports the performance of several human activities in real space, side by side with its being a substitute or competitor for other human activities performed in real space.

These geographic components and patterns of the Internet reflect the historical order of development of the Internet. The Internet has had a geographic user interface since its emergence in the 1990s, followed by its provision of information for the facilitation, upgrading and convenience of human spatial activities, eventually turning into an operational space by itself. We will begin our discussion with a short presentation of the Internet *per se* and the technologies for its universal use, followed by elaborations on each of its facets as a geographic technology.

The Internet is a communications system and technology for the storage and global transmission of information of all types – textual, audial, visual and audio-visual, including streaming. The Internet was originally invented in the US in 1969 as the ARPANET network, constituting an experimental alternative communications system for telephone services, developed for a potential replacement of the telephone system in case of nuclear disasters. It was originally experimented through a network connecting security headquarters with universities (Kellerman 2002), followed by the emergence of inter-university networks (e.g. BITNET and NSFNET). It took a long period of some 25 years of incubation and development for these early security and academic networks of communications and information to mature into a universally open and commercial entity, the Internet, in 1994. It took much less time, though, just seven years following its introduction, in 2001, for the Internet to be adopted by one half of Americans. Its current universal availability has been considered as the best example for the adoption of a technology for purposes completely different than those envisaged by its developers (Urry 2003, 63).

The Internet consists of two classes in terms of purposes and uses: information [cyber] space (mainly the World Wide Web (WWW or simply the Web)) and communications [cyber]space (mainly e-mail and Web 2.0 applications such as Facebook and Twitter) (Kellerman 2007). Information space refers to digital information sets or systems, consisting of information organized along spatial notions such as the Web, thus involving geographical metaphors such as sites, homes and navigation/surfing. Information cyberspace further refers to digital information sets at large, such as data archives and library catalogs (Fabrikant and Buttenfield 2001; Couclelis 1998). Search engines allow for an easy access to websites and files. The second class of the Internet is communications cyberspace, referring to the cyberspace of persons who communicate through numerous modes of communications. Communications [cyber]space is mostly inter- personal or shared by small groups, but it may also be widely accessible within social networking systems. The two digital spaces of information and communications are frequently interfolded, for example when e-mails are sent through an informative website. Such interfolded and even fused information and communications [cyber]spaces attest to the oneness of the Internet from its usage perspective.

The most dramatic trend in Internet use in recent years has been the rise of the mobile Internet, thus permitting access to both e-mail and the Web at any time and place, mainly through two leading technologies, Wi-Fi (wireless fidelity), introduced in 1991, and 3G/4G (third and fourth generations), introduced in 2001, both applied to both smartphones and laptops/tablets (Kellerman 2012). 'Broadband has become the stand- ard for Internet use' (Mossberger et al. 2013, 3), but basic mobile telephone technology was originally introduced as early as 1906, some 40 years prior to the introduction of the computer and almost 90 years prior to the introduction of the Internet. It was devel- oped by Lee de Forest, who claimed that 'it will be possible for businessmen, even while automobiling, to be kept in constant touch' (Agar 2003, 167).

The mobile phone has turned into the globally most widely diffused communications device, including developing countries, with a global penetration rate of some 95.5% in 2014 (90.2% in developing countries). However, the 2014 penetration rates of mobile broadband (globally 32% and in developed countries 83.7%) and the Internet (globally 40.4% and in developed countries 78.3%) present digital gaps between developed and developing countries (ITU 2015). Still, some 90% of the world's populated areas were covered by a mobile phone signal already in 2009, and as of 2002 the percentage popula- tion worldwide owning a mobile phone line has been higher than the percentage having a fixed one (ITU 2010, 2011). The global traffic of mobile data has doubled every year since 2008 (Allot Communications 2013), and a significant portion of this traffic stems from daily actions by users, other than for entertainment, which has become the most dominant producer of heavy traffic, given the volume of music and video clips (Kellerman 2010).

The fast growth rate of mobile broadband subscription has turned it into a dominant technology, but its adoption rate may reach a limit in the developing world, even if assuming future price drops for smartphones and broadband subscriptions, since most of the uses of the Internet require literacy. Castells (2009, 65) stated that mobile broadband is for the Internet what the electric grid has been for the provision of electric power, i.e. it permits universal distribution, but electricity use does not require literacy.

IMITATION OF REAL SPACE

The user interface of the Internet and its mode of operation by users imitate patterns and elements that typify the structure and use of physical space. These include at least two major facets: the Internet language of operation and the appearance and design of Internet screens.

Navigation and manipulation of cyberspace through the Internet involve a metaphorical spatial experience through the extensive use made of geographical language, symbols and tools, such as homepage, surfing, navigating, site and cursor. This continuous use of geographical language for Internet operations implies a constant discourse between real (material) and virtual (imagined) spaces through knowledge and information (vis-à-vis perception and cognition), as well as through spatial experience (real and imagined ones), thus accentuating the oneness of the material, the perceived and the imagined, as forms of social space (Wilson et al. 2013; see also Harvey 1989).

Spatial metaphors have turned out to be attractive for Internet users since they are well known to them from their daily lives and are simple to use (Schrag 1994). The growing placelessness of contemporary society, noted already in the 1970s (Relph 1976), has been coupled a generation later to an increased use of geographical language and symbols for the operation of digital communications. The spatial language for Internet operations has had to be kept highly homogenized and simplified in order for it to be comprehended by users with diversified cultures and languages, many of which are rich in spatial wording (Kellerman 1999).

The second major imitation of real space presented by Internet user interfaces is the Internet screen-spaces (ISSs), which constitute the visual interface between the Internet and its users, displayed on computer and smartphone screens (Kellerman 2015). 'Online interaction is currently dominated by visual interfaces, rather than aural, tactile, or olfactory interfaces', leading to the spatialization of non-spatial data (Zook et al. 2004, 159–160; see also Fabrikant 2000). Screen-spaces may consist of all possible visual presentation types: texts, pictures, maps, landscapes and combinations of these elements. ISSs are, therefore, virtual – in that they may visually present real space and material artifacts, and they are further cyberspatial – in their comprising a component of a digital communications medium.

ISSs, by their very nature, are not stable like printed virtual spaces, and they may disappear by pre-programmed commands or by users' actions. On the other hand, however, specific screen-spaces may be routinely and repetitively used by Internet subscribers. Internet users may find it difficult to cognize and eventually draw cognitive maps for the instantly appearing and disappearing virtual landscapes and informational screens. Whereas for real space, space and its maps are two completely separate entities, for ISSs, space and its maps may converge (Kwan 2001). Thus, 'cognitive communications cyberspaces are personally unique, and cannot be aggregated, whereas cognitive maps relating to a specific area may be compared and conclusions on a wider societal knowledge of an area drawn' (Kellerman 2014, 9).

As shown elsewhere (Kellerman 2015), Internet screen-spaces may be viewed as a category of image space that can be interpreted and analyzed using real space parameters, assuming that Web cyberspace can be considered as constituting a special form of social space. Thus, for instance, cyberspace on the Web constitutes a resource and a production force, similarly to real space, in its provision for online shopping. It can further be considered as text and as symbol for both individuals and organizations, and it may further be looked upon as constituting a landscape and even as a social value (see Dodge and Kitchin 2001 for detailed discussions).

MOBILITY OF INFORMATION THROUGH THE INTERNET

Since its inception, the Internet has gradually permitted an unrestricted and rather global transmission and receipt of information of all forms by its individual subscribers. Thus, for instance, the number of websites in 2015 was estimated as close to 1 billion (Internet Live Stats 2015), and the capacity of digital storage capacity at large surpassed analog back in 2002, or some eight years following the open introduction of the Internet (Wo 2011). Through this Internet feature of unrestricted access to information and its production, it has become possible for individuals to experience globalization on a continuous daily and routine basis, something which is unparalleled with other potential experiencing of globalization by individuals, such as pre-programmed television broadcasts or individual travel requiring preparations. As compared with the shipping of material products and the movement of humans, the immediate transmission of information by individuals through the Internet implies a most flexible movement similar to that of gas (Kellerman 2002).

This facet of Internet technology has changed some geographic dimensions of life for Internet users, notably for those who pay a fixed subscription rate. For instance, the Internet can be used for the transmission of information on a one-to-one basis through e-mail or chats, or from one-to-many through personal websites or social networking platforms. This transmission may include texts, pictures, audial conversations through VoIP (Voice over the Internet Protocol) or streaming clips or movies.

The most basic technological element in Internet communications is the transformation of information of all types into standard digital bits, or its very creation as such. This basic nature of information as a rather general entity is coupled by numerous telecommunications Internet technologies, led by routers, computers as servers and the transmission protocol of TCP/IP (Transmission Control Protocol/Internet Protocol). The rapid adoption of the Internet, as well as that of mobile telephony, has had to do with the prior partial existence of telecommunications infrastructures for its operation, such as fixed-line telephone systems, so that new connections to the system could be easily performed.

Of no less importance, though, has been the emergence of the Internet and mobile telephony at a time when these innovations constituted technologies and means for the support of the evolving information society, originally based on standalone computers such as PCs (personal computers), as well as on the digitization of the previously existing fixed-line telephone system. The information society, for its part, has placed a special emphasis on the production, processing, transmission and consumption of information (Wilson et al. 2013). This point is strongly demonstrated by the slow evolution of mobile telephony at the time of its original invention, early after the introduction of the telephone. Mobile telephony had to await its final development until the release of required wave spectrum in the late 1960s, when proper social and economic conditions emerged for such a long-awaited move by the American FCC (Federal Communication Commission).

At the socio-political end, personal telecommunications at large, and the Internet in particular, have been governed by an *open code*, which Lessig (2001, 246) considered as the 'heart of the Internet'. This open code permits unlicensed access for the production of Internet information, whether through the establishment of websites or through the writing of e-mail and chat messages. It further permits an open and free access to the consumption of Internet information, through the receipt of e-mails, as well as through accessing free-of-charge websites. The open code principle further permits the uncontrolled flows of information from any origin to any destination, unless controlled by governmental censorship, which is in partial or full effect in numerous countries (Warf 2013). The open nature of the Internet may be related to its origin in the US and the accent of American society on freedom of expression. The open code nature of the Internet has been questioned by numerous governmental forces of a regulatory nature, such as taxing authorities and copyright holders (Lessig 2001).

The open code nature of the Internet also allows for the innovation of information-related products, such as smartphone and PC applications, to be freely introduced and adopted for both the production and consumption of Internet information. All of these activities are unrestricted by a minimum or maximum age of users in countries which provide free Internet activities, so that the use of the Internet constitutes a completely informal activity, as compared, for example, with required driver licensing for free individual physical mobility (Kellerman 2006).

The open code nature of the Internet has had some additional expressions, for instance in the evolution of informal correspondence codes, the use of signs or icons for smiles, agreement, etc. By the very nature of the Internet as a mainly verbal communications system, literacy is much more required for its use than it is for driving, which is mainly based on road signs. Additional informal requirements for Internet use are some knowledge of basic computer operation and some knowledge of English, in order to permit access to information contained in over one-half of websites (see Hargittai 1999; W³Techs 2013).

The term 'Internetness' (Kellerman 2006) was proposed as referring to values, practices, norms and patterns within the three spheres of individuals, society and space, regarding the extensive adoption and use of the Internet. Culture and religion constitute leading social and informal dimensions which may, in some cases, influence the extent of use of the Internet, and its open code nature, notably when religious authorities attempt to restrict access to the system, or when they enforce censorships on its use. Furthermore, the use of the Internet is not only facilitated by its accessibility and affordability, but also by the capabilities of its users, as well as by their very choices of preferred uses (Kline 2013).

THE INTERNET AND PHYSICAL SPACE

The dissemination and integration of the Internet in physical space operations of almost all systems and for so many people worldwide have made it impossible to assume the operations and functioning of these systems and people without it. This also makes it difficult to elaborate here on all such activities. Thus, we will limit ourselves in the following discussions to just a selection of aspects and angles of Internet integration in physical space operations having a geographical significance: code/space; communications among individuals and between them and information; Internet and mobility; Internet and urban space; opportunities for individuals; and the Internet of things.

The heavy integration of the Internet into operations in physical space and the growing dependence of the latter on the Internet is part of a more general tendency of making the functioning of space operational through software of all kinds. This tendency was termed by Kitchin and Dodge (2011, 7) code/space: 'Code/space is quite literally constituted through software-mediated practices, wherein code is essential to the form, function and meaning of space'. One is tempted to view code as the invisible hand of technology, producing and managing not just one but rather two kinds of spaces: the real one and the virtual one of the Internet and communications.

Code may be recognized as constituting several dimensions simultaneously. It may be viewed as medium and means, mediating between economic and political systems, on the one hand, and daily human agency, on the other. Code development may, thus, respond simultaneously to customers' needs, as well as to motivations for profit and efficiency by business. Software may further be assessed as constituting a controlling element, for example in air and terrestrial traffic systems, as well as in manufacturing and for all kinds of information flows. Software can further turn into a kind of decontrolling means, when used, for example by hackers, to bring about malfunctioning of information and other systems.

The operation of software-based contemporary personal communications media by their users, notably fixed and mobile telephones and the Internet, is assumed to be rather convenient and user-friendly, and thus barrier-free. Once communications are initiated, the speed of reach of other places is normally instant. Moreover, users of communications media expect constant enhancements of communications speed through information technology innovations and continuously upgraded infrastructures. This speed of communications applies also to the access of websites to Internet users, permitting their instant search for information, for instance on transportation services of all types. It further permits organizations to establish their own, and rather closed, information systems, Intranets. The universal and easy access to information has increased the blurring of the traditional separation between home and work activities, leading Castells (2001, 234) to refer to *nomadic workers*. This trend emerged already when access to the Internet was made first through fixed PCs only (Kellerman 2006, 2012).

By the same token, however, portable Internet devices have made it possible for users to consume services without limitations of time and location in their routine daily life. It has become possible, therefore, for a person on the road to communicate with information sources available through the Web, while the locations of these sources may simultaneously change (e.g. through peering or hosting), as the using person moves. In other words, new sophisticated virtual mobilities permit us to act virtually without any regard to the location of the cities or countries from which we communicate, nor with regard to the internal spatial structures of cities from which we communicate.

Several mobile broadband applications, functioning through Internet infrastructure, have been specifically developed for users on the move. One such development is extended GPS (global positioning satellites) services, permitting road navigation on a global scale. Another such application is location-based services, providing commercial local information (e.g. on restaurants), transmitted to people's mobile phones, and changing with their movements. GPS components in smartphones permit not only access to locational information but also the other way around: the exposure of users' locations to LBS providers makes it potentially possible for providers to interfere with the privacy of users, even though they are located in public spaces (Gordon and deSouza e Silva 2011).

The wide adoption of mobile broadband implies instant access to cyberspace for its users, but it may further carry implications for their use and the meaning of urban physical space for smartphone users. The constant availability of GPS, Google Maps and LBS, even while walking or driving through unknown cities, or through unknown parts of known cities, implies an efficient movement of people through urban space aiming to reach specific addresses. It further implies a live and ongoing integration of virtual and real action spaces, with the virtual one guiding physical movements within real space. However, this efficient crossing of real space with the aid of tools located in virtual action space turns urban space into a kind of impediment rather than into a cultural urban exploration. Hence, this specific integration of virtual and real action spaces may reduce one's urban experience, in that cities turn into a mere mosaic of places of production and consumption, ignoring the traditional role of cities as providing residents and visitors with passive or active experiences, such as urban rhythms at different times of the day and the week (Allen 1999).

The growing convenience and speed of communications have resulted in the expansion of the extensibility and accessibility of users, making it possible for them to reach either new or veteran personal and business opportunities (see Janelle 1973; Adams 1995; Kwan 2001). The contemporary extended availability of access to the Internet, notably through personal fixed and mobile broadband, can be interpreted as providing enhanced locational opportunities of action for individuals, thus implying their wider self-extension. Communicating actors can reach people, information or virtual destinations and services, equally from wherever they are located at, or from any point of communications origin. However, even more important, from the perspective of locational opportunities, is the distanciation of human individual action, reaching nowadays a global extent (Giddens 1990). This refers to the increasing geographical spread of potential destinations for the retrieval and sending of information, and the ability of people to work, as well as to buy services globally, thus widening their locational opportunities and their action spaces. Cyberspace and the Internet have not merely emerged as a new world of information, but have further evolved into a dense concept of converging technological environments, human minds and personal motivations, as well as into a source for the generation of all kinds of human artifacts (Dodge and Kitchin 2001; Kellerman and Paradiso 2007).

Finally, let us return to our initial remarks on code/space. This tendency, as far as the Internet is concerned, is about to become of major significance in upcoming years, through the so-called emerging Internet of things, or the possible remote operation, monitoring and control of machinery and appliances through the Internet, for example, for environmental monitoring, for implanted medical devices and in households. This new line of Internet applications may involve regulatory and social changes as well as geographical ones, permitting, for instance, fewer visits to field machinery and equipment, as well as more flexible home visits (see Skarauskiene and Kalinauskas 2015).

THE INTERNET AS ACTION SPACE

In addition to its integration into the operations of real space which we just briefly presented, the Internet has further and rather gradually turned into a second action space for contemporary individuals, mainly in developed countries, side by side with their traditional and 'obvious' physical action space. In other words, Internet users may now perform numerous activities through the Web, activities which were traditionally performed solely in real space sites. This trend has affected an enormously wide array of uses and applications now being actually performed over the Internet. Internet users may now interact, also through mobile broadband connectivity, with a large variety of websites providing for major activities such as home-based work (telework), e-government, online shopping, online banking, travel online, e-learning and e-health. In addition, some applications and uses currently performed over the Internet could not be fully performed in real space before the maturing of the Internet. This applies mainly to the presentation of people's identity (Kellerman 2014).

This contemporary trend is based on the assumption that cyberspace may be manipulated so that it will become similar to real space in its constitution as a platform for human action. This very idea of viewing cyberspace in general as being somehow similar to real space in its very nature, as well as in its experiencing by users, is not novel. For instance, 'virtual environments contain much of the essential spatial information that is utilized by people in real environments' (Péruch et al. 2000, 115), and 'human behavior in cyberspace bears certain similarities with spatial behavior in the physical world' (Kwan 2001, 33). However, some differences between human perceptions of these two spaces still apply. For instance, 'what is near in physical space is often far in cyberspace, and vice versa' (Adams 1998, 93; see also Pickles 2004, 159).

Generally, the cyberspace enables and constrains its users in ways similar to physical space (Adams and Ghose 2003); 'cyberspace is hardly immaterial in that it is very much an embodied space' (Dodge 2001, 1), and from yet another viewpoint, 'information systems redefine and do not eliminate geography', and even more so, 'electronic space is embedded in, and often intertwines with, the physical space and place' (Li et al. 2001, 701). Thus, the Internet 'is shaped by, and reflects, the place-routed cultures in which it is produced and consumed' (Holloway and Valentine 2001, 153). Still, however, the Internet constitutes a 'different human experience of dwelling in the world; new articulations of near and far, present and absent, body and technology, self and environment' (Crang et al. 1999, 1). Thus, cyberspace has its own geography, it is symbol-sustained (Benedikt 1991; Batty 1997), and it has its own materiality (Kinsley 2014). Side by side with the particular identity of cyberspace, the very experiencing of the Web involves also a strong imprint of real space: 'Space isn't a mere metaphor. The rhetoric and semantics of the Web are those of space. More important, our *experience* of the Web is fundamentally spatial' (Weinberger 2002, 35).

The use of the Internet as second action space is growing rapidly, led by Scandinavian countries and the US, but this trend still has room for additional growth. For example, it was estimated that in 2012 some 88% of smartphone users in the US had travel reservation applications installed in their devices, followed by 79% of them having shopping applications installed in their smartphones (*The Economist* 2012). By the same token, in 2012 some 80% of US smartphone owners accessed retail content through their devices (ComScore 2012). However, the total value of online shopping reached only 4% of the total retail shopping in the US in 2009 (US Bureau of the Census 2012).

There are two exceptions to the general trends regarding the application of the Internet for daily activities. First, home-based work (telework), which is the most established activity offered online, existing already before the introduction of the Internet, has remained modest in its adoption, reaching only 4–5% of workers working mainly from home, thus demonstrating workers' and employers' preferences for traditional office work, possibly joined by complementary work on the move and at home. Second, social networking, which is used by most Internet subscribers worldwide, is led by adoption rates in developing countries, since it does not involve any economic activity and can be performed also from Internet cafés (Kellerman 2014).

The use of the Internet as second action space may potentially turn it into a complementary or substituting option for real space facilities, offering the same services. Potentially, at least, some change may emerge in the balance between services offered through facilities located in real space and within easy physical reach of potential customers, on the one hand, and virtual services provided solely through the Internet, or services located anywhere in real space and reached through the Internet, on the other. Such possible change would obviously be in favor of the virtual and at the expense of the real. It might further possibly bring about some structural changes in business facilities located in real space, making them more attractive to potential customers (Kellerman 2014).

The growing use of the Internet as second action space implies not just a widening consumption or use of the Internet by individuals, but also the widening and deepening of the production side of the information society, which consists, among other things, of the production and maintenance of enhanced websites, as well as the development and marketing of mobile phone applications, sold through virtual stores of mobile phone manufacturers. The availability of these applications fosters the consumption of more products and services by individual users of the Internet. Thus, a cycle of growing use, or demand, and growing supply of mobile phone Internet services has emerged.

CONCLUSION

Following a general exposition of the Internet, we discussed several of its dimensions and implications as a geographical technology: the Internet as an imitation of real space; the mobility of information through the Internet; the Internet and physical space operations; and, finally, the Internet as action space. Thus, we moved from the Internet as metaphor of real space to its major roles for the storage and instant transmission of information, and from there to its relationships with real space, as facilitating, operational and monitoring technology for real space activities, and finally to its possible competition with real space.

We will conclude this chapter by drawing some attention to social aspects and implications of the wide adoption of the Internet and its use via mobile devices. The most important contemporary dimension of the Internet is the socially growing instant access to broadband services. This widening and location-free access has several implications concerning contemporary society. First, growing virtual mobility may increase rather than decrease physical mobility, since one does not have to be tied to a desktop any more in order to instantly perform any Internet activity. Second, the speeding up of daily activities, whether for production, consumption or social communications, may reach now a higher level, since all communications and information media have become fully mobile, thus prompting continuous attention by users. Third, the blurring of separation between work/business and leisure which has typified the spheres of work and home in recent years may continue to intensify, since both work and social activities can be easily performed when away from both office and home. The blurring between leisure and work activities is further amplified, since for both activities the same equipment, software and channels are used (see Kellerman 2006).

Speed is one of the leading elements in the continuous development of Internet technologies and this has its roots in previous technological developments. Way before the introduction of the commercial Internet, Virilio (1983, 45) called our era *the age of the accelerator*, and this nature of contemporary society has been accentuated time and again, notably regarding car driving, or accelerated and personal physical mobility. 'Speed is the premier cultural icon of modern societies . . . Speed symbolizes manliness, progress, and dynamism' (Freund and Martin 1993, 89).

The use of the Internet implies not only the blurring of boundaries between work and home/leisure. At the sphere of social relationships, Licoppe (2004) also recognized an emerging pattern of continuous *connected relationships* through various media of electronic communications, so that 'the boundaries between absence and presence eventually get blurred' (p. 136). For the public sphere, the use of mobile phones may blur the distinction between the private and the public, as well as that between indoors and outdoors (Kopomaa 2000), into what Sheller (2004) termed *mobile publics*. Whereas telephones and computers were traditionally considered as devices to be used indoors and involving some privacy of communications, wirelessness has implied less privacy and a change of social boundaries (see Kellerman 2006). Furthermore, mobile surfing of the Internet accentuates placelessness, or less attachment to places of residence, through exposure to remote places while being physically located in a fixed place, or while being on the road. 'The contradictory experience of being somewhere and nowhere at the same time is perhaps the most obvious cognitive dissonance resulting from the use of the WWW' (Kwan 2001, 26; see also Kellerman 2002).

REFERENCES

Adams, P.C. 1995. A reconsideration of personal boundaries in space–time. *Annals of the Association of American Geographers* 85: 267–285.

Adams, P. 1998. Network topologies and virtual place. *Annals of the Association of American Geographers* 88: 88–106.

Adams, P. and R. Ghose. 2003. India.com: the construction of a space between. *Progress in Human Geography* 27: 414–437.

Agar, J. 2003. *Constant Touch: A Global History of the Mobile Phone.* Cambridge: Revolutions in Science.

Allen, J. 1999. Worlds within cities. In D. Massey, J. Allen and S. Pile (eds), *City Worlds*. pp. 53–98. London: Routledge.

Allot Communications. 2013. Allot mobile trends report 02/2013, accessed from http://www.allot.com/Press_Releases.html.

Batty, M. 1997. Virtual geography. *Futures* 29: 337–352.

Benedikt, M. 1991. Cyberspace: some proposals. In M. Benedikt (ed.), *Cyberspace: First Steps.* pp. 119–224. Cambridge, MA: MIT Press.

Castells, M. 2001. *The Internet Galaxy: Reflections on the Internet, Business, and Society*. New York: Oxford University Press.

Castells, M. 2009. *Communication Power*. Oxford: Oxford University Press.

ComScore. 2012. ComScore: 4 out of 5 smartphone owners use device to shop: Amazon is the most popular mobile retailer, accessed from http://techcrunch.com/2012/09/19/comscore-4-out-of-5-smartphone-owners-use-device-to-shop-amazon-most.

Couclelis, H. 1998. Worlds of information: the geographic metaphor in the visualization of complex information. *Cartography and Geographic Information Systems* 25: 209–220.

Crang, M., P. Crang and J. May. 1999. Introduction. In M. Crang, P. Crang and J. May (eds), *Virtual Geographies: Bodies, Space and Relations*. pp. 1–13. London: Routledge.

Dodge, M. 2001. Guest editorial. *Environment and Planning B: Planning and Design* 28: 1–2.

Dodge, M. and R. Kitchin. 2001. *Mapping Cyberspace*. London: Routledge.

Fabrikant, S. 2000. Spatialized browsing in large scale data archives. *Transactions in GIS* 4: 65–78.

Fabrikant, S. and B. Buttenfield. 2001. Formalizing semantic spaces for information access. *Annals of the Association of American Geographers* 91: 263–280.

Freund, P. and G. Martin. 1993. *The Ecology of the Automobile*. Montreal: Black Rose Books.

Giddens, A. 1990. *The Consequences of Modernity*. Cambridge: Polity Press.

Gordon, E. and A. de Souza e Silva. 2011. *Net Locality: Why Location Matters in a Networked World*. Chichester: Wiley-Blackwell.

Hargittai, E. 1999. Weaving the Western web: explaining differences in Internet connectivity among OECD countries. *Telecommunications Policy* 23: 701–718.

Harvey, D. 1989. *The Postmodern Condition*. Oxford: Blackwell.

Holloway, S.I. and G. Valentine. 2001. Placing cyberspace: processes of Americanization in British children's use of the Internet. *Area* 33: 153–160.

Internet Live Stats. 2015. Total number of websites, accessed from http://www.internetlivestats.com/total-number-of-websites/.

ITU. 2010. Statistics. International Telecommunication Union, accessed from http://www.itu.int/ITU-D/ict/statistics/material/graphs/movile_reg-09.jpg.

ITU. 2011. The world in 2011: ICT facts and figures. International Telecommunication Union, accessed from http://www.itu.int/ITUD/ict/facts/2011/material/ICTFactsFigures2011.pdf.

ITU. 2015. Aggregate data. International Telecommunication Union, accessed from http://www.itu.int/en/ITU-D/Statistics/Pages/stat/default.aspx.

Janelle, D. 1973. Measuring human extensibility in a shrinking world. *Journal of Geography* 72: 8–15.

Kellerman, A. 1999. Space and place in Internet information flows. *Netcom* 13: 25–35.

Kellerman, A. 2002. *The Internet on Earth: A Geography of Information*. London: Wiley.

Kellerman, A. 2006. *Personal Mobilities*. London: Routledge.

Kellerman, A. 2007. Cyberspace classification and cognition: Information and communications cyberspaces. *Journal of Urban Technology* 14: 5–32.

Kellerman, A. 2010. Mobile broadband services and the availability of instant access to cyberspace. *Environment and Planning A* 42: 2990–3005.

Kellerman, A. 2012. *Daily Spatial Mobilities: Physical and Virtual*. Burlington, VT: Ashgate.

Kellerman, A. 2014. *The Internet as Second Action Space*. London: Routledge.

Kellerman, A. 2015. Image spaces and the geography of Internet screen-space. *GeoJournal* 79; doi: 10.1007/s10708-015-9639-1.

Kellerman, A. and M. Paradiso. 2007. Geographical location in the information age: from destiny to opportunity? *GeoJournal* 70: 195–211.

Kinsley, S. 2014. The matter of 'virtual' geographies. *Progress in Human Geography* 38: 364–384.

Kitchin, R. and M. Dodge. 2011. *Code/Space*. Cambridge, MA: MIT Press.

Kline, D. 2013. *Technologies of Choice? ICTs, Development and the Capabilities Approach*. Cambridge, MA: MIT Press.

Kopomaa, T. 2000. *The City in Your Pocket: Birth of the Mobile Information Society*. Helsinki: Gaudeamus.

Kwan, M-P. 2001. Cyberspatial cognition and individual access to information: the behavioral foundation of cybergeography. *Environment and Planning B* 28: 21–37.

Lessig, L. 2001. *The Future of Ideas: The Fate of the Commons in a Connected World*. New York: Random House.

Li, F., J. Whalley and H. Williams. 2001. Between physical and electronic spaces: the implications for organizations in the networked economy. *Environment and Planning A* 33: 699–716.

Licoppe, C. 2004. 'Connected' presence: the emergence of a new repertoire for managing social relationships in a changing communication technoscape. *Environment and Planning D: Society and Space* 22: 135–156.

Mossberger, K., C. Tolbert and W. Franko. 2013. *Digital Cities: The Internet and the Geography of Opportunity*. New York: Oxford University Press.

Péruch, P., F. Gaunet, C. Thinus-Blanc and J. Loomis. 2000. Understanding and learning virtual spaces. In R. Kitchin and S. Freundschuh (eds), *Cognitive Mapping: Past, Present, and Future*. London: Routledge.

Pickles, J. 2004. *A History of Spaces: Cartographic Reason, Mapping and the Geo-coded World*. London: Routledge.

Relph, E. 1976. *Place and Placelessness*. London: Pion.

Schrag, Z. 1994. Navigating cyberspace – maps and agents: different uses of computer networks call for different interfaces. In G. Staple (ed.), *Telegeography 1994: Global Telecommunications Traffic*. Washington, DC: Telegeography Inc.

Sheller, M. 2004. Mobile publics: beyond the network perspective. *Environment and Planning D: Society and Space* 22: 39–52.

Skarauskiene, A. and M. Kalinauskas. 2015. The Internet of things: when reality meets expectations. *International Journal of Innovation and Learning* 17: 262–274.

The Economist. 2012. A sense of place, 27 October, p. 27.

Urry, J. 2003. *Global Complexity*. Cambridge: Polity.

US Bureau of the Census. 2012. Statistical Abstract of the United States 2012, Tables 1055 and 1159, accessed from http://www.census.gov/compendia/statab/2012/tables/12s1159.pdf.

Virilio, P. 1983. *Pure War*. New York: Semiotext(e).

W³Techs. 2013. Usage of contents languages for websites, accessed from http://w3techs.com/technologies/overview/content_language/all.

Warf, B. 2013. *Global Geographies of the Internet*. Dordrecht: Springer.

Weinberger, D. 2002. *Small Pieces Loosely Joined {a unified theory of the web}*. Cambridge, MA: Perseus.

Wilson, M., A. Kellerman and K. Corey. 2013. *Global Information Society: Knowledge, Mobility and Technology*. Lanham, MD: Rowman and Littlefield.

Wo, S. 2011. How much information is there in the world? *USCNews*, accessed from http://news.usc.edu/29360/how-much-information-is-there-in-the-world/.

Zook, M., M. Dodge, Y. Aoyama and A. Townsend. 2004. New digital geographies: information, communication and place. In S. Brunn, S. Cutter and J. Harrington (eds), *Geography and Technology*. pp. 155–176. Dordrecht: Kluwer.

10. Tuning in to the geographies of radio
Catherine Wilkinson

> Radio is a landscape, a place inhibited by heroes and villains.
> Vowell 1996

The literature on geography and radio is notably scarce. Although regrettable, the largely dormant study of radio in the geography literature is not unsurprising, owing to geopolitics' fixation with visual media (Pinkerton and Dodds 2009). Smith (1994) supports this point, highlighting geography's emphasis on landscapes over soundscapes. More recently, a number of scholars have presented soundscapes as relevant to geographical debate, moving away from discussions of space and place which have been embedded within visual epistemologies, towards a new subfield of sonic geographies (e.g. Boland 2010; Boyd and Duffy 2012; Gallagher and Prior 2014; Matless 2005). This chapter devotes attention to the portable soundscape of radio. Using Truax's (2000, 4) definition, a soundscape is 'the totality of all sounds within a location with an emphasis on the relationship between individual's or society's perception of, understanding of and interaction with the sonic environment'. In particular, this chapter presents the origins and historical context of radio, and theoretical perspectives to the study of radio by scholars across the globe. Radio is defined herein, in line with Pluskota (2015, 327), as the evolving traditional terrestrial-based broadcast model, which when combined with satellite radio and Internet technologies, make up the 'media listening experience' market we recognise today.

Radio's character as an invisible and perpetual medium makes it difficult to study. This has not dissuaded interdisciplinary scholarship from focussing on the history and cultural, political and economic impacts of radio. However, radio has not been given the study that it deserves, particularly owing to the burgeoning scholarship on television, which has attracted attention as a subject of historical analysis (Sterling 2009), and its impacts on children and other vulnerable groups (Hilmes 2002). As Hilmes (2002) points out, it was not until the rise of political talk radio in the 1980s that the medium began to receive due critical attention, the majority of which was from a sociological perspective. This has led some authors to contend that, even within academic media research, radio struggles to find a voice and definition (Johnson and Foote 2009). Others have argued that, uniting people electromagnetically, radio creates a set of intersecting communities, not only radio listeners, but also those who study radio as a technological, social, cultural and historical phenomenon (Squier 2003a). Geographers are one such community of scholars.

The remainder of this chapter is structured into four main parts. First, it outlines the origins and historical context of radio. Second, it highlights conceptual debates about how radio has been studied by geographers. Third, the spatial perspective of radio is outlined. Fourth, technological convergences in relation to radio are considered, with a particular focus on radio in the digital age.

A BRIEF HISTORY OF RADIO

Radio histories differ across countries and nations; radio in one country or locality is not the same as it is elsewhere – it is used differently in different places and at different times, and has different meanings associated with it (Tacchi 2000). What remains consistent is the way that radio functions as *more than* just a technology or a method of communication, but as a process through which we, as human beings, can represent and negotiate our desires and fears (Squier 2003a). This brief outline of the origins and historical context of radio does not focus on any particular geographical context, but provides a more general trajectory of the development of radio broadcasting.

Radio is a technology dating back to the nineteenth century, and has been heralded as comparable to 'the apparatus in the labs of Frankenstein movies' (Douglas 1999, 49) during this time. Even in its infancy, radio made an impact on the cultural imagination (Squier 2003a). In the 1920s, radio grew into the medium through which farmers followed prices, weather and other variables. In rural over and above urban life, radio became the key source of information and entertainment, and an invaluable connection to the rest of the world (Craig 2001; Ulin 2003). In 1927, the content of radio programmes was under examination within the context of localism, as a study of broadcasting in New York City concluded that radio was, at that time, 'used almost entirely as an entertainment device for the advertising of the radio itself, and of the business which provide[d] the programs' (Lichty and Topping 1976, 49).

The 1950s and 1960s witnessed the explosion of a youth radio market. With the flourishing of FM radio during this time, a new audience emerged, largely male and obsessed with 'high fidelity' (Squier 2003a, 13). Resultantly, radio became a battlefield between automated playlists of Top 40 hits and the non-conforming programming of college FM stations (Squier 2003a). By the 1970s and 1980s, the growth of talk radio instigated a new radio-based activism, as well as alternative models for male selfhood (Squier 2003a). By the end of the twentieth century, radio became a taken-for-granted medium, outshone by new technologies including computers, digital television and mobile telephones. This was only the beginning. Changes in radio technology provided a variety of new programming models: Internet radio, digital radio and low-power radio (Squier 2003a). In the process of constructing these new technologies, certain aspects of radio's potential were exiled, while others were enhanced, as I tease out later in this chapter.

CONCEPTUAL DEBATES

Radio has long been heralded as holding an 'unobtrusive presence in the geography of domestic space' (Moores 1988, 23). It is used to maintain or alter mood, it is emotionally evocative, comforting and a part of domestic soundscapes (Tacchi 2000). Other research (Cordeiro 2012) positioned radio as an 'everyday companion', assisting people to structure their routines and feeding their daily informational needs. With radio being personified as such, important geographical issues come to the fore concerning radio's ubiquitous nature, and hold resonance with Pinkerton and Dodds's (2009, 16) observation that 'radio listening whether in bed, while taking a shower or eating our breakfast is part of the daily fabric of many people around the world'. As Pinkerton and Dodds (2009) note, much literature now exists on how sound is entrenched in history, cultures, institutions, technologies and, indeed, geographies, and the consequences of this, although this is not specifically related to radio. Heeding Pinkerton and Dodds (2009), when geographical work is concentrated on radio, this tends to focus on large-scale geopolitical questions in relation to international, as opposed to small-scale, stations.

In their seminal paper, Pinkerton and Dodds (2009) note that extant research is dominated by the social and cultural in music, and that few researchers (cf. Power 2001) have explored in substantial empirical depth the geopolitical consequences of radio broadcasting, or listening more generally. Further, the authors note that radio scholarship has predominantly focussed on domestic radio environments in the Western world. The authors cite Liebes's (2006) examination of Israeli broadcasting. According to Liebes (2006), radio has the potential to create a number of acoustic spaces within which individual listeners, and indeed communities, can communicate their shared identities. The authors conclude by calling for further critical study of radio, sound and broadcasting/listener engagement within the geographical literature.

The most poignant example of such in extant literature is Vaillant's (2002) consideration of local radio in Chicago between 1921 and 1935. According to the author, during the 1920s, local broadcasting transcended a 'radio imaginary' or an 'imagined community', instead promoting face-to-face community life, for instance by encouraging listeners to participate in programmes or by organising community socials. As Vaillant (2002, 26) eloquently asserts:

> The airwaves became a neighbourhood as well as a metropolitan stage, and many listeners tuning to independent stations swelled with pleasure and pride at hearing music and cultural programs that acknowledge and validate their particular languages, histories, and cultural backgrounds for both the designated group and the larger audience to hear.

I quote Vaillant (2002) here because he evokes an image of radio broadcasting and listening as a community *per se*. Vaillant (2002) echoes Keough's (2010, 77) evaluation of community radio as 'more than just an FM frequency, but rather a meaningful place to station personnel and the listening community'. Through a case study approach, Vaillant (2002, 26) situates independent radio broadcasting and listening within a struggle among urban ethnic, immigrant and middle- and working-class Americans at this time to 'claim urban space, shape public culture, and define the contested terms of ethnic difference, racial differentiation, and Americanism'. This represents an important step in using radio as a means of empowerment and to give a voice to those deprived of one – those experiencing voice poverty (see Salazar 2009).

Other scholarship has explored the subversive power of radio. Douglas (2004) studies radio's power to sustain emotional states, to ignite the imagination and to enable expression of stigmatised parts of the self. Street (2012, 29) tells how radio producers quickly came to understand the capacity of the medium to cross barriers between fact and imagination, from naturalism to impressionism. The power of radio to do this has often been demonstrated in the 'other worldliness' of imaginings, an area where the pictorial nature of sound is particularly suggestive. By activating experiences aurally, radio taps an influential and deep connection to the human brain, as cognitive scientists are discovering (Squier 2003b). Radio's 'blindness' in this sense is not a disadvantage, but an opportunity to create a new kind of intimacy and individual fantasies (Munich 2003). McChesney (2001, v) heralded radio as 'the quintessential people's medium'. Certainly, establishing a bond with the listener has been positioned as essential to the success of audio programming (Johnson 2015). As Sauls (2011) points out, the dedication of radio broadcasting to connect to its listeners has been at the forefront since its inception. As is clear then, in line with Squier (2003a), scholarship has positioned radio as constructed by, and in its turn helping to shape, society and culture.

SPATIAL PERSPECTIVES

In terms of communication technologies, radio is the most local of the mass media (Hunteman 2003). Geographically bound by the physical properties of electromagnetic waves, radio was traditionally defined by its broadcast range (Hunteman 2003). The idea of community, so central to broadcast regulation, began to shift from its former definition as a local phenomenon to something that could extend across an entire nation. By including the listening audience in the content of radio shows, programmes made this new national audience central to radio entertainment. Audience participation programmes accelerated the process by which the new mass audience of radio came to stand in for the nation in general and 'the people' in particular (Loviglio 2002, 90). However, as Hilmes and Loviglio (2002) remind us, this shift towards togetherness, often heralded as a positive thing, was for some considered a threat to traditional notions of both community and individualism. This changing idea of community most widely impacted on understandings of community radio.

The alternative and community radio started in the turbulent 1960s and 1970s and provided a platform for local voices and concerns in hundreds of cities and towns globally. Reed and Hanson (2006, 214) define the community served by community stations as 'in a specific geographic area'. Others take a more networked view, asserting that community stations reach people who live in dispersed locations, yet have shared interests (e.g. Gumucio-Dagron 2013; McCain and Lowe 1990). Encompassing both strands, Coyer (2006) presents community radio as serving communities of interest, providing the examples of lesbians and gay men, blind and partially sighted people, Afro-Caribbeans and Asians and/or small geographic areas. Crucially, however, while being broad in appeal, these stations remain situated within the context of their local areas. Taken together then, in line with Squier (2003b), 'community radio' is generally understood to be local programming that serves the cultural, civic or informational needs of an audience that is either geographically or demographically limited.

As well as considering the wider community served, Leal (2009) devotes attention to the community of staff and volunteers working within the stations or, to borrow Gaynor and O'Brien's (2011, 31) term, the 'community within'. Leal (2009) finds that the 'community within' exhibits a considerable degree of diversity with regards to gender, age and cultural background, although predominantly including marginalised sections of the population, particularly the unemployed. Leal (2009, 156) refers to community radio stations as 'communities manifested from the discursive practices of individuals who share physical space and a similar social situation', depicting a community that has a geographical base, yet serves a common interest. These conceptualisations of community reflect a social network approach (e.g. Fabiansson 2006), whereby community is (dis) located – situated in, yet distinct from, the locale.

Other community radio research has recognised the contested nature of community. Using the case of Radio KC in Paarl, Western Cape, Davidson (2004) considers the steps that staff must take to understand the communities they serve. The importance of this lies in identifying programming content that includes stories and voices from within the community. From observations and participatory research, Davidson (2004) concludes that the concept of community extends beyond the geographical community outlined in the station's broadcast licence, being cognitive, normative and imagined. Interestingly, Davidson (2004) discovers that, while staff and volunteers at Radio KC came from within the geographical vicinity of the station, they did not perceive themselves as 'the community', rather they considered themselves as *serving* the community. In a study of XK FM, a community radio station in Platfontain, South Africa, Mhlanga (2009) finds heterogeneity in the community served by the station. Kanayama (2007) draws the same conclusion in a study of the role of community radio in Japan. These readings stand to counter understandings of community as homogenous which, as Young (1990) argues, typically privilege unity over difference.

During the 1930s and 1940s radio played a pivotal role in defining the national community. Radio was the only medium capable of addressing the entire country simultaneously, and its voice was both literal and figurative, consisting of an address to a national 'we' by radio networks, as well as providing a model of what a 'real American' should sound like (Russo 2002, 258). As Hilmes and Loviglio (2002) point out, radio waves and their impermeable mobility across social boundaries served as a symbol for national togetherness. This, of course, builds on Anderson's (1983) conception of an imagined community (see also Hilmes 1997 and Karpf 2013 who position radio this way). 'Imagined communities', as outlined by Anderson (1983), counter the idea that communities are defined by territory, claiming that communities are mental constructs.

Anderson's (1983, 4) point of departure is that nationality, 'nation-ness' and nationalism are cultural artefacts. Anderson (1983, 6) defines the nation as 'an imagined political community – and imagined as both inherently limited and sovereign'. He argues that it is imagined because nation members will never know their fellow-members, yet they carry with them the image of their communion. This is notwithstanding that there may be exploitation and inequality between fellow citizens. Despite the power of media in shaping ideas of the modern nation, Anderson (1983) positions geography as remaining important. To explain, the nation is imagined as 'limited' because it has finite boundaries, beyond which other nations lie. It is in this sense that Anderson (1983) touches on the importance of radio for the cultivation of nationalism in many colonial settings. In turn, as Spinelli (2003) argues, public radio imagines its audience, and how it conceives of the dynamics of the relationship between its producers and its consumers.

Drawing on Anderson (1983), Friedland (2001) argues that, when assuming an imagined community perspective, identities are formed from a collective sense of history and culture that meshes communities together. This has resonance with Levine (2008), who discusses community in relation to identities associated with music, graphic design, narrative history and other forms of culture. According to Levine (2008), people form clusters in communities, each of which has a distinct character. However, for Friedland (2001), the cognitive, moral and imagined aspects of community only cohere within the set of social structures that bind the community domain. Rose (1990) argues that, although communities may be imagined, they are not idealist, because such imaginings are grounded in specific social, economic and political circumstances. In sum then, following Loviglio and Hilmes (2013), radio is still a medium of local specificity and intimacy, but is increasingly defining its audience not through geography, but through cultural affinity.

RADIO IN THE DIGITAL AGE

Radio has survived numerous market threats: the phonograph, eight track, tape cassette and the compact disc. However, perhaps one of the most significant threats to radio, which has yet to be resolved today, is the one the Internet has introduced (Pluskota 2015). The digital transformation of radio began with streaming audio in the late 1990s, highlighted by the introduction of the iPod in 2001. Fast forward a few years and radio had evolved into something different: it has materialised, differentiated and de-spatialised (Loviglio and Hilmes 2013). The debut of podcasting in 2004 combined the mobility of the iPod with digital syndication software and web-based subscription. With the upsurge of Wifi, the entrance of the iPhone in 2007 and the rapid proliferation of smartphones, tablets, e-readers and digital radios, the concept of radio shifted rapidly (Loviglio and Hilmes 2013). However, at a time when music, news and talk can be heard via the Internet, satellite and MP3 sound files, and films and television programmes can be accessed via on-demand technologies, the future of radio as a distinct medium, and of broadcasting as a technological means, was considered no longer certain. Hilmes (2013) tells us that, in considering the future of radio, radio's death has been predicted as often as its survival. The conclusion that Hilmes (2013) makes in her chapter on the new materiality of radio is that radio has not only survived, but has revived – both as a creative medium and as a shared cultural experience.

   Although the reach of community stations is typically limited, there is increasing recognition that use of Internet technologies for online simulcasting is expanding this reach, making content available at the time of broadcast, and afterwards through downloads and podcasts. In accordance with Coyer (2005), Internet broadcasting offers a redefinition of community radio, away from geographical restrictions, to transnational broadcasting. As Sujoko (2011, 17) puts it, owing to technological advances in mobile phones and the Internet:

> Community radio is no longer broadcasting 'outwards' and downwards, from a central source of information. Instead, the messages exchanged are multi-sourced, constantly adjusting to and recognising their location(s), and so producing a consistent adaptability and negotiability, even as they rework the existing cultural perspectives of their community.

   For certain authors (Algan 2013; Rooke and Odame 2013), the Internet holds potential as an outlet where listeners can learn about, and interact with, station DJs, participate in competitions, find out about upcoming events and obtain local news bulletins. Certainly, Internet streaming challenged radio stations to think about the visible face they presented to the public (Hilmes 2013). However, the opportunities offered by the Internet, as envisioned by Sujoko (2011) and Rooke and Odame (2013), have been met with sharp critique. Hallett and Hintz (2010) argue that community radio's convergence online has created a widespread impression that the Internet is a viable alternative for frequency allocation, thus incorrectly implying that the access problem is solved. For instance, it is assumed that if somebody listening in a certain location cannot typically access a community radio station via FM, they can listen to the radio station online. However, this does not take into consideration factors such as the digital divide.

To interrogate this further, it is useful to consider Coyer's (2005) argument that the aspiration to create a community radio project, as a collective, transcends the technological means of distribution afforded by the Internet. Therefore, while the Internet has renovated public access to content, it fails to acknowledge people's social or political motivations to create their own media. Coyer (2005, 32) believes that 'the potential for endless possibilities within the digital arena cannot serve as a remedy to the issue of scarcity on the traditional dial'. Whilst the (relatively) limitless space available to broadcast online in some ways addresses the problem of restrictions, the author suggests that the Internet has its own limitations, including the digital divide and community access, arguing that the traditional format of radio is more intimate, despite there being fewer opportunities for interaction. If Coyer (2005) is correct, it is questionable whether there is any advantage, bar increased listening figures, to moving radio online. However, Coyer (2005, 40) confesses that even increased listener figures may be overly presumptuous: 'just because something is online doesn't mean anyone is listening'. Other authors, too, believe that debates have a myopic focus and question whether, owing to technological advancements, radio's locally distinctive identity is being demolished (e.g. Dunaway 1998). What is clear is that technology offers us the opportunity for connecting the radio listener to a community or a geographical space imaginatively and physically concurrently (Street 2012). Despite celebrating success for more than a century, Pluskota (2015) argues that, in order to remain competitive, the radio industry must consider changing its model to one that can compete with other listening options, particularly regarding listening choices, customisation and access. Moody (2010) complements yet extends this viewpoint, believing that, if radio is to innovate, it needs to engage with its origins in audio.

CONCLUSION

Despite a wide scope of studies looking at radio through a number of different lenses, including social, political, and cultural, few studies devote sufficient attention to radio for what it is, a sonic medium. Peters (2012, 1241), discussing pirate radio, considers how radio DJs on Radio Caroline[1] 'harnessed' the depth and dynamism of the sea to create unique audio experiences for listeners on land. Although not using the lens of sonic geographies explicitly, Peters (2012, 1241) is concerned with the management and manipulations of materiality and affect, particularly of the rough seas, which helped to form new 'cocomposed relations' on the airwaves. Through strategic acts to 'play up' the power of the sea, Radio Caroline listeners became 'seduced by . . . this motionful, dynamic, and uncertain hydroworld' (Peters 2012, 1249). Also devoting attention to sonic properties is Arkette's (2004) discussion of the ways in which different radio stations incorporate diverse modes of presentation. For instance, Radio 3 demonstrates 'presenters, in respectful tones and subdued inflections' conveying their knowledge of musical arte-facts to the listeners (Arkette 2004, 165). Meanwhile, Jazz FM's 'brand of aural image is centred on the often full-bodied and breathy voice of the presenter' (Arkette 2004, 165). Contrastingly, Capital and Virgin DJs work collectively, 'talking and cracking jokes amongst themselves in an attempt to dispel the image of radio as a unilateral disembod-ied voice' (Arkette 2004, 165). The author calls listeners to play a part in forming their acoustic communities, rather than considering radio an ambient landscape.

Certainly, there is demand for heightened scholarly attention to geographies of speech. Brickell (2013, 1), for example, has recently argued that geographers have failed to devote sustained attention to speech as a practice that 'provokes meanings in, and of, spaces'. This is analogous to Livingstone's (2007) argument that spaces of speech have been dis-regarded. Echoing this notion is Kanngieser's (2012) call for a geography of voice and a politics of speaking and listening, composing a sonic geography of voice, including study of tone and volume, amongst other qualities. The dominant critical discourse about the radio voice is mostly aesthetic – a voice is beautiful or ugly, grates or is resonant, has good elocution or poor (Karpf 2013). Considering voice in relation to radio is important, for as Street (2012, 7) tells us:

> It is a 'voice' that can show you pictures. It is this which, if we allow it, can open the world of response before understanding which may be found to be, in the end at the root of all true understanding after all.

Bainbridge and Yates (2013) also recognise that there is more research to be done in considering radio's deployment of voice as its predominant means of listener engage-ment. Central to this is to unite radio and sound which have typically, and somewhat bizarrely, been two separate areas of scholarship (Johnson 2015). I thus advocate that future studies of radio should adopt a sonic geographical perspective. This will enable exploration of radio's new sounds and forms, which comprise increasing creativity.

NOTE

1. Radio Caroline is a British radio station founded in 1964 to circumvent record companies' control of popular music broadcasting in the United Kingdom and the BBC's radio broadcasting monopoly (see Chapman, 1990).

REFERENCES

Algan, E. 2013. Youth, new media, and radio: mobile phone and local radio convergence in Turkey. In J. Loviglio and M. Hilmes (eds), *Radio's New Wave: Global Sound in the Digital Era*. New York: Routledge.
Anderson, B. 1983. *Imagined Communities: Reflections on the Origin and Spread of Nationalism*. London: Verso.
Arkette, S. 2004. Sounds like city. *Theory, Culture and Society* 21(1): 159–168.
Bainbridge, C. and C. Yates. 2013. The reparative spaces of radio. *The Radio Journal – International Studies in Broadcast & Audio Media* 11(1): 7–12.
Boland, P. 2010. Sonic geography, place and race in the formation of local identity: Liverpool and Scousers. *Geografiska Annaler: Series B, Human Geography* 92(1): 1–22.
Boyd, C. and M. Duffy. 2012. Sonic geographies of shifting bodies. *Interference: A Journal of Audio Culture* 1(2): 1–7.

Brickell, K. 2013. Towards geographies of speech: proverbial utterances of home in contemporary Vietnam. *Transactions of the Institute of British Geographers* 38(2): 207–220.

Chapman, R. 1990. The 1960s pirates: a comparative analysis of Radio London and Radio Caroline. *Popular Music* 9(2): 165–178.

Cordeiro, P. 2012. Radio becoming r@dio: convergence, interactivity and broadcasting trends in perspective. *Participations* 9(2): 492–510.

Coyer, K. 2005. Where the 'hyper local' and 'hyper global' meet: case study of Indymedia radio. *Westminster Papers in Communication and Culture* 2(1): 30–50.

Coyer, K. 2006. Community radio licensing and policy. *Global Media and Communication* 2(1): 129–134.

Craig, S. 2001. 'The farmer's friend': radio comes to rural America, 1920–1927. *Journal of Radio Studies* 8(2): 330–346.

Davidson, B. 2004. Mapping the radio KC community. *Ecquid Novi* 25(1): 43–60.

Douglas, S. 1999. *Listening In: Radio and the American Imagination, from Amos 'n' Andy and Edward R. Murrow to Wolfman Jack and Howard Stern.* New York: Times Books.

Douglas, S. 2004. *Listening In.* Minneapolis, MN: University of Minnesota Press.

Dunaway, D. 1998. Community radio at the beginning of the 21st century: commercialism vs. community power. *The Public* 5(2): 87–103.

Fabiansson, C. 2006. Being young in rural settings: young people's everyday community affiliations and trepidations. *Rural Society* 16(1): 47–61.

Friedland, L. 2001. Communication, community, and democracy: toward a theory of the communicatively integrated community. *Communication Research* 28(4): 358–391.

Gallagher, M. and J. Prior. 2014. Sonic geographies: exploring phonographic methods. *Progress in Human Geography* 38(2): 267–284.

Gaynor, N. and A. O'Brien. 2011. Community radio in Ireland: 'defeudalising' the public sphere? *Javnost – The Public* 18(3): 23–38.

Gumucio-Dagron, A. 2013. Look back for the future of community radio. *Media Development* 13(2): 6–12.

Hallett, L. and A. Hintz. 2010. Digital broadcasting – challenges and opportunities for European Community radio broadcasters. *Telematics and Informatics* 27(2): 151–161.

Hilmes, M. 1997. *Radio Voices: American Broadcasting, 1922–1952.* Minneapolis, MN: University of Minnesota Press.

Hilmes, M. 2002. Rethinking radio. In M. Hilmes and J. Loviglio (eds), *Radio Reader: Essays in the Cultural History of Radio.* New York: Routledge.

Hilmes, M. 2013. The new materiality of radio: sounds on screens. In J. Loviglio and M. Hilmes (eds), *Radio's New Wave: Global Sound in the Digital Era.* New York: Routledge.

Hilmes, M. and J. Loviglio. 2002. Introduction. *Radio Reader: Essays in the Cultural History of Radio.* New York: Routledge.

Hunteman, N. 2003. A promise diminished: the politics of low-power radio. In S. Squier (ed.), *Communities of The Air: Radio Century, Radio Culture.* Durham, NC: Duke University Press.

Johnson, P. 2015. Editor's remarks: the golden years, then and now. *Journal of Radio & Audio Media* 22(2): 142–147.

Johnson, P. and J. Foote. 2009. Alternative radio: other voices, other definitions, beyond the mainstream community as public interest convenience, and necessity. *Journal of Radio Studies* 7(2): 282–286.

Kanayama, T. 2007. Community ties and revitalization: the role of community radio in Japan. *Keio Communication Review* 29(1): 5–24.

Kanngieser, A. 2012. A sonic geography of voice: towards an affective politics. *Progress in Human Geography* 36(3): 336–353.

Karpf, A. 2013. The sound of home? Some thoughts on how the radio voice anchors, contains and sometimes pierces. *International Studies in Broadcast & Audio Media* 11(1): 59–73.

Keough, S. 2010. The importance of place in community radio broadcasting: a case study of WDVX, Knoxville, Tennessee. *Journal of Cultural Geography* 27(1): 77–98.

Leal, S. 2009. Community radio broadcasting in Brazil: action rationales and public space. *Radio Journal: International Studies in Broadcast & Audio Media* 7(2): 155–170.

Levine, P. 2008. A public voice for youth: the audience problem in digital media and civic education. In W. Lance Bennett (ed.), *Civic Life Online: Learning How Digital Media Can Engage Youth.* Cambridge, MA: MIT Press.

Lichty, L. and M. Topping. 1976. *American Broadcasting: A Source Book on the History of Radio and Television.* New York: Hastings House.

Liebes, T. 2006. Acoustic space: the role of radio in Israeli collective history. *Jewish History* 20(1): 69–90.

Livingstone, D. 2007. Science, site and speech: scientific knowledge and the spaces of rhetoric. *History of the Human Sciences* 20(2): 71–98.

Loviglio, J. 2002. Vox pop: network radio and the voice of the people. In M. Hilmes and J. Loviglio (eds), *Radio Reader: Essays in the Cultural History of Radio.* New York: Routledge.

Loviglio, J. and M. Hilmes. 2013. Introduction: making radio strange. In J. Loviglio and M. Hilmes (eds), *Radio's New Wave: Global Sound in the Digital Era*. New York: Routledge.

Matless, D. 2005. Sonic geography in a nature region. *Social & Cultural Geography* 6(5): 745–766.

McCain, T. and G. Lowe. 1990. Localism in Western European radio broadcasting: untangling the wireless. *Journal of Communication* 40(1): 86–101.

McChesney, R. 2001. Radio and the responsibility of radio scholars. *Journal of Radio Studies* 8(2): v–viii.

Mhlanga, B. 2009. The community in community radio: a case study of XK FM, interrogating issues of community participation, governance, and control. *Ecquid Novi: African Journalism* 30(1): 58–72.

Moody, R. 2010. Foreword. In J. Biewen and A. Dilworth (eds), *Reality Radio: Telling True Stories in Sound*. Chapel Hill, NC: University of North Carolina Press.

Moores, S. 1988. The box on the dresser: memories of early radio and everyday life. *Media, Culture & Society* 10(1): 23–40.

Munich, A. 2003. In the radio way: Elizabeth II, the female voice-over, and radio's imperial effects. In S. Squier (ed.), *Communities of The Air: Radio Century, Radio Culture*. Durham, NC: Duke University Press.

Peters, K. 2012. Manipulating material hydro-worlds: rethinking human and more-than-human relationality through offshore radio piracy. *Environment and Planning A* 44(5): 1241–1254.

Pinkerton, A. and K. Dodds. 2009. Radio geopolitics: broadcasting, listening and the struggle for acoustic spaces. *Progress in Human Geography* 33(1): 10–27.

Pluskota, J. 2015. The perfect technology: radio and mobility. *Journal of Radio & Audio Media* 22(2): 325–336.

Power, M. 2001. Acqui Lorenco Marques! radio colonization and cultural identity in colonial Mozambique. *Journal of Historical Geography* 26: 605–628.

Reed, M. and R. Hanson. 2006. Back to the future: Allegheny mountain radio and localism in West Virginia community radio. *Journal of Radio Studies* 13(2): 214–231.

Rooke, B. and H. Odame. 2013. 'I have a blog too?' Radio jocks and online blogging. *Journal of Radio & Audio Media* 20(1): 35–52.

Rose, G. 1990. Imagining poplar in the 1920s: contested concepts of community. *Journal of Historical Geography* 16(4): 425–437.

Russo, A. 2002. A dark(ened) figure on the airwaves: race, nation, and the Green Hornet. In M. Hilmes and J. Loviglio (eds), *Radio Reader: Essays in the Cultural History of Radio*. New York: Routledge.

Salazar, J. 2009. Self-determination in practice: the critical making of indigenous media. *Development in Practice* 19(4–5): 504–513.

Sauls, S. 2011. Locally and free: what broadcast radio still provides. *Journal of Radio & Audio Media* 18(2): 309–318.

Smith, S. 1994. Soundscape. *Area* 26(3): 232–240.

Spinelli, M. 2003. Not hearing poetry on public radio. In S. Squier (ed.), *Communities of The Air: Radio Century, Radio Culture*. Durham, NC: Duke University Press.

Squier, S. 2003a. Communities of the air: introducing the radio world. In S. Squier (ed.), *Communities of the Air: Radio Century, Radio Culture*. Durham, NC: Duke University Press.

Squier, S. 2003b. Wireless possibilities, posthuman possibilities: brain radio, community radio, radio Lazarus. In S. Squier (ed.), *Communities of The Air: Radio Century, Radio Culture*. Durham, NC: Duke University Press.

Sterling, C. 2009. The rise of radio studies: scholarly books over four decades. *Journal of Radio & Audio Media* 16(2): 229–250.

Street, S. 2012. *The Poetry of Radio – The Colour of Sound*. Abingdon: Routledge.

Sujoko, A. 2011. Talking culture: Indonesian community radio and the active audience. *Social Alternatives* 30(2): 16–20.

Tacchi, J. 2000. The need for radio theory in the digital age. *International Journal of Cultural Studies* 3: 289–298.

Truax, B. 2000. *Acoustic Communication*. Santa Barbara, CA: Greenwood.

Ulin, D. 2003. Science literacies: the mandate and complicity of popular science on the radio. In S. Squier (ed.), *Communities of the Air: Radio Century, Radio Culture*. Durham, NC: Duke University Press.

Vaillant, D. 2002. Sounds of whiteness: local radio, racial formation, and public culture in Chicago, 1921–1935. *American Quarterly* 54(1): 25–66.

Vowell, S. 1996. *Radio On: A Listener's Diary*. New York: St Martin's Griffin.

Young, I. 1990. The ideal of community and the politics of difference. In L. Nicholson (ed.), *Feminism/Postmodernism*. London: Routledge.

11. Eyes in the sky: satellites and geography
Barney Warf

The world's network of satellites and Earth stations comprises a critical, often over-looked, element in the global telecommunications infrastructure. Satellites are used extensively by telecommunications companies, multinational corporations, financial institutions and the global media to link far-flung operations, including international data transmissions, telephone networks, teleconferencing, broadband Internet and media sales of television and radio programs. They are widely used by the military, frequently deployed during catastrophes such as hurricanes or earthquakes, and essential to environ-mental monitoring (e.g. forest fires, glaciers). This technology has received remarkably little attention in the geographical literature despite the powerful ways in which it shapes the control, use and experience of space. While most treatments of this topic typically depict it in technologically deterministic, apolitical terms, satellites are deeply embedded in terrestrial political relations. Although satellites circulate in outer space, their origins and impacts occur very much on the ground.

There are several reasons for studying satellites, including their economic significance for global networks of capital and information and as an avenue through which com-modity relations have penetrated outer space, a domain once exclusively the province of the nation-state. Satellites have long been a major means by which national sovereignty has been both extended and contested above the Earth's atmosphere, and at times blur rigid national borders by facilitating the communication of information. Access to satel-lite technology both mimics and reinforces terrestrial relations of wealth and power, and the regulation of international satellite traffic – largely through the International Satellite Organization (Intelsat) and a series of regional organizations – mirrors the changing power-geometries of states in the contemporary world-system.

Politically, satellites have long comprised a major component of national espionage efforts and are deeply implicated in state monitoring of territory, forming an important tool in sustaining nationalist discourse and military strategy throughout. Perched high above the Earth, satellites constitute a panopticonic system of surveillance without walls. Satellites have been central to concerns over 'epistemic sovereignty' (Coombe 1996), the fact that some nation-states know much more about the territory of their rivals than the observed states may know about themselves. Throughout the Cold War, satellites were instrumental in the discursive scripting of geographic space, its ideological construction within hegemonic modes of understanding shared by politicians, military planners and the media that were typically infused with the indiscriminate 'othering' of the communist foe. In this light, satellites constitute an intertwined set of mutually reinforcing technical, political and ideological relationships. Cosgrove (1994) argued that, far from comprising politically neutral representations, space photography (such as the Apollo photographs) legitimated and sustained a discourse of 'one Earth' effectively embraced and encom-passed by one nation, the US. Thus, satellite imagery can be embedded conceptually within a long-standing Western tradition of ocularcentrism that privileges 'seeing' and the visual (Warf 2012), a form of knowledge that acquires an *a priori* aura of detached epistemological neutrality despite its manifestly political origins and impacts. From a feminist perspective, Litfin (1997) maintains that satellites reinforce masculinist and positivist norms of an all-seeing detached observer.

This chapter addresses the satellite industry from several angles. It opens with a brief note on terminology, defining satellites, Earth stations and their capabilities. Second, it offers an overview of the historical roots of the industry, stressing its military role during the Cold War. The third section delves into the global regulatory nature of the sector, emphasizing the key role played by Intelsat, as well as its more recent, smaller, competitors. In the fourth section, the implications of the neoliberal move to deregulation and privatization are addressed. Finally, it turns to two significant technological changes, competition from fiber optics cables and the potential of low-orbiting satellites.

A NOTE ON TERMINOLOGY

Satellites in orbit appear in a variety of sizes and degrees of technological sophistication (Maral and Bousquet 2002). Large satellites capable of handling international traffic sit 35,700 km (22,300 miles) high in geostationary orbits, which are by far the most valuable orbital slots because only in that narrow sliver of space do satellites and the Earth travel at the same speed relative to each other, making the satellite a stable target for signals transmitted upward from Earth stations (Frieden 1996). Large satellites in geostationary orbit can observe, collect and transmit data to (i.e. leave a 'footprint' over) roughly 40% of the Earth's surface (Figure 11.1). Because such orbital arcs are a scarce resource, their distribution is strictly controlled through international organizations (Smith 1987; Hart 1991; Hudson 1990). The cost of launching satellites and the fuel needed to maintain them in their proper orbit are also constraints to their economic viability. Satellites typically have a 15 year life span, primarily because they exhaust their available fuel, necessitating their eventual replacement by a new, frequently much improved, generation. From its vantage point, a broad-beam geostationary satellite covers so much area that only three or four are sufficient to provide global coverage. Because the cost of satellite transmission is not related to distance, it is commercially competitive in rural or low-density areas (e.g. remote islands), where high marginal costs dissuade other types of providers, particularly fiber optics providers (Giget 1994; Goldstein 1998; Warf 2006).

The terrestrial counterpart of the satellite is the Earth station. There are tens of millions of Earth stations located worldwide, ranging in size from 0.5 to 30 meters. The vast majority, however, can only receive information, not transmit it (i.e. downlink only). When microwave signals are sent over great lengths and become broadly diffused, Earth stations require large, powerful antennas to receive them. The world's 483 publicly owned Earth stations designed for international traffic are concentrated in the largest and wealthiest countries, particularly the US, which, with 70, has vastly more than any other state. Countries without these facilities (e.g. Afghanistan), or those with an insufficient number to satisfy domestic demand, must rely upon leased connections to other nations.

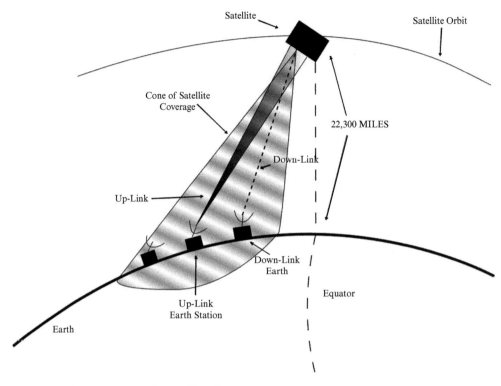

Figure 11.1   Dynamics of a satellite footprint

HISTORICAL ORIGINS OF THE SATELLITE INDUSTRY

The origins of the satellite industry lie in rocketry developed during World War II (Pelton 2013). Starting with the Soviet Union's launch of *Sputnik* in 1957, satellites played a key role in the militarization of space during the Cold War (Stares 1985; Burrows 1986; Richelson 1990; DeVorkin 1992). The first US satellite, *Explorer I*, was put into orbit one year later. In 1960, the CIA established the National Reconnaissance Office to operate an emerging satellite espionage capacity. The world's first spy satellites, a series of 95 launched under the Pentagon's Corona project, revolutionized Western understanding of the Soviet Union (Galloway 1972; Cloud 2001; Monmonier 2002). Draped in shrouds of secrecy (its images were not publicly released until 1995), the Corona reconnaissance system offered accurate images of the vast 'denied territory' of the Eurasian landmass (Day et al. 1998), allowing the location and magnitude of military facilities (especially long-range missile silos) to be pinpointed (Richelson 2001). The United Nations' Outer Space Treaty of 1967 attempted to limit the militarization of space, an inconvenience to military planners that has largely been cast aside. In 1969, American military satellites were coordinated in the Defense Satellite Program to provide continuous, comprehensive and seamless coverage of the entire Earth's surface (Gaffney 2000; Richelson 2001). Today, satellites form the basis for the US Defense Department's C³ (command, control and communications) infrastructure, by far the most sophisticated military intelligence and surveillance system in the world. In 2003, the US created its first military unit to defend satellites, the 614th Space Intelligence Squadron.

As the resolution and accuracy of military satellite imagery progressed over time, civilian uses followed accordingly. The visual power of satellites is usually expressed in terms of the length of the smallest feature that analysts can see when photographic processing is pushed to its limits. The exponential improvements in this capacity are evident in the changes from the early *Landsat* photos of the 1970s, with a resolution of 60 meters, to the current generation of spy satellites, which can identify objects 5 cm in length. When satellites were the exclusive purview of the military worldwide, nations jealously hid the abilities of their surveillance systems so that foes would be less likely to evade and counter them; today, commercial providers boast openly about their systems' powers of resolution.

Gradually, civilian applications assumed an increasingly more important role in the satellite industry's development. The first of these included the *Landsat* satellites launched by NASA in the 1970s, a process filled with bureaucratic politics, including the multiple constituencies within the scientific community involved, NASA's rivalry with the Defense Department, funding battles and the competing choices in technical design (Mack 1990). The major difference between military and civilian purposes involved the shift in emphasis from surveillance to communications; although the technology remains important in both respects for military purposes, in the civilian domain communications remains the dominant application. In 1962, the US launched AT&T's *Telstar*, the world's first communications satellite (Jansky and Jeruchim 1987). In the 1960s, NASA initiated the *Syncom* series, which allowed television companies to initiate satellite transmission of programs, ultimately leading to an explosion of home broadcasting in the 1990s (Inglis 1991). The first privately owned satellite, *Early Bird* (*Intelsat 1*), was launched in 1965 and operated by Western Union. The first Soviet civilian satellite, *Molniya*, launched in 1965, was intended for purely domestic communications, in contrast to the American *Early Bird*, which was designed for international purposes.

Since the late 1950s, the world has launched more than 5500 satellites, the vast majority of which were sent into orbit by the US, which launched 1854 before 2004, and the USSR/ Russia, which launched 3183. However, expanding commercial prospects gradually induced other countries to acquire the capacity to launch satellites. In 1973, Canada inaugurated the *Anik* satellite; France followed suit with the *Spot* and *Telecom* systems in the 1970s and 1980s, respectively; and in 1977, Italy launched *Sirio* and Japan its *Sakura* (Hudson 1990).

REGULATING THE GLOBAL SATELLITE INDUSTRY

International regulation has long been an important force shaping access to satellite technology, market structure, pricing and its uses over time (Smith 1987; Hudson 1990). The international regulation of telecommunications dates back to the International Telegraph Union in 1865, which later evolved into the United Nations' International Telecommunications Union (ITU). The United Nations' Outer Space Treaty, signed in 1967, constitutes the legal basis for extending international law into outer space, including principles governing its exploration and use for purposes of communication (Martinez 1985).

A key issue in international regulation concerns orbital 'parking spots', particularly the coveted 'geostationary gold', which are allocated by the ITU. The ITU as an arena in which satellite geopolitics unfold confronts a dilemma between states already occupying slots and newer entrants demanding their own, who often feel that early users enjoy an unfair advantage (Martinez 1985). Newer claimants, typically developing countries, call for abolition of the currently existing 'first come, first served' policy that favors economically advanced states. Such tensions reflect the fact that satellites have created a new resource – geostationary slots – but the world community lacks a means to allocate it with considerations in mind other than profitability. Because the total number of orbital slots is limited by the physics of rotation and problems of bandwidth spillover, the steady growth in satellites has generated increasingly crowded conditions, with fewer degrees of orbital arc between them. Within countries, national communications boards (in the case of the US, the Federal Communications Commission, or FCC) assign slots to private firms, generally via open auctions (Rothblatt 1982).

Closely associated with international regulation is the allocation of electromagnetic frequency spectra. This issue, the ITU's most successful area of jurisdictional competence, is accomplished through the World Administrative Radio Conference, which, through a series of international conferences and agreements, provides legal protection for the engineering lifetime of a satellite but does not allow states to enjoy unlimited control over a given frequency.

The ITU must also address the conflict between the technological capacities of satellites and national sovereignty that occurs when signal footprints exceed the borders of a target country (i.e. spillovers), the nature of which depends upon the size of the footprint and the capacity of receiving Earth stations. The globalization of satellite transmission of television and radio traffic also threatens to undermine national controls over information flows. This process has powerful cultural and ideological repercussions: for example, the world's largest media companies rely heavily on communications satellites to provide a largely homogenous diet of television and video programs around the world (Pool 1998; Negrine 2013). States that seek to restrict imports of foreign media – including much of the Muslim world and former Soviet bloc – have typically found it impossible to assert national controls over global flows of information beamed from above. Such 'externalities' point to the blurring of the distinctions between foreign and domestic policy that communications technology has accelerated in a period of rapid globalization.

Two distinct regulatory regimes may be identified in the political organization of the satellite industry. The first consists of the state-dominated system associated with Intelsat and complementary regional systems, which corresponded with the post-war Fordist boom, the Pax Americana and the Cold War. In the telecommunications industry this era witnessed national ownership of most telecommunications systems via government monopolies. The second, more recent, regulatory regime consists of the neoliberal, privatized and deregulated system of governance characteristic of the contemporary age of post-Fordist flexible production, one associated with the decline of Intelsat's monopoly, and mounting competition within the industry as well as from new technologies such as fiber optics. While no simple dichotomy can be said to exist, for each era contains elements of the other, these brief sketches serve to illustrate the complex ways in which the industry has evolved politically, economically and geographically. An examination of the organizations involved in regulating the satellite industry in each era sheds light on one of the multiple avenues through which terrestrial politics are projected to new heights, in this case quite literally.

By far the largest and most comprehensive organization involved in the regulation of global satellite traffic is Intelsat, a private organization headquartered in Washington, DC, with representatives from governments around the world (Kildow 1973; Akwule 1992). Intelsat is the largest and most comprehensive organization involved in the regulation of global satellite traffic, and owns and operates a fleet of 59 high-powered spacecraft in geosynchronous orbit over the Atlantic, Pacific and Indian oceans, far more than any other global or regional satellite system. Intelsat's large number (387) of international Earth stations (the only kind capable of handling international traffic) – more than 80% of the world's total – has long endowed it with near-monopoly status in this industry. The network boasts of being the only truly global satellite system. It is by far the world's leading provider of satellite services, including the transmission of one-half of all international phone calls (although this share has been rapidly declining). Today, Intelsat accounts for roughly 37% of global commercial satellite capacity, only 20% (or less) of the trans-Atlantic market and 12% (or less) of the trans-Pacific market.

In order to expedite the entry of private commercial providers in the satellite market, the US government initiated a regulatory framework conducive to the growing interests of private firms (Hudson 1990). In 1962, Congress passed the Communications Satellite Act, which established the privately owned Communications Satellite Corporation (Comsat), which enjoyed a monopoly over international satellite service to and from the US for the next generation (Galloway 1972). Comsat became the nucleus of Intelsat in 1964, with *Early Bird* providing its early fleet, and originally owned one-half of its shares; it remains the intermediary between Intelsat and the US government and telecommunications industry. Comsat policies are dictated by a collaborative effort of the FCC, State Department and National Telecommunications and Information Administration. The organization started with 19 signatories, all from the economically developed world, but excluding the Soviet bloc; today, 201 states are represented.

Intelsat is not regulated by any public body and is accountable only to its owner-members, which are stock holders (Hudson 1990). Political power is concentrated in the Board of Governors, in which voting power is proportionate to shares of ownership, with a dues structure similar to that of the United Nations. Not surprisingly, the US is the largest contributor, responsible for almost one-quarter of Intelsat's total revenues. Most international telephone calls (which generate two-thirds of its revenues) are routed through Intelsat's satellites, each of which carries tens of thousands of voice circuits, although international television is its most rapidly growing source of revenue. Other services include radio, integrated digital services, business-to-business services (e.g. orders, inventory control), financial data transmission, rural community services, news gathering and weather reports. Although it was initially erected to serve only international traffic, Intelsat began to offer domestic services as well, at Algeria's instigation in 1976. Today it remains by far the world's largest provider of domestic (as well as international) satellite services through leased transponders, and provides a means to regulate orbits and frequencies as well as for international pooling of financial resources and sharing of risk.

Table 11.1   World's 20 largest satellite carriers, 2014

Carrier	Headquarters	Satellites
Intelsat	Luxembourg	59
SES	Luxembourg	52
Eutelsat	France	37
Telesat Canada	Canada	13
Russian Satellite Comm.	Russia	11
Indian Space Research Org.	India	11
JSAT	Japan	8
Star One	Brazil	7
Arabsat	Saudi Arabia	6

Intelsat's infrastructure dwarfs that of any single state or organization. It owns and operates a fleet of 59 high-powered spacecraft in geosynchronous orbit over the Atlantic, Pacific and Indian oceans, more than any other network (Table 11.1). Intelsat's 433 Earth stations, which comprise 79% of the world's total capable of international traffic, give it a near-monopoly status. The distribution of Earth stations reflects international discrepancies in wealth and power between economically developed and underdeveloped nations.

While Intelsat's clients are primarily corporate, it maintains a non-profit side as well, a policy designed in part to ameliorate concerns among less developed nations that the organization serves only the interests of the West. The Intelsat signatory agreement calls for global averaging of prices between high-traffic routes to subsidize less profitable ones, a measure intended to ease the costs on poorer nations; thus all nations pay the same price, regardless of the actual service costs, an attempt to provide ubiquitous coverage (Akwule 1992). Project Access, started in 1987, grants free use of its satellites for educational, health and humanitarian communications. Similarly, Peacesat (the Pan Pacific Education and Communication Experiments by Satellite Project) provides health, education and research services over the Pacific Ocean using NASA satellites.

Because Intelsat's member states were allied with the US during the Cold War, it was long viewed by the Soviet Union as an instrument of US domination over the world telecommunications network (Hudson 1990). To counteract Intelsat's hegemony, the USSR developed its own system, Intersputnik, headquartered in Moscow. Although Intersputnik was proposed in 1968, it did not begin operation until 1974, and today consists of eight geostationary satellites (and several non-geostationary ones) and 32 Earth stations capable of international traffic. Traffic on the Intersputnik system, however, is only a fraction of that on Intelsat. During the 1980s, Intersputnik's members were limited to the USSR and client states in Eastern Europe as well as Iraq and Syria. With the collapse of the USSR in 1991, Intersputnik membership became more complicated as other nations sought access to its network, leading the US, Canada, Japan and China to join while the Central Asian republics dropped out. Thus, the cessation of the Cold War blurred conventional geographic divisions in the geopolitical regulatory order, producing new, more complex spatialities of belonging and exclusion.

In contrast to Intersputnik, which was intended to rival Intelsat, three other, smaller satellite systems were erected to complement the organization. Concerned that prevailing networks provided insufficient coverage of maritime regions, the International Maritime Organization, the UN agency concerned with safety at sea, initiated Inmarsat in 1979 to provide maritime satellite services (Doyle 1977). Because maritime communications are a low-volume and hence low-revenue market, they had been ignored by Intelsat, which catered to more lucrative corporate interests (Frieden 1996). Headquartered in London, Inmarsat began as a non-profit cooperative of 28 countries, but in 1998 was privatized as Inmarsat Ltd. Its network consists of 12 geostationary satellites and 23 Earth stations that assist aviation and shipping firms, governing, among other things, distress calls, ship-to-shore communications, ship-to-ship telephony, electronic data interchange and weather updates, and providing safety information about high-traffic regions and potential hazards. Much of its original maritime emphasis has given way to sales of services to governments around the globe.

The European Telecommunications Satellite Organization formed Eutelsat in 1977 to complement Intelsat services over the European continent and Russia, including television, telephony and data communications for computer networking. Coordinated by the European Space Agency and headquartered in Paris, Eutelsat wholesales its capacity (24 satellites and 23 Earth stations) to national carriers, which in turn sell it to retailers (Hudson 1990). In 2001 it was privatized as Eutelsat Communications.

Developing countries also complemented Intelsat with their own regional systems. In 1976, the Arab League formed the Arab Satellite Communications Organization, or Arabsat, a regional satellite system. With only 21 members, six geostationary satellites and 18 Earth stations, however, Arabsat is relatively small. Political differences among Arab nations have hindered its growth: for example, Egypt was barred from membership between 1979 and 1989 because it signed the Camp David Accords. Given the authoritarian control exerted by many Arab governments over flows of information, frequent prohibitions exist against imports of foreign satellite programs deemed immoral or decadent. Other, similar systems include the Regional African Satellite Communications System (Rascom), and in South America, the Condor Alliance of Andean Nations (Frieden 1996).

This brief overview demonstrates the key role of the nation-state in the changing regulatory organization of the industry, the contradictions between state and market imperatives, their changing boundaries and relative significance, and the intractably geographical nature of the industry. Despite the fact that satellites were designed to annihilate space via instantaneous communications, uneven social and spatial access to satellite technology has been a key feature of the industry since its inception. While satellites may be popularly conceived as 'above politics', in fact they have long been inundated with it. This dimension, however, changed after the Cold War and the subsequent neoliberal triumph of corporate interests, when a new regime of accumulation and regulation came into being.

SATELLITES IN THE POST-COLD WAR ERA: THE NEOLIBERAL TURN

During the late twentieth century, marked by the global transition to post-Fordism and the hegemony of neoliberalism, the satellite industry underwent enormous changes, including worldwide deregulation, mounting competition within the industry as well as with other technologies and rising numbers of corporate providers and applications as market-driven imperatives came to dominate. This transformation included a steady shift in the provision of telecommunications services worldwide from the public to the private sector, in which they are less frequently provided as part of national efforts to serve the public interest and increasingly allocated on an ability-to-pay basis.

Deregulation was a fundamental aspect of the emergence of the post-Keynesian state in its efforts to accommodate increasingly hypermobile, internationalized capital. The deregulation of the US satellite industry began with the 1972 Federal Communications Commission 'open skies' policy, which started an era of privately owned commercial satellites (Kinsley 1976). At the end of the Cold War, American government restrictions on the sale of satellite data and images were gradually lifted. In 1997, limitations were eliminated on international satellite service providers operating in the US, previously restricted in the amount and type of services that they could provide. The deregulation of telecommunications began in 1984 with the American dissolution of the monopoly long held by AT&T; other countries followed suit with the privatization of state-owned monopolies such as British Telecom, Deutsche Telekom or Nippon Telegraph and Telecommunications. Indeed, telecommunications has been a key industry in the global triumph of neoliberalism, as corporate capital became increasingly reliant upon the movement of electronic information (Salin 2000). Compared with the telephone industry, which began deregulation in the 1980s, the deregulation of the American satellite market began relatively early. In 1972, the FCC announced its 'open skies' policy, allowing any party to apply for orbital slots, effectively initiating the era of privately owned domestic commercial satellites (Kinsley 1976). In 1979, FCC restrictions were lifted on licensing requirements, minimum antenna diameters and the rate ceilings charged by common carriers. In 1985, the FCC allowed a private satellite carrier, Orionsat, to compete directly with Intelsat for the first time (Jansky and Jeruchim 1987). In 1997, limitations were abolished on international satellite service providers authorized by the US, which were previously restricted in the amount and type of public switched services they could provide. At the global level, this change was paralleled by the gradual demise of the Intelsat monopoly, the growth of new technologies such as fiber optics, and the proliferation of private carriers (Demac 1986). Thussu (2001) argues that privatization and deregulation can only widen the discrepancy between the world's information have and have-not nations, including, for example, the cessation of global price averaging among routes of differential profitability.

Deregulation was implemented in the context of dramatically shifting market conditions. Given the explosion of the global service economy, including international finance and producer services, and the increasingly information-intensive nature of production in general, the demand for satellite services grew rapidly. International trade in services, which relies heavily upon satellites, constitutes roughly one-quarter of the world's trade by value and one-third of US export revenues. Accordingly, total satellite traffic grew rapidly in the late twentieth century. Today, the growth of international telephony, the Internet and the increasingly global market for television broadcasting are driving forces behind the growth of civilian satellite communications traffic.

The unhampered flow of satellite traffic across national borders wreaks havoc with traditional notions of national sovereignty, a theme repeatedly noted in the literature on critical geopolitics. As Achilleas (2002, 37) notes, 'Television by satellite involves high political and legal stakes because of two underlying principles long considered to be antinomic:

freedom of information and sovereignty'. Some states argue that satellite traffic interferes with their internal affairs, as when Malaysia and Indonesia complain that Australian television portrays their governments in an unflattering light. Without legal protection of their freedom to transmit internationally, satellite providers would be limited to national sectors of space. While such rights are generally protected via the 1967 Outer Space Treaty, in practice numerous governments attempt to limit flows of information across their borders using signal interference. As privatization ended the monopoly that governments held for four decades on orbital espionage, many countries became fearful of the political and social repercussions of diffused surveillance capacities. For example, private Earth stations are banned in homes in China, Malaysia and Singapore, with varying degrees of success.

Globally, deregulation and enhanced competition took the form of the erosion in the quasi-monopoly position held by Intelsat. Indeed, distinct if imperfect parallels can be drawn between AT&T's status domestically and Intelsat's hegemony internationally. Some critics allege that the lack of competition in the satellite industry is to blame for its recent woes in light of the rise of fiber optics. In the post-Cold War era, Intelsat has been faced with two major sources of competition. The first stems from national satellite systems erected by individual nations. Indonesia, the first developing nation to use satellites for domestic connectivity, launched several generations of its *Palapa* ('Unity') satellites in the 1980s to provide services to all 27 provinces, and recently began to sell them to neighboring ASEAN countries as well. India initiated the Indian National Satellite System (Insat), primarily to provide low-cost educational services to remote villages, and in 1981 launched its first indigenous telecommunications satellite, the *Ariane Passenger Payload Experiment* (*APPLE*). In the 1990s India put into orbit three Insat satellites, thus joining the small group of nations that not only build but also launch these devices. Brazil's BrazilSat system leased its first satellite in 1974 and eventually launched three more, expanding its domestic network with 21 Earth stations, 17 of which are located in the Amazon River Basin. It has been supplemented by several private satellite providers. In the same vein, Mexico, China, France, Turkey and Thailand all initiated national satellite systems, some with their own launch capabilities. Unlike complementary networks such as Inmarsat or Eutelsat, national satellite systems are designed explicitly to bypass Intelsat.

The second source of new competition for Intelsat came from private satellite companies. Given the high fixed costs and barriers to entry into the satellite industry, particularly launch costs (Jussawala 1984; Snow 1987), few private firms took early advantage of the openings provided by deregulation. However, as entry costs have declined, the industry has become steadily commercialized (Achilleas 2002). Orionsat first filed an application with the FCC to compete with Intelsat in 1983, and had two satellites in operation by 1994, serving the trans-Atlantic market. The second private provider, Panamsat (Pan-American Satellite Corporation), was founded in 1988, with one satellite, by president Reynold Anselmo to 'bust open' the Intelsat monopoly in Latin America and the Caribbean, and became the first to challenge Intelsat on a global basis (Hudson 1990). The financial risks associated with such ventures are aptly illustrated by Motorola's Iridium venture, designed to take advantage of the explosive cellular phone market, which in 1999 was forced into bankruptcy by its clumsy and cost-ineffective technology, showing that the sky is not the limit. Nonetheless, the largest fixed satellite service providers, including Intelsat, reveal a diversity of companies with substantial revenues (Table 11.1). Aerospace firms such as Boeing have been key players in this process, frequently through consortia such as Skybridge and Teledesic. Lockheed Martin, for example, purchased Comsat for $2.7 billion in 1998, in the process acquiring 24% of Intelsat stock; it also purchased 15 geostationary orbital slots from Intersputnik. These systems represent the further intrusion and dominance of market imperatives in what had once been a sector aligned purely along the prerogatives of national security.

COMPETITION WITH FIBER OPTICS

Today, two technologies – satellites and fiber optics lines – almost exclusively constitute the only technologies deployed by the global telecommunications industry. The transmission capacities of both of these modalities grew rapidly in the late twentieth and early twenty-first centuries, a reflection of several intertwined trends in the global economy, deregulation, enormous technological change and the shifting nature of demand. Firms engaged in international traffic frequently employ both technologies, often simultaneously, either in the form of leased fiber circuits from shared corporate networks constructed by consortia or leased time from satellite companies.

Although they overlap to a great extent, satellite and fiber optics carriers exhibit an important degree of market segmentation. Economically, both reflect the typical cost structure of telecommunications, i.e. high fixed costs and barriers to entry and low marginal costs. However, firms offering these services serve overlapping, but slightly different markets: satellites overwhelmingly dominate mass media transmission, although fiber carriers have recently begun to invade this market (e.g. cable television). Fiber carriers are heavily favored by large corporations for data transmissions and by financial institutions for electronic funds transfer systems, in part because of the higher degrees of security and redundancy this medium offers (Warf 1995). Telecommunications markets that can be equally well served by both technologies, such as telephony and Internet traffic, are the focus of particularly intense competition. Television, long transmitted via satellite, has increasingly migrated to fiber (i.e. cable TV) and now comprises 25% of the traffic in that sector.

These two types of carriers are differentiated geographically as well: because their transmission costs are unrelated to distance, satellites are optimal for low-density areas such as rural regions and remote islands, where the relatively high marginal costs of fiber lines are not competitive. In contrast, fiber optics carriers prefer large metropolitan regions where dense concentrations of clients allow them to realize significant economies of scale and where frequency transmission congestion often plagues satellite transmissions. Satellites are ideal for point-to-area distribution networks, whereas fiber optic lines are preferable for point-to-point communications, especially when security is of great concern. In short, satellite and fiber optics services are quasi-substitutable, serving overlapping but not identical sectoral and geographic markets.

Satellites have been increasingly marginalized because of rapid technological changes in fiber optics. The introduction of integrated services digital networks greatly facilitated this process, as did the emergence of TCP-IP protocols and packet switching for the Internet. New repeaters introduced in the early 1990s, which pick up, amplify and transmit signals every 50 miles, further reduced costs and improved transmission quality. Optical fiber transmissions were greatly enhanced by the development of optical amplifiers, which simplified the electronic repeater, increased reliability, and by dense wavelength division multiplexing, making it possible to transmit multiple wavelengths over a single pair of fibers and raising maximum transmission capacities from 280 megabits per second in the 1980s to 12 terabits by 2008, a 42,800% increase. Moreover, the number of strands per fiber cable increased from two to eight pairs. The FLAG Atlantic-1 cable, which entered commercial service in June 2001, alone almost doubled trans-Atlantic fiber capacity. As a result, today, fiber optics carriers carry 95% of world-wide transmission capacity (up from 16% in 1988).

The satellite industry responded to the challenge posed by fiber with its own technical changes, including on-board processing, digital compression, dynamic routing through the satellite mesh, multibeam antennas and cheaper Earth stations. Even the industry's 800 pound gorilla, Intelsat, was forced to drop its prices significantly. Despite these changes, satellites failed to match the latest leaps in fiber optics capacity and can compete with transoceanic submarine cables only with enormous difficulty. As their competitive edge has eroded, satellite providers have been steadily forced to serve markets in low-density regions, relatively low-profit arenas compared with the lucrative high-volume, corporate data transmissions market.

The satellite industry has also staked much of its future on the rapid growth of wireless technologies and cellular phone markets, which have fostered the widespread deployment of small, low-orbiting satellites. The rapid growth of cellular traffic, however, has encouraged the deployment of small direct broadcast satellites positioned only a few hundred miles high in non-geosynchronous orbit, oriented to very small antennae and possessing relatively small footprints: when microwave signals are sent great lengths and become broadly diffused, as is the case in international traffic, Earth stations must have large and powerful antennas to receive them. Large numbers of cell phone users – more than 70% of the planet – have generated a steady growth in the demand for satellite services; today, three-quarters of all satellites launched serve the wireless telephony market. The growth of satellite television has also fueled the growth of direct broadcast satellites. Providers of such services, such as SkyTelevision, DirecTV, EchoStar and Sky Angel, are all privately owned corporations.

CONCLUDING COMMENTS

Satellites as both mechanisms of observation and communication reflect, and in turn feed back into, terrestrial politics in many ways. Born of Cold War rivalry, satellites played a key role in the militarization of space: the footprints of those in geostationary orbit create a spatial field of power subject to the gaze and manipulation of military planners. Although the military's role in the satellite industry has declined, it continues to remain an important segment distinct from civilian applications.

In civilian markets, satellites have long played a key role in international transmissions of voice, video and data traffic. Despite the enormous growth in capacity of fiber optics, which has come to dominate international telecommunications, satellites remain instrumental to the worldwide delivery of television and radio programs, particularly in remote regions, as well as the burgeoning wireless market.

To minimize externalities and allocate scarce resource such as prime orbital slots, global flows of information require international forms of regulation, which has occurred through a series of international and regional organizations, including the International Telecommunications Union. Martinez (1985, 131) notes 'To the extent that maldistributed access to the geostationary orbit – which favors the few space powers – mirrors the larger overall asymmetrical distribution of global power, the ITU is being politicized by technologically powerless states attempting to exploit a slightly more favorable decision-making structure.' The largest organization for the provision of international satellite services is Intelsat, traditionally perceived as a mechanism for the assertion of US hegemony over the industry during the Cold War. Several regional regulatory organizations (e.g. Eutelsat, Arabsat) testify to the continued local character of much of the industry.

In the context of digital neoliberalism, new national satellite systems and the worldwide deregulation of the industry have gradually eroded Intelsat's monopoly status and increasingly shifted control over the technology from the state to private capital, an important moment in the worldwide ascendancy of neoliberalism. The transition from instruments of military intelligence to commercial entities indicates the steady penetration of commodity relations into outer space. Throughout the world, long-standing government monopolies over access to international satellite facilities have given way to corporate carriers. This process extends into the commercialization of satellite imagery: with the transformation of the industry from agent of national security to privatized tool of profit-seeking firms, satellite images have increasingly become commodities in their own right. As the information collected and transmitted through satellites has increasingly fallen under a decentralized network of private agencies, the power of capital to allocate satellite resources has expanded while national security concerns, the industry's traditional *raison d'être*, have been progressively eclipsed, but not removed entirely.

Popular notions that telecommunications annihilate distance and render all places equally accessible may be dismissed immediately; despite claims that satellites provide equal levels of service over vast parts of the world's surface, in practice access to this technology is highly unevenly distributed among (and within) countries. In short, satellites, whether military or corporate, do not simply reflect the world's geopolitics, they are simultaneously *constitutive* of it, blurring the boundaries between Earth and space, the global and the local, the public and the private.

REFERENCES

Achilleas, P. 2002. Globalization and commercialization of satellite broadcasting: current issues. *Space Policy* 18: 37–43.
Akwule, R. 1992. *Global Telecommunications: The Technology, Administration, and Policies*. Boston, MA: Focal Press.
Burrows, W. 1986. *Deep Black: Space Espionage and National Security*. New York: Random House.
Cloud, J. 2001. Imaging the world in a barrel: CORONA and the clandestine convergence of the earth sciences. *Social Studies of Science* 31: 231–251.
Coombe, R. 1996. Authorial cartographies: mapping proprietary borders in a less-than-brave new world. *Stanford Law Review* 48: 1357–1366.
Cosgrove, D. 1994. Contested global visions: one-world, whole-earth, and the Apollo space photographs. *Annals of the Association of American Geographers* 84: 270–294.
Day, D., J. Logsdon and B. Latell (eds). 1998. *Eye in the Sky: The Story of the CORONA Spy Satellites*. Washington, DC: Smithsonian Institution Press.

Demac, D. (ed.). 1986. *Tracing New Orbits: Cooperation and Competition in Global Satellite Development*. New York: Columbia University Press.

DeVorkin, D. 1992. *Science with a Vengeance: How the Military Created the US Space Sciences after World War II*. New York: Springer.

Doyle, S. 1977. INMARSAT: the International Maritime Satellite Organization – origins and structure. *Journal of Space Law* 5(1): 45–64.

Frieden, R. 1996. The technology, law, and policy of international satellites. In *International Telecommunications Handbook*. pp. 187–206. Boston, MA: Artech House.

Gaffney, T. 2000. *Secret Spy Satellites: America's Eyes in Space*. New York: Enslow.

Galloway, J. 1972. *The Politics and Technology of Satellite Communications*. Lexington, MA: Lexington Books.

Giget, M. 1994. Economics of satellite communications in the context of intermodal competition. *Telecommunications Policy* 18: 478–492.

Goldstein, I. 1998. Stars of good omen: satellites in the global electronic marketplace. In D. Leebaert (ed.), *The Future of the Electronic Marketplace*. pp. 335–366. Cambridge, MA: MIT Press.

Hart, R. 1991. Orbit spectrum policy – evaluating proposals and regimes for outer space. *Telecommunications Policy* 15: 63–74.

Hudson, H. 1990. *Communication Satellites: Their Development and Impact*. New York: Free Press.

Inglis, A. 1991. *Satellite Technology: An Introduction*. Boston, MA: Focal Press.

Jansky, D. and M. Jeruchim. 1987. *Communication Satellites in Geostationary Orbit*. Norwood, MA: Artech.

Jussawala, M. 1984. The economic implications of satellite technology and the industrialization of space. *Telecommunications Policy* 8: 237–248.

Kildow, J. 1973. *INTELSAT: Policy-maker's Dilemma*. Lexington, MA: Lexington Books.

Kinsley, M. 1976. *Outer Space and Inner Sanctums: Government, Business, and Satellite Communication*. New York: Wiley.

Litfin, K. 1997. The gendered eye in the sky: a feminist perspective on Earth observation satellites. *Frontiers: A Journal of Women's Studies* 18: 26–47.

Mack, P. 1990. *Viewing the Earth: The Social Construction of the Landsat Satellite System*. Cambridge, MA: MIT Press.

Maral, G. and M. Bousquet. 2002. *Satellite Communications Systems: Systems, Techniques and Technology*. New York: Wiley.

Martinez, L. 1985. *Communication Satellites: Power Politics in Space*. Dedham, MA: Artech House.

Monmonier, M. 2002. *Spying with Maps: Surveillance Technologies and the Future of Privacy*. Chicago, IL: University of Chicago Press.

Negrine, R. (ed.). 2013. *Satellite Broadcasting: The Politics and Implications of the New Media*. London: Routledge.

Pelton, J. 2013. History of satellite communications. In J. Pelton, S. Madry and S. Camacho-Lara (eds), *Handbook of Satellite Applications*. pp. 27–66. New York: Springer.

Pool, I. 1998. Direct-broadcast satellites and cultural integrity. *Society* 35: 140–151.

Richelson, J. 1990. *America's Secret Eyes in Space: The U.S. Keyhole Spy Program*. New York: Harper and Row.

Richelson, J. 2001. *America's Space Sentinels: DSP Satellites and National Security*. Lawrence, KS: University Press of Kansas.

Rothblatt, M. 1982. Satellite communication and spectrum allocation. *American Journal of International Law* 76: 56–77.

Salin, P. 2000. *Satellite Communications Regulations in the Early 21st Century*. The Hague: Martinus Nijhoff.

Smith, M. III. 1987. The orbit/spectrum resource and the technology of satellite telecommunications: an overview. *Rutgers Computer & Technology Law Journal* 12: 285–311.

Snow, M. 1987. Economics of satellite communications. In J. Howland and J. Pelton (eds), *Satellites International*. London: Macmillan.

Stares, P. 1985. *The Militarization of Space: U.S. Policy, 1945–84*. Ithaca, NY: Cornell University Press.

Thussu, D. 2001. Lost in space. *Foreign Policy* 124: 70–71.

Warf, B. 1995. Telecommunications and the changing geographies of knowledge transmission in the late 20th century. *Urban Studies* 32: 361–378.

Warf, B. 2006. International competition between satellite and fiber optic carriers: a geographic perspective. *Professional Geographer* 58: 1–11.

Warf, B. 2012. Dethroning the view from above: toward a critical social analysis of satellite ocularcentrism. In L. Parks and J. Schwoch (eds), *Down to Earth*. pp. 42–60. Piscataway, NJ: Rutgers University Press.

12. The geography of mobile telephony
Jonathan C. Comer and Thomas A. Wikle

As a technological innovation, the cellular (or mobile) telephone has introduced dramatic change in the ways people interact. In a world of anywhere, anytime connectivity, communication has become personalized as calls to geographic places have been replaced by calls to a specific person. The utility and simplicity of mobile phone technology has contributed to its rapid growth. Nearly half the world's population owns or has access to a mobile telephone (GSMA 2015). In support of a growing number of mobile users are networks supported by hundreds of thousands of antenna sites scattered across cities and towns and increasingly pushing deeply into less accessible places. Mobile phone capabilities have also evolved. The newest generation of 'smartphones' offers the functionality of high-resolution cameras, MP3 players, GPS navigators and other devices. For the developing world, mobile connectivity offers Internet access where it has been limited or entirely absent. With its capacity to offer functions or services personalized to the needs of its user, the mobile phone has become a communications technology for the masses, ubiquitous in developed and less-developed countries and across social, cultural and demographic groups. In this chapter we explore the history, growth and impacts of mobile telephony on human patterns and processes with a focus on variation in subscriber densities across regions and countries.

Known in Germany as the *handy* and in China as the *shou-ja* (or hand machine), the mobile telephone began as a substitute for the traditional wired telephone. However, voice calling is just one of several functions supported by mobile phone technology as other phone functions have gained popularity. Today's phones are data portals, providing access to weather forecasts, stock market prices, traffic information and news. At the same time, text-based messaging (texting) and software applications (apps) have made mobile phones substitutes and in many ways replacements for desktop computers. In locations featuring third- or fourth-generation (3G or 4G) infrastructure, mobile phones support data-intensive functions ranging from interactive gaming to synchronous video communication and television or movie streaming.

Given their extraordinary rate of adoption, it may be useful to consider some ways in which mobile telephones have reshaped social and spatial patterns of communication. Extant literature suggests that mobile telephony's growth and expansion have influenced human society and behavior in at least four ways. First, the importance of physical place and distance has diminished (Graham and Marvin 1996; Plant 2001). Mobile devices lessen 'the need to tie particular activities to particular places at particular times' (Traxler 2011, 26). Communication previously tied to discrete nodes can take place nearly anywhere, including from within a moving vehicle.

A second impact can be seen in new forms of social interaction organized on the basis of common interests rather than geographic proximity (Graham 2000; Nyiri 2003; Adams 2005; Ling and Campbell 2009). Through voice calls, texting and social media access, mobile phones offer a substitute to face-to-face interaction. Peer-to-peer and peer-to-group exchanges can involve the exchange of photographs and videos transmitted in real time. As noted by Ling (2008), mobile communication has become more frequent and expressive. In addition, it has introduced new social maladies. For example, hyper-connectivity has contributed to anxiety among persons who are disconnected or out of contact with others, even for brief periods of time (Dery et al. 2014).

A third impact can be seen in increasingly blurred boundaries that previously separated home and work time and public versus private space. For some, mobile communication offers efficiencies, transforming the home-to-work commute and other idle periods into productive time for conversations with coworkers, clients and customers. Mobile connectivity has enabled work-related tasks to be carried out at the local coffee shop or while in line at the supermarket. However, 24/7 connectivity has invited job-related requests into family and personal time, lengthening the work day. Once-sacred evening, weekend and vacation time is increasingly subject to work-related intrusions (Cooper 2002; Traxler 2011). Mobility has also weakened boundaries that previously separated personal and public spaces. With the pay phone having all but disappeared, private conversations have invaded airport waiting areas, restaurants and hotel lobbies with libraries, museums and theaters responding to the scourge of electronic sounds and one-way conversations by posting signs limiting or forbidding mobile phone use.

Finally, mobile telephony has influenced interpersonal relationships and family dynamics. Face-to-face conversations are increasingly subject to interruptions from rings, chimes and beeps as social etiquette takes a backseat to an immediate need to answer a call or reply to a text message or social media posting. The need to remain connected is especially acute within younger generations, for whom a world without personal connectivity and communications mobility has never been known. For the generation known as the Millennials, mobile communication is a key element of personal identity and a focal point for socialization (Pew Research Center 2010). Among teenagers, mobile phone ownership represents independence and a venue for peer interaction outside adult control.

Along with reshaping social patterns of communication, mobile telephony and other wireless technologies have profoundly influenced the consumption and dissemination of information as newspaper and magazine publishers move away from print media in favor of the electronic distribution of stories and other news. In addition to text and photographs, news and magazine 'apps' offer short video clips often accessible after viewing a video-based advertisement. Along with consumption, mobile technology has impacted how news and other information are captured as photographs and videos taken by bystanders are increasingly featured in breaking news (Campbell and Park 2008). In some cases concealed phones have been used to record video evidence used in the prosecution of public officials and police officers.

Within countries with government-controlled media systems, mobile technology has contributed to a grassroots form of information sharing. Peer-to-peer or person-to-group exchange of information via text messaging has become a popular way to disseminate information in places having restrictions on the media or free speech. For example, mobile phones played a key role in 2001 protests when more than a million people took to the streets of Manila in opposition to Philippines President Joseph Estrada (Bociurkiw 2011), a phenomenon Rheingold (2003) has described as 'smart mobbing'. More recent examples of smart mobbing took place in 2005 on the eve of Spain's national elections, when it was rumored that government officials were withholding information about a terror attack (Suárez 2006), and when texting and voice calls helped in organizing demonstrations against unfair election practices in Kyrgyzstan (Howard 2010). In some countries mobile phones have offered a method for influencing policy, such as in China, where citizens can send text messages to members of the National People's Congress (Hachigian and Wu 2003).

EARLY MOBILE TELEPHONE SYSTEMS

A look back at early wireless communication offers a starting point for exploring the growth and diffusion of mobile telephone technology. The first wireless telephone calls were made more than 100 years ago when ship-to-shore radio enabled cruise ship passengers to place calls through the public telephone system. However, along with sound quality issues, the size and weight of ship-board radio equipment made the technology impractical for use in automobiles. The development of the integrated circuit following the Second World War offered new possibilities by reducing the size of radio components. In the early 1950s the first automobile radiotelephones (car telephones) were introduced in St Louis featuring three shared channels and a 50 mile range. Early car telephones were relatively unsophisticated, requiring the user to check for an open channel before asking a mobile operator for assistance in connecting to a wired telephone number. A subscriber paid a monthly service fee plus about 30–40 cents per call (US$3–4 in 2016 dollars). Despite the absence of privacy and the high cost for placing or receiving calls, the demand for car telephones grew rapidly, especially among metropolitan-based business users. During the late 1950s and 1960s car telephone systems spread to most large US cities and by 1964 subscribers could place outgoing calls without operator assistance.

Alongside expanding in US cities, car telephone systems became popular in European countries, with systems established in Sweden (1955), the UK (1959), West Germany (1959) and Finland (1970). By 1978 car telephone systems had been developed in much of Europe, including Austria, Spain, Luxembourg, the Netherlands, Italy and Switzerland (Garrard 1998). However, despite improvements in sound quality and declining operating costs, car telephone subscriptions remained expensive because of high demand and the limited bandwidth available for system expansion. For example, in the early 1960s more than 700 subscribers in New York City shared just 12 radiotelephone channels (Agar 2004). In addition, most car telephones could be used only within the geographic coverage defined by the network of radio repeaters located on high points such as tall buildings or nearby mountaintops.

In the early 1960s radio engineers began experimenting with methods that might reduce signal interference and increase radiotelephone subscriber capacity. Among the most promising concepts was a system calling for low-power (less than 1 Watt) mobile transceivers that could reach a network of geographically distributed base stations connected to the public telephone system. Communications zones around each base station formed overlapping areas of a few square miles, enabling frequencies to be reused among nonadjacent 'cells'. As a caller moved between cells, a computer handed off communications to the nearest base station on the basis of signal strength (Davis 1988). In 1973 American Telephone and Telegraph (AT&T) tested a cellular system in New York City and a short time later cellular networks were established in Chicago, Washington, DC and Baltimore. As subscriber numbers increased, cellular networks were extended along major transportation arteries, eventually reaching suburbs and smaller towns (Wikle 2002). Because of their size and weight the first cellular transceivers were restricted to use in vehicles. However, technological improvements soon brought reductions in bulkiness leading to the first truly portable telephone, nicknamed, the 'bag phone'. In 1983 Motorola unveiled its 28 ounce DynaTAC handset that was replaced by even smaller and lighter handsets produced by Samsung (Japan), Ericsson (Sweden) and Siemens (Germany).

Beyond mobility, cellular systems differ from wired telephone networks in important ways. Whereas the addition of each new subscriber to wired telephone service requires a physical connection, new cellular subscribers can be added by creating an electronic record. Networks can be expanded geographically through the addition of antenna sites, most often located on the tops of buildings in urban areas or mounted on poles or towers in suburban or rural regions (Wikle 2002). As subscriber density grows within an area, additional antennas can be added to expand the system's call capacity. Outside the United States, early cellular networks were developed in Japan, Sweden, Denmark, the UK, Norway, Finland and Saudi Arabia. Becoming operational in 1985, London's first cellular system featured a dozen antennas mounted on downtown office buildings and towers positioned along the M4 motorway.

WORLDWIDE DIFFUSION OF MOBILE TELEPHONY

As a consumer product, the mobile phone's popularity grew at an extraordinary rate. In just 35 years (1980–2015) networks were extended into nearly every populated region on Earth, including African countries with low per capita income such as Congo (Puro 2002; Comer and Wikle 2008). Whereas almost a century passed before the world reached a billion wire-lined phones, it took less than 10 years to reach the same number of cellular users (Donner 2008). A key factor in the growth and success of mobile telephony has been competition (Warf 2013). Whereas high infrastructure costs associated with wired telephone systems make it difficult for a geographic area to support more than one telephone provider, numerous and overlapping cellular networks in the same region are possible.

To understand the global expansion of mobile telephone technology it may be helpful to review terminology used for representing networks and subscriber density. The term 'provider' refers to public or private organizations that offer subscriber services while 'coverage area' is the geographic region of service as defined by infrastructure. Each provider offers access to infrastructure and services through a 'subscriber identity module' (SIM card) inserted within a phone or other device. Since each device (i.e. mobile phone, cellular computer modem, cellular-capable tablet computer) authenticates to a network through a separate SIM, it is possible for an individual to be associated with several subscriptions.

The ratio of SIM subscriptions to the population, or penetration, is a frequent benchmark for comparing mobile phone use across geographic areas. However, care is needed in the interpretation of penetration rates given the possibility of over-representing mobile telephone use in regions where multiple devices and subscriptions are maintained by a single subscriber. China serves as an example of a country with penetration rates that offer a misleading view of actual mobile phone use. Although having a penetration exceeding 100%, China's mobile phone users make up less than half (46%) of its population (ITU 2015). Actual user numbers are often lower because penetration doesn't account for persons within a population who wouldn't utilize a mobile phone or cellular-capable device (i.e. newborns, the very ill, incarcerated persons). In addition, penetration can underestimate the number of users where mobile phones are shared among two or more persons, a common practice in sub-Saharan Africa. Despite these limitations, penetration serves as a useful proxy for estimating the growth and expansion of mobile telephone use. Finally, the term 'depth' refers to features separating phones with basic talk and text functions from 'smart' devices that offer web-browsing, video streaming and other functions. Smartphones depend on the availability of third-generation (3G) and fourth-generation (4G) digital infrastructure. The United States offers an illustration of how depth can vary. About 90% of US adults owned a cellular phone in 2014 but only 64% of them were smartphones (Smith 2015).

MAPPING THE DIFFUSION OF MOBILE TELEPHONY

Previous studies that have looked at technology adoption and diffusion may be useful for characterizing the expansion of cellular networks and subscribers. For example, Rogers (1983) observed that the adoption of new technologies begins slowly following high development costs that eventually decline and level off as markets become saturated. Influences on technology adoption may include urbanization (Gatignon et al. 1989) and socioeconomic characteristics of populations, especially income (Takada and Jain 1991; Dekimpe et al. 1998; Rouvinen 2004; Donner 2008). Figure 12.1 provides a look at mobile phone penetration rates on a global scale using the United Nations M49 regionalization scheme. In 2005, mobile penetration rates were approximately four times higher in developed (82.1%) than in developing (22.9%) countries. However, while developing countries retained a clear lead in 2015, the gap narrowed considerably as a result of high growth rates in many developing countries (ITU 2015). Although a somewhat coarse measure of the digital divide, Figure 12.1 demonstrates that developing countries fall short of reaching connectivity levels common in the developed world, especially when taking into account infrastructure investments in fixed (land) lines. Compared with the average landline penetration in developed countries of 39%, developing countries achieved landline penetration rates of just 9.4% (ITU 2015).

Mobile telephone penetration can be illustrated through a regional look which shows that Europe's substantial lead in 2005 had disappeared by 2015 (Figure 12.2). Compared with Europe's penetration rate of 120%, the Commonwealth of Independent States (CIS, including Russia and former Soviet republics) reached a penetration rate of 140% over the same time period. Penetration rates in the Americas and Arab States were even lower (108%) than in Europe with Africa and Asia having the fewest subscriptions per capita with 2015 penetration rates below the world average of 96.8% (ITU 2015). It should be noted that some discrepancies in penetration rates may be the result of broad country groupings used by the ITU (2015).

Table 12.1 offers a summary of region-level indicators and their variability. Europe's relative homogeneity stands out with the lowest maximum, highest minimum and smallest range (and smallest interquartile range, or IQR) among its members coupled with the

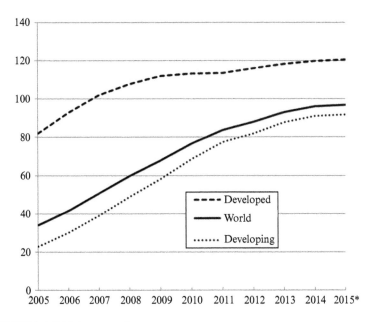

Source: ITU (2015). * 2015 data are estimates.

Figure 12.1   Cell phone penetration in the developed versus developing world, 2005–2015

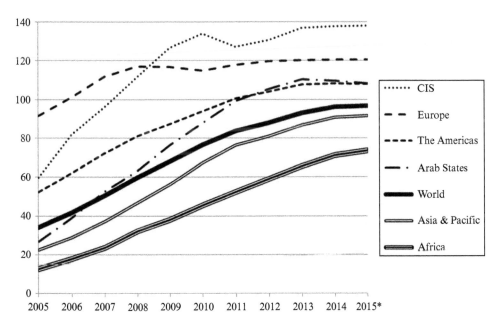

Source: ITU (2015). * 2015 data are estimates.

Figure 12.2   Cell phone penetration by world region, 2005–2015

Table 12.1   Cell phone penetration by world region, 2005–2015

Region	Maximum	Median	Minimum	Range	IQR
Asia and Pacific	322.6	92.3	11.2	311.4	61.8
Africa	210.4	72.3	6.4	204.0	45.4
Americas	179.8	106.6	22.5	157.3	48.9
Arab States	218.4	114.3	32.4	186.0	85.7
CIS	168.6	123.7	73.8	94.8	33.3
Europe	163.0	117.5	82.6	80.4	23.9

Source: ITU (2015).

second-highest median. At the same time, significant growth within Asia and African regions is revealed through large ranges; the highest and lowest penetration rates in Asia belong to Macao (China) and the Democratic People's Republic of Korea (North Korea). In Africa, Gabon leads with 210.4% while Eritrea has among the world's lowest penetration rates with just 6.4%. In 1999 only a few European countries had penetration rates exceeding 50%. By 2014 Europe, the United States, and some Arab states had reavched penetration rates exceeding 100% while those in Africa and Asia grew to 69 and 89%, respectively (ITU 2015).

Maps offer a method for visualizing the changing spatial patterns of mobile telephony. Figures 12.3–12.6 represent snapshots of penetration rates at four points in time: 2000, 2005, 2010 and 2014 (the last year of available data at the country level from the ITU). In lieu of maintaining the same range values for each year we elected to organize the data by quintiles. This facilitated comparisons showing each country's relative position or ranking at various times. Figure 12.3 reveals the strong advantage held by Europe and the Americas in 2000 and the very poor connectivity in northern and central Africa. Australia, New Zealand, Japan and South Korea stand out in the highest category. In comparison, Figure 12.4 reveals a shift in mobile telephony dominance between 2000 and 2005, showing a decline within the Americas relative to other regions with the United States, Canada and Mexico each dropping by one quintile. In contrast, CIS states and some African countries improved their relative positions. Although not visible on the map, the Cayman Islands experienced the largest percentage increase in penetration, moving from 25.7 to 166.5% and representing the world's highest penetrated rate in 2005. Twenty-one other countries also exceeded 100% in 2005, including several Caribbean nations, some European countries, and China's provinces of Hong Kong and Macao.

As shown in Figure 12.5, penetration rates continued to increase between 2005 and 2010 in Russia and the CIS while sliding further in North and South America. The United States, Canada, and Mexico again dropped to a lower quintile. At the same time, many European countries also experienced a decline with some falling out of the top 20%. With few exceptions, penetration rates in Africa seemed to stagnate during this period. Libya stood out as one of the few African countries to move into a higher category as its penetration increased from 35.7 to 180.4% between 2005 and 2010. The second largest gain was in Saudi Arabia, increasing from 57.4 to 189.2%. As shown in Figure 12.2, this is consistent with high growth rates associated with other Arab states during the late 2000s.

Figure 12.6 represents penetration rates for the last year (2014) for which country-level

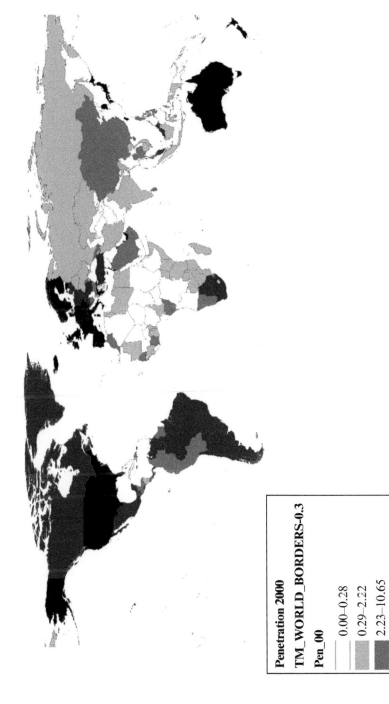

Penetration 2000

TM_WORLD_BORDERS-0.3

Pen_00

0.00–0.28
0.29–2.22
2.23–10.65
10.66–30.09
30.10–81.48

Source: ITU (2015).

Figure 12.3 Cell phone penetration by country, 2000

169

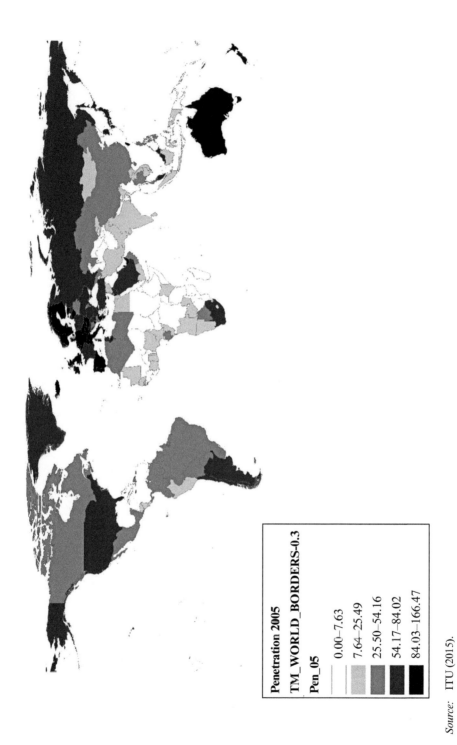

Penetration 2005

TM_WORLD_BORDERS-0.3

Pen_05

      0.00–7.63

      7.64–25.49

      25.50–54.16

      54.17–84.02

      84.03–166.47

Source: ITU (2015).

Figure 12.4   Cell phone penetration by country, 2005

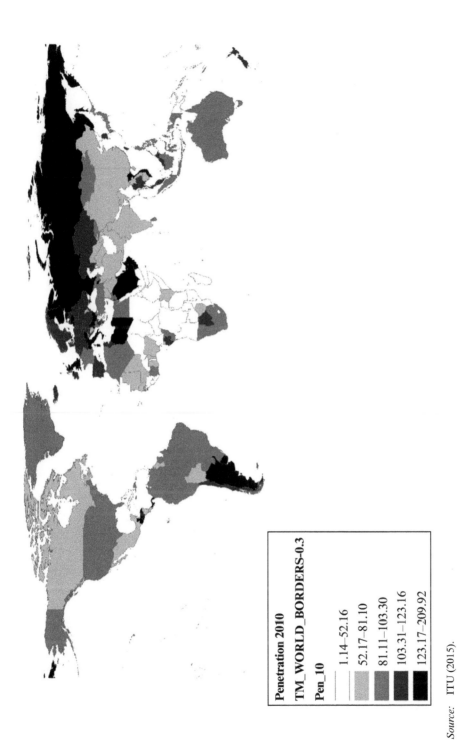

Penetration 2010

TM_WORLD_BORDERS-0.3

Pen_10

1.14–52.16
52.17–81.10
81.11–103.30
103.31–123.16
123.17–209.92

Source: ITU (2015).

Figure 12.5 Cell phone penetration by country, 2010

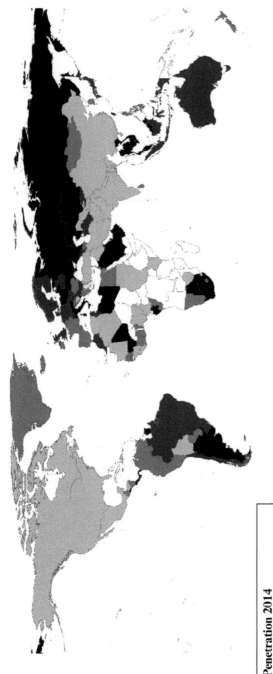

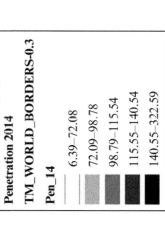

Penetration 2014

TM_WORLD_BORDERS-0.3

Pen_14

	6.39–72.08
	72.09–98.78
	98.79–115.54
	115.55–140.54
	140.55–322.59

Source: ITU (2015).

Figure 12.6 Cell phone penetration by country, 2014

data were available. Consistent with previous time periods, the United States dropped again, putting it into the second-lowest quintile along with Canada and Mexico. Most European countries remain in the second and third quintiles behind Russia and many CIS countries. Remarkably, some African countries jumped into the highest category, notably Botswana, Gabon and Mali. Within Asia, Macao experienced the most significant gain between 2010 and 2015, remaining the country with the world's highest penetration (322.6%).

FACTORS INFLUENCING RATES OF PENETRATION

As demonstrated in our analysis of penetration rates, the spread of mobile networks followed a different pathway when comparing developed and developing countries. For example, mobile telephony experienced early and rapid growth in European countries. Initiated in 1981 by Nordic Mobile Telephone (NMT), Europe's first cellular system brought direct dial calling to subscribers in Sweden, Norway and Finland. Five years later cellular networks were established in other Western European countries including Germany, Italy, the UK and France. With the exception of NMT's system that offered roaming, early European networks were plagued with compatibility problems preventing handsets from being used outside of country-based calling areas. For example, subscribers to the UK's Total Access Communications System were not able to initiate or receive calls on Italy's Radio Telephone Mobile System or France's RadioCom network. As a result of problems, European nations elected to migrate towards a digital system called Groupe Speciale Mobile (GSM) with continent-wide compatibility and six times more caller capacity than analogue technology being used in the United States. By 1993 GSM service was available in 22 Western European countries.

Compared with Western Europe, much of Eastern and Central Europe had a relatively poor communications infrastructure in the late 1980s. After the fall of Communism, former Soviet Bloc countries such as the Czech Republic, Poland and Hungary needed significant upgrades to telephone systems. In lieu of building additional wired infrastructure, telephone providers in these countries elected to construct third-generation (3G) cellular networks that were in many cases more advanced than systems used in elsewhere in Europe (Singh 1999; Steinmueller 2001), a process that has been described as 'technological leapfrogging'.

Despite an early role in developing mobile telephony, US cellular penetration lagged behind much of Europe and Asia. Contributing to slow growth was a substantial investment in wired telephone systems and a perception that mobile phones were a convenience to supplement, rather than replace, existing wired telephone services. Another factor was that the United States has large regions with relatively low population densities and few potential subscribers. In recent years US providers have significantly increased coverage within most rural areas while upgrading infrastructure to support 4G services. As a result of improved coverage and services, a growing number of US households have elected to abandon their landline telephones.

MOBILE TELEPHONY IN ASIA, LATIN AMERICA AND AFRICA

Compared with countries where mobile telephone systems were added in the presence of well-established wired telephone systems, the roll-out of mobile telephone systems in developing countries brought connectivity to places where it had previously been available only to the wealthy. For example, until the late 1980s it was rare for Chinese or Indian households to have telephone service with most calls made in locations having little privacy such as post offices. In recent years an expanding middle class in China and India has placed new demands on communications services. In 1987 China's first cellular network was established in Beijing and competition among mobile telephone providers began in 1994. As noted by Yu and Ting (2003), China's adoption of the cellular phone has become a symbol of the country's transition from a state-run to a market economy (Kshetri and Cheung 2002). By 2005 the number of Chinese cellular subscribers had reached nearly 400 million persons (ITU 2015). As in China, India's emerging middle class and its high population density facilitated rapid growth in cellular subscriptions. Mobile phones became tools enabling farmers to access information about crop prices and fishermen to find the most profitable markets (Rai 2000). With the acquisition of handsets sometimes serving as a barrier to mobile phone use, China and India have become destinations for used phones imported from Europe and the United States with an emerging cottage industry of local shops specializing in handset repair and reconditioning.

Although it was among the first to develop a cellular network, Japan's mobile telephone growth was initially slow because of the absence of competition and a requirement that subscribers lease mobile handsets from the country's cellular monopoly, Nippon Telegraph and Telephone (NTT). NTT's monopoly ended in 1994 with the entry of two new providers, Idou Tsushin and Daini. On a worldwide basis, Asia dominates smartphone penetration (all over 70%). Asian countries with the largest share of smartphones include the United Arab Emirates, South Korea, Saudi Arabia and Singapore (ITU 2015).

Latin America countries have a long history of poor-quality wired-telephone systems and long waiting lists for services (Lapuerta et al. 2003). Cellular systems first became available in Latin America in 1987 with most providers emulating the US analogue standard. In contrast to other regions, the possibility for competition did little to ease demand since most Latin American governments were unwilling to challenge long-standing wired telephone monopolies. By 1995 the average penetration rate within South American countries was just 1%. The introduction of competition in the late 1990s eventually lowered costs. For example, when the government of Chile allowed competition in 1998, prices dropped 40% and subscriptions doubled.

Mobile telephony in Africa followed a somewhat different pattern of expansion. In many African countries wired telephone penetration remains low (5%), not because of a lack of interest, but because of high infrastructure costs, especially in rural areas. The continent's first cellular networks were established in North African countries such as Tunisia (1985), Morocco (1988) and Algeria (1990). Elsewhere, implementation was slow. By 1992 only one cellular network was operating in Central Africa (Zaire) and by 1996 fewer than half of African countries had mobile telephone networks. As of 1999 only

10% of Africa's population, located mostly in the north and far south, had cellular coverage (Aker and Mbiti 2010).

Network expansion in many African countries has been hampered by difficulties in transporting equipment through remote or dangerous areas and by theft or vandalism (Garrard 1998). Across the continent wired telephone systems are also absent or affected by poor maintenance and unreliable sources of electrical power. In remote locations generators are often the only source of power for tower sites. Given the need for competition in lowering costs, some African governments have asked competing providers to share towers and other infrastructure. In a few cases governments themselves have been barriers to expansion where mobile telephone networks are taxed at higher rates compared with other services as a means of achieving short-term revenue goals.

The introduction of mobile telephone technology has played an important economic role in many African communities. For example, previously unconnected villages have benefitted from telephone, text and Internet services that enable business owners to gain access to commodity prices (James 2002; Aker and Mbiti 2010). For many Africans, the costs of purchasing a cellular handset and the difficulty of gaining access to credit have been barriers to cellular telephone use. To address this issue some companies have introduced low-cost handsets. For example, in 2015 Nokia began offering a US$30 handset featuring a color screen while Google unveiled an Android-based smartphone costing less than US$100 (Sahota 2014).

In contrast to much of the world where they are personalized communication devices, mobile phones in rural areas are often shared within family groups or even within villages. The phones themselves have introduced benefits beyond person-to-person communication. In countries without a well-functioning system of banks and credit, mobile money systems have evolved as alternatives to paying for goods and services with cash. For example, Kenyans use their mobile phones to pay their electricity bill, purchase bus tickets or shop at supermarkets. In other cases phones have become important tools for gaining access to health care. For example, a service in Tanzania and Nigeria called 'Mobile Baby' enables medical practitioners to utilize their mobile phone for sending ultrasound images and video clips from remote locations to physicians based in cities.

Despite progress elsewhere on the continent, much of sub-Saharan Africa remains outside cellular coverage. In 2014 sub-Saharan countries had an average penetration of just 39%. Some challenges to expansion include the relatively low density of potential subscribers and high maintenance costs. However, in spite of barriers, subscriptions in sub-Saharan countries continue to expand. In Kenya mobile telephone penetration grew 73.4% between 2010 and 2014. The growth in subscriptions was even higher in other countries such as Cameroon (75.1%) and Nigeria (77.8%). Remarkably, cellular systems operate even in African countries that have experienced civil unrest or that lack a stable central government such as Somalia, Liberia and the Congo.

CONCLUSION

Car telephone systems introduced in the United States, Japan and Europe in the 1950s served as precursors for today's mobile telephone networks. Like other communications technologies that have spread rapidly such as the printing press, wired telephone, radio and television broadcasting, and the Internet, the mobile telephone's impact on human patterns and process has been extraordinary. In less than 35 years mobile telephone technology has become available to half the world's population. In contrast to the diffusion of other technologies, such as the personal computer and Internet, the mobile telephone has expanded within populations that include men and women, younger and older persons, educated and illiterate populations, and within developed and less-developed countries. Most important, mobile telephones continue to evolve in form and function, adapting to new roles and offering an expanding range of services and benefits.

Driven by unique social and political circumstances, mobile telephone network and subscriber growth patterns have followed different pathways within developed and less-developed countries and regions. Wealth alone cannot explain the expansion of subscriptions as penetration has grown rapidly within countries with low per capita income. Competition has played a key role in the development and expansion of networks as users evaluate providers on the basis of coverage area and the quality, reliability and cost of services provided. Today, many developing countries have more technically advanced systems than their developed counterparts as first- and second-generation networks are bypassed in favor of more advanced systems. Within the developing world mobile phones have helped broaden trade networks while reducing costs for small businesses. At the same time they have played an important role in addressing the so-called 'digital divide' by offering Internet access in places where broadband networks have been absent. In spite of mobile telephony's rapid expansion throughout much of the world, cellular networks have yet to be established in a few isolated places. For example, fearing the impact of news and information that may challenge state positions, the leadership of North Korea maintains a moratorium on establishing a mobile telephone network for its own population. Not to be dissuaded, some North Koreans have purchased black market handsets to reach cellular towers across the border with China.

REFERENCES

Adams, P. 2005. *The Boundless Self: Communication in Physical and Virtual Spaces*. Syracuse, NY: Syracuse University Press.

Agar, J. 2004. *Constant Touch: A Global History of the Mobile Phone*. Cambridge: Icon Books.

Aker, J. and I. Mbiti. 2010. Mobile phones and economic development in Africa. *Journal of Economic Perspectives* 24(3): 207–232.

Bociurkiw, D. 2011. Revolution by cell phone. *Forbes.com*, 9 October 2010, accessed 13 January 2016 from http://www.forbes.com/asap/2001/0910/028.html.

Campbell, S. and Y.T. Park. 2008. Social implications of mobile telephony: the rise of personal communication society. *Sociology Compass* 2(2): 371–387.

Comer, J. and T. Wikle. 2008. Worldwide diffusion of the cellular telephone, 1995–2005. *Professional Geographer* 60(2): 252–269.

Cooper, G. 2002. The mutable mobile: social theory in the wireless world. In B. Brown, N. Green and R. Harper (eds), *Wireless World: Social and Interactional Aspects of the Mobile Age*. pp. 19–31. London: Springer.

Davis, J. 1988. Cellular mobile telephone services. In B. Guile and J. Quinn (eds), *Managing Innovation: Cases from the Services Industries*. pp. 144–164. Washington, DC: National Academy Press.

Dekimpe, M., P. Parker and M. Sarvary. 1998. Staged estimation of international diffusion models: an application to global cellular telephone adoption. *Technological Forecasting and Social Change* 57(1–2): 105–132.

Dery, K., D. Kolb and J. MacCormick. 2014. Working with connective flow: how smartphone use is evolving in practice. *European Journal of Information Systems* 23(5): 558–570.

Donner, J. 2008. Research approaches to mobile use in the developing world: a review of the literature. *The Information Society* 24(3): 140–159.

Garrard, G. 1998. *Cellular Communications: Worldwide Market Development*. Boston, MA: Artech House.

Gatignon, H., J. Eliashberg and T. Robertson. 1989. Modeling multinational diffusion patterns: an efficient methodology. *Marketing Science* 8(3): 231–247.

Graham, S. 2000. The end of geography or the explosion of place? Conceptualizing space, place and information technology. In M. Wilson and K. Corey (eds), *Information Tectonics: Space, Place, and Technology in an Electronic Age*. pp. 7–28. New York: Wiley.

Graham, S. and S. Marvin. 1996. *Telecommunications and the City: Electronic Spaces, Urban Places*. New York: Routledge.

GSMA. 2015. The mobile economy, accessed 13 January 2016 from http://www.gsmamobileeconomy.com/ GSMA_Global_Mobile_ Economy_Report_2015.pdf.

Hachigian, N. and L. Wu. 2003. *The Information Revolution in Asia*. Santa Monica, CA: Rand Corporation.

Howard, P. 2010. *The Digital Origins of Dictatorship and Democracy: Information Technology and Political Islam*. New York: Oxford University Press.

ITU. 2015. ICT statistics database. International Telecommunication Union, accessed 24 October 2015 from http://www.itu.int/en/ITU-D/Statistics/Pages/stat/default.aspx.

James, J. 2002. Information technology, transaction costs and patterns of globalization in developing countries. *Review of Social Economy* 60(4): 507–519.

Kshetri, N. and M.K. Cheung. 2002. What factors are driving China's mobile diffusion? *Electronic Markets* 12(1): 22–26.

Lapuerta, C., J. Benavides and S. Jorge. 2003. *Regulation and Competition in Mobile Telephony in Latin America*. Washington, DC: Inter-American Development Bank.

Ling, R. 2008. *New Tech, New Ties: How Mobile Communication is Reshaping Social Cohesion*. Cambridge, MA: MIT Press.

Ling, R. and S. Campbell. 2009. *The Reconstruction of Space and Time: Mobile Communication Practices*. New Brunswick, NJ: Transaction Publishers.

Nyiri, J. 2003. *Mobile Communication: Essays on Cognition and Community*. Vienna: Passagen.

Pew Research Center. 2010. Millennials: a portrait of Generation Next, accessed 13 January 2016 from http:// www.pewsocialtrends.org /files/2010/10/millennials-confident-connected-open-to-change.pdf.

Plant, S. 2001. *On the Mobile: The Effects of Mobile Telephones on Individual and Social Life*. Schaumberg, IL: Motorola.

Puro, J.-P. 2002. Finland: a mobile culture. In J. Katz and M. Aakhus (eds), *Perpetual Contact: Mobile Communication, Private Talk, Public Performance*. pp. 19–29. Cambridge: Cambridge University Press.

Rai, S. 2000. In rural India, a passage to wirelessness. *New York Times*, 4 August, pp. C1–C3.

Rheingold, H. 2003. *Smart Mobs: The Next Social Revolution*. Cambridge, MA: Perseus Publishers.

Rogers, E. 1983. *Diffusion of Innovations*. New York: Free Press.

Rouvinen, P. 2004. *Diffusion of Digital Mobile Telephony: Are Developing Countries Different?* Research Paper 2004/13, the United Nations University World Institute for Development Economics Research.

Sahota, D. 2014. Africa gets smart: continent prepares for device revolution. *Telecoms.com*, 9 January 2014, accessed 13 January 2016 from http://telecoms.com/ opinion/africa-gets-smart-continent-prepares-for-device-revolution/.

Singh, J. 1999. *Leapfrogging Development? The Political Economy of Telecommunications Restructuring*. Albany, NY: State University of New York Press.

Smith, A. 2015. U.S. smartphone use in 2015. Pew Research Center, accessed 14 December 2015 from http:// www.pewinternet.org/2015/04/01/us-smartphone-use-in-2015/.

Steinmueller, W. 2001. ICTs and the possibilities of leapfrogging by developing countries. *International Labour Review* 140(2): 193–210.

Suárez, S. 2006. Mobile democracy: text messages, voter turnout and the 2004 Spanish general election. *Representation* 42(2): 117–128.

Takada, H. and D. Jain. 1991. Cross-national analysis of diffusion of consumer durable goods in Pacific Rim countries. *Journal of Marketing* 55(2): 48–54.

Traxler, J. 2011. The mobile and mobility: information, organisations and systems. In J. Pokorný, V. Repa, K. Richta, W. Wojtkowski, H. Ling, C. Barry and M. Lang (eds), *Information Systems Development*. pp. 25–34. New York: Springer.

Warf, B. 2013. Geographies of global telephony in the age of the internet. *Geoforum* 45: 219–229.

Wikle, T. 2002. Cell tower proliferation in the United States. *Geographical Review* 92(1): 45–62.

Yu, L. and T.H. Ting. 2003. Culture and design for mobile phones in China. In J. Katz (ed.), *Machines that Become Us: The Social Context of Personal Communication Technology*. pp. 187–198. New Brunswick, NJ: Transaction.

13. Streaming services and the changing global geography of television
Ramon Lobato

On 6 January 2016, in a presentation at the Consumer Electronics Show in Las Vegas, Netflix CEO and co-founder Reed Hastings announced that his $50 billion company had just become a global television service. With the exception of China, North Korea, Crimea and Syria, Hastings declared, Netflix was now unblocked and operating in every country of the world, ending years of patchy regional availability. 'Today', said Hastings, 'you are witnessing the birth of a new global Internet TV network. With this launch, consumers around the world – from Singapore to St. Petersburg, from San Francisco to Sao Paulo – will be able to enjoy TV shows and movies simultaneously – no more waiting.'[1]

For anyone interested in the geography of television, this was a significant development. Ever since 2007, when Netflix unveiled a limited streaming service for its US customers as a sideline to its mail-order DVD rental business, there has been much speculation about how the company would leverage the global distribution capacity of the Internet to bring over-the-top subscription television to international viewers. These rumours were confirmed when the Netflix streaming service began rolling out internationally, first to Canada in 2010, then to Latin America in 2011, to various European countries in 2012–2013, and to Australia, New Zealand and Japan in 2015. During these years the service also attracted many out-of-region users, who used VPNs (virtual private networks) and DNS (domain name system) proxies to covertly access the service before Netflix was available in their country.

What can the case of Netflix tell us about the changing geography of television? And what can existing scholarship on the geography of television tell us about Netflix? This chapter will draw out some of these connections between new and old forms of TV, from a media-geographical perspective. Streaming services are a hybrid phenomenon that brings together two historically distinct but now interrelated technological systems. Streaming uses the infrastructure of telecommunications and the Internet, yet from the audience perspective, streaming is very much an extension of the television experience, and in terms of its commercial operations it relies upon the existing networks and relationships of TV production and distribution. So in streaming, two technological systems – television and telecommunications – now interact to produce a hybrid system with its own spatial logics.

Using Netflix as a case study,[2] this chapter explores the geography of streaming services by focusing on three issues – infrastructure, distribution and platform space. The aim here is not to provide a comprehensive account of these areas, each of which has its own extensive technical literature, but rather to explain how key concepts used with the interdisciplinary field of media geography can be productively revisited and rethought for the streaming age.

THE GEOGRAPHIES OF TELEVISION

Television as a medium has long been understood in terms of its capacity to compress, enchant, fill, expand and otherwise transform space. The fundamentally spatial power of the medium is referenced in the nineteenth-century usage of the precursor term 'tele-vision' – the transmission of images at a distance – which was understood as one of a suite of visionary spatial-extension technologies alongside telegraphy and telephony (Hay 2004).[3] Since the 1950s, a rich tradition of critical writing has explored how television brings 'there' over 'here', realizes time–space compression and fosters globalization – all fundamentally geographic concerns. From Marshall McLuhan's (1964) philosophy of electronic media and Raymond Williams's (1974) theory of mobile privatization, through to the 'situational geography' of communications theorist Joshua Meyrowitz (1985) and the influential writing of communications geographer Paul Adams (1992), we now have many generative models for understanding television as a spatial medium.

In extending this line of research to streaming, there are two key issues to keep in mind. The first is that television is not a singular phenomenon but a bundle of different transmission technologies, industrial practices, textual forms and audience experiences. Each of these elements has its own spatial logic, so a meaningful analysis must be attuned to these specificities. This is a point made by Lisa Parks (2007, 114) who, in an essay on cable TV infrastructure, suggests that 'if television technology is a historically shifting form and set of practices, then it is necessary to consider more carefully how the medium's content and form change with different distribution systems'. For Parks, this task involves 'being able to understand and differentiate its distribution mode from others, whether these be broadcast, satellite, web, or wireless – systems that have themselves emerged in places in different ways' (Parks 2007). It is therefore helpful to disaggregate TV into its constituent technologies for the purposes of analysis. Streaming has a different spatial logic from broadcast, which is different again from cable, satellite or mobile television.

The second issue is that, from an analytical perspective, any single form of television will have multiple geographies simultaneously. These would include, but are not limited to, an infrastructural geography of signal transmission and reception (e.g. Starks 2013); an industrial geography of commercial relationships, alliances, and competition (Christophers 2009); an experiential geography of domestic, public and mobile viewing (Spigel 1992); and a cultural geography involving traffic in images, ideas and representations (Morley and Robins 1995). To fully understand television as a spatial medium, we must be attuned to the co-existence and interactions of these various dimensions. With these qualifications in mind, let us now explore what a media-geographical analysis of streaming services might reveal.

STREAMING INFRASTRUCTURE

A useful point of departure is to consider the material properties of TV transmission technologies, in terms of their spatial reach, restrictions and dynamics. This is what we might call the infrastructural geography of television, and it is important because it determines the availability of TV culture for dispersed communities and viewers around the world.

As Parks reminds us, each TV technology has its own unique spatial logic. For example, the reach of broadcast TV is naturally limited by the finite reach of radio transmitters and their interaction with the topography of the landscape. In contrast, satellite has different characteristics: the footprint of a broad-beam satellite can cover up to 40% of the Earth's surface, providing signals to an unlimited number of receivers, and overcoming terrestrial transmission obstacles such as mountain ranges and oceans. This entails a different kind of reception experience as well as a different political economy, characterized by high start-up costs and expensive hardware but formidable reach (including easy cross-border transmission). Attention to the specificity of TV infrastructure can help to explain the diverse evolution of industries and viewing cultures around the world – for example, why some countries including the United States, the Netherlands and Singapore are predominantly cable TV markets, while in the Middle East and most of Europe satellite dominates.

Infrastructure also has more subtle effects on TV culture, as it feeds into the structure of programming and, by extension, audience reception. For example, satellite's numerous business-to-business applications have changed the industrial logic of television by accelerating the development of particular formats, genres and modes of address: direct transmission of programming between stations, from big-city TV studios to regional affiliates, led to the rise of national networks; the ability to beam content direct to international content partners created international made-for-TV spectacles (Our World, Live Aid) and global news networks (CNN); and the transnational footprints of commercial satellite operators have had profoundly disruptive and in some cases cosmopolitanizing effects on national broadcast cultures (Sakr 2001; Parks 2005).

Following this line of inquiry, let us now consider some of the infrastructural characteristics of Internet television. Streaming platforms, as 'over-the-top' video delivery services, are naturally reliant on telecommunications infrastructure – the vast networks of fibre and coaxial cable, copper telephone wires and satellite data links that form the Internet's underlying foundation. As Internet geographers have shown (Zook 2006; Warf 2013), these various layers of infrastructure are vital to how we experience the Internet, if at all, and their spatial arrangement both reflects and entrenches longer histories of political–economic advantage and exclusion. The location and reach of ISPs; connection speeds and reliability; the variety of services available in particular locations; typical scenarios of access (wired connections, cybercafés or mobiles); regulatory control and censorship; the relations between communications infrastructure and urbanization – these are all essentially spatial issues.

For streaming video applications, which require high download speeds, location is especially important. As an example, consider Internet connection speeds. The minimum bandwidth required to use Netflix and other streaming services is 0.5 mbps (megabits per second), and the recommended level is at least 3.0 mpbs. While Netflix automatically adjusts its resolution level upward or downward to match a customer's bandwidth, it cannot operate functionally below these thresholds. However, given that many of the world's largest emerging economies, including Indonesia, India and the Philippines, have sub-3.0 average connection speeds (Akamai 2015) – and much slower average speeds and higher costs via mobile – this effectively puts Netflix off-bounds for most of the users in these countries.[4]

Taking this into account, one sees how the global expansion of premium streaming services is unlikely – at least in the short to medium term – to extend very far beyond the global urban middle classes. In this sense, premium video streaming services are fundamentally different from basic Internet protocols, such as email, which have greater flexibility and capacity to cater for mobile-first/mobile-only and low-bandwidth users. The need for credit card payments also restricts the diffusion of subscription services.[5] So there is a spatial as well as an economic logic at work in determining the scale and extent of streaming take-up.

A second issue to consider when thinking about Internet TV infrastructure is the location of back-end services. For example, private data centres play a vital role in online video because they handle the traffic of major platforms. YouTube traffic relies on Google's global network of 14 data centres across the United States, Western Europe, South America and Asia (typically located in areas with cheap and reliable electricity supply). Another vital infrastructure element for streaming platforms is Content Delivery Networks (CDNs), third-party networks of distributed servers that reduce latency in streaming video transmission to make the experience faster and smoother (e.g. by caching popular video content in local servers closer to the end-user to reduce load times and buffering). CDNs such as Akamai and Amazon CloudFront have global networks of servers, concentrated mostly in the global North, which they use for local storage of popular content. In this sense they constitute an invisible but vital local delivery system for streaming infrastructure, one which has its own geography of location and service-provision.

Netflix is an interesting case here because it has been developing a custom-built infrastructure tailored to its needs and using both in-house and third-party systems. In the past, Netflix used to maintain its own network of data centres but after a major crash in 2008 it has been progressively moving its services to the public cloud. Its last data centre was closed a few years ago and Netflix now uses Amazon Web Services instead. In contrast, the Netflix CDN strategy has been different: it used to partner commercially with Akamai, Limelight and Level 3 for CDN services, but since 2010 has been investing in its own CDN system called OpenConnect. So we can see here different tendencies of centralization and decentralization at the level of infrastructure, which are typical of the information technology industries.

For users, this network infrastructure is mostly invisible, or becomes visible only when it fails; for this reason, following infrastructure theory, we can say that it is powerful precisely because of this invisibility (Star 1999; Parks and Starosielski 2015). Back-end infrastructure also inevitably has its own spatial logics of inclusion and exclusion. For example, we can expect Netflix's investments in its CDN infrastructure to primarily benefit established and emerging markets rather than peripheral areas with few current or potential subscribers. Studying these infrastructural issues reveals how not only the pleasures of streaming but also the irritation of buffering, drop-outs and pixellation are all spatially organized.

DISTRIBUTION AND LICENSING

A second strand of research on television geography concerns content distribution. Scholars working in this area have studied international programming *flows* – how various kinds of TV content reach audiences in different places, and the institutional structures that determine these flows. From this perspective, power is not just a question of transmission – the availability of a signal and receiving equipment – but also what is *on* the television (Christophers 2009; Steemers 2014). In other words, we are talking about cultural questions of diversity, accessibility and local representation.

Early work on international television distribution – its volume, intensity and political economy – emphasized the one-way flow of content from the 'West to the Rest', in line with the dominant cultural imperialism thesis of the time (e.g. Boyd-Barrett 1977). Later work has been more attuned to intra-regional and multi-directional flows, including regional TV markets within Latin America and Asia (Sinclair et al. 1995; Iwabuchi 2002; Straubhaar 2007). Research on the less obvious forms of international distribution – such as the global trade in TV formats and franchises (e.g. Moran 1998; Waisbord 2004) – has added complexity to these debates by showing the importance of industries in Israel, Korea, the Netherlands and other second-tier exporter nations. Another line of research focuses on the location of television's commercial institutions, including distributors, programme markets and trade fairs (Scott 2004; Moran 2009).

This body of work generates productive exchanges between media studies and geography. For example, Michael Curtin's (2003, 2007) diverse body of research into 'media capital', undertaken in various contexts including China and the United States, investigates how geographies of production, talent mobility, distribution and regulation interact. Likewise, the economic geographer Brett Christophers's work on the political and cultural economy of international television explicitly addresses the relationship between distribution, geography, technology and capital in order to explain how television and its underlying power-regimes are 'accommodated and contested in materially geographic ways', and further, 'how programming *itself* materializes in place, and indeed how "place" materializes in programming' (Christophers 2009, 3).

One aspect of streaming media that brings these concerns into sharp focus is digital content licensing. Streaming services partner with established TV producers and distributors to license their content and make it available on the platform. The terms of these licences vary, and may include fixed or percentage-based payments, exclusive or non-exclusive terms, or advertising revenue splits. In most cases commercial content is licensed on a territory-by-territory rather than a global basis, in line with standard practice in international TV sales. From a geographic perspective, then, these deals are interesting because they highlight conflict between different scales and spatial logics of distribution. As Peter Yu has described, the Internet, as a global, decentralized network, is structurally transnational in nature and frequently finds itself 'in a collision course with the territoriality principle in intellectual property law' (Yu 2015). In contrast, television's industrial practices are fundamentally shaped around territorial licensing and national copyright law, and streaming services must therefore be retrofitted with the location-aware capabilities required to enforce these boundaries, including geoblocking and geo-filtering (Trimble 2012; Yu 2015).

Current debates about Netflix licensing give an insight into these spatial conflicts. While Netflix produces its own 'Netflix originals' (*House of Cards*, *Orange is the New Black*) and licenses some titles on an exclusive, global basis (e.g. *Making a Murderer*), the majority of its content is licensed from the major US entertainment conglomerates, including Disney, Time Warner, Comcast, Fox, Viacom and CBS. For these conglomerates, Netflix is just one more buyer – albeit a large and formidable one – and should acquire content territory by territory in the usual fashion. Therefore, while Netflix has been pushing hard for global licensing terms which would enable it to run a standardized global service, in practice it must usually negotiate the patchwork of territorial licensing deals like every other buyer, and is constrained by the deals done with other TV broadcasters who may have paid handsomely for exclusive rights in particular territories.

Netflix describes this situation as being a 'prisoner of territorial licensing' (USNetflix tweet, 6 January 2016). From the user perspective this means the Netflix streaming catalogue varies significantly from country, expanding and contracting and morphing when accessed from different parts of the world, with user access to overseas catalogues restricted through geoblocking. This is a source of controversy for its users, who increasingly expect global availability of their favourite content and find the territorial licensing system mystifying. This tension manifests itself in a range of everyday irritations, as when Netflix subscribers travelling abroad find that favourite series have disappeared from the catalogue, or when pre-release buzz about a show reaches fans in a territory where that content is not licensed, and thus is unavailable.

These catalogue discrepancies constitute a geographic vector of difference, meaning that the experience of streaming is likely to vary from place to place. For example, in the United States the Netflix catalogue has around 7000 mostly American titles; in the UK, it is smaller but well stocked with American, British and international programming (including BBC shows), but in many of the peripheral territories where Netflix has just begun operating the choices are significantly reduced, and local-language subtitles are not available. In some countries, Netflix is even missing its signature shows, such as *House of Cards*, having previously sold the rights to local TV channels (Sadaqat 2016). So this ostensibly global service is in reality very patchy and unevenly localized.

Thailand is one of the new Netflix territories that went live on 6 January 2016. So far, the reaction from Thais has been mixed (*Bangkok Post* and AP 2016). Netflix Thailand has disappointed many because it has very few Thai titles available, and there is no Thai interface support, which means users must be able to read some English. Yet, owing to the vagaries of international licensing, Netflix Thailand includes some premium content unavailable in the United States, such as *The Godfather* and *Better Call Saul*, which has pleased some viewers. There are also local regulatory priorities at work that will shape how the service is experienced within Thailand. At the time of writing, it is expected that Thai authorities will seek to censor or block some of Netflix's more racy content, which would likely fall foul of classification rules.[6] Alternatively Netflix may seek to 'follow cinematic practice by pixelating smoking, drinking and bloody violence, as well as censor nude scenes' (*Bangkok Post* and AP 2016). From this example we can see how an ostensibly global digital service is differentiated by multiple layers of industrial and regulatory contingency. From the end-user perspective, Netflix Thailand will be experienced as a rather different service from its other local iterations.

As global consumers become more familiar with streaming platforms, their awareness of the geographic disparities of content licensing and regulation grows. Through online forums, how-to sites and international catalogue comparator services (such as Flixsearch), streaming fans around the world are becoming more savvy about the limited nature of many local catalogues and the better options elsewhere. This has led to widespread take-up of circumvention and unblocking services, many of which are custom-built for accessing international streaming sites (Lobato and Meese 2016). In recent years an array of unblocking tools and workarounds – including hundreds of consumer-oriented VPN services (such as Private Internet Access, Hotspot Shield, Hide. Me, BlackVPN) and cheap DNS proxy services (Getflix, Unblock US) – have become popular with users as circumvention tools for accessing foreign Netflix sites, along with other kinds of geoblocked and government-censored content and services. While data on the use of circumvention services is unreliable, existing research paints a broad picture of substantial and growing global use (e.g. Roberts 2010; Global Web Index 2014).

As we can see from these examples, the distribution geographies of Netflix and other streaming services are complex and fast-changing. For our purposes there are two take-away lessons here. The first is that the cultural geography of video streaming is, for better or worse, inextricably linked to the economic geography of TV licensing and the legal institutions of territorial copyright. Any claim about the 'global' nature of Internet video would therefore need to be tempered with a more sober analysis of how these various geographies intersect, and the social costs and benefits.

Second, in looking at the issue of geoblocking, circumvention and proxy/VPN use, we can observe different degrees of *mobility* on the part of viewers, with a small number able to easily cross digital borders and access programming meant for viewers in other territories, and a much larger number remaining effectively immobile. Given the current popularity of circumvention tools – which are by now established as an integral part of the Internet TV landscape, and will remain popular with users so long as catalogue and pricing discrepancies between countries exist – questions arise as to who can use these tools and who cannot. In other words, differential access to digital mobility needs to be understood as closely related to class, education and technological competency.[7] This dialectic between mobility and immobility is likely to become a more significant feature of television culture in the years ahead.

PLATFORM SPACE

So far we have explored two aspects of streaming, infrastructure and licensing, and discussed their spatial logics. We have also begun to consider differences and continuities vis-à-vis older forms of television. In so doing, we have been investigating foundational structures associated with what Paul Adams (2007), in his taxonomy of communication geography, calls 'media-in-space' – flows of content across space.

However, there are other spatial dimensions of television, and of streaming, that we can analyse. One final dimension is what Adams calls 'space-in-media', or the way media experiences and technologies are themselves spatially organized. In Adams's words, 'communications are a kind of space – that is, a structured realm of interaction that both enables and constrains occupants in particular ways' (Adams 2010, 46), so it is helpful to consider how these communication and media-spaces are experienced from the user perspective, in terms of their internal spatiality. As anyone who has used Netflix, YouTube or Hulu will know, the interfaces of streaming services are organized in particular ways, using various kinds of motifs and elements designed as navigational aids – playlists, charts, categories, 'verticals', 'horizontals', and so on. As such, these design features invite particular kinds of movement around the site and particular modes of searching, discovery and access. The resulting 'platform spaces' require analysis because they have an agency of their own. No media-space is neutral. It is therefore important to consider how today's streaming platforms are spatially structured, and to what effect.

Various strands of thought in media scholarship are useful when considering this topic. Within television studies, for example, there is a long tradition of textual and ideological analysis that investigates how the experience of television viewing is organized around particular spaces of production, such as the TV studio, and their corresponding ways of seeing (think of the frontality of the sitcom set and the newsreader desk). As we move further into the digital age, these ways of experiencing televisual space are overlaid with the space of the platform – understood as a meta-structure that plays an active role in determining how we select, experience and understand the materials available on those services, which will in turn have their own internal spatial-aesthetic politics.

Geographers, including Kitchin and Dodge (2011), and media scholars, including José Van Dijck (2013), have written eloquently about the nature of digital platforms. Their work draws our attention to the structural capabilities and affordances of particular kinds of digital systems, determined at the level of code, and the way they shape user behaviour and subjectivity in subtle yet powerful ways. One of the recurring themes in this literature is the way that platforms work to classify and organize various kinds of content, making some things discoverable and other things obscure. As the cultural sociologist David Beer (2013: 62) writes in a critical synthesis of this literature, digital platforms present us with the need to consider 'how classificatory processes work to order culture on commercial, organizational, informal and everyday levels'. Central to these processes is algorithmic filtering, understood as a technique for shaping consumption in the context of complexity. As Beer (2013, 63) writes, 'we can only imagine the density of algorithmic processes and the complex ways that they are now a part of the ordering, structuring and sorting of culture'.

Consider the example of YouTube (Figure 13.1). When we visit the site or use a YouTube mobile app, we are presented with a carefully structured screen-space that encourages certain kinds of interactions. YouTube's design has evolved over the years to become less like a traditional TV interface and more like a customizable, interactive database tailored to the interests of the individual user. Its purpose is to provide a constantly updated smorgasbord of content to keep users coming back to the site, paying attention and generating advertising revenue. Across this platform YouTube and its parent company Google carefully prioritize certain kinds of strategically important partner content. So while the platform is a space of interaction, it is also a commercial space in which every inch of screen real-estate is valuable.

To the left of the YouTube landing page, in PC view, we find links to our Library, Subscribed Channels, and various other sections. At the top, a search bar invites us to enter a query. The rest of the screen is filled with video links, organized into categories such as Recommended, Watch It Again and Popular Right Now, and populated algorithmically with content according to our user history. These are arranged in horizontal layers which can be expanded or dismissed, so that the viewer can partially customize the space according to their own preferences. Because this content is all personally customized, no two unique users will experience the same YouTube front page. Like most streaming sites, YouTube has distinctly different interfaces for mobile, tablet and computer views, which adds to the chameleonic nature of the platform.

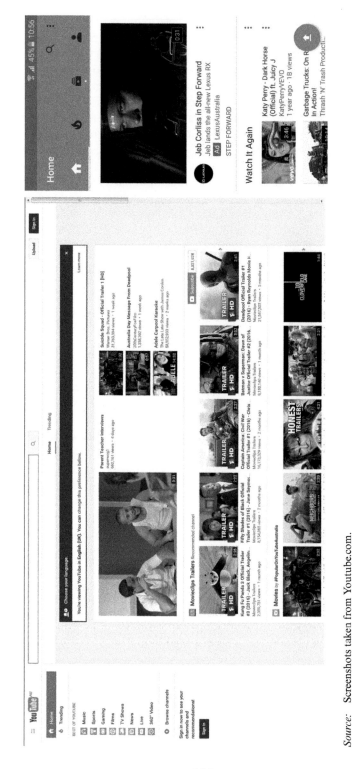

Source: Screenshots taken from Youtube.com.

Figure 13.1 YouTube platform space, on PC and mobile, as of January 2016

Design features like this are significant from a media-geographic point of view because they involve sorting, rearrangement and presentation of materials within the bounded but potentially infinite space of the browser window, tablet or mobile phone screen. While it has some parallels in media history, today's platform culture is in many respects unprecedented because of the level of personalization involved. Platform space raises new conceptual issues for researchers, who must constantly update and refine their approaches, for example, the modes of spatial analysis used in pioneering studies of cyber-space (e.g. Shields 1996). Users do not navigate contemporary streaming services in the same way as they moved through early MUDs, Geocities sites or Friendster. The norms, practices and design principles are constantly changing. We are in the realm of what Tarleton Gillespie (2010) calls 'the politics of platforms'.

Netflix again provides a rich case study to explore these issues, because its highly sophisticated platform design is so central to the user experience (Figure 13.2). Netflix is rightly famous for its personalization: every video selection that appears on the home-page is the result of intricate algorithmic calculation based on user-submitted data (movie ratings and viewing history), collaborative filtering (predictions based on other people's activities) and manual coding of films for all conceivable metadata points, from character types to endings (Madrigal 2014). As Amanda Lotz (2014, 74–75) notes, the recommendation algorithm has a structural purpose within the vast consumption-space of Netflix, in the sense it 'provide[s] subscribers with a different paradigm for thinking about and organizing viewing behavior, and one that substantially challenges the long dominant, linear, "what's on" proposition'. As Lotz argues, 'Personalized queues in combination with recommendation algorithms [become] valuable tools for *navigating an environment of post-network programming abundance*' (p. 79, emphasis added).

Drilling down into the Netflix interface, we can identify some design features that help to create this particular spatialized experience of abundance. Like YouTube, the Netflix interface is organized into horizontal strips of video links organized around particular categories, most of which are chosen on a user-customized basis from a list of thousands of potential genres, from broad categories (Romantic Comedies) through to hyper-specific micro-genres (Imaginative Time Travel Movies from the 1980s, Emotional Fight-the-System Documentaries; Madrigal 2014). This smorgasbord of content is arranged into movie-like reels of colour that slide off the right side of the page, suggest-ing an infinite variety of choices. In this way the viewer is positioned as the sovereign navigator-user of an endless archive of screen content. Such design choices are carefully constructed to conceal limitations in what is in fact a finite Netflix catalogue, as discussed above. Crucially, all these choices are informed by location, based on user-entered cus-tomer profile data and the IP address of the user. From the kinds of categories promoted on the front page (telenovelas in Mexico, but not in Malta), through to the range of inter-face languages, subtitles and soundtracks, location is an integral input into the platform's algorithmic calculations and its appearance to the user.

One interesting aspect of the Netflix platform space is how it frames the emergent practice of streaming by positioning it in relation to older screen technologies. Until 2015 the Netflix site had a light background and the video links were formatted in vertical, DVD cover-style boxes, so that the overall effect was reminiscent of a video store. Now, the background is dark – as in a movie theatre – and the DVD covers have been rear-ranged into a horizontal format suggesting frames on a celluloid film strip. This extensive

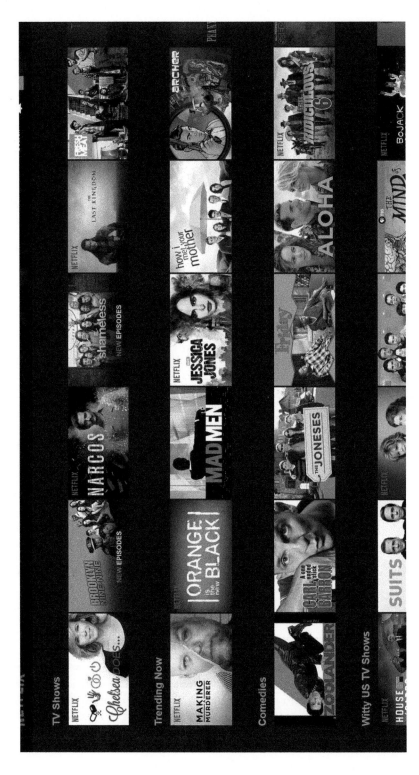

Source: Screenshot taken from Netflix.com.

Figure 13.2   The Netflix front page, organized into horizontal strips (below the fold view)

and expensive site update seems designed not only to make the service as tablet-friendly as possible, hence the shift to horizontal format, but also to discursively reposition the site within the pantheon of older media technologies by moving the idea of Netflix away from video-store and DVD culture – surely a fading memory for most of its users – and realigning the service with that most resilient medium, cinema. Interestingly, the iconography of television is nowhere to be found in this platform space: there are no screens, remote controls, advertisements or schedules. Even though the idea of TV is central to Netflix's commercial ambitions – recall Hastings' boasting of Netflix as 'a new global Internet TV network' – and TV programming represents a massive proportion of the Netflix catalogue, the TV experience does not seem to be central to how Netflix wishes its users to imagine streaming. Perhaps this is due to the degraded nature of the televisual medium, and Netflix's related desire to market itself as a premium service. In any case, it is surely one of the ironies of Internet television that its referent medium is being simultaneously effaced, reified, integrated and remediated through the emergence of streaming services.

CONCLUSION

In this chapter we have taken a spatial journey through Netflix, as a way into the wider field of television geography in the age of streaming. Starting with the infrastructural politics of the Internet and ending with the pleasures of platform space, we have approached the topic of streaming from a number of different angles that, it is hoped, together provide useful entry points into what is undoubtedly a challenging and fast-changing topic. It is no accident that we have finished with the user experience, because it is where all the issues canvassed in this chapter become visible and tangible: the geography of Internet infrastructure and connection speeds as the ultimate gatekeeper for access to video-on-demand; the variability of digital licensing, and its underlying economic and legal structures, as constraints on content availability; the drive towards personalization as a key logic of digital media (and the corresponding movement away from a monolithic idea of the global market to a model of located, niche audiences); the spatial organization and politics of platforms; the use of recommendation as a strategy for navigating the 'environment of abundance'; and so on. As a key site for thinking about streaming culture, the case of Netflix touches on all of these fundamentally *spatial* issues.

I began this chapter by asking what streaming services can teach us about the geography of television, and what the geography of television can teach us about streaming. Perhaps the major lesson here is that streaming introduces another, spatially differentiated layer of inclusion and exclusion onto the existing palimpsest of television technologies. As we have seen from the various examples and controversies described above, complex questions of cultural geography are emerging, in which the boundaries of transmission, service-provision, distribution and regulation are mobile – but also *incommensurate with one another*. Internet geography, as the underlying foundation for streaming, exists alongside broadcast and pay TV but has distinct spatial logics. Overlaying this infrastructural geography of access is a political–economic geography of distribution and licensing, which has a separate logic again. Then on top of this there is the always-evolving interactive geography of the platform itself. Countries, regions and individual users are differentially advantaged and disadvantaged across these various criteria. Some vectors of difference, such as territorial licensing, are contiguous with national borders while others, such as access to circumvention know-how (VPNs/proxies), are fundamentally transnational in nature. Each of these vectors of social, technological and infrastructural difference are vast areas of investigation in their own right, and further research is needed to understand their interaction.

NOTES

1. Hastings' remarks, circulated via a Netflix press release, were widely reported in international media. The original press release ('Netflix Is Now Available Around the World', 6 January 2016) is available at http://pr.netflix.com.
2. The Netflix model of browser-based subscription TV is just one kind of video streaming, which co-exists with other subscription sites (Amazon Prime, Hulu Plus, Mubi), ad-supported free sites (YouTube, DailyMotion), catch-up TV (iPlayer, iView) and pirate streaming services (Popcorn Time and its various forks). Netflix is a unique case because it is the only subscription service with a truly global brand, and the only one to have become a cultural phenomenon in its own right.
3. As Hay notes, 'the idea of television emerged as a response to spatial questions and to modern ways of imagining and representing geography and mobility' (Hay 2004, 974).
4. The spatial distribution of Netflix subscribers is impossible to establish with certainty because Netflix does not release country-by-country subscriber numbers. However, as of late 2015, from the available information – based on Netflix's disclosures about the ratio of US to international customers as of 2015 in various investor reports – it appears Netflix is still a US-centric service in terms of its usage, with two-thirds of its subscriber base located in the United States.
5. Fidel Rodríguez (2016) notes that, even though Netflix's much-discussed 2015 expansion to Cuba made it one of the first American companies to do business in the island in recent years, no Cubans can actually use the service because they do not have credit cards and Netflix does not accept local payment (not to mention the obvious income-level and Internet speed disparities). Any discussion of international availability of Netflix, and of streaming in general, must therefore be textured with this kind of detail.
6. Blocking of subscription streaming services by local authorities is already starting to happen in a number of countries, and will probably increase in the future. In Indonesia, for example, Netflix has been blocked for obscenity.
7. The appetite and know-how required for this mobility is itself unevenly distributed, but like other forms of mobility – such as air travel – it is growing, and the barriers to access are decreasing as various user-friendly apps and free 'geo-dodging' tools become available. The efficacy of these border-crossing workaround tools may wane as Netflix begins to crack down on 'geo-dodging', as it has recently promised.

REFERENCES

Adams, P. 1992. Television as gathering place. *Annals of the Association of American Geographers* 82(1): 117–135.

Adams, P. 2007. *Geographies of Media and Communication: A Critical Introduction*. Chichester: Wiley-Blackwell.

Adams, P. 2010. A taxonomy for communication geography. *Progress in Human Geography* 35(1): 37–57.

Akamai. 2015. *State of the Internet Report, Q3 2015*. Cambridge: Akamai, accessed 10 January 2016 from http://stateoftheinternet.com.

Bangkok Post and AP. 2016. After long wait, Netflix launches in Thailand, with caveats. *Bangkok Post*, 7 January, accessed from http://www.bangkokpost.com/news/general/818868/after-long-wait-netflix-launches-in-thailand-with-caveats.

Beer, D. 2013. *Popular Culture and New Media: The Politics of Circulation*, Basingstoke: Palgrave Macmillan.

Boyd-Barrett, O. 1977. Media imperialism: towards an international framework for the analysis of media systems. In J. Curran, M. Gurevitch and J. Woollacott (eds), *Mass Communication and Society*. pp. 116–135. London: Edward Arnold.

Christophers, B. 2009. *Envisioning Media Power: On Capital and Geographies of Television*. Lanham, MD: Rowman and Littlefield.

Curtin, M. 2003. Media capital: towards the study of spatial flows. *International Journal of Cultural Studies* 6(2): 202–228.

Curtin, M. 2007. *Playing to the World's Biggest Audience: The Globalization of Chinese Film and TV.* Berkeley, CA: University of California Press.

Gillespie, T. 2010. The politics of 'platforms'. *New Media and Society* 12(3): 1–19.

Global Web Index. 2014. *The Missing Billion: How Web Analytics is Wiping the Emerging World off The Map. Industry Whitepaper.* London: Global Web Index.

Hay, J. 2004. Television and geography. In H. Newcomb (ed.), *Encyclopedia of Television*, 2nd ed. pp. 974–979. London: Routledge.

Iwabuchi, K. 2002. *Recentering Globalization: Popular Culture and Japanese Transnationalism.* Durham, NC: Duke University Press.

Kitchin, R. and M. Dodge. 2011. *Code/space: Software and Everyday Life.* Cambridge, MA: MIT Press.

Lobato, R. and J. Meese (eds). 2016. *Geoblocking and Global Video Culture.* Amsterdam: Institute of Network Cultures.

Lotz, A. 2014. *The Television will be Revolutionized*, 2nd ed. New York: NYU Press.

Madrigal, A. 2014. How Netflix reverse engineered Hollywood. *The Atlantic*, 2 January 2014, accessed from http:/ www.theatlantic.com/technology/archive/2014/01/how-netflix-reverse-engineered-hollywood/282679/.

McLuhan, M. 1964. *Understanding Media: The Extensions of Man.* New York: McGraw-Hill.

Meyrowitz, J. 1985. *No Sense of Place: The Impact of Electronic Media on Social Behavior.* New York: Oxford University Press.

Moran, A. 1998. *Copycat Television: Globalisation, Program Formats and National Identity.* Luton: University of Luton Press.

Moran, A. 2009. *TV Formats Worldwide: Localizing Global Programs.* Bristol: Intellect Books.

Morley, D. and K. Robins. 1995. *Spaces of Identity: Global Media, Electronic Landscapes, and Cultural Boundaries.* London: Routledge.

Parks, L. 2005. *Cultures in Orbit: Satellites and the Televisual.* Durham, NC: Duke University Press.

Parks, L. 2007. Where the cable ends. In S. Banet-Weiser, C. Chris and A. Frietas (eds), *Cable Visions: Television Beyond Broadcasting.* pp. 103–126. New York: New York University Press.

Parks, L. and N. Starosielski. 2015. *Signal Traffic: Critical Studies of Media Infrastructures.* Urbana, IL: University of Illinois Press.

Roberts, H. 2010. *2010 Circumvention Tool Usage Report.* Cambridge, MA: Berkman Center for Internet & Society.

Rodríguez, F. 2016. Cuba: videos to the left – circumvention practices and audiovisual ecologies. In R. Lobato and J. Meese (eds), *Geoblocking and Global Video Culture.* pp. 178–189. Amsterdam: Institute of Network Cultures.

Sadaqat, R. 2016. Patience! You're gonna get more titles in UAE: Netflix. *Khaleej Times*, 13 January 2016, accessed 29 January 2016 from http://www.khaleejtimes.com/business/local/patience-youre-gonna-get-more-titles-in-uae-netflix.

Sakr, N. 2001. *Satellite Realms: Transnational Television, Globalization and the Middle East.* London: I.B. Tauris.

Scott, A. 2004. *On Hollywood: The Place, The Industry.* Princeton, NJ: Princeton University Press.

Shields, R. (ed.). 1996. *Cultures of Internet: Virtual Spaces, Real Histories, Living Bodies.* London: Sage.

Sinclair, J., L. Jacka and S. Cunningham (eds). 1995. *New Patterns in Global Television: Peripheral Vision.* New York: Oxford University Press.

Spigel, L. 1992. *Make Room for TV: Television and the Family Ideal in Postwar America.* Chicago, IL: University of Chicago Press.

Star, S. 1999. The ethnography of infrastructure. *American Behavioral Scientist* 43(3): 377–391.

Starks, M. 2013. *The Digital Television Revolution: Origins to Outcomes.* Basingstoke: Palgrave Macmillan.

Steemers, J. 2014. Selling television: addressing transformations in the international distribution of television content. *Media Industries Journal* 1(1): 44–49.

Straubhaar, J. 2007. *World Television: From Global to Local.* Thousand Oaks, CA: Sage.

Trimble, M. 2012. The future of cybertravel: legal implications of the evasion of geolocation. *Fordham Intellectual Property, Media & Entertainment Law Journal* 22(3): 567–657.

van Dijck, J. 2013. *The Culture of Connectivity: A Critical History of Social Media.* New York: Oxford University Press.

Waisbord, S. 2004. McTV: understanding the global popularity of television formats. *Television and New Media* 5(4): 359–383.

Warf, B. 2013. *Global Geographies of the Internet.* Dordecht: Springer.

Williams, R. 1974. *Television: Technology and Cultural Form.* London: Fontana.

Yu, P. 2015. Towards the seamless global distribution of cloud content. In A. Cheung and R. Weber (eds),

Privacy and Legal Issues in Cloud Computing. pp. 180–213. Cheltenham, UK and Northampton, MA, USA: Edward Elgar Publishing.

Zook, M. 2006. The geographies of the internet. *Annual Review of Information Science and Technology* 40(1): 53–78.

PART IV

TRANSPORTATION
TECHNOLOGIES

14. Automobility in space and time
Aaron Golub and Aaron Johnson

Close to a billion cars roam the roads of the Earth. Over the past century, cars have shaped and reshaped their surroundings – they facilitate high-speed movement, explode the geography of opportunity for drivers and open up new territories for residence, work and recreation; at the same time they 'consume' large amounts of space with their insatiable requirements for street space, parking lots, fueling stations, limited-access freeways, off-ramps, and so on. Disconnected from the car itself are the numerous upstream and downstream systems that are required to make the car function – from the thousands of oil wells and pipelines around the planet, to iron and aluminum mines, to the factories, suppliers, dealers, service stations and millions of miles of roadways, parking lots, traffic signals and traffic engineers, and the vast environmental impacts, in turn, of all of these systems. We can see how automobility is not merely a cultural pattern, a set of vehicles or infrastructures, a type of urban or suburban form, or an industrial sector or two. Automobility is a system, or indeed, a system of systems. As in most systems there are involved multiple actors and institutions covering a wide range of geographies both virtual and real. At play are the social geographies of race, class and gender, of exclusion and segregation, the geographies of production and consumption, and the geographies of extraction, pollution, disposal and climate change all at varying scales, from local to global. The automobile spreads its tentacles over the entire Earth and implicates each driver in a wide expanse of social and environmental costs and benefits both seen and unseen.

In this chapter we explore the various overlapping and interlocking geographies, both spatial and social, of automobility. The chapter integrates data and literature from the fields of urban geography, urban studies and planning, and transportation engineering and policy studies. It will synthesize an introduction to the idea of automobility, the configurations of its subsystems, its supporting institutions and actors, and the array of external effects that automobility creates. It will conclude with a brief look forward as we witness for the first time in history a marked departure from the continued increase in automobile use and ownership trends. We will focus on automobility in the United States, but similar systems are found worldwide – see for example, the work of Paterson (2007) on automobility politics in the UK, and Sperling and Gordon (2009) for a global review of automobility trends and challenges.

DEFINING AUTOMOBILITY

Automobility, as a complex system is hard to define as a single concept. This complexity is perhaps best captured in this passage from a seminal piece on the subject by Sheller and Urry (2000):

Automobility is a complex amalgam of interlocking machines, social practices and ways of dwelling, not in a stationary home, but in a mobile, semi-privatized and hugely dangerous capsule. . . . we argue that automobility has reshaped citizenship and the public sphere via the mobilization of modern civil societies. . . . In particular we argue that civil society should be reconceptualized as a 'civil society of automobility', a civil society of quasi-objects, or 'car-drivers' and 'car-passengers', along with disenfranchised 'pedestrians' and others not-in-cars, those that suffer a kind of Lakanian 'lack'. There is not a civil society of separate human subjects which can be conceived of as autonomous from these all-conquering mechanic complexes. (p. 739)

The system of automobility, therefore is both freeing and unsettling (Sheller and Urry 2000). It is freeing to those who can command such a system for their personal mobility benefits, and unsettling to those who may, out of choice or force or poverty, be excluded from this system. Exclusion from it, unfortunately, does not shield one from its effects; the civil society of automobility, as described above, transforms many, nearly all, social institutions into ones optimized for automobility. Those excluded therefore not only experience the comparably more constricted movement and freedom, they also bear the externalized costs of automobility, while their own mobility may be degraded by the optimization of the transportation systems for automobility. That is, one is not just neutrally left out, one is penalized for non-participation as the geographies and configurations of nearly everything reflect this automobilization. Geographies become fragmented, privatized, divided off, favoring those with automobiles over those without:

If *urbanization* leads to the intensification of human habitats, the concentration of places in space, and the unification of condensed temporal flows, then automobilization, by contrast, leads to the extension of human habitats, dispersal of places across space, the opportunities to escape certain locales and to form new sociologies, and the fragmentation of temporal flows, especially through suburbanization. (Sheller and Urry 2000, 742)

The system of automobility is freeing but also coercive. It ironically forces its subjects into flexibility – rather than everything coordinated and convenient for access by foot, horse, carriage, train, it is forced to dispersal, and thus are its mobile subjects. One is hard-pressed to find the slow and coordinated city – remnants of urbanizations past in old city centers which have mostly also been reconfigured to accommodate automobility. This mobility-focused approach creates significant external costs and unintended consequences (Cervero 1996; Freund and Martin 1993). The size and extent of the roads and parking needed to support such an approach become a hindrance to the use of modes of transportation other than the automobile (Chester et al. 2015). As activities spread out and adapt to automobility and land use and zoning, parking, right of way design, all reconfigured and optimized for automobility, travel by other modes degenerates. One-quarter of the land area of Los Angeles is designated for automobiles (streets, freeways, parking); a bus rider, cyclist or pedestrian must navigate that extra area to reach destinations, while most are forced to use the automobile or suffer the inconvenience of not driving (Chester et al. 2015; see also Weinberger 2012). This creates a self-reinforcing cycle where those being forced into automobility then become its constituents, unable or unwilling to place limitations on the system upon which they now depend.

Automobility as a system unsurprisingly reinforces and exacerbates other social processes, such as racial and class exclusion and segregation (Golub et al. 2013). While urban and regional segregation before the automobile operated at fine, block-level or even building-level scales, automobility facilitated segregation at much wider spatial scales. Automobility at once exploded the geographies of regional growth and opportunities for middle-class whites (i.e. 'white flight') and allowed for the confinement and abandonment of urban working-class and minority communities (Golub et al. 2013; Soja 1989; Pulido 2000; Self 2003). The infrastructure of automobility (freeways, off-ramps, interchanges, etc.) further burdened these same abandoned urban communities with noise, blight and pollution, while discriminatory regional housing and employment markets limited the mobility of these communities; they were trapped by automobility.

The idea of automobility as a system of systems, where drivers are 'subjects' in a complex and fluid operation more complex than that of its components stems from the work of Urry and others central to the 'mobility turn' in sociological research (see Urry 2007). These mobility systems, which include automobility, are 'assemblages of humans, objects, technologies, and scripts that contingently produce durability and stability of mobility. Such hybrid assemblages can roam the countrysides and cities, remaking land-scapes and townscapes through their movement' (Urry 2007, 48). This lens was developed as a paradigm for research and for deciphering social systems. For the interested reader, there are a range of important works in this literature, including the work of Urry, Thrift, Sheller, Featherstone and Gartman, and especially the edited volume by Featherstone et al. (2005).

THE DIMENSIONS OF AUTOMOBILITY

Nothing defines urban passenger transportation in the United States more than the auto-mobile. The dependence on automobiles for passenger travel is extreme. The number of vehicles has long surpassed the number of households and the number of licensed drivers (Pisarski and Polzin 2015). The share of trips to work in the United States made by auto-mobile is around 88%; the rates of using alternatives such as public transit, bicycling and walking are just a few percent each. This is a marked departure from other developed countries: in the United States, only 7% of all urban travel is made by foot or bicycling, lower than the next lowest, Canada, at 12%, and far below the rates in England (16%), France (28%) and Germany (34%) (Pucher and Dijkstra 2003).

Over half of the US population lives in the suburbs, up from 23% in 1950, and only about 25% live in traditional central cities (Pisarski and Polzin 2015, 4). Post-war urban development characterized by land uses that are segregated by use, highly dispersed and often surrounded by parking lots, walls and other road infrastructure make walking or cycling inconvenient and unsafe. This built environment all but prohibits traveling by modes other than the automobile, thereby creating what is sometimes termed 'automo-bile dependence'. The net effects are profound: US cities consume an average of 2.4 MJ of transportation energy for every dollar of regional product, ranging from a high of 3.1 MJ/$ in Phoenix to lows of 1.7 and 1.8 MJ/$ in Washington, DC and New York, respec-tively. In comparison, European cities consume around 0.8 MJ of energy per dollar of product (Newman and Kenworthy 1999).

While there are a few cities where 'automobile dependence' is less acute, this trend holds broadly across every region of the country and in cities old and new. In each, a complex set of public and private investments and urban management practices is reproduced: from the construction of roads, freeways and bridges, traffic signals and management systems, parking and other infrastructures, to the legal and bureaucratic systems governing driver licensing, vehicle titling, registration and environmental checks to the policing and roadway emergency response services, to the private activities in vehicle sales, insurance and maintenance and the transport, storage and distribution of fuels. In total, these systems account for about 11% of US employment (Hill et al. 2010) and 17% of household expenditures (Bureau of Labor Statistics 2015). These systems are replicated in each and every city, region and state around the United States, with very few exceptions. Considering the significantly different histories of development from state to state and city to city – places born as rural extractive outposts or teeming urban ports and industrial sites – the present-day uniformity of automobile dependence is a testament to the power of this planning regime.

The United States was not always this way. It was an early innovator in public transit systems and had a healthy transit ridership until the 1920s. The electric streetcar was invented in Richmond, Virginia in 1888, and by 1910, most cities of any significant size had trolley systems with the total track mileage peaking at 45,000 miles (about 3 times the size of the interstate highway system) in 1918 (Sawers 1984, 228).

The imposition of automobility upon existing urban systems created hybrid geographies of slow, dense central places that needed retrofitting, alongside far-flung suburban development, edge cities and decentralized places of work, recreation and residences (Muller 2004). Following Muller's typology and superimposing approximated historical boundaries of western Baltimore City and County as an example, one can observe the vastly different forms of settlement and infrastructure (Figure 14.1a). The early city was small and compact. In the 1900s, with the development of the streetcar, it began to reach out into the greater county area. By the 2000s, activities were scattered across the map as the interstate system; the Baltimore beltway and its various tributaries exploded the geographies of settlement. The west sub-center of Woodlawn, for example, built out in the mid to late 1990s, exemplifies the edge cities developed along interstates and beltways around many urban areas across the country. Several resulting transportation outcomes are shown in Figures 14.1b–d. One can observe the differences in auto ownership, 'vehicle miles traveled' (VMT) and transportation expenses between the compact urban core and its suburbs.

ORIGINS AND EVOLUTIONS OF AUTOMOBILITY

The origins of automobility, automobile dependence and automobile systems in the United States have fascinated scholars of urban development and technology for some time and are the subject of a large body of academic and popular literature. For excellent broad overviews of the history, see Flink (1970), Norton (2008), St Clair (1986), Jones (2008) and McShane (1995). Here, we make a brief review of that literature by highlighting six representative theoretical threads developed to explain the growth of automobility in the United States. Although at times these threads are fairly distinct, in many ways they overlap, inform or reinforce each other.

The first thread involves the esthetic movement towards automobiles and

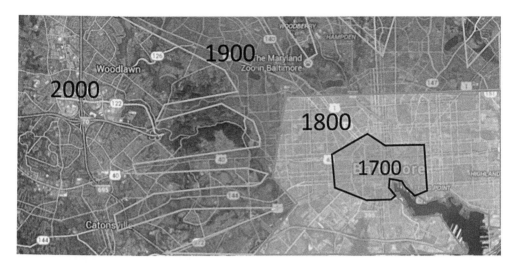

Source: Muller (2004) and Google (2015).

14.1a   Historical contours of urbanization in Baltimore City and West Baltimore County

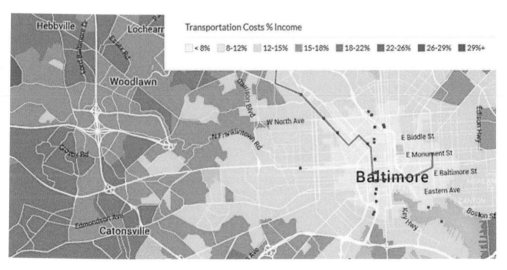

Source: Google (2015) and Center for Neighborhood Technologies (2015).

14.1b   Transportation costs as percentages of regional typical income

Figure 14.1   West Baltimore city and its Baltimore County hinterlands, illustrating the evolution of regional growth and the impact of automobility

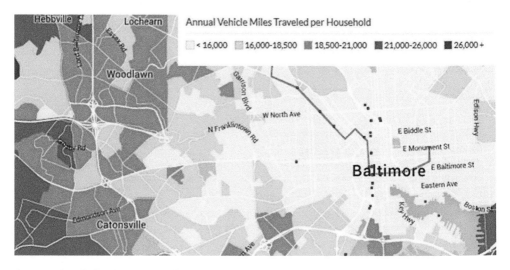

Source: Google (2015) and Center for Neighborhood Technologies (2015).

14.1c   Annual vehicle miles traveled per household

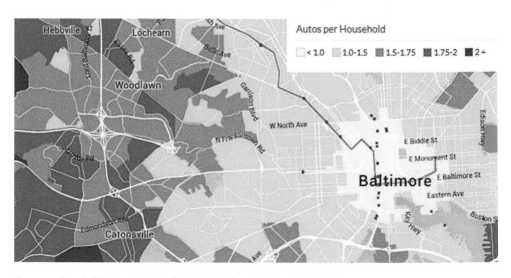

Source: Google (2015) and Center for Neighborhood Technologies (2015).

14.1d   Automobiles per household

Figure 14.1   (continued)

suburbanization: urban planners' perception of the day that suburbs and the automobile were morally, environmentally, aesthetically and managerially superior to urban living (e.g. Foster 1981; and McShane 1995). The next set of theories posits that the operating environment for public transportation, such as operating franchises, became too repressive to allow buses and streetcars to compete with the automobile, which was being supported by an array of subsidies (e.g. Bianco 1997, 1999). A third line of thought places the impetus of development on the automobile technology itself: as urban regions expanded, the automobile and freeway were better suited to this new environment, and were naturally chosen by utility-conscious consumers and voters in the free market of urban planning options (e.g. Meyer and Gomez-Ibanez 1981). A fourth thread of 'conspiracy theories' concerns automobile manufacturers' schemes to control public transit properties and stifle their growth (e.g. Snell 1974). A fifth line of research has shown how weaker rail-based industrial actors lost out to automobile-based actors in the arena of urban transportation development: the automobile industry was more effective at steering public policy in its favor (e.g. Whitt and Yago 1985; Whitt 1982; Yago 1984). A sixth line of thought argues that the planning bias toward automobiles reflected the Fordist political-economic regime emphasizing the mass consumption processes of suburbanization, where automobile-related industries played an important supporting role (e.g. Aglietta 2000; Sawers 1984; Ashton 1984; Baran and Sweezy 1968; Harvey 1989).

Not seeking explicitly to explain the mechanisms of growth of automobility, Gartman (2004) projects on this history an exploration of the evolution of cultural 'logics' of automobility, in three phases. Each phase, he argues, expanded and entrenched automobility while also revealing limits and contradictions to that process. The first phase was marked by an ideal of class distinction, the second by mass consumption and mass conformity while fostering an illusion of choice and individuality, and the third extended the individualization to the production process, creating the possibility for numerous subcultures and niches of automobility, again amplifying illusions of choice. Gartman (2004, 193) concludes:

> Each stage of the automobile has ultimately floundered due to the inability of this thing to satisfy human needs, to provide identity in sheet metal and autonomy in movement. So the contradictions pile up from one stage to the next, intensified and exacerbated but not solved.

We now turn to describing the systems which make up automobility.

THE AUTOMOBILITY 'SYSTEM'

Describing automobility as a system means that it is not enough to focus on the roads, a particular public policy, the behavior of the driver or the price of gasoline, to fully grasp what shapes and steers the automobile and its effects on society. As a complex 'assemblage', automobility systems in the United States are driven by a complex set of historical and institutional factors giving current practices great momentum and resistance to change (Geels et al. 2011; Urry 2004). Its complexity results from the combination of numerous histories, cultural norms, institutions (formal and informal), expectations and practices. Urban transportation is shaped and reshaped, produced and consumed across several groups of actors. Understanding automobility systems in the United States means that one must consider the needs of these various actors, how they interact with each other, and how they respond to demands for change. We must understand that a web of actors benefits from the current system. In this section, we briefly consider these different actors and how they interact. This discussion will then lead us into our final section, where we discuss strategies for change.

The Individual and the Household

The individual and household sit at the most micro-level of activity, making daily decisions about how to travel and less regular decisions about home location or vehicle ownership. These decisions are made mostly on rationally minimizing travel times and maximizing convenience while also balancing other decisions about racial and class differentiation, home value preservation and school quality (also tied to class maintenance). Individuals and households also engage with larger cultural forces and car ownership in general, and ownership of a specific vehicle in particular, is often a powerful sign of an individual's identity in our society. Automobile ownership is seen as a symbol of status, patriotism and of belonging to and supporting mainstream society (Paterson 2007; Gartman 2004; Flink 1975). These cultural contexts can become significant shapers of decisions regarding automobiles.

Planners and Developers

Urban planning emerged as an important force in the process of urban development in the United States just as automobiles were growing in significance. Many early planners felt that suburban areas offered a better quality of life compared with the crowded and dirty industrial cities of the time (Foster 1981). To this day, much professionalized urban planning practice reproduces the suburban, automobile-oriented models and therefore reinforces the systems of automobility. Since this is often what the (mostly driving) public and local governments request, this is what planners deliver.

Developers reproduce the automobility model, not out of a particular preference, but mostly because that is what seems to be the least risky investment (Levine and Inam 2004; Levine 2005). Banks are more likely to lend construction loans to build traditional automobility-oriented developments in suburban areas, and developers will find it easier to develop such sites on the edges of cities compared with dealing with potential neighborhood conflict and higher or unpredictable construction costs in urban 'infill' sites.

City, County and State Governments

Cities and county governments are intimately tied to automobilities, as they construct and manage the backbone of the automobility system – the local roadways that connect home and neighborhoods to arterials and regional and interstate freeways. They also manage the traffic signaling systems which allow the free flow of traffic through those local roadways. State governments have a special role in automobility because they are tasked with overseeing the design and construction of the state and interstate highway systems. Most states also collect their own gasoline taxes, mostly used for investment in roads and freeways as well as managing the drivers and vehicle licensing systems.

The Federal Government

The federal government has an important role for supporting automobility, as well as supporting alternatives to the automobile. Federal funds have long been used to support automobile use, first used to build roads in significant amounts during the 1920s and the 1930s New Deal stimulus package. The 1956 Interstate Highway Act solidified support into a set of financing systems, based on the national gasoline tax and federal planning support, to build the national network of interstate highways we have today (Rose 1990). Federal policies have also been important in managing automobile use. These include regulations to control safety, pollution from automobiles and fuel economy standards that all automobile companies must follow. Federal funds also support public transportation systems, although in small amounts compared with the monies spent on roads.

INDUSTRIAL STRUCTURES: OIL AND AUTOMOBILES

The petroleum and automobile industrial sectors are among the most heavily concentrated in the entire United States economy – relatively few companies account for nearly all of their industry's production. This concentration means that they can together easily coordinate their concerns, influence public policy and shape consumer demands. We must see urban transportation systems' use and dependence on petroleum and automobiles as being tied directly into the effects of the oil and automobile-related industrial pillars.

As automobile use grew, our economy became increasingly dependent on petroleum, and those firms who could supply it would become increasingly powerful. On the eve of World War II, the United States contained over 80% of the world's automobiles and consumed 65% of the world's oil produced (Philip 1994, 36–37) and thus control of global oil production was a key policy of the US government and military. Using the Middle East as an example, by the 1950s, five US oil companies, together with two European firms, controlled nearly all of the Middle Eastern oil resources. This oil dependence has put the United States and its allies into a subordinate position in Middle Eastern affairs, forcing them to support dictatorships, invasions and proxy wars to maintain the relatively smooth flow of oil, arguably contributing to the hatred of the United States and the West by many groups there today.

Automobile manufacturers became the focus of the emerging mass production economy, riding the wave of public investments in freeways and suburbia, and overcoming competition from transportation alternatives, such as streetcars. The 1956 Interstate Highway Act, passed by a federal commission with ample automobile-related representatives, solidified the course toward automobile dependence, guaranteeing financing and planning support for freeways (Rose 1990). The suburb and the single-family home, the freeway and the automobile, along with the oil and other resources needed to power them all, became the centerpieces of the post-war mass-consumption society, supported heavily by a variety of federal policies (Harvey 1989). Surrounding these larger pillar industrial actors sit countless small companies – automobile parts suppliers, independently owned car dealers, service stations and repair shops, along with other related sectors such as automobile insurance companies, drive-through restaurants and suburban homebuilders.

EXTERNAL EFFECTS

Automobility extends beyond the mere infrastructure, the drivers and the systems which create it – it also creates a range of direct and indirect detrimental effects on a range of systems at various geographic scales. We briefly review those here.

Traffic Fatalities and Injuries

In the United States, around 3000 people – roughly the same number that perished during the 11 September 2001 terrorist attacks – die every month on the nation's roadways from traffic crashes and have been dying at that rate for the past 700 months (ca. 60 years). For those who survive crashes, there is pain, suffering and thousands of hours of lost work time and cost of physical rehabilitation, etc. Together, traffic fatalities and injuries cost US society deeply, with estimates ranging between $46 and $161 billion per year (Delucchi and McCubbin 2010).

Detrimental Health Impacts

Studies have shown that mobility systems significantly impact peoples' activity levels, and in turn, their health. The lack of safe, walkable neighborhoods and barriers in neighborhoods created by transportation infrastructure (such as busy streets or freeways) lead to low rates of cycling and walking. This lack of activity is linked to higher body mass indexes (e.g. obesity), poorer health indicators (Frank et al. 2006; Keegan and O'Mahony 2003) and consequently additional health costs for the society.

Reduced Social Time Budgets and Productivity

While in good traffic conditions, driving is normally the fastest way to travel in US cities, during rush-hour the average traveler can suffer from long delays, which negatively affects family life and social relations. At a value of $10 per hour, these delays are estimated to cost between $63 and $246 billion per year (Delucchi and McCubbin 2010).

Local Air Pollution

In the US environmental legislation like the Clean Air Act, enacted in 1970, has reduced tailpipe pollution emissions by around 99% for most pollutants. However, large increases in driving mean that local air pollution remains a national problem. More than 120 million Americans live in counties that fail at least one of the National Ambient Air Quality Standards, imposing a health cost burden of around $60 billion per year (Environmental Protection Agency 2010; Parry et al. 2007).

Greenhouse Gas Emissions

Greenhouse gasses in the atmosphere manage the planet's greenhouse process, whereby the global climate is regulated. Most transportation systems other than bicycles burn fuel which create greenhouse gas emissions such as carbon dioxide and methane. In the United States, transportation is responsible for about one-third of the national greenhouse gas emissions, imposing a total cost of around $9 billion per year (Environmental Protection Agency 2011; Parry et al. 2007).

Over-exploitation of Non-renewable Resources

Cars and light trucks use a large amount of non-renewable steel, glass, rubber and other materials. Data from 2001 showed that automobile production in the United States consumed 14% of the national consumption of steel, 32% of its aluminum, 31% of its iron and 68% of its rubber (McAlinden et al. 2003, 21–23). Around 10 million automobiles are retired and junked every year with the majority of the built-in resources and embedded energy lost, worth around $3 billion.

Social Inequality, Exclusion and Isolation

Planning a mobility system around the need to own and operate a personal vehicle means that, for those who are unable to do so, the system will be poorly configured. In most metropolitan areas in the United States, for example, around 25% of the population is too old, too young or not able to afford an automobile and therefore they can become isolated and excluded from the mainstream of society (Taylor and Ong 1995; Lucas 2012). For example, in many central cities where low-income populations lack access to automobiles, a lack of access to healthy food and grocery options results in what is known as a 'food desert' (United States Department of Agriculture 2009). Furthermore, as mentioned earlier, transportation systems have been used to segregate or reinforce existing segregation in some cities (Golub et al. 2013; Mohl 1993).

Contamination of Habitats

Negative environmental impacts occur throughout the petroleum supply chain – from spills and flares at the local sites of extraction, to spills and toxic pollution emissions at ports and refineries, to local service stations where fuels can cause groundwater contamination. Roughly 10 million gallons of fuel are spilled into US waters every year (Etkin 2001). This does not include the large spills such as the Gulf (aka *Deepwater Horizon*) spill in 2010 of around 170 million gallons, or the *Exxon Valdez* spill in 1989 of 11 million gallons. Worldwide, more than 3 billion gallons have been spilled into waters since 1970, with typical annual environmental damages costing around $3 billion (Parry et al. 2007).

Costs of Petroleum Dependence

In the United States around half of the country's petroleum needs are imported from other countries, resulting in significant costs, estimated to be between $7 and $30 billion per year (Delucchi and McCubbin 2010), from lack of flexibility in the economy to respond to changes in price. The non-competitive structure of the oil industry has resulted in artificially high prices, with costs estimated to exceed $8 trillion since 1970 (Davis et al. 2010). The costs of US military presence in locations of strategic importance to the oil industry amount to between $6 and $60 billion per year (Davis et al. 2010).

FUTURES: A DECLINE OF AUTOMOBILITY?

For the first time in history, the secular trends of increasing car use and ownership are showing signs of stagnation and reversal (Millard-Ball and Schipper 2011). Between 1960 and 2010, the number of households with no car at all decreased dramatically, from 23% of total households to 9%, while the share of households with two or more vehicles increased (Oak Ridge National Laboratory 2014). Altering course, the overall car owner- ship rate has ticked down since 2010, while the number of licensed drivers per household began declining in the mid-2000s (see below).

Overall licensure rates have also declined slightly, although more profound changes can be seen in individual age cohorts. Figure 14.2 compares the licensure rates for age cohorts over three time points – 1980, 2000 and 2012. The decline in licensure rates for the younger cohorts is significant; 20–24-year-olds today are 25% less likely to have a driver's license than the same age group 30 years ago (a drop of almost 15% out of 55%). A similar pattern holds for all of the age groups younger than 40. Following these trends forward another decade or two suggests a significantly lower licensure rate for the overall population.

We may already be seeing the initial effects of the decline in licensing and vehicle ownership. Since the mid-2000s, there has been a significant reversal in long-term travel trends as measured by VMT, a measure of total vehicle travel (automobile and trucks, but mostly consisting of automobile travel; Figure 14.3). Note that stagnation began in the mid-2000s, notably before the high gasoline prices of 2008. Along with stagnation in automobile use is a rapid increase in the use of alternatives such as walking, cycling and public transit; between 2000 and 2012, the number of workers grew by 9% and driving alone to work grew by 10% while public transit use grew by 17%, motorcycling by 123%, bicycling by 61% and working at home by 43% (ORNL 2014).

Explanations of these declines in car usage and ownership have focused around increased virtual mobility (e.g. social media, mobile phones and home delivery), shifting cultural understanding of automobiles, decreased economic opportunities and increases in alternative modes of mobility (Cohen 2012; Pew Research Center 2006; Goodwin 2012; McDonald 2015). Across demographics groups there are trends of decreased car ownership and usage, but young adults and low-income communities are more affected than the general population (Blumenberg et al. 2012; Blumenberg and Pierce 2013). Household income and employment status have been found to have a significant impact on VMT per year, but those aged 15–26, as compared with 27–61, have been found to be more impacted than older adults (Blumenberg et al. 2012).

The effects of virtual mobility on automobility are not yet clear (Goodwin 2012), but among young adults this has been found to explain around 35–50% of the decrease in car usage and ownership (McDonald 2015). Increasingly, young adults

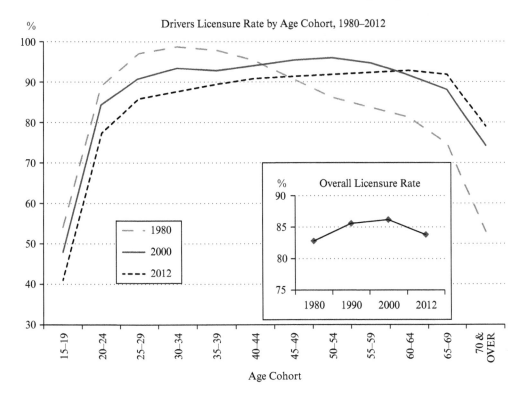

Source: US Department of Transportation, Federal Highway Administration. Highway Statistics, Table d220.

Figure 14.2 Driver licensing declines are signs of a shift away from automobility

value cars differently than older adults, and view them less as a source of independence and more as a burden and liability. Young adults also prefer to live in walkable urban areas, which would decrease their dependence on cars (Pew Research Center 2006). Decreased economic opportunities and earnings have been found to explain 10–25% of the decrease in car usage in young adults (McDonald 2015; Blumenberg et al. 2012).

Along with shifting cultural views of car ownership, young adults are also highly likely to use shared mobility systems such as car-share and bike-share programs (Goodwin 2012). These programs are seen as low-cost alternatives to owning and maintaining a car although ironically many in the low-income community experience barriers to using these programs (Kodransky and Lewenstein 2014). These barriers result from a lack of sharing systems' physical proximity to low-income neighborhoods, a lack of access to smart phones needed to run the applications, a lack of access to funds for up-front membership cost or user fees, and many residents are unbanked and lack access to a debt or credit card. There is also a lack of information and cultural difference in understanding how these programs work (Kodransky and Lewenstein 2014).

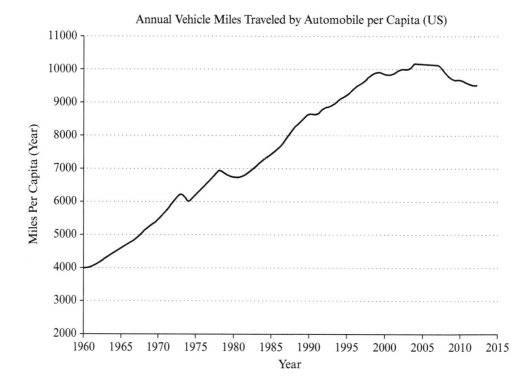

Sources: US Department of Transportation, Federal Highway Administration, 'Traffic Volume Trends' and US Census Bureau, 'Population Estimates'.

Figure 14.3   Vehicle miles traveled since 1960

REFERENCES

Aglietta, M. 2000. *A Theory of Capitalist Regulation: The US Experience*. London: Verso.
Ashton, P. 1984. Urbanization and the dynamics of suburban development under capitalism. In W. Tabb and L. Sawers (eds), *Marxism and the Metropolis*. New York: Oxford University Press.
Baran, P. and P. Sweezy. 1968. *Monopoly Capital: An Essay on the American Economic and Social Order*. New York: Monthly Review Press.
Bianco, M. 1997. The decline of transit: a corporate conspiracy or failure of public policy? The case of Portland, Oregon. *Journal of Policy History* 9(4): 450–474.
Bianco, M. 1999. Technological innovation and the rise and fall of urban mass transit. *Journal of Urban History* 25(3): 348–378.
Blumenberg, E. and G. Pierce. 2013. *Multimodal Travel and the Poor. Evidence from the 2009 National Household Travel Survey*. Los Angeles, CA: University of California.
Blumenberg, E., B. Taylor, M. Smart, K. Ralph, M. Wander and S. Brumbagh. 2012. *What's Youth Got to Do with It? Exploring the Travel Behavior of Teens and Young Adults*. Los Angeles, CA: University of California.
Bureau of Labor Statistics. 2015. Consumer expenditures (annual) news release, accessed from http://www.bls.gov/news.release/cesan.htm.
Center for Neighborhood Technologies. 2015. Housing Plus Transportation Index, accessed from http://www.cnt.org/tools/housing-and-transportation-affordability-index.

Cervero, R. 1996. Paradigm shift: from automobility to accessibility planning. Working Paper, IURD, Berkeley, CA.

Chester, M., A. Fraser, J. Matute, C. Flower and R. Pendyala. 2015. Parking infrastructure: a constraint on or opportunity for urban redevelopment? A study of Los Angeles County parking supply and growth. *Journal of the American Planning Association* 81(4): 268–286.

Cohen, M. 2012. The future of automobile society: a socio-technical transitions perspective. *Technology Analysis and Strategic Management* 24(4): 377–390.

Davis, S., S. Diegel and R. Boundy. 2010. *Transportation Energy Data Book: Edition 29*. ORNL-6985. Oak Ridge, TN: Oak Ridge National Laboratory, accessed from http://info.ornl.gov/sites/publications/files/pub24318.pdf.

Delucchi, M. and D. McCubbin. 2010. External costs of transport in the U.S. In A. de Palma, R. Lindsey, E. Quinet and R. Vickerman (eds), *Handbook of Transport Economics*. Cheltenham, UK and Northampton, MA, USA: Edward Elgar Publishing.

Environmental Protection Agency. 2010. Our nation's air – status and trends through 2008, accessed from http://www.epa.gov/airtrends/2010/index.html.

Environmental Protection Agency. 2011. 2011 Draft U.S. greenhouse gas inventory report, accessed from http://www.epa.gov/climatechange/emissions/usinventoryreport.html.

Etkin, D. 2001. Analysis of oil spill trends in the United States and worldwide. Presentation to 2001 International Oil Spills Conference, accessed from http://www.environmental-research.com/publications/pdf/spill_statistics/paper4.pdf.

Featherstone, M., N. Thrift and J. Urry (eds). 2005. *Automobilities*. Thousand Oaks, CA: Sage.

Flink, J. 1970. *America Adopts the Automobile, 1895–1910*. Cambridge, MA: MIT Press.

Flink, J. 1975. *Car Culture*. Cambridge, MA: MIT Press.

Foster, M. 1981. *From Streetcar to Superhighway*. Philadelphia, PA: Temple University Press.

Frank, L., J. Sallis, T. Conway, J. Chapman, B. Saelens and W. Bachman. 2006. Many pathways from land use to health: associations between neighborhood walkability and active transportation, body mass index, and air quality. *Journal of the American Planning Association* 72(1): 75–87.

Freund, P. and G. Martin. 1993. *The Ecology of the Automobile*. Montreal: Black Rose Books.

Gartman, D. 2004. Three ages of the automobile: the cultural logics of the car. *Theory, Culture and Society* 21(4–5): 169–195.

Geels, F., R. Kemp, G. Dudley and G. Lyons. 2011. *Automobility in Transition? A Socio-technical Analysis of Sustainable Transport*. London: Routledge.

Golub, A., R. Marcantonio and T. Sanchez. 2013. Race, space and struggles for mobility: transportation impacts on African-Americans in Oakland and the East Bay. *Urban Geography* 34(5): 699–728.

Goodwin, P. 2012. *Peak Travel, Peak Car and the Future of Automobility: Evidence, Unsolved Issues, Policy Implications and a Research Agenda*. Paris: International Transport Forum.

Harvey, D. 1989. *The Condition of Postmodernity*. Oxford: Blackwell.

Hill, K., D. Menk and A. Cooper. 2010. *Contribution of the Automotive Industry to the Economies of all Fifty States and the United States*. Washington, DC: Center for Automotive Research.

Jones, D. 2008. *Mass Motorization and Mass Transit: An American History and Policy Analysis*. Bloomington, IN: Indiana University Press.

Keegan, O. and M. O'Mahony. 2003. Modifying pedestrian behaviour. *Transportation Research Part A: Policy and Practice* 37(10): 889–901.

Kodransky, M. and G. Lewenstein. 2014. *Connecting Low-income People to Opportunity with Shared Mobility*. New York: Institute of Transportation and Development Policy.

Levine, J. 2005. *Zoned Out: Regulation, Markets, and Choices in Transportation and Metropolitan Land Use*. Washington, DC: Resources for the Future Press.

Levine, J. and A. Inam. 2004. The market for transportation–land use integration: do developers want smarter growth than regulations allow? *Transportation* 31: 409–427.

Lucas, K. 2012. Transport and social exclusion: where are we now? *Transport Policy* 20: 105–113.

McAlinden, S., K. Hill and B. Swiecki. 2003. Economic contribution of the automotive industry to the U.S. economy – an update. Center for Automotive Research, accessed from http://www.cargroup.org/?module=Publicationsandevent=ViewandpubID=57.

McDonald, N. 2015. Are millennials really the 'go-nowhere' generation? *Journal of the American Planning Association* 81(2): 90–103.

McShane, C. 1995. *Down the Asphalt Path: The Automobile and the American City*. New York: Columbia University Press.

Meyer, J. and J. Gomez-Ibanez. 1981. *Autos Transit and Cities*. Cambridge, MA: Harvard University Press.

Millard-Ball, A. and L. Schipper. 2011. Are we reaching peak travel? Trends in passenger transport in eight industrialized countries. *Transport Reviews* 31(3): 357–378.

Mohl, R. 1993. Race and space in the modern city: Interstate-95 and the Black community in Miami. In

A. Hirsch and R. Mohl (eds), *Urban Policy in Twentieth-Century America*. pp. 100–158. New Brunswick, NJ: Rutgers University Press.

Muller, P. 2004. Transportation and urban form: stages in the spatial evolution of the American metropolis. In S. Hanson and G. Giuliano (eds), *The Geography of Urban Transportation*. pp. 59–89. New York: Guilford Press.

Newman, P. and J. Kenworthy. 1999. *Sustainability and Cities: Overcoming Automobile Dependence*. Washington, DC: Island Press.

Norton, P. 2008. *Fighting Traffic*. Cambridge, MA: MIT Press.

Oak Ridge National Laboratory. 2014. *Transportation Energy Data Book: Edition 33*, accessed from cta.ornl. gov/data/spreadsheets.shtml.

Parry, I., M. Walls and W. Harrington. 2007. Automobile externalities and policies. *Journal of Economic Literature* 45(2): 373–399.

Paterson, M. 2007. *Automobile Politics: Ecology and Cultural Political Economy*. Cambridge: Cambridge University Press.

Pew Research Center. 2006. Americans and their cars: is the romance on the skids. A social trends report, accessed from http://www.pewsocialtrends.org/2006/08/01/americans-and-their-cars-is-the-romance-on-the-skids.

Philip, G. 1994. *The Political Economy of International Oil*. Edinburgh: Edinburgh University Press.

Pisarski, E.A. and S. Polzin. 2015. *Commuting in America 2013: The National Report on Commuting Patterns and Trends*. Washington, DC: American Association of State Highway and Transportation Officials.

Pucher, J. and L. Dijkstra. 2003. Promoting safe walking and cycling to improve public health: lessons from the Netherlands and Germany. *American Journal of Public Health* 93(9): 1509–1516.

Pulido, L. 2000. Rethinking environmental racism: white privilege and urban development in Southern California. *Annals of the Association of American Geographers* 90(1): 12–40.

Rose, M. 1990. *Interstate Express Highway Politics, 1939 to 1989*. Knoxville, TN: University of Tennessee Press.

Sawers, L. 1984. The political economy of urban transportation: an interpretive essay. In W. Tabb and L. Sawers (eds), *Marxism and the Metropolis*. New York: Oxford University Press.

Self, R. 2003. *American Babylon: Race and the Struggle for Postwar Oakland*. Princeton, NJ: Princeton University Press.

Sheller, M. and J. Urry. 2000. The city and the car. *International Journal of Urban and Regional Research* 24(4): 737–757.

Snell, B. 1974. American ground transport; a proposal for restructuring the automobile, truck, bus, and rail industries. Presented to the Subcommittee on Antitrust and Monopoly of the Committee on the Judiciary, US Senate, 26 February 1974. Washington, DC: US Government Printing. Office.

Soja, E. 1989. *Postmodern Geographies: The Reassertion of Space in Critical Social Theory*. New York: Verso.

Sperling, D. and D. Gordon. 2009. *Two Billion Cars: Driving toward Sustainability*. Oxford: Oxford University Press.

St Clair, D. 1986. *The Motorization of American Cities*. New York: Praeger.

Taylor, B. and P. Ong. 1995. Spatial mismatch or automobile mismatch? An examination of race, residence and commuting in US metropolitan areas. *Urban Studies* 32(9): 1453–1473.

United States Department of Agriculture. 2009. Access to affordable and nutritious food – measuring and understanding food deserts and their consequences – Report to Congress, accessed from http://www.ers.usda. gov/media/242654/ap036_reportsummary_1_.pdf.

Urry, J. 2004. The 'system' of automobility. *Theory, Culture and Society* 21(4–5): 25–39.

Urry, J. 2007. *Mobilities*. London: Polity.

Weinberger, R. 2012. Death by a thousand curb-cuts: evidence on the effect of minimum parking requirements on the choice to drive. *Transport Policy* 20: 93–102.

Whitt, J. 1982. *Urban Elites and Mass Transportation: The Dialectics of Power*. Princeton, NJ: Princeton University Press.

Whitt, J. and G. Yago. 1985. Corporate strategies and the decline of transit in U.S. cities. *Urban Affairs Quarterly* 21(1): 37–65.

Yago, G. 1984. *The Decline of Transit: Urban Transportation in German and U.S. Cities, 1900–1970*. New York: Cambridge University Press.

15. Air transport: speed, global connectivity and time–space convergence

Andrew R. Goetz

No mode of commercial transport is as technologically advanced as air transport. Reaching a typical cruising speed of 500 knots, or 575 miles (920 km) per hour, aviation is by far the fastest mode of commercial transport.[1] Conventional high-speed rail, the next-fastest mode, can attain top speeds of approximately 300 miles (480 km) per hour, roughly half the speed of commercial aircraft.[2] As part of the larger aerospace industry, including aircraft manufacturing, navigation, traffic control, safety systems and airport development, air transport reflects a very high degree of technological sophistication.

Technological development has been both cause and effect of changes in the global air transport environment. As a result of its progressive capability to move people and goods at very high speeds, air transport has greatly affected patterns of spatial interaction, encouraging connections between places separated by large distances. The rate of technological improvement in air transport's speed and reliability has resulted in increasing international connections and contributed to accelerated processes of globalization. The evolution of air transport from a highly regulated transportation service provider to an increasingly liberalized, market-driven globalized industry has accompanied air transport's technological maturation.

Even though air transport is global in scale, its evolution and development have been concentrated in more developed regions, especially North America and Europe. Airlines based in the United States, including American, United, Delta and Southwest, and in Europe, including Lufthansa, Air France–KLM, International Airlines Group (British Airways and Iberia), Ryanair and easyJet are among the largest in the world. In 2012, nearly 60% of all flights originated from an airport in North America or Europe (Bowen 2014). Airports in Atlanta, Los Angeles, London, Chicago, Paris, Dallas, Frankfurt and Amsterdam continue to be among the world's largest based on passenger traffic.

Yet over the recent period, air traffic and airlines in Asia-Pacific and the Middle East have increased dramatically. Airlines based in the Asia-Pacific region now account for more than a quarter of all passengers carried worldwide (O'Connor and Fuellhart 2014). Today, airports in Beijing, Tokyo, Hong Kong, Jakarta and Guangzhou are among the largest in the world. Even more recently, air transport in the Middle East region, specifically in Dubai with Emirates Airline and Istanbul with Turkish Airlines, has experienced explosive growth (Alkaabi 2014). Air traffic in Latin America and Africa is also increasing, underscoring continued global growth in air passenger traffic.

This chapter first provides an historical overview of air transport, with an emphasis on major technological advances. Some theoretical perspectives, conceptual issues and political debates inherent in the development of air transport are presented next, followed by an examination of recent geographic patterns and trends in airline and airport activity.

HISTORICAL OVERVIEW OF AIR TRAVEL DEVELOPMENT

It is surprising to many contemporary observers that the technology of heavier-than-air powered flight[3] is barely over 100 years old, and solely a product of the twentieth century. It is widely recognized that the modern aerial age began with the Wright Brothers' flight in *Kitty Hawk*, North Carolina, United States, on 17 December 1903. This inaugural flight lasted only 12 seconds but marked the first time that 'a machine carrying a man had raised by its own power into the air in full flight, had sailed forward without reduction in speed, and had finally landed at a point as high as that from which it started' (Wright 1913, 12).

From this early beginning, a legion of aspiring aviators began to improve the technology and pushed the frontier to set new records for speed and distance. In 1909, Louis Bleriot became the first to cross the English Channel in a heavier-than-air aircraft (Budd 2014). The first air passenger service started in 1914 between Tampa and St Petersburg, Florida, United States, using flying boats (aircraft that landed on water) but lasted only a few months. Aviation technology developed rapidly as a result of its application to military purposes during the First World War. By 1917, in addition to fighters and bombers, aircraft were being used to transport armed services mail in countries throughout Europe (Davies 2011). The US Post Office began an air mail service between Washington and New York in 1918. After the war, John Alcock and Arthur Whitten Brown became the first to make a non-stop crossing of the Atlantic Ocean, having flown from Newfoundland, Canada to Ireland in 1919 (Dierikx 2008). The first regularly scheduled daily international air passenger service commenced in 1919 between London and Paris, leading to the need to establish regulations for international air service, codified by the Paris Convention of 1919. The signatories to this agreement recognized the 'complete and exclusive sovereignty' of states over the air space above their territories, but also the 'freedom of innocent passage' through the airspace of other contracting states (Budd 2014).

During the 1920s, companies in the United States focused on developing air mail routes supported by contracts from the US Post Office, while nationally supported airlines in Europe focused on passenger operations to link major cities and overseas empires to their homelands (Davies 2011). The United States pioneered air navigation through the development of lighted airways using high-intensity beacons to guide pilots along prescribed routes at night, as well as a national system of airfields. To underscore the importance of ground facilities, airline promoter Clement Keys insightfully observed that 'ninety percent of aviation is on the ground' (Davies 2011, 7). In Europe, airlines such as Imperial Airways (UK), Deutsche Lufthansa (Germany), KLM (Netherlands), Sabena (Belgium) and Air France (France) were created and established as national carriers. After the US Air Commerce Act of 1926 and Charles Lindbergh's transatlantic solo flight in 1927, the prospects for passenger air travel in the United States were improved. Smaller airlines in the United States merged and consolidated by the 1930s to form the 'Big Four' of American, Eastern, TWA and United that each developed transcontinental trunk routes. Led by entrepreneur Juan Trippe, Pan American Airways was established as the 'chosen instrument' to pioneer international air routes from the United States, starting first with Latin America, and then eventually to Europe and Asia. US aircraft manufacturers, such as Douglas and Boeing, produced early prototypes including the Boeing

247 and Douglas DC-2 and DC-3 that became workhorses of passenger air transport. Entered into service in 1934, the 14-seat Douglas DC-2 reached speeds of 165 mph, while the 21-seat Douglas DC-3 became the flagship of every major US carrier and many other of the world's airlines by the late 1930s (Davies 2011).

Prior to the Second World War, European and US airlines had established global air services stretching around the world. Because of the range limitations of aircraft and the need for frequent refueling, air routes operated in a linear configuration with numerous intermediate stops. Most intercontinental air services still operated with flying boats because ground-based aircraft landings required airfields, or at least air strips, with navigational aids, and these were difficult to develop in unfamiliar territory. Imperial Airways (UK) had pioneered routes to India, in Africa from Egypt to South Africa, and to Australia (jointly with Qantas Empire Airways) to link its colonial territories to London. KLM (Netherlands) had developed a route from Europe to the Dutch East Indies, while Air France extended into French North and West Africa and Aeropostale (France) into South America over the South Atlantic. Aeroflot (Soviet Union) developed a large domestic network stretching from the Baltic to the Pacific. Pan American (United States) had reached New Zealand, Australia, the Philippines and Africa in addition to its Latin American and European services (Davies 2011).

Aviation technology experienced a quantum leap as a result of the Second World War and the military necessities for advanced bombers, fighters and high-speed transport planes. In 1939, fighter aircraft reached speeds of 350 mph, but by 1945, piston-engine fighters were topping out at 450 mph and the first jet-powered prototype fighters surpassed 500 mph (Hartley 2014). While the role of aviation during the First World War was relatively marginal, its role in the Second World War only 20 years later was central to military strategy and the eventual outcome of the war. Commercial airlines largely took a hiatus during the war, since pilots, aircraft and fuel were needed for the war effort. However, the war experience and aviation technology advances laid the groundwork for a massive expansion in the postwar period.

Anticipating the need to organize the postwar commercial aviation environment, representatives from 52 states met in Chicago in 1944 to create a new international regulatory framework. The Chicago Convention of 1944 recognized the so-called 'five freedoms' of the air, thus leading to a framework which individual states could use to negotiate air service agreements (Table 15.1). While consensus was readily achieved for the first two freedoms (right to fly over territory of another country and right to land for refueling and maintenance), the other freedoms were more contentious. The third and fourth freedoms refer to the right to transport passengers or cargo between the home country and a foreign country, while the fifth freedom refers to carrying passengers or freight between two foreign countries by a flight that originates in the home country. The United States was eager to pursue a more open multilateral regime that would allow all five freedoms for all signatory states. European nations, most notably the UK, preferred a more restrictive approach that would require negotiated bilateral agreements concerning the relative amount of air service provided between two countries based on the principle of reciprocity. The ensuing Bermuda I bilateral air service agreement in 1946 between the United States and the UK included which city-pairs could be served and the specific airlines designated to provide service, as well as rules concerning capacity levels, fares and freight rates, and became the basis for other international bilateral agreements (Bowen

Table 15.1   Freedoms of the air

First Freedom	Airline of one country can fly over the territory of another country
Second Freedom	Airline of one country can stop in another country for refueling and maintenance but not to pick up or drop off passengers or cargo
Third Freedom	Airline can carry passengers or cargo from its own country to a foreign country
Fourth Freedom	Airline can carry passengers or cargo from a foreign country to its own country
Fifth Freedom (or 'beyond rights')	Airline can carry passengers or cargo between two foreign countries provided that the flight originates or terminates in its own country
Sixth Freedom	Airline can carry passengers or cargo between two foreign countries via its own country (combination of third and fourth freedoms)
Seventh Freedom	Airline can operate 'stand alone' service between two foreign countries entirely outside of its own country
Eighth Freedom (or 'consecutive' cabotage rights)	Airline can carry passengers or cargo between two domestic points in another country provided that the flight originates or terminates in its own country
Ninth Freedom (or 'stand alone' cabotage rights)	Airline can carry passengers or cargo between two domestic points in another country without continuing service to its own country

Note: The first five freedoms were identified at the 1944 Chicago Convention. The other freedoms have been added since then.

Sources: Bowen (2010); Debbage (2014).

2010; Debbage 2014). The International Air Transport Association was established in 1945 as the fare-setting and global trade organization for the international airline industry, while the International Civil Aviation Organization was founded in 1947 as an agency of the United Nations principally concerned with the rules and procedures of international air navigation. The Chicago Convention and the formation of the International Air Transport Association and International Civil Aviation Organization set the stage for postwar commercial air transport expansion.

Advances in military aviation technology were quickly applied to civilian use after the war, and the air transport industry grew dramatically. The innovation of the pressurized compartment, pioneered by the US-manufactured Lockheed Constellation and entered into service by TWA on its transcontinental routes in 1944, improved passenger comfort while increasing speed and range (Davies 2011). Turboprop aircraft were introduced commercially in 1950, but would soon be eclipsed by jet aircraft. In 1952, the first commercial passenger jet, the British-built de Havilland Comet 1, was entered into service by BOAC on its London–Johannesburg route (Budd 2014). It also was entered into service from London to India and the Far East, reaching Tokyo by 1953. The de Havilland Comet 1 operated at an average speed of 458 mph, which was almost twice as fast as piston-engine passenger aircraft at the time, and reduced the journey time between long-distance city-pairs from days to hours (Davies 2011; Budd 2014). Besides improvements in speed, jet aircraft improved fuel efficiency,[4] range and reliability, while reducing vibration on the airframe, resulting in a smoother ride. The large number of intermediate stops previously required on long-distance routes for refueling was reduced dramatically, contributing greatly to the time savings. The improved reliability of jet engines also reduced the time needed for overhauls and maintenance, allowing the aircraft to keep flying more frequently and for a longer lifespan. Although the de Havilland Comet 1 was grounded in 1954 owing to several high-profile accidents caused by the previously unknown problem of 'metal fatigue', aircraft companies learned from this experience and began producing more airworthy jets. By the late 1950s, Pan Am had launched the Boeing 707, which quickly became the industry standard in competition with the de Havilland Comet 4 and the Douglas DC-8. The Soviet Union also developed the Tupolev Tu-104 in 1956, which was deployed in Aeroflot's long-distance route network. Pan Am started the first round-the-world service with the Boeing 707 in 1960, and every other major airline in the world began to acquire jet aircraft for their long-distance services.

Air transport expanded greatly in the 1960s and 1970s with the onset of the 'jet age'. The speed, reliability and comfort of jet aircraft, together with creative marketing from the major airlines, led to air transport's dominance of long-haul passenger travel. By 1960, air transport had eclipsed ocean liners as the preferred mode of travel over the Atlantic, while in the United States, airlines overtook the railroads to become the largest carriers of intercity passengers. In the 1960s, 13 of the 25 largest airlines in the world were based in North America, including the Big Four US domestic airlines (United, American, TWA and Eastern) and Pan Am, the airline with the largest international service. Another nine of the largest airlines were based in Europe, including Air France, BOAC and KLM. In 1961, 15 of the world's 20 top passenger airports were located in North America, with four located in Europe (Bowen 2010). US aircraft manufacturers, led by Boeing and Douglas, accounted for more than 90% of the world's airliner capacity in the 1960s (Davies 2011). The Boeing 737 was entered into service in 1968, and has since become the world's most widely used aircraft.

A major technological improvement occurred with the introduction of the Boeing 747, launched by Pan Am in 1970, the first wide-bodied 'jumbo-jet' that greatly increased passenger capacity and fuel efficiency. Other aircraft manufacturers followed suit, including the McDonnell-Douglas DC-10 and the Lockheed L-1011 TriStar, but neither was able to compete successfully with the 747. In Europe, the Airbus consortium was created in the late 1960s, involving French, British and German manufacturers in an effort to compete more effectively with Boeing and the other American companies. Airbus focused its initial efforts on the A-300, designed as a wide-bodied short-haul aircraft that was better suited to the geography of European air markets.

Another technological breakthrough was the development of supersonic air transport and its commercial application in the form of the Concorde jointly developed by the British Aircraft Corporation and the French company Aerospatiale.[5] The Concorde started commercial service in 1976 with Air France flying from Paris to Rio de Janeiro and British Airways[6] from London to Bahrain. One of the major problems with supersonic air transport was the sonic boom that was generated once the aircraft exceeded the speed of sound. This created a major noise problem over populated areas and relegated the use of the Concorde to only trans-oceanic flights. The high costs of Concorde production and operation led to extremely high fares (e.g. $10,000 round-trip ticket between London and New York), which greatly limited demand even though the speed of service reduced flight times significantly. The Concorde's fuel efficiency was very poor, requiring a ton of aircraft fuel for every seat carried across the Atlantic (Davies 2011). The

Concorde was permanently grounded in 2003 owing to safety concerns after an Air France take-off crash in 2000, and for economic reasons owing to extremely high costs and low passenger demand.

Technology has had an important role to play in the development of the air freight industry. The movement of cargo by air has been a feature of commercial aviation since the introduction of regularly scheduled air mail delivery in the 1910s. In addition to carrying luggage, passenger airlines also shipped cargo in the bellies of their aircraft. By the 1970s, the containerization revolution in freight transport came to the airline industry in the form of custom-designed containers for aircraft, and the beginning of all-cargo airline companies. Federal Express (later shortened to FedEx) commenced a dedicated air cargo service in 1973, focusing on small package delivery coordinated with its own ground transport that provided fast, reliable, door-to-door service. FedEx developed a hub-and-spoke network conducting overnight flights through its Memphis hub, delivering packages the following day. FedEx pioneered the use of information technology in tracking its shipments, such as bar code labels and scanners, computerized information systems, electronic communications in its vehicles and hand-held tracking devices. This business model proved to be extremely successful, revolutionizing the movement of packages, and has been adopted by competitors such as United Parcel Service (UPS), DHL and TNT (Gilbert and Perl 2010). Air cargo services emphasize the shipment of lighter-weight, higher-value and perishable commodities, such as electronic items, medical devices, fruits and vegetables, and cut flowers. Today, while the global air freight industry carries only 2% of all goods measured by weight, it carries over 40% of goods measured by value (Bowen and Rodrigue 2016).

By the late 1970s, the global airline industry had grown substantially and reached a stage of maturity wherein economists and policy-makers began to question the desirability of continued economic regulation or state ownership of airlines. In the United States, the Civil Aeronautics Board (CAB) had controlled airline route entry and exit, fares, mergers and acquisitions since the 1938 Civil Aeronautics Act was enacted during the Depression. After many studies and much policy debate, the United States promulgated the Airline Deregulation Act of 1978, which allowed US airlines to make decisions on entry, exit and fares, and transferred authority over mergers to the US Department of Justice upon termination of the CAB in 1985. The United States also adopted an 'open skies' policy for international air service and, as part of a wider neoliberal policy agenda including structural adjustment programs, encouraged other countries to follow suit. The European Union liberalized its air transport industry in a succession of three packages in 1987, 1990 and 1993, with full liberalization starting in 1997. Privatization and liberalization policies were adopted to varying degrees in Asia, Latin America, the Middle East and Africa. What ensued was a change in the market structure of the airline industry around the world whereby many state-owned airlines were privatized, many new entrants, including low-cost carriers, started airline service, mergers resulted in a smaller number of very large airlines, and global strategic alliances created mega-airline groupings. During this time, air passenger traffic has continued to increase dramatically, air fares on average have been reduced and the geographic extent of markets served has expanded greatly.

Concerns with aviation security date back to the 1960s and 1970s when airline hijackings became a problematic trend, being followed by numerous terrorist incidents on airlines, including the infamous 11 September 2001 attacks in the United States. Because of its high-profile role in the global economy and the symbolic importance of national flag carriers, air transport has become a target for terrorists, thus necessitating more intrusive and technologically enhanced security systems. In addition to metal detectors, which have been used in airports since the 1970s, new security technologies for passenger and baggage screening such as explosive trace detection, X-ray and backscatter X-ray machines, and CTX machines have been adopted by several countries after the 11 September attacks.

While the global airline industry has been restructuring, the major aircraft manufacturers have been developing more sophisticated technology. Boeing and Airbus have emerged as the two dominant commercial aircraft manufacturers in the world, now accounting for over two-thirds of the world's jetliners, including nearly all of the large civil aircraft (greater than 100 seats) fleet. Together, they and their component suppliers have been responsible for several principal technological advancements over the last several decades. The non-stop range of aircraft has been extended from approximately 6000 miles (10,000 km) in 1970 to over 9000 miles (15,000 km) today. While range has increased substantially, cruising speeds have remained about the same (up to 600 mph/1000 kph) because higher cruising speeds result in greater fuel consumption and thus higher operating costs. As a prodigious consumer of petroleum and increasing contributor of global greenhouse gases, air transport is developing technological improvements in aircraft engines and airframes to increase fuel efficiency and reduce harmful atmospheric emissions to try to achieve a more sustainable profile. Passenger seating capacity has increased on some aircraft, such as the Airbus A-380, which can accommodate over 500 passengers, but the Boeing 747 has increased its capacity only marginally from approximately 360–400 passengers since its launch in 1970. Furthermore, the most widely used aircraft in the world, such as the Boeing 737 and Airbus A320, are in the 100–150 seat range.

There have been tremendous advances in improving the safety and navigation features of aircraft, including more reliable engines, automated flight controls and enhanced warning and traffic control systems. The US Federal Aviation Administration is moving forward with its Next Generation (NextGen) Air Transportation System that is replacing its current radar-based air traffic control system with a satellite-based one using Global Positioning Systems technology. The US Federal Aviation Administration (2016) expects NextGen to deliver $134 billion in direct airline, industry and passenger benefits by 2030 as a result of shorter routes, increased air space capacity, fuel savings and reduced emissions. Taken together, these advancements have greatly improved the efficiency, reliability and safety of commercial air transport, and hold the promise of continued progress.

THEORETICAL PERSPECTIVES, CONCEPTUAL ISSUES AND DEBATES

Technology and Time–Space Convergence

Since its inception in 1903, the most significant aspect of air transport's technological development has been the global impact of its increased speed of travel. No other mode of commercial transport is faster than air travel. It is able to operate over both land and water, and can provide direct connections between cities around the world in a span of hours. Air transport has greatly accelerated the process of time–space convergence, or the shrinkage in the amount of time it takes to connect two places (Janelle 1968, 1969). In 1830, it took 6 months' sailing time to reach Bombay from London, by 1929 it took 7 days by piston-engine aircraft with overnight stops en route, and by 1976 it took only 9 hours on a non-stop jet flight (Bowen 2010). As a result, air transport has been a key enabling technology for the process of globalization, or the increasing geographical scale and intensification of economic, social, political and cultural interactions over extended distances (Janelle and Beuthe 1997). The degree to which the world has become increasingly interconnected would simply not have been possible without air transport.

At the same time, however, it is important to realize that the process of time–space convergence has not been geographically uniform. As Knowles (2006, 420) has pointed out, 'Reductions in time accessibility favour the largest transport hubs which draw closer to each other whilst smaller centres and less developed areas become relatively more peripheral'. As a result of air transport, London, for example, has been drawn closer to major European and global cities in space–time, while being relatively more difficult to reach for smaller cities and towns on Britain (Figure 15.1). Zook and Brunn (2006) examined time–space convergence using global airline geographies and noted that, while time and distance are still roughly correlated, there are numerous places that do not conform to this expectation. They referenced the work of Sheppard (2002, 308) on the concept of positionality as a 'relational construct' that captures the 'shifting, asymmetric, and path-dependent ways in which the future of places depends on their interdependencies with other places'. Sheppard also proposed 'the metaphor of "wormholes" to capture the non-Cartesian character of globalization in which physically distant places are tightly connected with one another when their positionality is closely overlapped' (Zook and Brunn 2006, 475). They also add 'the contrasting metaphor of "travelscarps" to refer to the inversions of proximities as places that are nearby in physical space become relatively more distant from one another' (Zook and Brunn 2006, 475). Examples of wormholes include the relatively fast, frequent and inexpensive connections between global cities such as London, New York and Tokyo, even though they are separated by large distances. Likewise, travelscarps exist for cities such as Minsk, Tirana and Volgograd even though they are relatively close in physical proximity to major centers in Europe. The unevenness of time–space convergence supports the contention that globalization has changed the economic playing field but has not created a 'flat world' of equal connectivity as Friedmann (2005) and others have proposed.

The effects of technology, market demand, liberalization and geography are clearly evident in O'Connor's (1995) model of international airline network evolution (Figure 15.2). In Stage 1, typical of airline networks in the 1930s and 1940s, long linear routes were established between major destinations with numerous intermediate stops owing to the technological limits of aircraft range at the time. The early 'Kangaroo Route' service in 1935 between London and Australia required over 30 intermediate stops and took over 12 days (Davies 2011). In Stage 2, technological developments in the 1950s and 1960s that increased the range and speed of aircraft meant that some intermediate stops could be bypassed, focusing on larger cities along the route. By 1960, the time required for the London–Sydney service was reduced to just over 34 hours with only eight intermediate

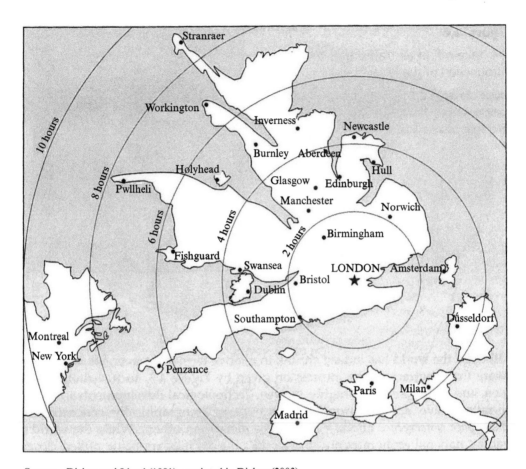

Source: Dicken and Lloyd (1981), reprinted in Dicken (2003).

Figure 15.1   The unevenness of time–space convergence

stops. Stage 3 signaled the development of international hub airports and the development of local feeder services in the 1970s and 1980s as technology and market geography began to favor specific intermediate locations within the network. Longer-range aircraft and the liberalization of air transport through more 'open skies' bilateral agreements encouraged multiple airlines to begin service using sixth freedom rights that allowed airlines from intermediate countries to carry passengers on a one-stop service between long-distance destinations. On the London–Sydney route, which had previously been dominated by British and Australian airlines, carriers from intermediate locations initiated connecting service through emerging hubs such as Singapore, Hong Kong and Bangkok, which had developed their own feeder networks in Southeast Asia. The fourth stage illustrates the expansion of major hub airports and the emergence of a principal axis shift where changing geographies of demand and increasing liberalization created new air transport service opportunities in the 1990s and 2000s. Air service along the east–west London–Sydney

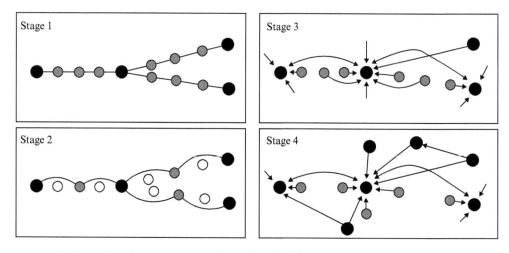

Source: O'Connor (1995), reprinted in Derudder and Witlox (2014).

Figure 15.2 Evolution of airline networks in Southeast Asia following technological advances, market demand and liberalization

route has been supplemented by increasing north–south traffic between Australia and Asia, strengthening hubs in Singapore, Hong Kong and Bangkok. Additional airlines, such as Emirates (UAE), now serve passengers between Europe, East Asia and Australia through their emerging hub at Dubai (Derudder and Witlox 2014).

Globalization and Strategic Alliances

A product of both technological development and liberalization policies, globalization has become one of the major forces in contemporary air transport. As a space-adjusting technology, air transport has contributed to the emergence of an increasingly globalized society. Herbert Callison, former Senior Vice-President of Delta Airlines, defined globalization as 'the unfettered freedom to sell and distribute a company's goods and/or services in all regions of the world. A truly global environment would foster vigorous competition *without regard to the national origin* of the provider – everyone would be in everyone else's backyard' (emphasis added). While this hyper-globalist position represents a particular unreconstituted view of globalization processes, their actual manifestation has involved reterritorialization rather than deterritorialization as the distinctiveness of national financial spaces is eroded but not necessarily removed. Individual states remain insistent upon sovereignty and control over vital national assets and concerns (Goetz and Graham 2004).

As a result, airlines have responded to globalization by creating and expanding global airline alliances. As Brian Graham (1995, 101) noted, 'the world-wide response to liberalization and deregulation has been the formation of strategic alliances at regional, national, and global scales'. While it has become easier for individual airlines to provide their own service in markets around the world, international service is still regulated by bilateral agreements that contain significant restrictions. For example, most states do not allow cabotage (eighth and ninth freedoms) rights for foreign airlines to operate

Table 15.2   Largest airlines in global alliances

Star Alliance	SkyTeam	oneWorld
United Airlines (United States)	Delta Airlines (United States)	American Airlines (United States)
Lufthansa (Germany)	China Southern (China)	IAG Group (UK/Spain)
Air China (China)	Air France/KLM (France/Neth)	Qantas Group (Australia)
Singapore Airlines (Sing)	China Eastern (China)	Cathay Pacific (Hong Kong)

service between domestic city-pairs, although these rights are permitted among countries within the European Union. There are also restrictions on foreign ownership of domestic airlines, such as in the United States, where effective and actual control of a US air carrier must be held by US citizens. Yet even in cases where airlines can operate services in foreign countries, they have chosen to partner with airlines based in other countries and regions to provide coordinated global service through marketing and operational alliances. Specific advantages include joint booking systems, sharing of frequent flyer programs, optimization of connections, and sharing of airport facilities and maintenance (Bowen and Rodrigue 2016). Today, most of the largest airlines in the world are affiliated with one of the three major global alliances – Star Alliance, SkyTeam or oneWorld (Table 15.2). These three alliances as of 2015 have grown to include over 60 airlines and now account for over 40% of global seat capacity (Bowen 2014).

DEREGULATION/LIBERALIZATION AND THE RISE OF LOW-COST CARRIERS

It has been nearly 40 years since the United States deregulated its airline industry, and many other countries have since privatized their state-owned airlines and liberalized their industries. In the debates leading up to the decision to deregulate airlines in the United States, there was considerable discussion about the degree to which air transport had matured. Certainly technology had advanced greatly since the 1930s when the system of regulation by the CAB was instituted. It was also felt that the economic structure of the industry had matured, and that the airline industry should no longer be considered a monopolistic 'public utility' in need of regulation, but that it would become a highly competitive industry with many new entrants if deregulated. A number of economic studies showed that unregulated intrastate airlines in California and Texas were able to offer lower fares and increased service in comparison with regulated airlines. Additional studies also found that there were no significant economies of scale or barriers to entry in the airline industry, thus making it possible for new entrants to compete successfully. Others suggested that the theory of contestable markets, wherein just the potential *threat* of entry would cause firms to avoid monopoly pricing, would apply to the airline industry (Goetz and Vowles 2009). It has since been shown that the airline industry is still characterized by large economies of size (including economies of scale, scope and network density), large barriers to entry and the inapplicability of theoretical contestability (Goetz 2002).

These conflicting theoretical arguments and findings have resulted in a mixed bag of outcomes from the US experience with airline deregulation. Most observers agree that US deregulation has been a success, particularly in lowering average fares, providing more flights, increasing carrier efficiency and encouraging new entrant competition, while maintaining a good safety record (Transportation Research Board 1991, 1999; US Department of Transportation 2000; US Government Accountability Office 2006; Morrison and Winston 2008). On the negative side, many airlines have encountered wide swings in profitability, with losses being much larger than gains. This financial turbulence has resulted in increasing instability in industry structure and employment, while service quality has declined. Fares and service for smaller cities, shorter-haul routes and more concentrated markets (where single-carrier domination persists) have also been negatively affected (Goetz and Vowles 2009).

One of the most significant outcomes of deregulation and liberalization has been the rise of low-cost carriers (LCCs; also called discount or 'no-frills' airlines) and their competition with full-service carriers (FSCs; also called network, 'flag' or 'legacy' airlines). Low-cost airlines emerged in the United States after the commencement of deregulation, led by upstarts such as PeoplExpress, which did not survive, and by former intrastate carrier Southwest Airlines which has since become one of the largest and most consistently profitable airlines in the world. LCCs differ from FSCs since they tend to be smaller in size, have lower costs, offer lower fares, offer fewer passenger services, have higher levels of aircraft utilization, use point-to-point route networks in addition to hub-and-spoke networks, have regional rather than global networks and utilize more secondary airports rather than primary airports in multi-airport regions (Table 15.3). The low-cost airline revolution in the United States has proceeded through several waves, or phases, of expansion and retrenchment, with new entrants rising and falling owing to economic conditions and responses from the incumbent full-service airlines.

The pattern of low-cost airline entry followed by FSC response has suggested a model of LCC evolution (Figure 15.3). After the onset of deregulation or liberalization, many

Table 15.3   Full-service carriers versus low-cost carriers

Full-service carriers	Low-cost carriers
Large carriers	Medium-sized carriers
Relatively high costs	Lower costs
Hub-and-spoke networks with coordinated scheduling	Mix of point-to-point and hub-and-spoke route networks
Primary airports in multi-airport regions	Primary and secondary airports in multi-airport regions
Different classes of service	One class of service
Aircraft utilization is lower (less time in the air)	Higher levels of aircraft utilization (more time in the air)
Large number of global and regional markets through extensive alliances	Regional markets with no or limited partnerships
Travel search engines and airline web sites	Online ticketing on airline web site only

Source: Bowen (2010); Doganis (2006).

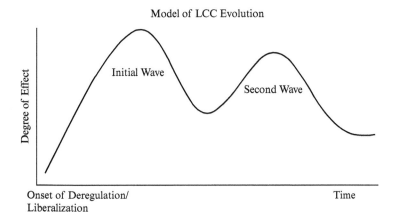

Model of LCC Evolution

Figure 15.3   A model of low-cost carrier evolution

new entrant LCCs are attracted into the industry and have a dramatic effect on lowering average fares and altering market shares. Many new entrants, however, do not survive the initial wave and are forced out owing to overexpansion, mismanagement and/or the responses of the major FSCs. After a period of consolidation and increasing fares, another round of LCC entry begins but the second wave produces a smaller number of new entrants and their impacts are not as large as the initial wave. Most of the second wave new entrants do not survive, and the major FSCs and surviving LCCs consolidate into larger airlines that expand their dominance over the market. While it is too early to tell whether this model will apply to other regions, such as the European Union or Asia-Pacific, that have been experiencing LCC growth, this model does characterize the US experience with LCCs thus far.

RECENT GEOGRAPHIC PATTERNS AND TRENDS

The geographic patterns of global air transport activity have been changing as a reflection of technological, economic, social, cultural, demographic and political dynamics as well as challenges from financial, security and environmental concerns. The airline industry has always been highly sensitive to economic fluctuations, but the first decade of the twenty-first century witnessed several catastrophes that resulted in massive financial losses for the world's airlines. The turmoil in the 2001–2006 period coincided with the effects of the 9/11 terrorist attacks in the United States, wars in Afghanistan and Iraq, heightened security measures, the SARS outbreak and rising fuel prices, which together resulted in huge financial losses for the global airline industry, amounting to over US$42 billion. Another period of severe financial turmoil occurred as a result of sharply rising fuel prices and the fallout from the global economic crisis of 2008–2009, resulting in over US$30 billion in losses to airlines (Goetz and Budd 2014). The global airline industry has stabilized since 2010, but has faced mixed financial results owing to the uneven recovery around the world. During the 2014–2016 period, airlines have benefitted greatly from sharply falling petroleum prices, although petroleum resource availability and pricing instability remain long-term concerns.

Air transport has been and continues to be concentrated in the developed regions of the world, especially the United States, Europe and Asia-Pacific. Most of the largest airlines and airports are based in these three regions, with a few notable exceptions such as rapidly rising Emirates Airline and Dubai International Airport in the Middle East region.

Airlines

US and European full-service carriers are still among the largest in the world, but have maintained their positions because of megamergers and the formation of airline holding companies (Table 15.4). In the United States, the major FSCs have undergone several waves of consolidation, including the disappearance of former industry giants Pan Am and Eastern Airlines in the early 1990s, and TWA in the early 2000s. A more recent round of consolidation, including Delta's merger with Northwest in 2011, United's merger with Continental in 2012, and American's merger with US Airways in 2014, has resulted in these three megacarriers maintaining their ranking among the top four airlines in the

Table 15.4   *Largest passenger airlines in the world, ranked by number of passengers and passenger-kilometers, 2014*

Rank	Airline	Passengers (millions)	Rank (1998 Rank)	Airline	Passenger-kilometers (billions)
1	Delta (United States)	129.4	1 (3)	Delta (United States)	290.9
2	Southwest (United States)	129.1	2 (1)	United (United States)	287.5
3	China Southern (China)	100.7	3 (48)	Emirates (UAE)	230.9
4	United (United States)	90.4	4 (2)	American (United States)	208.0
5	American (United States)	87.8	5 (35)	China Southern (China)	166.1
6	Ryanair (Ireland)	86.4	6 (15)	Southwest (United States)	162.4
7	China Eastern (China)	66.2	7 (9)	Lufthansa (Germany)	143.4
8	easyJet (UK)	62.3	8 (4)	British Airways (UK)	137.2
9	Lufthansa (Germany)	59.8	9 (8)	Air France (France)	134.5
10	Air China (China)	54.6	10 (42)	Air China (China)	112.2

Notes: (1) Air France/KLM is an airline holding company composed of Air France (France) and KLM (Netherlands) and together they carried 87.3 million passengers in 2014. (2) IAG Group is an airline holding company composed of British Airways (UK), Iberia (Spain), Air Lingus (Ireland) and Vueling (Spain) and together these airlines carried 77.3 million passengers in 2014. (3) LATAM Airlines Group is an airline holding company composed of LAN Airlines (Chile), TAM (Brazil) and several other Latin American airlines, and together they carried 67.8 million passengers in 2014.

Source: International Air Transport Association (2014); Bowen (2014).

Table 15.5   Largest freight airlines in the world, ranked by freight ton-kilometers

Rank	Airline	Ton-kilometers (billions)
1	FedEx (United States)	16.0
2	Emirates (UAE)	11.2
3	UPS (United States)	10.9
4	Cathay Pacific (Hong Kong)	9.5
5	Korean (South Korea)	8.1
6	Lufthansa (Germany)	7.1
7	Singapore (Singapore)	6.0
8	Qatar (Qatar)	6.0
9	Cargolux (Luxembourg)	5.8
10	Air China (China)	5.3

Source: International Air Transport Association (2014).

world based on passenger-kilometers. Likewise, the formation of the airline holding companies Air France (France)/KLM (Netherlands) and IAG, composed of British Airways (UK), Iberia (Spain), Air Lingus (Ireland), and Vueling (Spain), created the fifth and sixth largest airline groups in the world respectively based on revenues.[7] Lufthansa (Germany) remains in the top 10 for both passenger and freight activity. Based on ton-kilometers of freight carried, US-based cargo airlines FedEx and UPS are still among the world leaders while Lufthansa and CargoLux are the European representatives among the top 10 (Table 15.5).

The impact of LCCs is readily apparent through the remarkable growth of Southwest Airlines (United States), now the second largest airline in the world based on passengers carried. Both Southwest Airlines and jetBlue (United States) have carved out a substantial market share at many US airports, and altogether LCCs account for nearly 30% of seat capacity in the United States (Bowen 2014; Tierney 2014). Air transport liberalization in the European Union has resulted in the dramatic rise of LCCs, especially Ryanair (Ireland) (ranked sixth in the world based on passengers carried) and easyJet (UK) (ranked eighth), while reducing market shares for many flag carriers. Expansion of the EU to include countries in Central and Eastern Europe together with expanded LCC service there and in other more peripheral regions have altered geographic patterns of mobility, tourism, development and migration throughout Europe (Dobruszkes 2014; Goetz and Budd 2014).

The largest growth over the last 20 years has been for airlines in the Asia-Pacific and Middle East regions. Based on passenger-kilometer rankings, Emirates is now the third largest passenger airline in the world (ranked 48 in 1998), China Southern is fifth (ranked 35 in 1998), and Air China is tenth (ranked 42 in 1998) (International Air Transport Association 2014; Bowen 2014). Based on the number of passengers carried, China Southern, China Eastern and Air China are each in the top 10. The extremely rapid growth of air transport in China is a direct reflection of its rising position in the global economy, its large domestic market and its enormous investments in airport and air transport infrastructure (Goetz and Budd 2014). In addition to China, the other big story of growth belongs to the Gulf region of the Middle East, especially Emirates

Airline (Alkaabi 2014). The phenomenal growth of Emirates can be attributed to an ambitious strategy to serve international traffic mainly between Europe and Asia through its geographically well-positioned hub airport in Dubai. Emirates is controlled by a state-owned corporation and benefits from protective domestic state policies, but it has taken advantage of a liberalizing international air transport environment that favors open skies and expansive sixth-freedom rights (Goetz and Budd 2014). Based on ton-kilometers of freight carried, Emirates, Cathay Pacific (Hong Kong), Korean Airlines (South Korea), Singapore and Qatar are among the top 10 carriers.

In other world regions, liberalization and privatization processes in Latin America have led to numerous bankruptcies and subsequent consolidations (Lipovich 2014). This resulted in the disappearance of many formerly state-owned airlines and the emergence of multinational Latin American airline consortia such as LATAM [composed of LAN (Chile) and TAM (Brazil)] and Avianca [composed of Avianca (Colombia), TACA (El Salvador) and other subsidiaries]. At the same time, some Latin American countries, such as Argentina, Bolivia, Ecuador and Venezuela, have re-nationalized, created or strengthened new state airlines as a reaction to previous liberalization policies. While the smallest of the major world air transport regions, the air transport market in Africa is one of the fastest growing. Most airlines in Africa remain state-owned or controlled, including the 'big five' of Royal Air Maroc, Egyptair, Ethiopian, Kenya Airways (privatized in 1996, but the Kenyan government still holds the largest ownership share) and South African Airways (Pirie 2014; Goetz and Budd 2014).

Airports

The geography of global air transport is reflected in the rankings of international airports based on total passenger and freight volumes (Tables 15.6 and 15.7). While there is a general correlation between metropolitan area population size and airport passengers, certain cities are more important in air transport provision than their populations would suggest. Cities that serve as airline passenger or freight hubs, as well as important command-and-control centers of the global economy, tend to have the largest air activity in the world (Goetz 2015).

In 2014, Atlanta Hartsfield-Jackson International, the major hub for Delta Airlines, was the largest passenger airport in the world, serving over 96 million passengers, followed by Beijing Capital, London Heathrow, Tokyo Haneda and Los Angeles International. Of the top 30 passenger airports in 2014, 12 were located in the United States, 10 in Asia-Pacific, five in Europe, two in the Middle East and one in Latin America. Although many US airports continue to handle increasing passenger volumes, their world rankings today are lower, indicating a relative decline. In 1993, the United States accounted for the four largest, seven of the top 10, and 18 of the top 30 passenger airports in the world. Many Asia-Pacific passenger airports, especially those in Beijing, Jakarta, Guangzhou, Singapore, Shanghai, Kuala Lumpur and Bangkok, have grown substantially as a reflection of increased economic importance of the region. In the Middle East, Dubai has experienced a meteoric rise to become the sixth largest passenger airport in the world, and Istanbul is now the thirteenth largest. European passenger airports have generally maintained their relative position, with London Heathrow, Paris Charles de Gaulle, Frankfurt and Amsterdam Schiphol among the largest.

Table 15.6   Top 30 passenger airports by terminal passengers handled, 2014 and 1993

Rank	2014		1993	
	Airport	Passengers (millions)	Airport	Passengers (millions)
1	Atlanta Hartsfield-Jackson International	96.2	Chicago O'Hare International	65.1
2	Beijing Capital International	86.1	Dallas Fort Worth International	49.7
3	London Heathrow	73.4	Los Angeles International	47.8
4	Tokyo Haneda International	72.8	Atlanta Hartsfield International	47.7
5	Los Angeles International	70.7	London Heathrow	47.6
6	Dubai International	70.5	Tokyo Haneda International	41.5
7	Chicago O'Hare International	70.0	Denver Stapleton	32.6
8	Paris Charles de Gaulle	63.8	San Francisco International	32.0
9	Dallas Fort Worth International	63.5	Frankfurt Rheim	31.9
10	Hong Kong Chek Lap Kok International	63.1	Miami International	28.7
11	Frankfurt	59.6	New York JFK International	26.8
12	Jakarta Soekarno-Hatta International	57.0	Newark International	25.8
13	Istanbul Ataturk International	56.8	Paris Charles de Gaulle	25.7
14	Amsterdam Schiphol	55.0	Paris Orly	25.3
15	Guangzhou Baiyun International	54.8	Hong Kong Kai Tak	24.4
16	Singapore Changi	54.1	Detroit Metro Wayne Co.	24.2
17	New York JFK International	53.6	Boston Logan International	24.0
18	Denver International	53.5	Phoenix Sky Harbor International	23.5
19	Shanghai Pudong International	51.7	Minneapolis/St Paul International	23.5
20	Kuala Lumpur International	48.9	Osaka International	23.3
21	San Francisco International	47.1	Seoul Kimpo International	22.6
22	Bangkok Suvarnabhumi	46.5	Las Vegas McCarran International	22.5
23	Seoul Inchon	45.6	Honolulu International	22.1
24	Charlotte Douglas International	44.3	Orlando International	21.5
25	Las Vegas McCarran International	42.9	Amsterdam Schiphol	20.8
26	Phoenix Sky Harbor International	42.1	Toronto Pearson International	20.5
27	Madrid Barajas	41.8	Houston Intercontinental	20.3
28	Houston Bush Intercontinental	41.2	London Gatwick	20.1
29	Miami International	40.9	Tokyo Narita	20.0
30	Sao Paulo Guarulhos International	39.8	St. Louis Lambert International	19.9

Source: Airports Council International (1994, 2015).

Table 15.7   Top 20 freight airports by tons carried, 2014

Rank	Airport	Tons carried (millions)
1	Hong Kong Chek Lap Kok International	4.41
2	Memphis International	4.26
3	Shanghai Pudong International	3.18
4	Seoul Inchon	2.56
5	Anchorage Stevens International	2.48
6	Dubai International	2.37
7	Louisville International	2.29
8	Tokyo Narita International	2.13
9	Frankfurt	2.13
10	Taipei Taoyuan	2.09
11	Miami International	2.00
12	Paris Charles de Gaulle	1.89
13	Singapore Changi	1.88
14	Beijing Capital International	1.83
15	Los Angeles International	1.82
16	Chicago O'Hare International	1.67
17	Amsterdam Schiphol	1.67
18	London Heathrow	1.59
19	Guangzhou Baiyun International	1.45
20	New York JFK International	1.32

Source: Airports Council International (2015).

On the freight side, airports in the Asia-Pacific region held five of the top 10 positions, with Hong Kong, Shanghai, Seoul, Tokyo and Taipei leading the way. Each of these cities is a major trade center, not only for airports but also seaports, and they are experiencing strong economic growth. The leading freight airports in the United States are the major hubs for all-cargo airlines: Memphis (FedEx), Anchorage (FedEx and UPS) and Louisville (UPS). Memphis and Louisville are centrally located hubs within the continental United States, while Anchorage's intermediate location along the great circle flight route between the US mainland and the Asia-Pacific region explains its important air cargo function. Dubai's recent air transport growth includes freight as well, and it joins Hong Kong and Tokyo as the only cities that rank in the top 10 for both air passenger and freight activity. Frankfurt and Paris are the leading freight airports in Europe.

CONCLUSIONS

Air transport has been and continues to be a major space-adjusting technology, bringing cities and regions closer together in time–space. The effects of air transport technology have been manifested unevenly with more developed regions such as the United States, Europe and more recently Asia-Pacific leading commercial aviation development and experiencing more and better connections. The pace of technological change has also been uneven with rapid advances in aircraft speed during the 1950s and 1960s, and improvements in efficiency, safety, security and navigation technology in more recent years.

Since air transport is itself a product of remarkable technological advancements from the twentieth century, it is expected that the twenty-first century should yield significant improvements, especially in the continued applications of information technology and greater operational efficiencies. Each generation of aircraft has been able to fly farther and more safely, using less fuel per passenger-mile (Goetz and Budd 2014). The effects of improved air transport technologies will continue to be a major focus of geographical research.

NOTES

1. Current air service does not include supersonic air transport, such as the Concorde, which attained maximum cruise speeds of 1340 miles (2158 km) per hour, before it was grounded in 2003 owing to safety and environmental concerns.
2. This does not include magnetic levitation (maglev) technology, which has attained maximum speeds of 375 miles (600 km) per hour in tests.
3. Hot-air balloons became a viable technology in the late eighteenth century, but they lacked propulsion and an effective means of directional control (Budd 2014).
4. Early jet engines were not as fuel efficient as piston engines but had progressed considerably by the late 1950s with the development of turbofan engines (Bowen 2010).
5. The Soviet Union also developed the Tupolev Tu-144 supersonic airliner with which Aeroflot operated a short-lived passenger service from Moscow to Alma Ata in Kazakhstan in 1977–1978.
6. British Airways resulted from a merger between BOAC and British European Airways in 1972.
7. IATA compiles passenger and freight statistics based on the individual airlines, not the airline groups.

REFERENCES

Airports Council International. 1994. *Annual Airport Traffic Statistics, 1993*. Geneva: Airports Council International.

Airports Council International. 2015. *Annual Airport Traffic Statistics, 2014*. Geneva: Airports Council International.

Alkaabi, K. 2014. Geographies of Middle Eastern air transport. In A. Goetz and L. Budd (eds.), *The Geographies of Air Transport*. Farnham: Ashgate.

Bowen, J. 2010. *The Economic Geography of Air Transportation: Space, Time and the Freedom of the Sky*. London: Routledge.

Bowen, J. 2014. The economic geography of air transport. In A. Goetz and L. Budd (eds.), *The Geographies of Air Transport*. Farnham: Ashgate.

Bowen, J. and J-P. Rodrigue. 2016. Air transport. In *The Geography of Transport Systems*, accessed 6 March 2016 from https://people.hofstra.edu/geotrans/eng/ch3en/conc3en/ch3c5en.html.

Budd, L. 2014. The historical geographies of air transport. In A. Goetz and L. Budd (eds.), *The Geographies of Air Transport*. Farnham: Ashgate.

Davies, R. 2011. *Airlines of the Jet Age: A History*. Washington, DC: Smithsonian Institution Scholarly Press.

Debbage, K. 2014. The geopolitics of air transport. In A. Goetz and L. Budd (eds.), *The Geographies of Air Transport*. Farnham: Ashgate.

Derudder, B. and F. Witlox. 2014. Global cities and air transport. In A. Goetz and L. Budd (eds.), *The Geographies of Air Transport*. Farnham: Ashgate.

Dicken, P. 2003. *Global Shift: Reshaping the Global Economic Map in the 21st Century*, 4th edn. New York: Guilford.

Dicken, P. and P. Lloyd. 1981. *Modern Western Society*. London: Harper and Row.

Dierikx, M. 2008. *Clipping the Clouds: How Air Travel Changed the World*. Westport, CT: Praeger.

Dobruszkes, F. 2014. Geographies of European air transport. In A. Goetz and L. Budd (eds.), *The Geographies of Air Transport*. Farnham: Ashgate.

Doganis, R. 2006. *The Airline Business*, 2nd edn. New York: Routledge.

Friedmann, T. 2005. *The World is Flat: A Brief History of the Twenty-First Century.* New York: Picador/Farrar, Straus and Giroux.

Gilbert, R. and A. Perl. 2010. *Transport Revolutions: Moving People and Freight without Oil.* London: Earthscan.

Goetz, A. 2002. Deregulation, competition, and antitrust implications in the U.S. airline industry. *Journal of Transport Geography* 10(1): 1–19.

Goetz, A. 2015. The expansion of large international hub airports. In R. Hickman, M. Givoni, D. Bonilla and D. Banister (eds.), *Handbook on Transport and Development.* Cheltenham, UK and Northampton, MA, USA: Edward Elgar Publishing.

Goetz, A. and L. Budd. 2014. Conclusion. In A. Goetz and L. Budd (eds.), *The Geographies of Air Transport.* Farnham: Ashgate.

Goetz, A. and B. Graham. 2004. Air transport globalization, liberalization and sustainability: post-2001 policy dynamics in the United States and Europe. *Journal of Transport Geography* 12(4): 265–276.

Goetz, A. and T. Vowles. 2009. The good, the bad, and the ugly: 30 years of U.S. airline deregulation. *Journal of Transport Geography* 17(4): 251–263.

Graham, B. 1995. *Geography and Air Transport.* Chichester: John Wiley and Sons.

Hartley, K. 2014. *The Political Economy of Aerospace Industries: A Key Driver of Growth and International Competitiveness?* Cheltenham, UK and Northampton, MA, USA: Edward Elgar Publishing.

International Air Transport Association. 2014. *World Air Transport Statistics*, 59th edn.

Janelle, D. 1968. Central place development in a time–space framework. *Professional Geographer* 20: 5–10.

Janelle, D. 1969. Spatial reorganization: a model and concept. *Annals of the Association of American Geographers* 59: 348–364.

Janelle, D. and M. Beuthe. 1997. Globalization and research issues in transportation. *Journal of Transport Geography* 5: 199–206.

Knowles, R. 2006. Transport shaping space: differential collapse in time–space. *Journal of Transport Geography* 6: 407–425.

Lipovich, G. 2014. Geographies of Latin American air transport. In A. Goetz and L. Budd (eds.), *The Geographies of Air Transport.* Farnham: Ashgate.

Morrison, S. and C. Winston. 2008. The state of airline competition and prospective mergers. Statement for 'Competition in the Airline Industry' Hearing before the Judiciary Committee Antitrust Task Force, United States House of Representatives, Washington, DC, 24 April.

O'Connor, K. 1995. Airport development in Southeast Asia. *Journal of Transport Geography* 3(4): 269–279.

O'Connor, K. and K. Fuellhart. 2014. Air transport geographies of the Asia-Pacific. In A. Goetz and L. Budd (eds.), *The Geographies of Air Transport.* Farnham: Ashgate.

Pirie, G. 2014. Geographies of air transport in Africa: aviation's 'last frontier'. In A. Goetz and L. Budd (eds.), *The Geographies of Air Transport.* Farnham: Ashgate.

Sheppard, E. 2002. The spaces and times of globalization: place, scale, networks, and positionality. *Economic Geography* 78: 307–330.

Tierney, S. 2014. Geographies of air transport in North America. In A. Goetz and L. Budd (eds.), *The Geographies of Air Transport.* Farnham: Ashgate.

Transportation Research Board. 1991. *Special Report 230: Winds of Change: Domestic Air Transport Since Deregulation.* Washington, DC: National Research Council.

Transportation Research Board. 1999. *Special Report 255: Entry and Competition in the U.S. Airline Industry – Issues and Opportunities.* Washington, DC: National Research Council.

US Department of Transportation. 2000. *Changes in Pricing since Deregulation.* Washington, DC: Office of Aviation Analysis.

US Federal Aviation Administration. 2016. Next Generation Air Transportation System (NextGen), accessed 14 March 2016 from https://www.faa.gov/nextgen/.

US Government Accountability Office. 2006. *Airline Deregulation: Reregulating the Airline Industry Would Likely Reverse Consumer Benefits and Not Save Airline Pensions.* GAO-06-630, Washington, DC: GAO.

Wright, O. 1913. How we made the first flight. *Flying* 2(11): 10–12, 35–36.

Zook, M. and S. Brunn. 2006. From podes to antipodes: positionalities and global airline geographies. *Annals of the Association of American Geographers* 96: 471–490.

16. Drones in human geography
Thomas Birtchnell

Standing on the ferry's deck we wait to depart. The bright springtime sun illuminates the sheer plateau of sand on the close seafloor in the aperture between the starboard side and the moorings of the jetty. Accompanying children comment on the possibility of a stingray or shark sighting, unlikely inhabitants of the underwater desert below us. A shadow darting across the sand causes the ferry's passengers to look up. A square resembling the CPU and fans of the interior of a laptop computer hovers above the glint of a lens. The aircraft is obvious and foreboding. Such omnipresence in such an unlikely venue takes the ferry's occupants by surprise. The drone flits backwards and forwards, rising and lowering indecisively, as if browsing for consumables in a supermarket aisle. A knowledgeable onlooker classifies the unidentified flying object as a quad-copter drone. 'It's filming us with its Go-Pro camera right now'. Children begin to scan the encircling residential buildings for the remote controllers of this lingering technology if indeed the object did lack sentience. There is something unsettling about the flagrant, probing observations of the drone as it banks and eddies in the wind. Sophisticated computing allows it to remain stable and aloft, regardless of the coastal air currents. Odd is the sense of violation. The probability of our activities being watched was high already. The CCTV along the foreshore, long-range camera lenses of tourists and ever-present cameras on smart phones guarantee the saturation of our privacy in surveillance. The brashness of this breach causes silence to descend. 'Are they allowed to do that? There are children here.' The drone draws closer and then circles away as if to respond to the concern of the question through some kind of bizarre airborne sign language. Adults look away unable to offer sure wisdom in the face of the spectacle of childish delight. Eventually the drone darts into an open window on the top floor of an otherwise nondescript apartment, revealing the owners of the drone who had no doubt been watching us as well as the beautiful vista of the harbor. 'Can I have a drone for my birthday' a child asks?

With the widespread awareness of unmanned ('uncrewed' is a more gender neutral term) aerial vehicles (UAVs) or drones' videogame-like roles in conflict zones, the idea that they offer more benefit to society than disbenefit smacks of sophistry. Robots, and drones in particular, are emerging as a geographical topic of interest (Del Casino 2015). However, the global market for non-military drones has ballooned into a $2.5 billion industry, one that is growing 15–20% annually (Dillow 2014). The spread of drones into public spaces is initiating the question: have we got them wrong? Indeed, 'many of the legal, normative and ethical issues that come out of the modern drone debate are ones that have come before' (Carvin 2015, 137). The proverb 'drones don't kill people, people kill people' could be apt here. Certainly advocates of drones for military purposes liken them to wristwatches rather than killer robots (Allenby 2015).

Geopolitics aside, the spread of airborne cameras to all and sundry is certainly a compelling reason for geographical inquiry. The ubiquity of public surveillance through the spread of communicative technologies – that is, smart phones, digital cameras and

CCTV – takes on new heights with the entry into the market of consumer drones along-side professional airborne data capture devices for journalism, surveys and many other purposes. Drones are not necessarily an all-new technology, in fact they are a combination of existing and ancestral ones. Simply put, a 'drone is a computer that flies, and has the capability to take action in the real world' (Clarke 2014b, 259). The novelty is a geographical one: an aerial viewpoint in three-dimensional space is massified by the marketization of autonomous aircraft.

Foresight to the year 2050 imagines drones as a commonplace in the aerial vista of 'smart cities' alongside birds (Hill 2015). For the present-day urban-dweller the ramifications of ubiquitous drones are far from foreknown and in some respects alarming. Robotics expert Illah Reza Nourbakhsh imagines in 2040 the invention of a benign swarming robot able to detect eye contact and respond by fluttering around the head of the onlooker. The botigami transforms Central Park in just a year from a social hub of laughter and relaxation to a place people hurry through either wearing very dark sunglasses or stooped over, petrified of making eye contact with one of the 'Robot Smog' (Nourbakhsh 2013).

Accompanying the entry into public space of drones are all sorts of questions about privacy and its invasion by unwanted eyes. Some of these issues have been summarized for academic audiences elsewhere (Valavanis and Vachtsevanos 2015). Such concerns pale in comparison to those experienced in 'drone warfare' as these autonomous robots provide previously unimagined options for reconnaissance, killing and dismemberment in conflict zones. However, the non-military uses of drones are far from benign in many cases and there are many social and geographical consequences.

In this chapter I review different geographical perspectives on the drone. Academics are already spearheading a range of applications for drones and the multiple institutional uses mean they are testbeds for wider social ramifications (Morris 2015). In the second section I review the technical varieties and different affordances of various models of drone. In the third section, I consider the thread of engagement by geographers with the military aspects of drones, spanning geopolitics, human rights, justice and conflict issues. I also consider the civil and commercial applications of drones. In the fourth section I consider an exercise with a mid-range drone by a human geography department. Finally, I bring together a geographical perspective on the consequences of drones for human geographers.

DEFINING DRONES

The term 'drone' is contentious in itself with advocates of domestic commercial products unwilling to adopt or support the term in marketing (industry leader Skycatch uses the terms UAV; Freeman and Freeland 2016). However, this position has softened recently with some leading companies offering 'drones' or 'mini-drones', for instance GoPro. Hence the term drone is adopted in this handbook entry to reflect both military and civil technologies. It is important to understand gradations between full pilot control and full autonomy in drone technology. There are also technical distinctions to be made between fixed, rotary and even flapping wing aircraft. There has been a flood of rotary-wing aircraft into consumer markets, which have mainstreamized drones to some extent. While the majority of such consumer-level heli- or quad-copters on the market operate much in the same manner as a remote control kit aircraft or car – that is, the pilot is the decision-maker with the onboard computer ensuring stabilization in-flight – other drones are able to follow a predetermined flight-path within a boundary set by global positioning system (GPS) markers and even respond to unforeseen incidents, whether caused by the weather or other vehicles in the air. Computing power is perhaps the major difference, with fixed-wing unmanned units requiring far more processor speed owing to the complexities of landing and tracking of as well as controlling flight for winged craft. More esoteric innovations include multi-modal 'caged' robots able to fly and roll into complex structures – these are applicable currently to offworld applications rather than worldly ones (Floreano and Wood 2015).

Aside from consumer applications (primarily leisure or amateur photography), drone companies are already diversifying their interests into commercial ones, including agriculture, photography, surveillance, search and rescue, construction and ecological study. For instance, US start-up 3DR's product line includes a single plane-style drone, four copter drones and a unique consumer drone: a quadcopter able to follow a predetermined flight-path and stabilize the capture of visual data. The company's platforms capture aerial imagery for consumer interests and data analysis, including mapping, surveying and 3D modeling. The increasing demand for drones inspired the company founder Chris Anderson to note that 'hardware is the new software' (Stuart and Anderson 2015, 102). In this handbook entry a core focus is the arrival of mid-range drones with advanced data capture features into the purview of human geographers.

DRONE HISTORY

An awareness of UAVs caught the public imagination of the British, and their American allies, in the Second World War with the deployment of the V1 Flying Bomb, a type of early cruise missile. The noise the original instance of the aircraft made was likened to an insect in flight, perhaps reflected in the popular German name for them: *Maikäfer* – maybug or cockchafer. Indeed, the well-known novelist H.G. Wells noted in his short story 'The reign of Uya the lion' for *A Story of the Stone Age* series in the *Idler* magazine in 1897 that 'now and again a cockchafer would drone through the air' so this association between the word 'drone' and the insect-like noise and nature of the V1 'doodlebug', as British slang came to class it, is not as whimsical as one might first think (Wells 2008). The UAVs were also known as 'buzz bombs' in the UK and the United States because once the buzzing stopped they would fall from the air and unleash the devastating effect of their payload. A humorous account from *The Rotarian* magazine in 1944 (in the US edition) illustrates this link between the sound of the V1 and the drone of an insect for both British and American audiences. A Rotary Club talk on bee-keeping in Southern England is rudely interrupted by the

> well-known drone of a 'doodle-bug' flying over the town. Suddenly the drone ceased – and crash!!! The room rocked, the members were shaken. A Rotarian got to his feet, and in a calm voice asked: 'Can the speaker state whether it is true that bee stings cure rheumatism?' (Soar 1944, 62)

Insect-like too was the inhuman flight of the pilotless planes and the apparently indiscriminate targets they hit. Perhaps because of the droning sound of the V1 and the unthinking, bee-like nature of its flight, the term 'drone' soon entered the lexicon for a pilotless plane or guided missile. For instance, a 1947 article in *Life* magazine imagines future 'superfast, light drone aircraft' to carry atomic missiles over the Arctic (Murphy 1947, 61), and a 1950 article from *Popular Mechanics* describes a 'robot aircraft' or 'drone plane' laying a wire 'faster and more safely than a crew of soldiers in a front-line area' (*Popular Mechanics* 1950, 96).

The improvement of radio-controlled aircraft saw a surge of interest in military uses for drone aircraft beyond bombing. A patent titled 'Locating objects viewed by remote television camera' filed in 1956 by Arthur Stocker demonstrates the way thinking was going with his proposal: 'a drone aircraft flying behind the enemy line carries a TV camera . . . An advantage of such an arrangement is that a human pilot is not exposed to the danger of enemy gunfire' (Stocker 1962, unpaginated). So it was that the term 'drone' became synonymous with light unpiloted aircraft primarily for military applications. In the next section I summarize the human geographical engagement with drones in warfare and border security.

DRONES IN CONFLICT SPACES

The use of drones by the military has rapidly escalated since 2000 – a 140-fold increase in 12 years in the United States (Hall 2015). The embodied experience of drone operators in their 'presence at a distance' is a key focus for geographers concerned with geopolitics and spatial violence. Since military drones such as the Reaper or Predator are deployed to 'loiter' for upwards of 30 hours at a time, multiple people become a part of the UAV assemblage in order to facilitate its more-than-human vision capabilities (Williams 2011). The distance of the drone from its operators combined with unhuman capabilities unsettles observers of the complex relationship between humans and technologies in theaters of war. The rise in the deployment of drones for military applications is a non-human, or post-human, phenomenon (Cudworth and Hobden 2015).

The topic that has garnered most academic attention is the use of drones as surveillance or weapon systems in warfare. According to geographer Peter Adey (2014), drones are a part of a dispositif of security in recent times wherein drone operators are immersed in an affective atmosphere including technologies, (juridical) decisions, materials, military personnel, pilots, intelligence gatherers and lawyers. The atmosphere of the drone 'assemblage' simplified is distinctive from that of traditional aerial combat or conventional battlefield warfare. The materiality of the drone has also been understood as 'a moment in an energetic politics of plastic and metals, resource wars, parts commodity chains, visual infrastructures, military installations in the mountains of Utah and Pachir Wa Agam, silence and height, absent presence, geographical information systems (GIS), video affects, guilt and rationalized culpability, and so on' beyond simply the materiality of the 'solid drone' (Jackson and Fannin 2011, 437). For Adey the attention to the failures of drones in conflict spaces – that is, killing or maiming civilians rather than enemy combatants – is attributable to the disjunction between affective atmospheres. There is much evidence of the military awareness of the necessity to emulate conflict atmospheres in training and deploying drone operators with simulations adopting 'cultural settings' in set designs and various accessories. However, quite a degree of decision-making is left to the drone itself, which automates airstrikes through a 'kill-chain': find, fix, track, target, engage and assess (Gregory 2011). The extended network of these kill chains has compelled human geographers to talk of a 'Predator Empire' involving and expanding geography of drone bases in and around areas of concern (Shaw 2013).

The gradual shift from peopled to unpeopled aerial vehicles is indicative of the benefits to nation-states of death from above and at a distance. Human geographers point out that facile reassurances from politicians that drone operations are clean, surgical and precise are dubitable owing to the complexity of subject targeting in comparison to the limitations of artificial intelligence (Adey et al. 2011). Yet there is also a burgeoning need to alter or revise legislation on the ethical conventions of war in light of automation (Plaw and Reis 2015). For human geographers, drones offer a form of symbolic violence wherein they do not instigate protective or incisive measures but rather serve to mark boundaries through 'socio-spatial transformations' as one Kafka-inspired reading puts it (Nasir 2015). The dronification of warfare represents the production of a 'boundless battle space' that is heavily computerized, involving software modeling of collateral (civilian) damage (deaths) and the generation of calculated flightpaths via geotagging and real-time target location data. A paradox here is the incredible individualization of the 'enemy' through data capture and awareness at the same time as the anonymization of the drone operator as combatant (Shaw and Akhter 2014). Moreover, there is a deeper concern about the racialization of algorithms and the 'necropolitics' of this kind of warfare (Allinson 2015).

DRONES IN CIVIL SPACES

Aside from the military deployment of drones in conflict spaces there is a growing response in geography to their demobilization and domestication. Partly, this is no doubt due to the more positive media response towards UAVs highlighting their domestic use and not their military one (Freeman and Freeland 2016). The regulatory framework for drones is following that of firearms with supporters arguing for their potential in self-defense systems on the one hand and on the other hand their altruistic benefits as helpers and assistants for dull, dirty and dangerous jobs (Boucher 2014). Civil markets are now embracing various forms of drone from lowbrow applications such as delivery of pizza and textbooks to highbrow ones such as weed control and wildfire management.

Observers of the progress of drone technologies point out that other major ubiquitous technologies trace their origins back to military investment and use, for instance the Internet (Jacobstein 2013). A key factor in the debate around the introduction of UAVs into society is their legislation and the unforeseen consequences that could emerge from their ubiquity. Potentially dangerous technologies are already popular in civil space, with the most serious being perhaps the automobile. Discussions of autonomous automobiles (or more colloquially driverless cars) sing from the same hymn sheet as those on autonomous aircraft. There are a number of legislative hurdles to the introduction of drones into civil airspace: threats to human safety through malicious or accidental impact; collisions with other airborne vehicles; and the lack of natural and economic controls over problems including weather conditions, technical interference, irresponsible usage and infrastructure shortfalls (Clarke and Moses 2014).

The threats to human safety are real and possibly catastrophic. The use of drones for malicious purposes by criminals, terrorists or protestors is already a source of risk assessment for governments and notable instances of security breaches include one landing in 2014 on the White House lawn (*The Economist* 2015). The second area of concern is the accidental or benign threats to safety. Drones are already an issue around landmarks where they are launched for amateur photography and film. Open source crowd mapping involving geovisualized data points from aerial surveillance technologies also has sociocultural implications as the innovation of HarrassMap – a Cairo-based interactive online mapping interface for reporting and mapping incidents of sexual harassment anonymously and in real time, in Egypt – demonstrates, in attempting to secure public space for women (Grove 2015).

There are a number of generic application categories in load delivery and surveillance (Clarke 2014a, 239). Both of these categories carry with them unpredictable and confusing legislative hurdles and challenges. First, drones are useful in consensual cargo carriage point-to-point. An example of this is a recent trial by the company Amazon of near-proximity freight deliveries under a designated weight being administered by a drone (Jabour 2013). Benefits are freedom from traffic congestion and terrain not to mention the speed of delivery. Beyond the advantages of shifting consumer freight quickly and cheaply, as the crow flies, there are also niche uses of point-to-point cargo, particularly medical products such as blood and pharmaceuticals to hospitals, mass casualty events and offshore vessels (Thiels et al. 2015).

Scholarly thought is being given to the operational features of this freight system as an alternative to a delivery truck visiting all customers. One model imagines UAVs delivering to all eligible customers within its flight range while the delivery truck serves customers with large parcels or those out of range. A second model imagines a UAV and truck working together with the UAV collecting freight from either the depot or the truck itself (Murray and Chu 2015).

One issue with utilizing drones for delivery is the balance of privacy with necessity. A study with public focus groups in the UK and Italy understood distribution applications as being in the same ballpark as recreational ones, and too easily misused for breaching privacy or criminality (Boucher 2015). The backlash against Google Glasses – that is, an augmented reality technology with image and video capture capabilities able to overlay data visualizations over the user's eyes – as 'creepy' was due to suspicions of covert data capture (Tene and Polonetsky 2013).

Drones are also applicable to consensual cargo carriage point-to-area. Disaster recovery is one area applicable to this function wherein UAVs provide connectivity to emergency management field workers (Tuna et al. 2014). UAVs also have agricultural functions in spraying pesticides, as widely used in Japan by Yamaha Corporation: 2400 drones cover 1 million hectares of rice fields (Freeman and Freeland 2014).

Second, surveillance applications are bound to occur, both licit and illicit. Journalists are early adopters of drones for alternative viewpoints and greater opportunities for access to subjects (*The Economist* 2014). Much public space is already surveilled through CCTV and through inadvertent image capture owing to smart phones and other handheld devices now carried by the majority of people in the Global North's cities. The key factor in drone surveillance is a geographical one: UAVs provide access to three-dimensional space. Besides military engagement is their use in border 'protection' and sovereignty geopolitics. Drones, as autonomous, interoperational technologies, constitute a part of the 'smart border' of the modern nation-state (Pötzsch 2015).

There is some evidence that the use of drones for crime fighting and deterrence is socially acceptable as long as the trade-off between security and privacy is not too marked (West and Bowman 2016). Others are highly critical of permanent and perpetual policing through an all-seeing panoptic flying eye (Neocleous 2013). Supporters cite the life-saving potential of the technology. In the United States, five national bills structure the legal deployment of drones in civil spaces (Straub 2014). UAVs for information gathering require a warrant and their use as a weapon is prohibited, particularly to kill a US citizen. Only a few exceptional cases allow for data collection for law enforcement.

One possibility for drones in civil use is in nature conservation (Sandbrook 2015). The provision of high-quality image and video data to the public in near real time would certainly support many conservation efforts and campaigns, for instance combatting illegal hunting of white rhino in Africa. There are, however, social issues – both positive and negative – with privacy, safety, data security and psychological wellbeing with conservation drones when in proximity to human settlements or property.

A similar application to conservation is eco-tourism, particularly in areas in isolation from humans (King 2014). Online tourism is a possibility here, with viewers able to conduct wildlife photography and videography from anywhere in the world, either in proximity at specially prepared eco-lodges or in their own homes far away from the sites.

Aside from eco-tourism there is virtual (or voyeur, depending on how you see it) tourism. One option is to blend virtual reality (VR) with real-time live data streams from UAVs so that tourists are able to move around and interact with the location via a 3D VR visor (Mirk and Hlavacs 2014). Drones could also imaginably allow people to tour areas in the Global South where legislation is more lax and corporeal dangers prevent physical tours. Such uses could be construed as a further stage of techno-colonialism through aviation (Purtschert 2016). While many would rightly balk at the suggestion of sending drones into 'lost' tribes in the Amazon to gather anthropological data, there are currently initiatives underway to work with, and even provide drones to, indigenous people so they can survey deforestation and incursions into their land (Winkler-Schor 2015).

Drones are already proving to be powerful tools in surveillance of natural disasters and emergency situations. In effect they offer the crowdsourcing of urban surveillance and put an array of sensors in the hands of the public to initiate environmental justice themselves (Monahan and Mokos 2013). With infrared cameras UAVs are able to assist the assessment and combatting of wildfires in collaboration with firefighters (Hodson 2014). One key area that comes to mind is in emergency responses to wildfires in the critical period between the danger passing and prior to researchers gaining access. Indeed, a member of the public used an octocopter drone to capture footage of firefighters in the Blue Mountains World Heritage National Park in Australia without the approval of Civil Aviation Safety Regulations (Holton et al. 2015). Authorities argued that flying a remotely piloted aircraft in the same airspace as firefighting helicopters and aeroplanes creates a real risk of a mid-air collision.

Drones already supply Volunteered Geographic Information to popular websites with geotagging features including Instagram and Flickr. In these images the geographic position is captured and the location can be visualized. Dedicated sites are on the way with the 'dronestagram project' a notable forerunner in the general public's acceptance of drone-based aerial mapping (Hochmair and Zielstra 2015).

RESEARCH WITH DRONES

Many human geographers are now familiar with the lower-cost rotary wing drones, such as octocopters, since they are readily available on the consumer market and are useful for journalism courses. Human geographers also have an awareness of the military fixed-wing drones, such as Predators and Reapers, capable of long-distance flight, artificial intelligence and weapon deployment owing to the ethical and moral consequences of remote conflict. There is not, however, a widespread awareness of the mid-range suite of fixed-wing drones, costing tens of thousands of dollars, available for complex data capture applications as well as high-resolution image and video.

These drones are dissimilar to octocopters, which generally require constant remote control, because they sport artificial intelligence, modular design and autonomous flight via an autopilot using stabilizing systems and onboard GPS in order to fulfill predestined flight missions.

In order to gain experience with this level of drones I organized with colleagues at the University of Wollongong a flight demonstration for staff and students in the Geography and Sustainable Communities (see Birtchnell and Gibson 2015). The suppliers at Ultimate Positioning Group kindly provided the demonstration of the Sensefly eBee, a 700 gram professional mapping drone able to cover 12 square km in a single flight with no flying skills required. Scientists who utilize orthomosaic and 3D models are familiar with this technology for extracting volume data, measuring distances and taking cross-sections.

Apart from scientific applications (civil engineers, biologists and geologists also attended the demonstration), drones such as the eBee offer a method for capturing social data from strategic vantage points and documenting material arrangements, time-uses and emergency responses in areas difficult to access by researchers. Some of our staff expressed their interest in experimenting with the deployment of mid-range drones in these kinds of settings to record social data.

Beyond considerations of whether academic researchers would be even allowed to fly drones over peopled areas there is the question of what human geographers would do with the data gleaned from such exercises. The possibilities are beguiling. 3D models involving very many points of data provide a window into land use in real time (Figure 16.1).

CONCLUSION

So is it possible to imagine virtuous drones in geographical contexts (Kennedy and Rogers 2015)? In this handbook entry I have reviewed the latest geographical aspects of drone technologies and highlighted the gradual but steady adoption of them in civil society at least in the case of entry-level technologies. There are significant hurdles to the ubiquity of drones with more sophisticated data capture instruments, notwithstanding their

Figure 16.1   3D model of University of Wollongong playing field

association with military injustice and 'killer robots'. Octocopter drones already capture data at many events on private and public land involving large numbers of people: music festivals, sporting events, political rallies and protests. It remains to be seen whether the commercialization of drones for distribution, journalism and surveillance will lead to clemency for their use in social research.

One area demanding further attention is the ubiquity of more complex drones with artificial intelligence capabilities that extend beyond airborne stabilization to finding, fixing, tracking, targeting, engaging and assessing subjects for research rather than termination. The scope to capture social actions from the air would represent a paradigm shift in the social sciences akin to the adoption of these technologies for ecological surveys of wildlife (Gonzalez et al. 2016). Imaginable applications include the social practices of mobility, crowd behavior, disaster response and management, and informal land use and infrastructure developments. Certainly there will be concern about the abuse of such technologies and public debates and roundtables will need to include human geographers as consultants. One practical application already foreseeable is in GIS and those human geographers with expertise in this area can expect to see much greater resolution in their studies as satellite mapping data are complemented by aerial surveys conducted in real time. Aerial sensor platforms for social phenomena currently remain on the 'bucket lists' of many GIS practitioners, however, until proper legislative frameworks emerge around the use and application of drones in public space.

REFERENCES

Adey, P. 2014. Security atmospheres or the crystallisation of worlds. *Environment and Planning D: Society and Space* 32(5): 834–851.

Adey, P., M. Whitehead and A. Williams. 2011. Introduction: air-target: distance, reach and the politics of verticality. *Theory, Culture & Society* 28(7–8): 173–187.

Allenby, B. 2015. Emerging technologies and the future of humanity. *Bulletin of the Atomic Scientists* 71(6): 29–38.

Allinson, J. 2015. The necropolitics of drones. *International Political Sociology* 9(2): 113–127.

Birtchnell, T. and C. Gibson. 2015. Less talk more drone: social research with UAVs. *Journal of Geography in Higher Education* 39(1): 182–189.

Boucher, P. 2014. Domesticating the drone: the demilitarisation of unmanned aircraft for civil markets. *Science and Engineering Ethics* 21(6): 1393–1412.

Boucher, P. 2015. 'You wouldn't have your granny using them': drawing boundaries between acceptable and unacceptable applications of civil drones. *Science and Engineering Ethics* published online. DOI 10.1007/s11948-015-9720-7.

Carvin, S. 2015. Getting drones wrong. *International Journal of Human Rights* 19(2): 127–141.

Clarke, R. 2014a. Understanding the drone epidemic. *Computer Law & Security Review* 30(3): 230–246.

Clarke, R. 2014b. What drones inherit from their ancestors. *Computer Law & Security Review* 30(3): 247–262.

Clarke, R. and L. Moses. 2014. The regulation of civilian drones' impacts on public safety. *Computer Law & Security Review* 30(3): 263–285.

Cudworth, E. and S. Hobden. 2015. The posthuman way of war. *Security Dialogue* 46(6): 513–529.

Del Casino, V. 2015. Social geographies II: robots. *Progress in Human Geography* DOI: 10.1177/0309132515618807.

Dillow, C. 2014. Get ready for drone nation. *Fortune* 170(6): 134–145.

Floreano, D. and R. Wood. 2015. Science, technology and the future of small autonomous drones. *Nature* 521: 460–466.

Freeman, P. and R. Freeland. 2014. Politics and technology: U.S. polices restricting unmanned aerial systems in agriculture. *Food Policy* 49(Part 1): 302–311.

Freeman, P. and R. Freeland. 2016. Media framing the reception of unmanned aerial vehicles in the United States of America. *Technology in Society* 44: 23–29.

Gonzalez, L., G. Montes, E. Puig, S. Johnson, K. Mengersen and K. Gaston. 2016. Unmanned aerial vehicles (UAVs) and artificial intelligence revolutionizing wildlife monitoring and conservation. *Sensors* 16(1): 97.

Gregory, D. 2011. From a view to a kill: drones and late modern war. *Theory, Culture & Society* 28(7–8): 188–215.

Grove, N. 2015. The cartographic ambiguities of HarassMap: crowdmapping security and sexual violence in Egypt. *Security Dialogue* 46(4): 345–364.

Hall, A. 2015. Drones: public interest, public choice, and the expansion of unmanned aerial vehicles. *Peace Economics, Peace Science, & Public Policy* 21(2): 273–300.

Hill, D. 2015. The street as platform: how digital dynamics shape the physical city. *Architectural Design* 85(4): 62–67.

Hochmair, H. and D. Zielstra. 2015. Analysing user contribution patterns of drone pictures to the dronestagram photo sharing portal. *Journal of Spatial Science* 60(1): 79–98.

Hodson, H. 2014. Enter the drone zones. *New Scientist* 223(2987): 19–20.

Holton, A., S. Lawson and C. Love. 2015. Unmanned aerial vehicles. *Journalism Practice* 9(5): 634–650.

Jabour, B. 2013. Amazon trialling the use of drones for parcel delivery, accessed 2 December 2013 from http://www.theguardian.com/technology/2013/dec/02/amazon-trialling-the-use-of-drones-for-parcel-delivery.

Jackson, M. and M. Fannin (2011). Letting geography fall where it may –aerographies address the elemental. *Environment and Planning D: Society and Space* 29(3): 435–444.

Jacobstein, N. 2013. Drones: a 360 degree view. *World Policy Journal* 30(3): 14–19.

Kennedy, C. and J. Rogers. 2015. Virtuous drones? *International Journal of Human Rights* 19(2): 211–227.

King, L. 2014. Will drones revolutionise ecotourism? *Journal of Ecotourism* 13(1): 85–92.

Mirk, D. and H. Hlavacs. 2014. Using drones for virtual tourism. In D. Reidsma, I. Choi and R. Bargar (eds), *Proceedings of the Intelligent Technologies for Interactive Entertainment: 6th International Conference, INTETAIN 2014*, Chicago, IL, 9–11 July 2014. pp. 144–147. New York: Springer.

Monahan, T. and J. Mokos. 2013. Crowdsourcing urban surveillance: the development of homeland security markets for environmental sensor networks. *Geoforum* 49: 279–288.

Morris, L. 2015. On or coming to your campus soon: drones. *Innovative Higher Education* 40(3): 187–188.

Murphy, C. 1947. *The Polar Concept: It is Revolutionizing Polar Strategy*. New York: Time.

Murray, C. and A. Chu. 2015. The flying sidekick traveling salesman problem: optimization of drone-assisted parcel delivery. *Transportation Research Part C: Emerging Technologies* 54: 86–109.

Nasir, M. 2015. Droning, zoning and organizing: Kafkaesque reflections on the nomos of the earth in the northwestern tribal belt of Pakistan. *Space and Polity* 19(3): 273–292.

Neocleous, M. 2013. Air power as police power. *Environment and Planning D: Society and Space* 31(4): 578–593.

Nourbakhsh, I. 2013. *Robot Futures*. Cambridge, MA: MIT Press.

Plaw, A. and J. Reis. 2015. Learning to live with drones: answering Jeremy Waldron and the neutralist critique. *Journal of Military Ethics* 14(2): 128–145.

Popular Mechanics. 1950. Drone plane lays wire. *Popular Mechanics* 94: 96–97.

Pötzsch, H. 2015. The emergence of iborder: bordering bodies, networks, and machines. *Environment and Planning D: Society and Space* 33(1): 101–118.

Purtschert, P. 2016. Aviation skills, manly adventures and imperial tears: the Dhaulagiri expedition and Switzerland's techno-colonialism. *National Identities* 18(1): 53–69.

Sandbrook, C. 2015. The social implications of using drones for biodiversity conservation. *Ambio* 44(4): 636–647.

Shaw, I. 2013. Predator epire: the geopolitics of US drone warfare. *Geopolitics* 18(3): 536–559.

Shaw, I. and M. Akhter. 2014. The dronification of state violence. *Critical Asian Studies* 46(2): 211–234.

Soar, H. 1944. *My Favourite Story*. Chicago, IL: The Rotary Club.

Stocker, A. 1962. Locating objects viewed by remote television camera. Google Patents. USPTO. United States, RCA Corp.

Straub, J. 2014. Unmanned aerial systems: consideration of the use of force for law enforcement applications. *Technology in Society* 39: 100–109.

Stuart, T. and C. Anderson. 2015. 3D robotics: disrupting the drone market. *California Management Review* 57(2): 91–112.

Tene, O. and J. Polonetsky. 2013. A theory of creepy: technology, privacy and shifting social norms. *Yale Journal of Law and Technology* 59: 59–102.

The Economist. 2014. Eyes in the skies; Drone journalism. *The Economist*, 29 March, accessed from http://www.economist.com/news/international/21599800-drones-often-make-news-they-have-started-gathering-it-too-eyes-skies.

The Economist. 2015. Dealing with rogue drones: copping a 'copter. *The Economist*, 2 May, accessed from http://www.economist.com/news/science-and-technology/21650071-hands-criminals-small-drones-could-be-menace-now-time.

Thiels, C., J. Aho, S. Zietlow and D. Jenkins. 2015. Use of unmanned aerial vehicles for medical product transport. *Air Medical Journal* 34(2): 104–108.

Tuna, G., B. Nefzi and G. Conte. 2014. Unmanned aerial vehicle-aided communications system for disaster recovery. *Journal of Network and Computer Applications* 41: 27–36.

Valavanis, K. and G. Vachtsevanos. 2015. Introduction. In K. Valavanis and G. Vachtsevanos (eds), *Handbook of Unmanned Aerial Vehicles*. pp. 3–4. Dordrecht: Springer.

Wells, H.G. 2008. *Collected Short Stories*. Sweden: Ulwencreutz Media.

West, J. and J. Bowman. 2016. The domestic use of drones: an ethical analysis of surveillance issues. *Public Administration Review* 76(4): 649–659.

Williams, A. 2011. Enabling persistent presence? Performing the embodied geopolitics of the Unmanned Aerial Vehicle assemblage. *Political Geography* 30(7): 381–390.

Winkler-Schor, S. 2015. *An Eye in the Sky: How Eco-Drone Monitoring and Successful Conservation Methodology Can Mitigate Deforestation in the Tropics*. Undergraduate thesis, University of Washington, Seattle, WA.

17. Geography of railroads
Linna Li and Becky P. Y. Loo

Railroads refer to the transport system that conveys passengers and freight by moving vehicles on rail tracks (Loo and Li 2015). In addition to the conventional forms of heavy rails, modern railroads can also be light rails, monorails or electromagnetic rails (Rodrigue et al. 2013). Passenger railroads can be intercity or intracity, whereas freight railroads are usually intercity rails that support long-distance transport. The general forms of intercity railroads include conventional intercity rail with speed less than 200 km/h and high-speed rail (HSR) with speed above 200 km/h (UIC 2015a). The speed of intracity rail transit systems is usually lower, but with shorter distance and higher frequency in the form of commuter rails, light rails, metros and trams. Generally, railroads have several advantages, such as large capacity, low transport cost, good safety and reliability and high-energy efficiency, but they are often less flexible and more capital intensive.

This chapter aims to discuss the changing geography of railroads by analyzing their global distribution and geographical implications. In the following part, we briefly outline the origins and history of railroads and highlight the major contemporary issues and debates about the revival of railroads. In the third part, the global distribution of both passenger and freight railroads is presented. In the fourth part, some geographical implications of railroads are analyzed. Finally, some comments on the future of railroads are made.

CONCEPTUAL ISSUES AND DEBATES

Railroads first appeared in Europe for mining purposes before the Industrial Revolution. They expanded rapidly after the application of steam locomotives in the early 19th century. Thereafter, the technology has evolved from steam trains to diesel trains and electric trains. The impact of railroads on regional development has been closely examined. On the one hand, railroads facilitated the industrialization process in the developed countries (Owen 1987) and played an important role in stimulating the economic development at the national or even continental scale (Rostow 1960). On the other hand, this view was thrown into doubt by Fogel's (1979) social savings argument, which shows that the contribution of railroads to economic development was quite weak in the United States. Nonetheless, the significance of railroads varies among different countries, and railroads do not have the same impact everywhere and under all circumstances (Loo 2015).

Globally, the distribution of railroads is uneven, and the pattern is strongly associated with several factors. First, the technical characteristics of railroads require relatively flat topography and are constructed mostly on level ground. Second, the operations of railroads need to be supported by large volumes of passengers and freight; hence, railroads are only economically feasible in areas with higher densities of population and economic activities. Moreover, the competition from other transport modes can influence the railroad network directly. In the areas where automobiles and air transport are well developed, it is usually difficult to develop and maintain railroads as a dominant transport mode. Hence, a geographical perspective is needed to analyze the key issues and debates about railroads.

At the international scale, borders between countries impose increased transport costs on international trade; some argue that borders can reduce trade by 20–50% (Anderson and van Wincoop 2003). Transport is an important means of reducing the effects of borders and unifying the economic space (Vickerman 2015). When combining cross-border and domestic transport systems, it leads to reduced trade costs and increased foreign direct investment, which induces higher economic growth and stimulates the institutional development for regional integration (Fujimura 2004). However, the development of cross-border transport infrastructure has to overcome many obstacles. For instance, railroads, as a land transport mode, are usually operated by domestic companies, with each country's own track gauge, electrification system, signaling system and working practices. As international railroads need to overcome the interoperability problems, there are not many successful cases of international railroads (Vickerman 2007).

At the national scale, the huge investment for constructing, operating and maintaining railroads imposes great pressures on the stakeholders. Traditionally, the public sector has taken the lead in developing railroads. In recent years, however, it was found that private sector involvement may be more efficient than the public sector and can also relieve the great financial burden on governments. Thus, introducing the private sector to the investment and operation of railroads has become a trend worldwide (Black 2015). In general, the public sector has the advantages of regulating the service quality and upholding social equity, whereas the private sector has the advantages of higher economic efficiency (Li and Loo 2015). In different countries, the forms of private sector participation in the railroad systems are different. The major forms of public–private partnership include service contracts, management contracts, lease arrangements, concessions, Build-Own-Operate-Transfer (BOOT), Build-Operate-Transfer (BOT), Build-Own-Operate (BOO), reverse BOOT (the public entity builds the infrastructure and progressively transfers it to the private sector), joint ownership or mixed companies, and outright sale or divestiture (Panayotou 1998). With these diverse strategies, the private sector can participate in both the construction and the provision of services of railroads.

At the local scale, despite the wide benefits of railroads in stimulating economic development, providing employment opportunities and making local communities more livable, railroad projects are challenged by high financial costs. There are several ways to improve the financial efficiency, such as making financial feasibility studies in the planning stage of railroad projects, adopting the 'Rail and Property' model to support railway construction and operation costs, and enhancing intermodalism with ports and other transport modes as well as railway mergers and acquisitions with other transport operators (Loo 2015).

At the terminal scale, the planning, construction and operation of mega rail terminals is of great importance (Loo 2015). For passenger railroads, transit-oriented development has received much attention. The mixed land use around a transit stop gives potential opportunities to integrate railway development and land use. Accordingly, rail stations are no longer seen as access points to the rail system only but also as places with many other activities like residential, commercial and industrial services (Reusser et al. 2008). The primary development area of a railway station usually refers to the area within 5–10 minutes' walk of the station (Yin et al. 2015). For freight railroads, the concept of 'Thruport' is put forward to describe the facilities that are 'designed to handle high volume of transmodal rail shipments' (Rodrigue 2008). These rail-based transshipment terminals are important in promoting intermodality between railroads and other transport modes, especially marine shipping.

However, there was a significant decline of railroads in developed countries in the middle of the 20th century, when automobiles and airplanes became more important (Shaw and Docherty 2009). This trend was also recorded in many developing countries, with passengers and freight shifting from railroads to automobiles and airplanes. However, the great reliance on automobiles and airplanes caused many problems, including urban sprawl, nonrenewable energy consumption, air pollution, noise, traffic congestion and injuries. In particular, the increasing greenhouse gas emissions and climate change have received much attention (Loo and Li 2012). In comparison with automobiles and airplanes, railroads can have much lower greenhouse gas emission intensity per passenger-km.

Recently, there seems to be a railroad renaissance (Loo 2015). Accordingly, the renewed interest in railways for passenger travel brings opportunities for reversing the modal shift trend from railroads to automobiles and airplanes. A fine balance between fulfilling mobility needs and reducing environmental impacts can be achieved. Many experiences show that HSR, with its advantages being high speed, comfort and convenience, can attract passenger travel from both automobiles and airplanes. For instance, the operation of HSR has attracted air passengers and increased the railway market share for the routes of Paris–Lyon and Madrid–Seville in Europe (European Commission and Directorate General Transport 1998). It is expected that HSRs will play a more important role in the future transport systems not only in developed countries but also in developing countries.

GLOBAL DISTRIBUTION OF RAILROADS

Railroad Networks Distribution

The distribution of railroad networks varies greatly by world region. According to the database of the UIC (2014), the total length of the world's railroads reached 1,029,230 km at the end of 2013. The regional share of the Americas was the largest (37%), followed by Europe (26%), Asia Oceania and the Middle East (22%), and then the Russian Federation (8%) and Africa (7%). However, the growth rate in the past five years was the highest in Africa (34.3%), followed by Asia Oceania and Middle East (2.7%) and Europe (0.5%). In the Americas and the Russian Federation, the length of railroads actually decreased, with negative growth rates (−1.4 and −1.1%, respectively). In other words, the railroad network is still the most developed in the Americas and Europe but it is growing most rapidly in Africa and Asia. With respect to the railroad network density, it was the densest in some European countries, such as Belgium (11.6 km per 100 km^2), Germany (9.4) and Luxembourg (9.2), as well as a few Asian countries, such as Japan (5.3) and Korea (3.7). In countries with large areas (over 3 million km^2), it was the densest in the USA (2.02) and India (1.96), followed by China (0.69), Russia (0.49) and Brazil (0.35), and lowest in Canada (0.12) and Australia (0.02).

Since the HSR technology was introduced in Japan in the 1960s, many other countries have developed HSR systems, such as France, Germany, Spain, the UK, Italy and China. Based on the statistics of UIC (2015b), there is currently a total of 22,954 km HSR in operation, 12,754 km under construction and 18,841 km planned in different parts of the world. The length of HSR under construction and planned is more than that of the current HSR in operation. In 2014, the lengths of HSR in Europe, Asia and other countries were 7351, 15,241 and 362 km, respectively (UIC 2015b). Compared with the dense HSR network in Europe, Japan and China, there is only one HSR line in the USA between Boston and Washington, DC. This means that the revival of railroads has a specific geographical distribution, mainly in Europe and Asia, rather than North America. By 2014, China had the longest HSR in the world, with 11,132 km in operation.

Distribution of Passengers and Freight Railroads

In terms of rail passenger transport measured in passenger-km, the regional shares of Europe, Russian Federation, Africa, the Americas, and Asia Oceania and the Middle East in the world total (2865 billion passenger-km) in 2013 were 17, 5, 1, 1 and 76%, respectively (UIC 2014). Their growth rates in the past five years were −2, −18, −20, 59 and 11%, respectively (UIC 2014). This shows that the passenger rail traffic is highly concentrated in Asia Oceania and the Middle East. Europe has the second largest passenger-km share but its growth rate is much lower than Asia. Although the passenger rail traffic share is very low in the Americas, it has shown a very high growth rate recently. Both the Russian Federation and Africa have low passenger travel shares and negative growth rates (Loo and Li 2015). Also, the number of passenger-km by HSR was highly concentrated in Asia and Europe, with shares of 69 and 31%, respectively. In Asia, the HSR passenger-km in China and Japan accounted for 40 and 22% of the world HSR passenger-km, respectively. In Europe, the HSR passenger-km mainly existed in France, Germany, Italy, Spain, the UK and Sweden.

In terms of rail freight transport measured in tonne-km, the regional shares of Europe, the Russian Federation, Africa, the Americas, and Asia Oceania and the Middle East in the world total (9789 billion tonne-km) in 2013 were 6, 23, 1, 33 and 37%, respectively (UIC 2014). Their growth rates in the past five years were −39, 5, 1, −8 and 4%, respectively (UIC 2014). Thus, the regional disparity of freight transport by railroads shows a different pattern. The concentration of freight rail traffic is found in Asia Oceania and the Middle East, and the growth rate in the Asia is even lower than that in Russia. The Americas have the second largest tonne-km share, but negative growth rate. In Europe, not only its share of rail tonne-km is low but its growth rate is also negative.

SPATIAL AND GEOGRAPHICAL IMPLICATIONS OF RAILROADS

International Railroads: Promoting Regional Integration

As mentioned, cross-border railroads can reduce the transport costs imposed by borders and facilitate regional integration. One example is the Trans-European Rail Network, which is part of the Trans-European Transport Network (TEN-T) program implemented in Europe. This program aims to reduce disparity of accessibility between different Member States of the European Union, remove the transport bottlenecks to facilitate the consolidation of a single market, and overcome the barriers of incompatible railroad standards of different countries. Over the past 20 years, the European Commission has been engaged in restructuring a 'Single European Railway Area' (European Commission 2011). The main strategies include: open the domestic rail passenger market to competition; achieve a single vehicle type authorization and a single railway safety certification; develop an integrated freight corridor management approach; and ensure effective access to railway infrastructure by separating infrastructure management from service provision (European Union 2014). Although the length of railroads in Europe decreased from 237,671 km in 1990 to 215,734 km in 2012, more than 70% of the railroads in Europe are now using the standard track gauge (1435 mm) (European Union 2014). These railroads are utilized for both national and international transport. For passenger travel, domestic transport is still dominant, accounting for more than 90% of the total transport in most countries in 2013, except in Luxembourg, where 28% of its total rail passenger transport is international transport (Eurostat 2014a).

For freight railroads, the share of international transport is much larger, with an average share of 40% in all the European countries (Eurostat 2014b). In particular, the share is high in countries located along some key freight corridors, such as Latvia (90%) and Estonia (87%) at the border between Europe and Russia. The shares in the Netherlands, Luxembourg and Belgium also reach 85, 87 and 70%, respectively. Despite the wider use of railroads in international passenger and freight transport in Europe, there are still barriers for an integrated European railway network, such as the lack of technical standardization, the slow implementation of common EU legislation and border delays for international travel (Maurer et al. 2010).

By building an integrated rail network, the European Union aims to improve the accessibility of periphery areas and reduce regional disparity. The regions connected by the Trans-European Rail Network have gained in accessibility and enjoyed economic development. For example, Lyon and Lille in France experienced higher economic growth induced by the HSR (Hall 2009); some small and intermediate cities with HSR stations in Spain had much higher growth rate and housing investment than other local cities (Garmendia et al. 2012a, b); the cities within an hour of HSR travel distance from London, UK were also found to attract some economic activities from London (Chen and Hall 2011). However, some scholars argue that the railway system in Europe has mainly increased the economic activities in the core regions (Vickerman 1997; Ross 1994). The main reason is probably that the transportation infrastructure itself cannot guarantee economic growth in peripheral countries, and other aspects of regional policies are needed to facilitate economic development (Dall'Erba 2003; Vickerman 1995).

National Railroads: Changing Institutions and Governance

Currently, there are 136 countries with national railroad systems in the world (US Central Intelligence Agency 2015). Although different countries may have different railroad institutions, the general trend is to open up the government monopoly and separate the ownership and management of railroads (Black 2015). There are several representative national railroad institution models around the world.

In the United States, railroads have always been operated by private companies and regulated by government regulations to prevent monopoly and maintain open access at the same time. However, with the competition from automobiles and airplanes, about one-third of the railroad companies in the US faced bankruptcy by 1960. Thus, the government began to reform the industry. After enacting the Railroad Revitalization and Regulatory Reform Act in 1976 and the Staggers Rail Act in 1980, railroad companies obtained more freedom in setting price, merging and abandoning lines. Currently, the rail industry in the United States is concentrating on freight transport and competing successfully with other transport modes by low operation costs and making a profit (Slack 2013).

In Europe, railway development and operation were traditionally based on public ownership and operation, owing to the large investment and operating costs. However, government control of railways is increasingly being viewed as inefficient, quasi-monopoly and unresponsive to the market. Many European countries aim to improve the efficiency and competitiveness of the passenger rail system by reforming railway governance: some have completely separated the infrastructure management from rail operations and placed it in a state-owned organization, such as in Sweden and the UK; others have retained the rail infrastructure and train operations within the same holding company but as separate divisions, such as in Germany, Italy and Greece; some have separated infrastructure management from operations while still maintaining a monopoly train operating company that retains control of many functions, such as in France (Nash 2008). By reforming the railroad institution, the market has an opportunity to participate in rail operation.

In Japan, there are several milestones. The first milestone was at the beginning of the 20th century, when the major private railways were nationalized. The second was after the Second World War, when Japanese National Railways (JNR) was set up as a public corporation. The third was in 1987 when the state-owned JNR Company was split up and privatized into six regional passenger railway companies and one national cargo company. This reform has greatly improved the economic efficiency and transport efficiency of railroads in Japan by reducing labor costs and encouraging more business initiatives. In addition, other railway operators are also allowed (Aoki et al. 2000).

In China, railroad reform has just started. Before 2013, the railroads in China were owned and operated by the government under the former Ministry of Railways (Wu and Nash 2000). In 2013, the Ministry of Railways was split into the National Railway Administration, responsible for making railway technology standard and supervising railway safety operation and railway construction projects, and China Railway Corporation, responsible for construction and management. Although the operation of railroads has been separated from the government organization, political instead of economic power still dominates (Lee 2008). State-owned governance has the advantages of fulfilling social responsibility, but it may lack incentives to improve service and pursue profits.

Local Railroads: Balancing Wider Benefits and Financial Costs

Railroads are typically financially risky investment projects in light of the high capital and operating costs. The construction of railway systems requires heavy capital investment, i.e. planning and land costs, infrastructure building costs and superstructure costs; and there is no guarantee of sound financial returns in a short period because of its high maintenance and operating costs. Hence, the planning and investment of rail transport often receive much attention and debate. For instance, it is argued that there should be socio-economic benchmarks for evaluating cities for building or extending metro systems, such as population size and income level (Loo and Cheng 2010). For other railway systems, careful planning and feasible studies to contrast benefits and costs should be undertaken before the construction.

Despite the huge capital costs, there are many ways to improve the financial efficiency. For instance, some parts of the Shinkansen in Japan obtained high revenue despite the higher construction cost because of higher population density, closer stops and higher frequency, as well as lower price compared with airfares (Lee 2007). In 2004, the daily passenger volume per km was 319 in Japan, much higher than in France (38 passengers per km) and Germany (79 passengers per km). In France, some of the TGV system has much higher financial efficiency than other countries owing to the lower construction cost. Its lower construction costs are related to advanced technology, urban form and terrain, the eschewal of mixed use (affecting gradient and curve parameters), and the use of existing tracks built for conventional trains to access urban centers, avoiding expensive urban construction. The construction cost of TGV is only 10 million Euros per km, much lower than that of the Shinkansen (35–41 million Euros per km). These two cases have shown that both high traffic demand and low construction cost are important for the financial efficiency of railways. This is supported by the argument of Vickerman (1997) that a HSR between two city centers requires the passenger demand to reach between 12 and 15 million per year so that it can compensate the cost.

The construction and operation model of railways can also influence their financial efficiency. For instance, the MTR of Hong Kong is commercially operated, unlike many other railway systems operated by the public sector. By the 'Rail and Property' development model, the MTR Corporation gets more than half of its income from property development that supports its rail transit operation (Cervero and Murakami 2009). Non-fare income from the property above the metro stations keeps the metro system financially sustainable without government subsidy and provides the citizens with high-quality services at affordable fares. This experience may enlighten other railway systems around the world, especially in cities with high population densities and rapid economic growth (Tang et al. 2004).

Railroad Terminals: Transit-oriented Development and Intermodality

For passenger rail terminals, there is evidence of the impacts of railways on the station-area development. In Japan, a case study about Tokyo shows that railway service may bring more economic activities, such as real estate, retail and commerce around the stations (Calimente 2012). However, it is argued that development around the stations, especially newly built ones, is closely related to their accessibility to the city center (Sands 1993). In Germany, the railway station Kassel-Wilhelmshohe has managed to attract many hotels, offices and retail activities around it (Sands 1993). By adopting the strategy of transit-oriented development, the land use structure around the railway stations changes (Cervero and Landis 1997), and the land and property value increases

(McDonald and Osuji 1995). However, studies show that not every catchment area of a railway station gets more development. For instance, the Le Creusot and Haute-Picardie stations along the TGV in France did not attract more business activities, mainly owing to their low passenger demand and service frequency (Yin et al. 2015). Therefore, passenger demand and accessibility of the station is important for its catchment development. Among the different kinds of economic activities, the location and prices for offices around the station area are most widely discussed. One study in the Netherlands concluded that, for different types of railway stations, their attractions for office location are different. The stations with international HSR services can attract international offices around the stations; however, domestic railway stations are less likely to attract offices, because of the short domestic distances in the Netherlands (Willigers and Van Wee 2011). Another study identifies several factors influencing the office prices in the area around railway stations. These factors include regional economy, image, regional rail accessibility and urban car accessibility, land development structure, public support, national accessibility and the 'clustering' effect (de Jong 2007). Actually, the property price change around the railway station can be complicated and hard to predict, because it is affected by multiple factors. Some studies show that the effect of railway stations can make the commercial property prices increase within a short range, and make the residential property prices increase over a long distance, especially for commuter railway stations (Debrezion et al. 2007). However, some other studies do not find any significant effect of the impact of railway on the housing prices in Tainan, Taiwan (Andersson et al. 2010).

The main functions of freight rail terminals are loading/unloading the freight from ships or road vehicles and transferring the goods wagons to local sidings. They are usually equipped with a lot of storage, loading sidings and loading/unloading facilities, and thus need larger spaces than the passenger rail terminals and are often located on the outskirts of cities (Rodrigue et al. 2013). However, the impact of freight railroads on the station-area development has changed over time. In the early period, the rail terminals tended to attract various kinds of industries and some even became important industrial areas. As the industrial structure changes and the location of industry depends less on the railroads, some rail terminals are relocated or abandoned by the end of the 20th century (Rodrigue 2015). There are many cases of the redevelopment of old freight rail terminals and their surrounding areas, especially in Europe, with its long history of railroads. Another trend is for the freight rail terminals to be incorporated into different transport supply chains and integrated with shipping or highways, especially port terminals, which has broadened the hinterlands and improved the efficiency of railroads.

CONCLUSIONS AND LOOKING FORWARD

The Future of Passenger Railroads

People's mobility is determined by fixed time and money budgets. It is predicted that future mobility of the world population will increase as income grows, and people tend to choose faster modes of transport within a fixed travel time budget (Schafer and Victor 2000). The past few decades have seen a significant decline in conventional railroads in passenger travel in almost all regions in the world. However, recent experiences in many countries show that the HSRs have challenged both the automobiles and the airplanes in medium-distance travel. For instance, HSR often dominates the travel market within 350 km travel distance in European city pairs, and competed with air transport from 350 to 1000 km (European Commission and Directorate General Transport 1998). Currently, the HSR network worldwide has reached 22,954 km and there are another 31,596 km under construction or planned before 2025 (UIC 2015b). It is expected that, in the future, HSRs may contribute more to intercity travel in various world regions. Yet the development of HSR is constrained by some regional geographical characteristics, such as high population density, appropriate topography and weather conditions, thus it is more popular in Asia and Europe but less developed in the Americas, Russia and Africa. Therefore, the renaissance of passenger railroads needs to be understood in regional contexts.

The role of railroads in urban transport is also important, since it can accommodate a large capacity of urban travel and reduce urban traffic congestion and pollution. In the past two decades, many cities in the world have built rail transit systems, such as metros, light rails and tramways (Goetz 2012). Currently, there are 162 metro systems, 94 light rail systems and 342 tramway systems in operation in 79 countries worldwide; meanwhile, there are 24 metros, 11 light rails and 28 tramways under construction and 13 metros, 9 light rails and 21 tramways planned (Light Rail Transit Association 2015). Since rail transit systems need large travel demand to support them, they are also determined by the geographical characteristics of the cities, such as urban population densities and land use characteristics.

In addition, despite the wider economic, social and environmental benefits of railroads, they are only part of the transport system, which comprises road, air, water and railroads. In the future, it will be necessary to integrate railroads with other transport modes to promote an integrated transport system to support peoples' mobility.

The Future of Freight Railroads

Since freight transport by railroads has the advantages of energy efficiency and lower carbon intensity (McKinnon 2007), it should be further promoted. Railway containerization has become a worldwide trend and many railroad stations have developed into land-based load centers (Loo and Liu 2005). However, railroads lack flexibility and may only be used for long-distance freight transport; hence, intermodal transport between railroads, road and water transport is important for sustainability (Mathisen and Hanssen 2014). In an intermodal transport system, the strengths of different transport modes can be utilized to minimize economic costs and energy consumption. For railroads, besides the conventional intermodality between railroad and marine transport to connect a seaport and its hinterland, intermodal rail–truck freight transport has also become important (Morlok and Spasovic 1994). It requires railroads and road transport to undertake long-haul and short-haul transport, respectively, by synchronizing the schedules between these two modes, using standardized load units, and coordinating between multiple actors of transport management (Bontekoning et al. 2004). Different kinds of innovations in transport technologies are developed to facilitate the intermodal activity (Bontekoning and Priemus 2004). For instance, 'Thruport' is a facility to handle high volumes of transmodal rail shipments and is applied successfully in North America. It has the advantages of reconciling time and flows in rail freight distribution (Rodrigue 2008). In the future, inter-modal transport between railroads and other transport modes still needs to be improved, such as through introducing new procedures or technologies to intermodal transport infrastructure, further standardizing vehicles between transport modes and countries, developing information systems for integrated transport, etc. (Janic 2001).

Based on the changing geography of railroads, this chapter argues that the revival of railroads has a spatial dimension, with passenger railroads concentrating in Asia and Europe and freight railroads concentrating in the Americas. By analyzing the spatial and geographical implications of railroads at different spatial levels, railroads can play an important role in sustainable transport worldwide. However, railroads still have their many limitations; thus, they have to be integrated with other transport modes to support both intermodal passenger and freight transport in the future.

REFERENCES

Anderson, J. and E. van Wincoop. 2003. Gravity with gravitas: a solution to the border puzzle. *American Economic Review* 93: 170–192.

Andersson, D., O.F. Shyr and J. Fu. 2010. Does high-speed rail accessibility influence residential property prices? Hedonic estimates from southern Taiwan. *Journal of Transport Geography* 18(1): 166–174.

Aoki, E., M. Imashiro, S. Kato and Y. Wakuda. 2000. *A History of Japanese Railways 1872–1999*. Tokyo: East Japan Railway Culture Foundation.

Black, J. 2015. National railway system. In B.P.Y. Loo and C. Comtois (eds.), *Sustainable Railway Futures: Issues and Challenges*. Farnham: Ashgate.

Bontekoning, Y. and H. Priemus. 2004. Breakthrough innovations in intermodal freight transport. *Transportation Planning and Technology* 27(5): 335–345.

Bontekoning, Y., C. Macharis and J. Trip. 2004. Is a new applied transportation research field emerging? A review of intermodal rail–truck freight transport literature. *Transportation Research Part A: Policy and Practice* 38(1): 1–34.

Calimente, J. 2012. Rail integrated communities in Tokyo. *Journal of Transport and Land Use* 5(1): 19–32.

Cervero, R. and J. Landis. 1997. Twenty years of the Bay Area Rapid Transit System: land use and development impacts. *Transportation Research Part A: Policy and Practice* 31(4): 309–333.

Cervero, R. and J. Murakami. 2009. Rail and property development in Hong Kong: experiences and extensions. *Urban Studies* 46(10): 2019–2043.

Chen, C. and P. Hall. 2011. The impacts of high-speed trains on British economic geography: a study of the UK's Inter-City 125/225 and its effects. *Journal of Transport Geography* 19: 689–704.

Dall'Erba, S. 2003. European regional development policies: history and current issues. University of Illinois EUC Working Paper, Vol. 2, No. 4.

de Jong, M. 2007. *Attractiveness of HST Locations* (Master of Urban Planning). Universiteit van Amsterdam, Amsterdam.

Debrezion, G., E. Pels and P. Rietveld. 2007. The impact of railway stations on residential and commercial property value: a meta-analysis. *Journal of Real Estate Finance and Economics* 35(2): 161–180.

European Commission. 2011. Roadmap to a Single European Transport Area – towards a competitive and resource efficient transport system, accessed 20 June 2015 from http://ec.europa.eu/transport/themes/strategies/doc/2011_white_paper/white_paper_com(2011)_144_en.pdf.

European Commission and Directorate General Transport. 1998. *Cost 318: Interactions between High-speed Rail and Air Passenger Transport: Final Report*. Luxembourg: Office for Official Publications of the European Communities.

European Union. 2014. *EU Transport in Figures: Statistical Pocketbook 2014*. Luxembourg: Publication Office of the European Union.

Eurostat. 2014a. Railway passenger transport statistics – quarterly and annual data, accessed 20 June 2015 from http://ec.europa.eu/eurostat/statistics-explained/index.php/Railway_passenger_transport_statistics_-_quarterly_and_annual_data.

Eurostat. 2014b. Railway freight transport statistics, accessed 20 June 2015 from http://ec.europa.eu/eurostat/statistics-explained/index.php/Railway_freight_transport_statistics.

Fogel, R. 1979. Notes on the social saving controversy. *Journal of Economic History* 39(1): 1–54.

Fujimura, M. 2004. Cross-border transport infrastructure, regional integration and development. ADB Institute Discussion Paper No.16.

Garmendia, M., C. Ribalaygua and J. Urena. 2012a. High-speed rail: implication for cities. *Cities* 29: 526–531.

Garmendia, M., V. Romero, J. Urena, J. Coronado and R. Vickerman. 2012b. High-speed rail opportunities around metropolitan regions: Madrid and London. *Journal of Infrastructure Systems* 18(4): 305–313.

Goetz, A. 2012. Introduction to the Special Section on rail transit systems and high speed rail. *Journal of Transport Geography* 22: 219–220.

Hall, P. 2009. Magic carpets and seamless webs: opportunities and constraints for high-speed trains in Europe. *Built Environment* 35: 59–69.

Janic, M. 2001. Integrated transport systems in the European Union: an overview of some recent developments. *Transport Reviews* 21(4): 469–497.

Lee, J. 2008. Railway governance and power structure in China. *International Journal of Railway* 1(4): 129–133.

Lee, Y.S. 2007. A study of the development and issues concerning high speed rail (HSR). Working Paper. Oxford University: Korea Railroad Research Institude and Transport Studies Unit.

Li, L. and B.P.Y. Loo. 2015. The promotion of social equity through railways. In B.P.Y. Loo and C. Comtois (eds.), *Sustainable Railway Futures: Issues and Challenges*. Aldershot: Ashgate.

Light Rail Transit Association. 2015. A complete listing of light rail, light railway, tramway and metro systems throughout the world, accessed 30 June 2015 from http://www.lrta.org/world/worldind.html.

Loo, B.P.Y. 2015. Prospects for sustainable railways. In B.P.Y. Loo and C. Comtois (eds.), *Sustainable Railway Futures: Issues and Challenges*. Aldershot: Ashgate.

Loo, B.P.Y. and A.H.T. Cheng. 2010. Are there useful yardsticks of population size and income level for building metro systems? Some worldwide evidence. *Cities* 27(5): 299–306.

Loo, B.P.Y. and L. Li. 2012. Carbon dioxide emissions from passenger transport in China since 1949: implications for developing sustainable transport. *Energy Policy* 50: 464–476.

Loo, B.P.Y. and L. Li. 2015. Rail transport (passenger). In D. Richardson, M. Castree, F. Goodchild, A.L. Kobayashi, W. Liu and R. Marton (eds.), *The International Encyclopedia of Geography: People, the Earth, Environment, and Technology*. Malden, MA: Wiley-Blackwell.

Loo, B.P.Y. and K. Liu. 2005. A geographical analysis of potential railway load centers in China. *Professional Geographer* 57(4): 558–579.

Mathisen, T. and T. Hanssen. 2014. The academic literature on intermodal freight transport. *Transportation Research Procedia* 3: 611–620.

Maurer, H., A. Burgess, P. Hilferink, E. Kroes and T. Whiteing. 2010. Effects of the EU rail liberalisation on international rail passenger transport. In European Transport Conference, 2010, accessed from http://citeseerx.ist.psu.edu/viewdoc/download?doi=10.1.1.681.2964&rep=rep1&type=pdf.

McDonald, J. and C. Osuji. 1995. The effect of anticipated transportation improvement on residential land values. *Regional Science and Urban Economics* 25(3): 261–278.

McKinnon, A. 2007. CO_2 emissions from freight transport: an analysis of UK data, accessed from http://www.greenlogistics.org/SiteResources/d82cc048-4b92-4c2a-a014-af1eea7d76d0_CO2%20Emissions%20from%20Freight%20Transport%20-%20An%20Analysis%20of%20UK%20Data.pdf.

Morlok, E. and L. Spasovic. 1994. Redesigning rail-truck intermodal drayage operations for enhanced service and cost performance. *Journal of the Transportation Research Forum* 34(1): 16–31.

Nash, C. 2008. Passenger railway reform in the last 20 years – European experience reconsidered. *Research in Transportation Economics* 22(1): 61–70.

Owen, W. 1987. A global overview. In *Transportation and World Development*. Baltimore, MD: Johns Hopkins University Press.

Panayotou, T. 1998. *The Role of the Private Sector in Sustainable Infrastructure Development*. Cambridge, MA: Harvard Institute for International Development.

Reusser, D., P. Loukopoulos, M. Stauffacher and R. Scholz. 2008. Classifying railway stations for sustainable transitions–balancing node and place functions. *Journal of Transport Geography* 16(3): 191–202.

Rodrigue, J.-P. 2008. The thruport concept and transmodal rail freight distribution in North America. *Journal of Transport Geography* 16(4): 233–246.

Rodrigue, J.-P. 2015. Structuring effects of rail terminals. In B.P.Y. Loo and C. Comtois (eds.), *Sustainable Railway Futures: Issues and Challenges*. Aldershot: Ashgate.

Rodrigue, J.-P., C. Comtois and B. Slack. 2013. *The Geography of Transport Systems*. London: Routledge.

Ross, J. 1994. High-speed rail: catalyst for European integration? *Journal of Common Market Studies* 32: 191–214.

Rostow, W. 1960. *The Stages of Economic Growth: A Non-Communist Manifesto*. New York: Cambridge University Press.

Sands, B. 1993. The development effects of high-speed rail stations and implications for California. Working paper, Institute of Urban and Regional Development, University of California, Berkeley, CA.

Schafer, A. and D. Victor. 2000. The future mobility of the world population. *Transportation Research Part A: Policy and Practice* 34(3): 171–205.

Shaw, J. and I. Docherty. 2009. Transport: railways. In N. Thrift and R. Kitchin (eds.), *International Encyclopedia of Human Geography*. pp. 91–99. Amsterdam: Elsevier.

Slack, B. 2013. Transport planning and policy. In J.-P. Rodrigue, C. Comtois and B. Slack (eds.), *The Geography of Transport Systems*. New York: Routledge.

Tang, B.S., Y.H. Chiang, A. Baldwin and C.W. Yeung. 2004. *Study of the Integrated Rail-Property Development Model in Hong Kong*. Hong Kong: Hong Kong Polytechnic University.

UIC. 2014. Synopsis 2013, accessed 18 June 2015 from http://www.uic.org/spip.php?article3299.

UIC. 2015a. General definitions of highspeed, accessed 16 June 2015 from http://www.uic.org/spip.php?article971.

UIC. 2015b. High Speed Lines in the World, accessed 18 June 2015 from http://www.uic.org/IMG/pdf/20140901_high_speed_lines_in_the_world.pdf.

US Central Intelligence Agency. 2015. The World Factbook, accessed 24 June 2015 from https://www.cia.gov/library/publications/the-world-factbook/rankorder/2121rank.html.

Vickerman, R. 1995. The regional impacts of trans-European networks. *The Annals of Regional Science* 29(2): 237–254.

Vickerman, R. 1997. High-speed rail in Europe: experience and issues for future development. *Annals of Regional Science* 31: 21–38.

Vickerman, R. 2007. Policy implications of dynamic globalized freight flows in North America. In M. Maggio and R. Stough (eds.), *Globalized Freight Transport: Intermodality, E-commerce, Logistics and Sustainability*. Cheltenham, UK and Northampton, MA, USA: Edward Elgar Publishing.

Vickerman, R. 2015. Railways and borders: the international dimension. In B.P.Y. Loo and C. Comtois (eds.), *Sustainable Railway Futures: Issues and Challenges*. Farnham: Ashgate.

Willigers, J. and B. Van Wee. 2011. High-speed rail and office location choices. A stated choice experiment for the Netherlands. *Journal of Transport Geography* 19(4): 745–754.

Wu, J.H. and C. Nash. 2000. Railway reform in China. *Transport Reviews* 20(1): 25–48.

Yin, M., L. Bertolini and J. Duan. 2015. The effects of the high-speed railway on urban development: international experience and potential implications for China. *Progress in Planning* 98: 1–52.

18. Ports and maritime technology
Jean-Paul Rodrigue

ANCIENT TECHNOLOGY, NEW WORLD

Maritime transport involves maritime shipping and the ports operations, an industry that is extensively globalized, having assets in every market. Since ancient times, maritime transportation has been linked with the scale and scope of trade and its commercial activities. While in the past trade was mostly conducted out of convenience and profit to supply goods, many of which not essential, trade became increasingly a factor of economic benefit covering a whole range of goods from the essential to the trivial. The origins of maritime shipping can be traced to around 3200 BCE when small Egyptian coastal and river sail ships began to be used. By 1200 BCE these routes reached as far as Sumatra, but volumes were limited to the vagaries of sailing technologies. Ships were not designed to effectively face the challenges of long-distance navigation and physical elements such as sea currents and wind patterns. Phoenician ships supported the first main trade routes along the Mediterranean from 1200 BCE to 800 BCE, setting the stage for Greek and Roman trade networks from 600 and 200 BCE, respectively. Between the first and sixth centuries, ships were sailing between the Red Sea and India, aided by summer monsoon winds. From the ninth century, these maritime routes became controlled by the Arab traders. Asia also saw the development of its own regional trade networks with Chinese merchants frequenting the South China Sea and the Indian Ocean by the tenth century. Most of early maritime shipping activity focused around the Mediterranean, the northern Indian Ocean and Pacific Asia. From the fifteenth century these networks would further expand into the North Atlantic and the Caribbean.

Although sailing was used for millennia to transport people and cargoes, the capacity of sail ships rarely exceeded 100 tons and these ships were not designed for deep-sea travel. For instance, the Dhow was an ancient coastal sailing ship found through the Middle East and South Asia and supported Arab trade during the feudal era. Through their innovation in maritime technology, European colonial powers, mainly Spain, Portugal, England, the Netherlands and France, would be the first to establish a true global maritime trade network. In the fifteenth century, new ship designs started to emerge, able to carry larger quantities of cargo over longer distances. The caravel was the first major breakthrough in European maritime technology, used by Spain and Portugal from the thirteenth century. These early caravels were small, three-masted vessels with a crew of five or six and were about 50 tons in size. They were used as fishing boats or coastal cargo ships since navigation could not permit high sea faring until the middle of the fourteenth century.

From the 1430s to the 1530s caravels were used for trade and exploration, being between 100 and 200 tons in size. One of the first effective cargo sailing vessels was the carrack, which could carry 1500 tons at about 10 km/hour. From the fifteenth to the seventeenth centuries, the carrack became the linchpin of long-distance maritime trade. By the seventeenth century, it was gradually replaced by the galleon, which being on average of even lower capacity, was much more maneuverable and cost effective. Still, some galleons exceeded 1500 tons. Sailing technology reached its peak efficiency by the nineteenth century, when clipper ships were introduced that were able to carry a good quantity of cargo with effective distances of about 700 km per day. Still, the effective daily travel distance for maritime transport can vary considerably depending on prevailing wind and sea current conditions, which is a major factor in the effectiveness of sail ships.

With the development of the steam engine in the mid-nineteenth century, trade networks expanded considerably with maritime technology much less constrained by dominant wind and sea current patterns. The mechanization of maritime technology was supported by improved civil engineering capabilities, enabling extensive commercial harbors and artificial canals to be built. With the opening of the Suez Canal (1868), the second half of the nineteenth century saw an intensification of maritime trade between Europe and Asia. The completion of the Panama Canal in 1914 led to additional trade opportunities in the Americas. The lock size also became a de facto standard in maritime shipping that would endure until late in the twentieth century. During that century maritime transport grew exponentially as globalization and seaborne trade became interrelated (Stopford 2009). These trade relations are also influenced by the existing maritime shipping capacity with bigger ships and port development projects taking place across the world (Talley 2009). There is thus a level of reciprocity between trade and maritime shipping capabilities. In the first decade of the twenty-first century, seaborne trade accounted for 90% of global trade in terms of volume and 70% in terms of value. Maritime shipping remains one of the most globalized industries in terms of ownership and operations (Pinder and Slack 2004). It is, however, the development of containerization that can be considered the most significant technology impacting ports and maritime shipping.

THE HUMBLE HERO: THE CONTAINER

Simple Idea, Complex Ramifications

Known as the 'silent humble hero' of the global economy (Levinson 2006), containerization is practically synonymous with globalization. Few other technologies had such wide ramifications for the geography of production, consumption and distribution (Fremont 2007). A wide variety of goods can be carried in a container, such as toys, apparel, car parts, televisions, meat, fruits, coffee and lumber. A container is a large standard-size metal box into which cargo is packed for shipment aboard specially configured transport modes. The reference size is the 20-foot box of 20 feet long, 8 feet 6 inches feet high and 8 feet wide. This represents one Twenty-foot Equivalent Unit (TEU), a measure commonly used to refer to the level of container activity. 'Hi cube' containers are also common and they are 1 foot higher (9 feet 6 inches) than the standard container height. Containers are designed to be moved with common handling equipment, enabling quick intermodal transfers between ships, railcars, truck chassis and barges using a minimum of labor. The container serves as the load unit rather than the cargo contained therein.

The idea behind containerization is rather simple since the container remains a box with a latching system. Its inventor, Malcom McClean, purposefully did not patent his invention so that the container could diffuse rapidly. However, from the concept ideas designed in the 1930s, we have to wait until the 1950s for the first commercial application of the container and until the 1980s to see the beginning of its massive diffusion as a support to global trade. Therefore, the simple concept of the container came with massive ramifications on the design and engineering of transportation modes and terminals, the organization of supply chains, location and economic activities as well as the corporations carrying them (Song and Panayides 2012).

A Transport and Logistical Unit

Within global production networks the container is concomitantly a transport, production and distribution unit. Irrespective of what it carries, the container is a transport unit requiring modes, terminals, infrastructure and equipment to undertake modal and intermodal operations. As a transport unit, the container thus requires an extensive technical and technological system. As a production unit the container carries inputs, intermediary goods and outputs (final goods). With the container becoming the standard transport unit for international transportation, many production segments have embedded the container as a production planning unit with inputs and outputs considered as containerized batches. Industrial capacity is as such a function of intermodal capacity. The container became a distribution unit as well, leading to radical changes in freight distribution with a switch to time-based management strategies. The shorter the transit time, the lower the inventory level, which can result in significant cost reductions. The fact that the container is also its own warehousing unit has led to new distribution strategies where the modes as well as the terminals can be part of inventory management systems.

Keeping it Cool: Refrigerated Containers

The transport of refrigerated goods by container represents a market where this technology has substantial impacts. Land, sea and air modes all have different technologies and operations for keeping food fresh throughout the transport chain. Refrigerated containers, also known as reefers, account for a growing share of the refrigerated cargo being transported around the world. While in 1980, 33% of the refrigerated transport capacity in maritime shipping was containerized, this share rapidly climbed to 72% in 2013. The reefer has become a common temperature-controlled transport unit used to insure load integrity since it can accommodate a wide range of temperature settings and accordingly a wide range of temperature-sensitive cargoes. Also, it is a versatile unit able to carry around 20–25 tons of refrigerated cargo and is fully compatible with the intermodal transport system, which implies a high level of accessibility to markets around the world. Refrigerated containers account for more than 50% of all the refrigerated cargo transported in the world, with the banana alone accounting for 20% of all seaborne reefers trade.

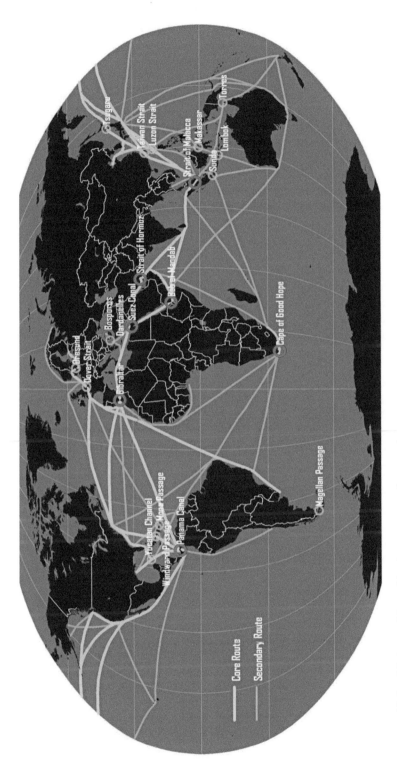

Figure 18.1   Main maritime shipping routes and strategic passages

MARITIME SHIPPING AND THE APPEAL OF MASSIFICATION

Maritime Circulation

Maritime transportation operates on its own space, which is geographical by its physical attributes, strategic by its control and commercial by its usage (Kaluza et al. 2010). While geographical considerations tend to be constant in time (with the exception of the seasonality of weather patterns), strategic and especially commercial considerations are much more dynamic. As such, geography confers a stability to the structure of global shipping networks that are forced to pass through specific strategic passages with the primary passages being the most important (Figure 18.1). Without them there would be limited cost-effective maritime shipping alternatives, which would seriously impair global trade. Among those are the Panama Canal, the Suez Canal, the Strait of Hormuz and the Strait of Malacca, which are key locations in global trade. Secondary passages support maritime routes that have alternatives, but would still involve a notable detour. These include the Magellan Passage, the Dover Strait, the Sunda Strait and the Taiwan Strait. The Suez and Panama canals are among the best examples of massive achievements in maritime technology. In spite of both being above a century old, both remain critical bottlenecks in global trade.

The physical and operational characteristics of strategic passages dictate the volume and capacity of ships transiting through them. Capacity can be measured in deadweight tons (dwt), which is the mass a ship can carry, and in TEUs, which is the volume a containership can carry. The capacity of a passage in dwt does not relate effectively to the capacity in TEU. For instance, a 220,000 dwt containership would have a capacity of about 20,000 TEU and a draft of 16.5 m, while a similar tanker would have a draft reaching 21 m. The reason behind this gap lies in the cargo density, also known as the stowage factor, which is the ratio between cargo volume and cargo weight. For containers the average cargo density is about 3 m^3 per ton, while for oil and iron ore this ratio is 1.1 and 0.44 m^3 per ton, respectively.

The capacity of a strategic passage is also a function of the number of transits it can handle. The Panama Canal can handle about 45 transits per day, while this figure is about 100 (78 before the expansion) per day for the Suez Canal. The capacity of the Strait of Malacca is difficult to assess, but about 210 commercial ships are transiting each day. By handling about 30% of the global maritime trade, Malacca is the most important passage in the global economy, followed by Suez (15%) and Panama (5%).

Mega Maritime Structures: Panama and Suez Canals

The continuous growth of global trade since the 1990s has placed additional pressures on two major artificial chokepoints of global trade, the Panama and the Suez canals, to handle a growing number of ships in a timely and predictable manner. This became increasingly apparent as a growing share of the global containership fleet reached a size beyond the capacity of the Panama Canal. The usage of these 'post-Panamax' ships along the Pacific Asia/Suez Canal/Mediterranean routes as well as the development of the North American rail land bridge using double-stacked container trains competed with the canal as an intermediate location in global maritime shipping.

A decision to expand the Panama Canal was reached in 2006 by the Panamanian government. The expansion is a US$6 billion project that involves building a new set of locks on both the Atlantic and Pacific sides of the canal to support a depth of 60 feet, a width of 190 feet and a length of 1400 feet, which would accommodate ships up to 12,500 TEU depending on their load configuration. The dredging of access channels as well as the widening of several sections of the existing canal is also required. This would allow larger bulk vessels to pass through the canal, thus permitting new opportunities for shipping services. A new containership class, dubbed New-Panamax, will be created to add to the existing Panamax ship class.

The Suez Canal also underwent significant improvements in recent years. With additional deepening and widening projects, the depth of the canal reached 22.5 m in 2001 and 24 m in 2008, to accommodate the largest Post-Panamax containerships. In 2014, a new expansion project that increased the capacity to about 100 transits per day took place. The 'New Suez Canal' was inaugurated in August 2015 at a cost of over US$8 billion. This expansion included a new 35 km section enabling the canal to transit ships in both directions at once. Prior to this expansion, convoys were organized since a section of the canal could only handle ships in one direction at a time. The expansion substantially improved the operational capabilities of the canal by reducing waiting and transit times in addition to improving capacity. The expanded Suez Canal has draft limitations similar to those of the Strait of Malacca.

THE PUSH TOWARDS ECONOMIES OF SCALE

Maritime transportation is the most energy-efficient mode, but bunker fuel accounts for about 50% of its operating costs. Fluctuations in energy prices have a direct impact on operations with limited options to absorb higher energy costs. The outcome is a greater reliance on economies of scale to mitigate those costs. The high-energy-price environment that prevailed from the mid-2000s to the mid-2010s incited maritime to adopt slow-steaming (reducing speed) practices for many of its services. Yet maritime transportation only accounts for 2.3% of greenhouse gas emissions attributed to fossil fuel consumption. The global market is dominated by a few maritime shipping companies in close relationship with global terminal operators, some of which are parent companies. The system is oligopolistic (10 dominant shipping companies and four global terminal operators), but highly competitive (Slack and Fremont 2006). A powerful economic force in maritime shipping in the last two decades has been economies of scale leading to larger containerships, placing pressures on port terminal and inland freight distribution to cope.

The principle of economies of scale is fundamental to the economics of maritime transportation as the larger the ship, the lower the cost per unit transported (Cullinane and Khanna 2000). This trend has been particularly apparent in bulk and containerized shipping. For instance, the evolution of containerization, as indicated by the size of the largest available containership, has been a stepwise process (Figure 18.2). Changes are rather sudden and correspond to the introduction of a new class of containership by a shipping company (Maersk Line tended to be the main early mover), quickly followed by others. In a sense, each of the steps represents a technological evolution with improved ship and engine design.

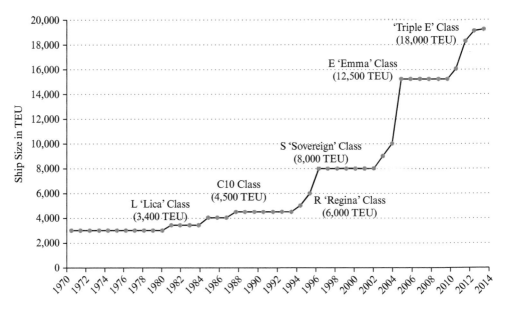

Figure 18.2   The largest available containership, 1970–2015 (in TEUs)

A new class generally takes the name of its first ship put in commercial use. There are variations concerning how many containers can be carried by a containership depending on the method of calculation. For instance, for the Emma class, a ship can carry about 15,200 TEUs of containers if they are all empty, which represents all of the available container slots. If all of the carried containers are loaded with an average load of 14 tons per container, then about 11,000 TEUs can be carried (25% less). The official capacity figures used is 12,500 TEUs, which considers that containerships carry a mix of loaded and empty containers, but containerships are usually able to carry slightly more.

Since the 1990s, three substantial steps have taken place in the evolution of containership sizes. The first involved a jump from 4000 to 8000 TEUs, effectively moving beyond the 'Panamax' threshold of around 5000 TEU, an important operational limitation in maritime shipping. Most modern container ports were designed to handle ships of the Panamax class since this standard has been around for a century.

The second step took place in the 2000s to reach the 12,500 TEU level, which is essentially a 'New-Panamax' class when the extended Panama Canal will come online in 2016. However, this 12,500 TEU capacity is assessed from a ship fully loaded with containers averaging 14 tons each. Using a load configuration that includes empties as well can result in effective capacities of 15,200 TEU for an E class ship.

A third step is unfolding. In 2011, China Shipping Container Lines took delivery of the first containerships whose theoretical capacities are 14,100 TEUs, setting a new landmark above the 14,000 threshold. However, the range of ports that can accommodate these ships is limited, so their service configuration is scheduled along the East Asia–Europe route. Later in 2011, Maersk announced the order of a new generation of 18,000 TEU ships, dubbed 'Triple E' vessels, the first delivered in 2013. The 18,000 TEU

Table 18.1 Technical characteristics of the world's three main strategic passages

	Panama/expanded	Suez	Malacca
Minimum depth	12 m (40 feet)/15.2 (50 feet)	24 m (78 feet)	21 m (68 feet)
Length	64 km	193 km	800 km
Standard	Panamax/New-Panamax	Suez-max	Malacca-max
Ship capacity	65,000 dwt/119,000 dwt	240,000 dwt	240,000 dwt
TEU equivalent	4500/12,500	22,000	20,000
Capacity (transits per day)	45/57	100	500 (approximate)
Transits (day | year)	36 | 14,000	47 | 17,000	210 | 75,000
Average transit time	16.5 h	14 h	20 h
Share of global maritime trade	About 5%	About 15%	About 30%

threshold was overtaken in 2014 when ships of more than 19,000 were introduced. Ship sizes are likely to level around 20,000 TEU, which is close to the limit of the Suez Canal and the Strait of Malacca (Table 18.1).

From a maritime shipper's perspective, using larger containerships is a straightforward process as it conveys economies of scale and thus lowers costs per TEU carried. From a port terminal perspective, this places intense pressures in terms of infrastructure investments, namely portainers. Several converging factors underline that further economies of scale in maritime shipping are unlikely to unfold within the foreseeable future, or at least would come at a high cost. The more economies of scale are applied, the lower the number of ports able to handle such ships, which limits commercial options and accessibility. Economies of scale involve higher costs for inland operations as a large quantity of containers arrives at once and must be handled effectively to maintain an adequate level of service. Economies of scale require higher levels of capital intensiveness in infrastructure and equipment (ships, portainers, terminal facilities) that is prone to risk.

Larger quantities of containers in circulation have incited maritime shipping companies to rely more on transshipment hubs to connect different regions of the world. This went on a par with the growing share of transshipments in regard to the totality of maritime containerized traffic, from around 11% in 1980, to 19% in 1990, to about 28% in 2012. The challenge is no longer about building larger ships, but about finding paying cargo to fill the ships. It is therefore a possibility that the optimal size of a containership would be in the 8000–10,000 TEUs range. Containerization may be entering a phase of maturity where growth prospects are more limited, which is not because of technological constraints, but because of commercial and physical limitations of maritime passages and ports.

PORTS: BETWEEN FORELANDS AND HINTERLANDS

The Complexity of Port Infrastructure and Superstructure

Port activities cover a wide range of functions, each adding a layer of complexity and technological expertise. For instance, providing maritime access involves activities such

Table 18.2   Main components of port technology

Land	Land acquisition (purchase or lease)
	Land reclamation projects
Maritime access	Access channel dredging
	Sea locks and breakwaters
	Vessel traffic service and ship movement information networks
	Light buoys and navigational aids
Port infrastructure	Internal locks; docks, quays, jetties, piers, berths
	Harbor basin dredging
Port superstructure	Pavement
	Warehouses, sheds
	Cranes and gantries and other mobile/semi-mobile equipment
	Terminal and office buildings
	Public utilities (sewage, water supply, electricity)
Infrastructure links	Railways, roads, canals, tunnels and bridges within the port area
Port maintenance	Maintenance dredging; infra- and superstructure
Port services	Cargo handling (stevedoring, storage, stowage)
	Nautical services (pilotage, towage, mooring)
	Other services (firefighting, water and electricity supply, security, bunkering, pollution control, etc.)
Port planning	Promoting logistics and industrial areas
	Marketing to existing and potential users
	Planning of infrastructure and superstructure developments

as the dredging of access channels, the construction of breakwaters (if required), the setting of vessel traffic services to help navigation to approach and navigate within the port as well as buoys and other navigation aids. All of these activities fall under different branches of port technology (Table 18.2). Infrastructure investments (civil engineering) and developments are usually done by port authorities while superstructure investments (mechanical engineering) are done by terminal operators, many of which are private companies leasing terminals. All these activities are necessary for the planning, development and operation of a port and are part of its incurred costs.

Terminal Automation

Since its inception, containerization was a highly mechanized activity, inciting investment in a large array of equipment types, including portainers and gantry cranes. Paradoxically, the level of automation remained relatively low as all the equipment was manually operated. Adding capacity usually involved adding equipment, but the coordination between the different elements of port terminal operation required increasing effort. Terminal automation involves the transformation of mechanized operations requiring a human operator to different systems where operations are fully or partially automated, often remotely. Automation is particularly suitable when labor and land costs are high, which gives an incentive to move towards more efficient alternatives, not necessarily to increase the level of port throughput, but to increase terminal productivity. Therefore, a significant constraint to automation remains the existing terminal cost structure and congestion level, which is co-dependent on the terminal's geographical setting. For instance, several Chinese container ports have limited incentive to automate since their labor cost is much lower; increasing throughput was done by adding labor and equipment. For North America and Europe, port labor costs are much higher, and so the push towards terminal automation is more prevalent.

Automation can be applied to three intermodal operations within the terminal. The first concerns loading and unloading operations. Since ships are getting bigger, the distance containers have to travel between the ship and the shore (or vice versa) has increased. To maintain the productivity (e.g. number of movements per hours), crane operations need to be faster, which cannot be effectively achieved without automation. Cranes able to handle several (two or four) containers at once are also introduced, many of which are automatically and remotely controlled. The second intermodal operation relates to yard management. Automated facilities can stack containers higher and with more density, which requires rail-mounted gantries or automated stacking cranes. Automated vehicles are also introduced to carry containers from cranes to stacking yards, increasing the throughput of both. The third intermodal operation involves the interface between the terminal and inland transport systems, which is usually improved with automated terminal gates where inbound or outbound containers can be rapidly scanned and allocated a slot for pick-up or delivery at the terminal. This increases the number of movements a truck can perform per day and lessen the average time spent at the terminal or waiting to access the terminal. In the coming years and with technological improvements, additional terminal automation is expected and should become the norm.

Ports as Logistics Clusters

Although specific economic activities (ship building, warehouses, offices of shipping agents, heavy industries, etc.) near ports have existed since the Industrial Revolution, the renewed role of ports as gateways to global trade has favored a convergence of logistics activities (Robinson 2002). Port-centric logistics zones have been planned in proximity to many port terminal facilities and almost every new port terminal project is usually accompanied by related logistics zones projects. Such zones support freight distribution activities directly related to maritime shipping. From a freight distribution perspective, inventory management is improved since containers can be picked up or dropped at the terminal facility. The added security that a port-centric logistics zone offers is also a positive factor, particularly in developing countries where cargo theft is more prevalent. Port-centric logistics zones do, however, have some drawbacks, particularly since they involve higher land costs and potentially more restrictive labor regulations if they are within the jurisdiction of dock workers. They also lock the shipping options of its customers to the port, which may not be the most suitable or may shift if shipping lines decide to change the configuration of their services.

Port authorities tend to be proactive in the development of port-centric logistics since this provides added value and helps diversify their involvement in regional freight distribution. Satellite terminal facilities supporting port activities can also be developed, such as off-dock rail yards and empty container depots. Port-centric logistics zones can be export-oriented or import-oriented depending on the trade structure they are embedded with. Most of China's special economic zones can be considered as export-oriented port-centric logistics zones. Dubai has emerged as a major logistics hub in conjunction with massive investments in its port terminal (Jebel Ali) and the development of several major logistics zones in proximity. In North America and Europe, many port authorities have been proactive in the development of logistics zones. For instance, Savannah, Georgia is a gateway that has experienced significant development in import-oriented port-centric logistical activities with the growth of all-water services between Pacific Asia and the East Coast through the Panama Canal.

Pushing Massification in the Port Hinterland

Economies of scale in maritime shipping have pushed for improvements in port capacity and throughput, which in turn push for improvement in inland freight distribution. Freight flows on the foreland and hinterland are not taking place at the same momentum, particularly since economies of scale have been more effectively applied on the foreland than on the hinterland (Figure 18.3). In light of an increasing massification of containerized freight loads, and while the ultimate goal remains atomization (individual containers delivered to freight owners), the insertion of an intermediate hub can in some circumstances act as a mitigation strategy (Fleming and Hayuth 1994). Larger containerships can call at intermediate hubs with high capacity and frequency services. Through feedering, ports serviced through an intermediate hub can have smaller feeder ships (e.g. Panamax class) calling more frequently.

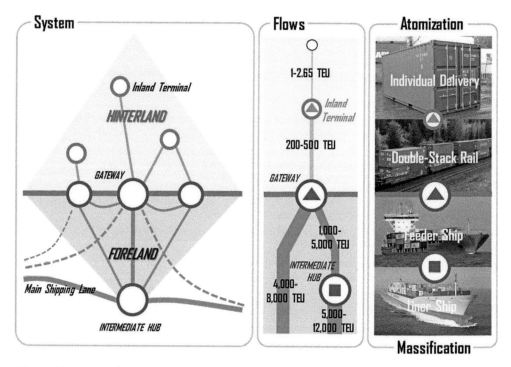

Figure 18.3   Port foreland and hinterland: from massification to atomization

There are also site constraints, environmental factors or market potentials that may limit the volumes generated by the hinterlands of some ports. The growth of hinterland traffic, if only supported by drayage (truck) operations, leads to increasing diseconomies such as congestion and capacity limitations. With this, the inland transport system involves higher transport costs and unreliability in freight distribution. The use of inland terminals serviced by rail or barge services can help cope with these diseconomies through a massification of several inland flows. Under such circumstances, many ports around the world have been actively involved with the development of their hinterland connectivity (Notteboom and Rodrigue 2005; Roso and Lumsden 2010). As such, technological developments on the maritime and port sides are inciting technological changes in how inland destinations are serviced.

CONCLUSION: TECHNOLOGY AND THE RATIONALIZATION OF MARITIME SHIPPING

Similar to most transport modes, technological development in maritime shipping focused on improving the capacity of modes and terminals, their energy efficiency, their performance and their safety. The ongoing rationalization of ports and shipping networks, in part driven by economies of scale, may lead to the setting of a global system of maritime freight distribution composed of a circum-equatorial route, transoceanic services and north–south connectors interwoven by intermediate hubs (Rodrigue 2010). Among the drivers that are most likely to impact this industry in the future are security issues, real estate pressures around terminal facilities, economies of scale, environmental concerns, the integration of information technologies, progresses in terminal automation and finding capital for financing.

As a greater share of the global population lives in urban areas, additional pressures are felt on port terminal facilities that find themselves with limited room for expansion. New sites are therefore located further away from existing activity centers and facing pressures to improve productivity. Hinterland transportation has also expanded with the setting of gateways and their corridors, helping accommodate global trade through intermodal options (e.g. rail and barge). Supply chain integration supported by information technologies has been beneficial for global freight distribution and will continue to do so in the future with a better utilization of existing port, ship and inland assets. Still, many of the drivers that have propelled growth in the shipping and port industries are being questioned.

Since containerized maritime transportation is strongly derived from the level of consumption of final goods, technological changes in the manufacturing sector are likely to place pressures on future growth potentials. Several manufacturing sectors are becoming less dependent on basic inputs costs such as labor and land; automation and robotization are emerging rapidly. Thus, there is a distinct possibility that there may be less offshoring (peak globalization) and therefore less growth incentives for maritime shipping. Like many economic sectors, technological changes can both be a factor of growth and rationalization and ports and maritime transportation continuously feeling these forces.

REFERENCES

Cullinane, K. and M. Khanna. 2000. Economies of scale in large containerships. *Journal of Transport Economics and Policy* 33: 185–207.

Fleming, D. and Y. Hayuth. 1994. Spatial characteristics of transportation hubs: centrality and intermediacy. *Journal of Transport Geography* 2(1): 3–18.

Fremont, A. 2007. *Le monde en boîtes. Conteurisation et mondialisation.* Paris: Les collections de l'Inrets.

Kaluza, P., A. Kölzsch, M. Gastner and B. Blasius. 2010. The complex network of global cargo ship movements. *Journal of the Royal Society Interface* 74(8): 1093–1103.

Levinson, M. 2006. *The Box: How the Shipping Container Made the World Smaller and the World Economy Bigger.* Princeton: Princeton University Press.

Notteboom, T. and J-P. Rodrigue. 2005. Port regionalization: towards a new phase in port development. *Maritime Policy and Management* 32(3): 297–313.

Pinder, D. and B. Slack (eds.). 2004. *Shipping and Ports in the Twenty-first Century: Globalisation, Technological Change and the Environment.* London: Routledge.

Robinson, R. 2002. Ports as elements in value-driven chain systems: the new paradigm. *Maritime Policy and Management* 29(3): 241–255.

Rodrigue, J-P. 2010. Maritime transportation: drivers for the shipping and port industries. International Transport Forum, Forum Paper 2010-2, Organization for Economic Cooperation and Development.

Roso, V. and K. Lumsden. 2010. A review of dry ports. *Maritime Economics & Logistics* 12(2): 196–213.

Slack, B. and A. Fremont. 2005. Transformation of port terminal operations: from the local to the global. *Transport Reviews* 25(1): 117–130.

Song, D-W. and P. Panayides. 2012. *Maritime Logistics: Contemporary Issues.* Wagon Lane, Bingley, UK: Emerald Group.

Stopford, M. 2009. *Maritime Economics*, 3rd edn. London: Routledge.

Talley, W. 2009. *Port Economics.* New York: Routledge.

PART V

ENERGY TECHNOLOGIES

19. Assessing the spatial, economic and environmental implications of biorefining technologies: insights from North America
Kirby E. Calvert, Jamie D. Stephen, M.J. Blair, Laura Cabral, Ryan E. Baxter and Warren E. Mabee

Fossil fuels, including oil, natural gas and coal, currently provide almost 80% of global primary energy supply. The use of fossil fuels extends well beyond the energy sector. On average, approximately 5% of the total output from a petroleum refinery is used to manufacture the plastics, paints and basic polymers that constitute the fabric of our material existence, while 4% of the output from a petroleum refinery is used in road construction (asphalt) and building envelopes (roof shingles) (Canadian Fuels Association 2013). The structural elements of our surroundings created by heat-intensive processes, especially steel and concrete, have expanded in scale and scope owing to access to energy-dense and easily distributed coal supplies (Smil 2015). Natural gas is the primary input into nitrogen fertilizer production, which has underpinned modern agricultural economies and rising yields.

Biomass that is not currently being used in traditional markets such as food, animal feed or lumber and pulp production can displace fossil fuels as the primary feedstock in many of these markets. Liquid biofuels such as bioethanol or biodiesel can replace gasoline or diesel in the transport sector. Solid biofuels such as wood pellets can replace coal or natural gas in heat and electricity sectors. Ethanol can be used as an alternative feedstock to ethane in ethylene production and, in turn, plastic manufacturing. The vision of facilitating sweeping displacement of fossil fuels across multiple sectors is captured in the concept of an 'advanced bioeconomy'.

Central to the bioeconomy vision is a biorefinery, a term that has found traction since Hines (1980) compared the conversion of crops with fuels with the oil refinery. As the analogy implies, biorefining is the application of various technological processes to convert biomass feedstock from agriculture, forestry or urban waste-streams into a combination of energy, fuels and chemicals (Huang et al. 2008). In contrast to traditional bioeconomies that use biomass directly, such as wood or agricultural residues for cooking and heating, biorefineries process biomass into a higher value, cleaner burning and more efficient fuel prior to distribution. These technologies have been the focus of significant public investments into research, development and commercialization of clean energy technologies, especially in developed nations (REN21 2015). Recently, investments have shifted away from food-based biorefineries (e.g. corn-to-ethanol; soy-to-diesel) towards advanced biorefineries which process lignocellulosic materials that are inedible to humans, such as straw and wood. This shift is in response to concerns about poor environmental performance and escalating food prices associated with food-based biorefineries.

Although advanced biorefineries can alleviate some of the concerns around food-based biorefining, dramatic reconfigurations to social and society–environment relations are inevitable. The development and implementation of advanced biorefineries opens up new markets for biomass, thereby providing a new production pressure on, and opportunity for, rural, resource-based economies. Land and resources once considered 'waste' or 'marginal' are increasingly valued for their commercial energy potential. In response, land markets and land tenure systems are changing at the local and global scales, while new forms of international trade (and, therefore global economic and environmental governance concerns) are emerging. The purpose of this chapter is to assess and conceptualize these opportunities, challenges, and implications.

The chapter focuses attention on the (changing) spatial patterns of technology implementation as well as the (changing) spatial organization of supply chains, otherwise known as the 'space-economies', of biorefining. The assessment builds on insights from environmental economic geography (EEG). As a complement to economic geography which has traditionally treated the environment as 'extra-economic' and has focused on social institutions and infrastructures, EEG attends specifically to the economy–environment interface when studying technological innovations (Patchell and Hayter 2013; Kedron 2015). Bridge (2008) identifies three primary lines of inquiry within EEG: (a) the influence of environmental processes on economic decision-making; (b) the environmental impacts of economic processes as they relate to livelihoods; and (c) how environmental governance reconfigures economic relations, with particular attention to social and environmental justice. As such, EEG holds strong conceptual and analytical ties with recent advances in land change science (see Munroe et al. 2014), and political–industrial ecology (see Newell and Cousins 2014), and builds on decades of work in political ecology and the 'ecological turn' in science and technology studies.

Technologies are a promising analytical entry point into understanding the ways in which environment and economies are co-constituted. Indeed, technologies are by definition an economic artefact, an economic actor and an economic tool; along with institutions value systems they are the primary means by which scarcity is managed and flows of materials and energy are exchanged across social and ecological systems. Conversely, then, it makes sense to apply EEG to the study of technology. The EEG lens provides a framework through which to question how and in what ways technologies connect and reconfigure economy–environment relations.

Through an analysis of the emerging space-economies of biofuel technologies, this chapter will make two conceptual contributions under the EEG framework. First, the chapter responds to a sympathetic critique by Bridge (2008, 79) that 'surprisingly few researchers have examined the extent to which the environmental qualities [hereafter "materiality"] of the commodity . . . affect the way [supply] chains work'. In what follows, the materiality of the resource base – in terms of quantity, quality, form, spatial distribution and timing – is shown to participate (albeit ambivalently, or unintentionally) in the fabrication of biorefining space-economies. There is a clear interdependence between the materiality of the resource base and processes and outcomes of technology siting and supply chain development. Second, the chapter demonstrates that the three lines of EEG inquiry identified above are not mutually exclusive. Indeed, the ways in which biorefining technologies incorporate into, and appropriate, (certain kinds of) environments is shaped at the nexus of resource materialities (line 1), technology capacities and designs (line 2) and environmental governance (line 3).

The chapter unfolds in three parts. First we describe the main pathways and products that constitute biorefining, with an emphasis on advanced biorefineries. Second, the space-economies of biofuels and biorefining are assessed across three themes: regional development; supply chain organization; and land-use. Third, this assessment is based on an empirical survey of existing patterns of biorefining using publically available documents, as well as spatially explicit techno-economic modelling conducted by the authors. The chapter focuses on technologies that process lignocellulosic materials (e.g. woody biomass and agricultural waste), although some comparison with grain-to-ethanol production is included. Finally, the chapter concludes with a discussion of avenues for future research.

BIOREFINING: TECHNO-ECONOMIC PERSPECTIVES

There is no formal typology by which to classify the myriad of biorefining technologies, processes or products. Common ways to classify biorefineries are according to: (a) feedstock types (first generation using grains or sugarcane; second generation using lignocellulosic material; third generation using microorganisms such as algae); (b) the primary product being generated (the type of fuel and/or chemical); and (c) the resource sector into which it is integrated (forestry or agricultural; see Cherubini et al. 2009). Here, we differentiate primarily by the technology or the bioconversion platform that is used to convert raw biomass into a higher-value product. The exact composition and titling of different platforms vary depending upon the author or agency. As such, the differentiations presented in Figure 19.1 should not be considered definitive. Rather, we use this conceptual model to present four basic platforms that are distinguished by the technology that is used to deliver combinations of energy, fuels and products.

The most basic pathway defined in the figure is the biomass-to-energy platform. Here, lignocellulosic biomass is converted into a form that is more convenient for transport and storage, often for combustion in order to deliver heat or electricity (although possibly for more advanced processing by way of the technologies described below). The most common forms of processed biomass are wood pellets, which have both industrial and residential use, and briquettes, which are most commonly used for cooking and space heating purposes. Torrefaction is a pyrolytic process where pellets or briquettes are heated in the absence of air. It has been proposed to improve the quality of the solid biofuel product; the process reduces the hemicellulose portion within the product (and thus increases the overall energy content) and plasticizes the lignin at the surface, which increases strength and reduces moisture absorption. Through torrefaction, the most volatile components of the biomass are combusted, leaving a product with a higher carbon and energy content.

The other three platforms shown in Figure 19.1 are more complex and are capable of delivering a wider range of products, including energy. The biochemical platform is designed to recover intermediate products from biomass, such as sugars and lignin. Generally speaking, this platform uses acid or enzymatic hydrolysis to release sugars from biomass. Starchy biomass is an ideal feedstock for bioconversion because it comprises a

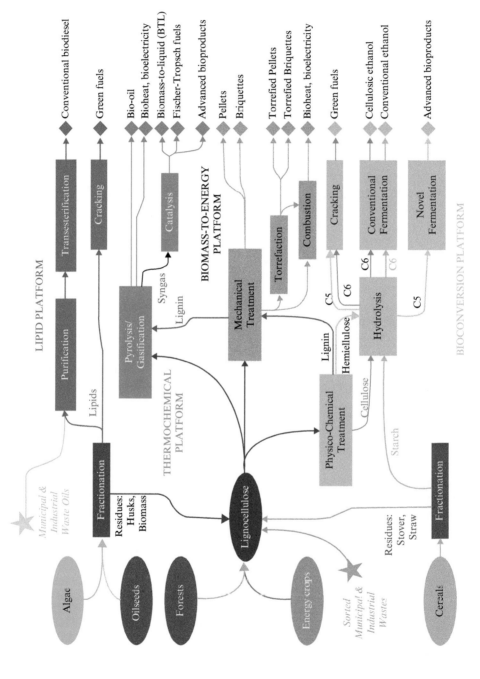

Figure 19.1   Biorefining platforms for processing biomass into fuels, energy and other products

single sugar (glucose), which simplifies process design and complexity. Sugar can then be fermented to ethanol or directly to hydrocarbons and other chemicals, and the non-sugar components of the biomass (mostly lignin) can be combusted for energy, or used as the basis for other products. In essence, this platform mimics physiological processes – such as ruminant digestion systems – that convert biomass into basic sugars that can be processed into higher-value products.

By comparison with the biochemical approach, the thermochemical platform utilizes pyrolysis – combustion in the absence of oxygen under different temperatures – or gasification – combustion with limited volumes of oxygen – to produce valuable products including syngas, bio-oils (a.k.a. pyrolysis oil), and solid, high-carbon char. These products can be recovered and further processed into fuels (i.e. ethanol and synthetic diesel), chemicals and plastics, with a portion of the product stream used to generate energy supporting the process (Zhong and Wei 2004). Pyrolysis and gasification are often referred to as 'robust' technologies that can be used to process heterogeneous feedstock, including woody biomass, and even pre- or post-consumer waste (Demirbas 2003; Yaman 2004). In essence, biorefineries are a set of technologies that reproduce the geologic processes that converted living biomass into the fossil resources we know today (i.e. extreme temperature and pressure). In many cases, excess heat is produced as a part of thermochemical conversion processes and this excess heat can be used as a co-product in useful applications such as electricity generation.

The lipid platform utilizes oils and fats from a wide variety of sources, including grains (e.g. soy, rapeseed, canola), tree crops (e.g. palm), animal slaughter by-products, consumer or industrial wastes (e.g. cooking oils) and algae (Pinzi et al. 2014). Conventional biodiesel production involves a transesterification of triglycerides contained within these oils to produce biodiesel (fatty acid methyl esters or less commonly ethyl esters) and glycerol, but the lack of other co-products has meant that the economics of the biodiesel industry has been a challenge. New pathways to produce 'renewable hydrocarbons', which in turn can be used to develop chemicals and fuels, have recently been proposed (Asomaning et al. 2014). The advantage of the latter pathway is that the 'green fuels' developed using this approach would be chemically indistinguishable from petroleum-based gasoline, diesel or jet fuel, removing blend restrictions and allowing existing infrastructure to be employed. These are typically referred to as 'drop-in' biofuels.

From a private investment perspective, the preferred technology and product of choice is influenced by market pricing, product supply and demand dynamics, processing margins, proximity to market, feedstock material properties, feedstock availability and policy supports. With the exception of a few low-hanging fruit opportunities, especially for the biomass-to-energy platform (e.g. niche markets in home heating), biofuels and biochemicals are not widely economically competitive. As such, recent investment decisions have been influenced by government policies and public funds (Mabee 2013; Kedron 2015). Policy support has come in the form of tax incentives, above-market price feed-in tariffs (bioelectricity), blending mandates (biofuels), import tariffs, support of research and development, loan guarantees, feedstock subsidization, preferential treatment under carbon pricing regimes and direct public investment in, or grants to, companies and demonstration/commercial facilities.

ENVIRONMENTAL ECONOMIC GEOGRAPHIES OF BIOREFINING

Regional Economic Perspectives: from Agriculture to Forestry

Although technologies exist to convert woody material derived from forests to biofuels, policy and investment decisions surrounding biorefining are predominantly materialized in traditional agricultural centres. In Canada, for instance, more than 99% of bioethanol is produced from agricultural feedstock, and 100% of biodiesel produced in 2015 was agriculture-based (USDA 2015). As of 2015, Abengoa, POET-DSM, Dupont and Beta Renewables established the first commercial-scale advanced (lignocellulosic) ethanol biorefineries (ultimately 80–120 million litres per year) in high-yield agriculture zones in the US corn belt. Each of these facilities is designed to convert cellulose (crop residues delivered as bales) transported no more than 100 km, and have ethanol as a primary product with lignin and other by-products used to generate process steam or energy.

On one hand this is a consequence of existing technical capacity. The science and technology of producing liquid biofuels from grain is well known, since the constituent polymers from which fuels and chemicals can be derived are more easily accessible. Moreover, supply chains and infrastructure are already in place to collect biomass in agricultural centres. On the other hand, the focus on agricultural centres is a consequence of political–economic preferences. Agricultural-based biorefining develops new markets for regional agricultural commodities and can stimulate investment in ailing rural economies. This provides an opportunity for governments to achieve a win–win outcome in economic and environmental governance. By way of an example, the Canadian government changed its environmental governance approach and reinstated taxes on imported ethanol (mostly from Brazil) in order to raise money to stimulate the development of a domestic ethanol industry to mitigate the impacts of increasingly imbalanced trade relations in the corn and corn-products sector (Calvert and Simandan 2015). In other words, favourable resource materialities, environmental governance objectives and economic growth objectives converged to stimulate policy and technology innovations in North America's grain-rich regions.

Developing technologies that can process woody material will increase the geographic range of the biorefinery sector, bringing sustainable sources of energy and materials as well as economic development into new forestry regions (Solomon 2008). The appeal of the forest-based biorefinery is especially strong given the recent downturn in pulp and paper production, especially in the forestry regions of North America (Blair et al. 2013). The primary technical challenge for the forest-based biorefinery is that, unlike starch, constituent polymers of wood serve a structural function and are resistant to decomposition (Sjöström 2013). This is especially the case for biochemical applications, since lignocellulose contains both five- and six-carbon sugars, which are difficult to ferment simultaneously since the most common natural strains of yeast metabolize one or the other, but rarely both. Mixing yeasts is often not possible, because they favour different operating conditions (temperature, moisture content, etc.). As such, centuries of knowledge about grain-to-fuel conversion do not translate directly into these areas.

Biorefining technologies have had to evolve in order to expand their geographic scope to forestry regions. The biochemical platform has evolved to include several stages that can fractionate and recover the various structural and chemical components of lignocellulosic biomass, including fibre, both five- and six-carbon sugars, and aromatic compounds from the lignin and extractives, which creates the potential for additional chemical co-products. Alternative platforms to the biochemical pathway are also being pursued. The thermochemical platform is typically preferred for woody feedstocks owing to their resistance to fractionation, but this platform is not yet competitive with 1G biofuels or fossil fuels.

Although there are some R&D-level thermochemical facilities currently operational (see Mabee 2013), wood pellets are currently the most common processed fuel from forestry centres. In fact, Rentech recently abandoned a thermochemical-to-liquid fuels project in Northern Ontario after unsuccessful attempts to scale up their technology, despite considerable public investment, and now provides pellets in the region for electricity generation. Furthermore, wood pellets are traded internationally in significant volumes – primarily from the interior of British Columbia, where vast quantities of dead wood are available owing to a recent pine beetle outbreak, and from plantations in the southeast United States, destined for Scandinavia and the UK to displace fossil fuels in the electricity and district heating sector in order to reduce carbon tax liabilities.

In addition to technical challenges in processing woody material, very few bioconversion/biofuel companies have strong connections to the forest sector. In contrast, the agricultural sector in Ontario and the mid-West had ties to the fuel value chain prior to, but surely strengthened by, the implementation of ethanol blending mandates. Since the early 20th century, for example, a group of farmers' cooperatives under 'United Co-operatives of Ontario' has operated fuel retail outlets, acting as a 'buying group' for rural Ontario. In 1991, UCO partnered with Suncor to form UPI Energy and in 1994 UCO sold their shares to GROWMARK, a large agricultural cooperative out of Illinois. These ties do not exist in the forestry sector.

Some regions provide opportunity to leverage a 'mixed feedstock' stream that includes both forestry and agricultural biomass (Calvert and Mabee 2014). South-eastern Ontario, for example, covers over 5.2 million ha or approximately 5% of the total land base of the Canadian province of Ontario. Moving from the southeast corner towards the northwest corner of the region, there is a clear shift from predominantly small urban centres with suburban and agricultural land covers to predominantly forested land managed within five separate forest management units. Forested regions have varying degrees of recreational land-use (e.g. cottages, resorts and parks) and decreasing amounts of agriculture as one travels to the north and west. As expected, biomass availability also mimics this transition (Figure 19.2). The region contains a transitional area, i.e. the area over which forestry and agricultural activities meet. At this transition zone, a hypothetical facility would be able to access a wider range of feedstock types from both forestry and agricultural operations, delivered at different times of the year. Locating along the transition zone not only ensures that feedstock can be procured at the lowest possible point along the supply–cost curve (Figure 19.3), but also insulates the facility from localized disruptions in the availability, composition and cost of supply related to weather or market competition.

From a technological perspective, encouraging 'feedstock-agnostic' technologies that can process lignocellulosic fibre from various source options is necessary in order to make this siting and procurement strategy viable. A heterogeneous feedstock is difficult

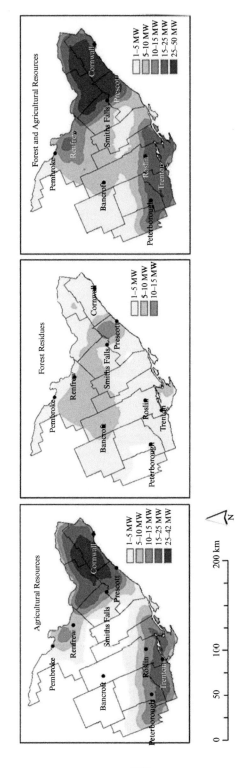

Figure 19.2   Spatial distribution density of potential energy production in a region with mixed agriculture and forestry operations

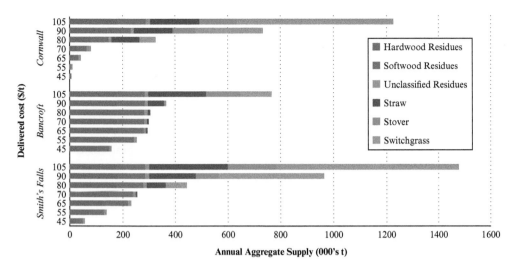

Note: Forest residues are generally the cheapest source of biomass, but limited in terms of their spatial distribution, timing and production potential.

Figure 19.3   *Supply–cost curve and feedstock mix at the transition zone between forestry and agricultural activities (Smith's Falls) in an agricultural centre (Cornwall) and a forestry centre (Bancroft)*

to process on an industrial scale, and would most likely require some version of the thermochemical platform (Demirbas 2003; Yaman 2004). Pyrolysis, for example, converts the solid feedstock into a combination of solid char, gas and liquid, the relative breakdown of which is dependent on operating temperature/pressure, feedstock properties and biomass residence (exposure) time. Another option is pelletization, in which case a blend of feedstock is converted into a relatively homogenous product that meets industry and regulatory standards for moisture and ash content. A third option to process mixed feedstock streams is biomass-based industrial (bio-industrial) parks. Conceptually speaking, a bio-industrial park is an agglomeration or cluster of industrial activities focused on the conversion of biomass feedstock into one or a combination of energy, fuels and products (e.g. platform chemicals) within separate but interconnected industrial processes. Using this concept, multi-biomass feedstock streams could be processed within multiple optimized units (i.e. parameterized to manage specific ash and lignin contents) rather than a single unit as feedstock is separated at the factory gate and fed into dedicated but parallel lines. The challenge with the latter model, however, is achieving economies of scale, an issue we take up next.

Supply Chain Perspectives I: Greenfield Development and Centralized versus Decentralized Plants

Some literature describes biorefineries in terms of their spatial relation with feedstocks along two lines. *Centralized* facilities, also referred to as 'integrated', 'stand-alone' or 'conventional design' biorefineries, are commonly described as single facilities which refine raw biomass taken directly from the field (Towers et al. 2007; Mesfun and Toffolo 2015; Stuart and El-Halwagi 2012). In contrast, a *distributed* configuration involves the transport of raw biomass to a 'depot' or 'terminal' for collection and/or pre-processing, before shipment to a second facility for final processing into one or more commodity outputs (Cafferty et al. 2014; Lamers et al. 2015).

The spatial organization of the biorefining supply chain is critical to environmental and economic performance (Nguyen et al. 2014). Scale economies are especially important for processing lignocellulosic biomass, which suggests that large, centralized processing facilities will be the most likely model of development for advanced biorefineries. Indeed, Dunnett et al. (2008) find limited potential for decentralized processing of lignocellulosic material into ethanol, and recent estimates by Stephen et al. (2012) suggest that very large facilities, approaching 1 billion litres of ethanol production per year, could take advantage of economies of scale and achieve significant cost reduction. Nguyen et al. (2014), however, show that decentralized pre-processing of 'intermediate' products (e.g. wood chips, hydrolysed biomass) along rail lines connected to a centralized processing facility has potential to reduce life-cycle costs and emissions. Decentralized intermediate production allows initial processing stages to be located near the resource while shifting the final processing steps closer to market (i.e. urban centres or shipping ports).

Understanding optimal spatial models of biorefinery implementation requires careful attention to the availability and spatial distribution of resources, as well as the desired product. According to recent modelling work conducted by the authors, biorefining technologies with high capital costs per unit of feedstock processed, such as cellulosic ethanol, benefit more from increasing scale than technologies with relatively low capital costs and poor economies of scale, such as wood pellets (Stephen et al. 2014). In other words, pellet production is more economical at small scale while cellulosic ethanol and bioelectricity production is more economical at large scale. This is because capital and non-feedstock operating cost (e.g. maintenance, labour) economies of scale exceed feedstock delivery cost diseconomies of scale for cellulosic ethanol. Conversely, the higher capital expenditure scaling factor for wood pellets results in feedstock transportation cost diseconomies of scale exceeding capital and operating cost economies of scale. Decentralization can potentially lower feedstock costs which, as seen in Figure 19.4, are the highest cost for the large wood pellet facilities. The impact of (de)centralization on process economics and environmental performance is the subject of continued research and debate surrounding biofuel production and biorefining, and is in many cases context dependent.

That said, even the most centralized biofuel production chain will be much more decentralized relative to conventional fossil fuel production chains. The low energy density and low bulk density of raw biomass prohibit feedstock transport over the continental distances that are typical in a fossil-fuel supply chain, and biomass resources are spatially distributed, spanning vast geographic areas, whereas fossil fuels mostly exist in spatially explicit subterranean pools that extend vertically. On average, approximately 1.5 tonnes of biomass, grown above ground, are required in order to replace the energy equivalent of 1 tonne of coal, which is recovered from subterranean deposits. Replacing one unit of petroleum and natural gas requires as much as 2–4 tonnes of biomass which must be transported above ground rather than below ground. In other words, each petroleum refinery will need to be replaced by no fewer than seven biorefineries, probably scattered across a region rather than centralized in a single location in order to

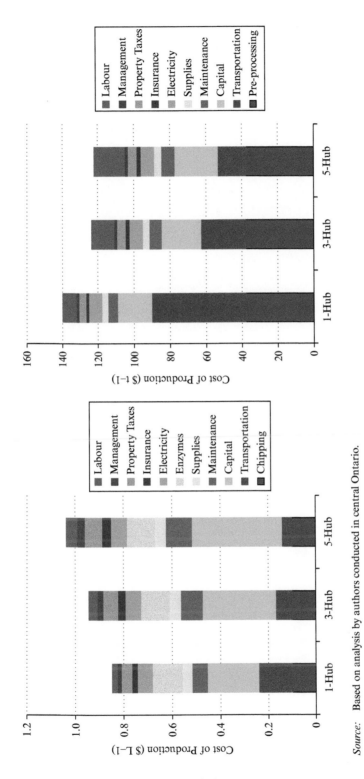

Source: Based on analysis by authors conducted in central Ontario.

Figure 19.4 Impacts of decentralization on cost of production for forestry-based ethanol (left) and wood pellets (right)

minimize transport distances between extraction and processing. This means a higher number of communities hosting industrial-scale biomass processing and a re-surfacing of infrastructure and activity associated with the fuel production chain. Indeed, the largest biomass processing facility in the world is approximately seven times smaller than the average petroleum processing facility in the United States (see Figure 19.5).

A recent study suggests that biorefinery integration within existing infrastructure results in better economic and social performance when compared with either a focus

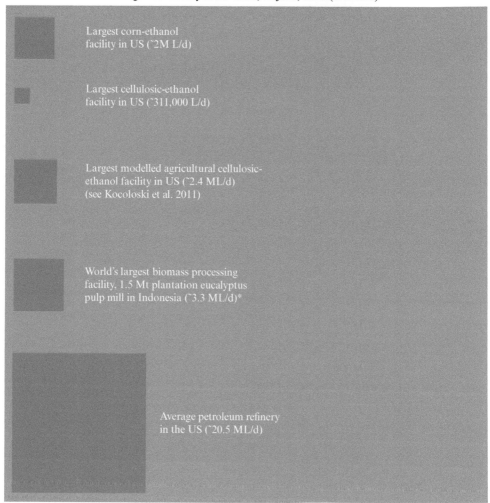

Largest oil refinery in the world, Gujarat, India (196ML/d)

Largest corn-ethanol facility in US (˜2M L/d)

Largest cellulosic-ethanol facility in US (˜311,000 L/d)

Largest modelled agricultural cellulosic-ethanol facility in US (˜2.4 ML/d) (see Kocoloski et al. 2011)

World's largest biomass processing facility, 1.5 Mt plantation eucalyptus pulp mill in Indonesia (˜3.3 ML/d)*

Average petroleum refinery in the US (˜20.5 ML/d)

Note: Sizes are proportional based on total feedstock throughput on a giga-Joule basis, and numbers represent daily production in litres (L) or million litres (ML) per day. It is assumed that woody feedstock could be converted to ethanol at a rate of 400 L/tonne.

Figure 19.5   Comparison of existing and anticipated biofuel production systems with existing petroleum refineries

on traditional wood and paper products or opening new, independent biorefineries (FPAC 2010). This is especially the case given a spate of pulp and paper mill closures that have left infrastructure idle. Here, again, the challenge is greater in forestry centres than in agricultural centres. Shuttered mills are not at a scale sufficient to achieve scale economies necessary for advanced biorefining into liquid fuels. It is therefore likely that 'first movers' in the area of wood-to-biofuel would integrate into the supply chain of a *cluster* of existing mills rather than a single mill. Recent work by the authors (see Blair et al. 2013) identified seven potential clusters of mills in Canada that could supply a first-mover forest-based biorefinery – even given relatively strict criteria and assumptions (e.g. each cluster is owned by a single operator). At each cluster, however, even utilizing 30% of the current residue production (which is optimistic, given that much of the residue is currently being used for internal energy) would not be sufficient to achieve scale economies necessary to support a centralized biorefinery. Further research (Blair et al. 2016) has shown that these clusters can reduce technology and market risk if they follow a decentralized pathway, whereby mill clusters produce intermediate commodities that could either be sent to a centralized biorefinery located closer to consumers or shipped internationally for final processing.

There is also a social dimension to consider when deciding to leverage existing infrastructure. Eaton et al. (2014, 228), for example, demonstrate that 'national socio-technical imaginaries', i.e. a 'collective vision of a feasible, desirable future social order, provided by technological projects', are framed in different ways by different individuals according to lived experiences and remembered histories within the locales that might host said technological projects. Simply stated, biorefineries receive more support within traditional 'resource towns' where extraction and management of forest resources have shaped livelihoods and lifestyles for generations. Familiarity with biomass processing is critical given the intensive land management, processing and trucking activity associated that is required to displace even modest amounts of petroleum, as described above.

Land-use Perspectives: Open-loop versus Closed-loop Biorefineries

Biorefineries operate as part of 'open-loop' or 'closed-loop' systems. Open-loop systems derive biomass from the waste-stream of some other primary process, e.g. municipal solid waste, agricultural residues or forestry residues. In these cases, the biorefinery directly integrates into, and rearranges, established resource flows and land-use practices. The environmental impacts of open-loop systems are nonetheless important to consider. All existing cellulosic-ethanol facilities operating in agricultural centres are examples of open-loop systems. Given the added pressure on agricultural residues, considerable agronomic and soil science research is being conducted to determine residue removal rates that are sustainable (i.e. do not compromise long-term soil fertility or other niche markets such as animal bedding), although it is widely believed that a removal rate of between 30 and 50% is appropriate depending on initial soil quality (Calvert and Mabee 2014). Similar concerns underpin decisions to utilize forestry residues more intensively.

The opportunities provided by open-loop systems, however, are limited by the availability of the feedstock. Replacing a meaningful amount of fossil fuels with biomass will require the development of closed-loop systems, wherein biomass is derived from production activities that are conducted exclusively for energy production. Closed-loop systems with dedicated biomass plantations (hereafter energy crops) increase the amount of biomass within proximity to the facility since the full harvest goes to the facility. Pellets or ethanol from fields of switchgrass or willow are examples of such systems.

Concerns about the land area that is required to produce energy crops underpin debates about the sustainability of closed-loop biorefining (Lynd et al. 2007). By way of example, consider the conversion of a refinery recently purchased by Delta Airlines to service its Philadelphia flights. This very modest-sized refinery, at 52,000 bbl/day of jet fuel, would require the biomass output from 15 farms in the surrounding region on a daily basis, even under very optimistic assumptions about field-level biomass yield and plant-level ethanol yield.

Current land-use debates feature two issues: (a) biophysical impacts such as habitat change and carbon sequestration as land is converted into energy crop production (Fargione et al. 2008); and (b) social impacts such as competition with food prices and large-scale land transfers from local subsistence to multinational corporations (i.e. land grabbing) (Nalepa and Bauer 2012; Baka 2014). Further complicating the sustainability and acceptance of energy crop production is the potential for indirect land-use change, where land may be converted elsewhere to compensate for food crop production that was displaced by energy crop production (Searchinger et al. 2008).

It is widely believed that the land-use and food system impacts of biomass production can be mitigated if energy crops are grown on so-called *marginal land*. Definitions of marginal land emphasize either biophysical or economic criteria. Definitions that are purely economic include 'land of marginal productivity' (Bryngelsson and Lindgren 2013), '[land of] low productivity and reduced economic return' (Kang et al. 2013) and 'economically marginal land' (Shortall 2013). Definitions that emphasize biophysical criteria include land with specific 'soil quality, erodibility and ecological sensitivity' (Kang et al. 2013), and the vague criterion of 'lower quality land' (Shortall 2013). These definitions must be applied cautiously, because context matters. For example, a parcel that is shown to be marginal because it supports inadequate yields of corn in the mid-west United States may be shown to be relatively good cropland (and not marginal) in areas where overall yields are lower (Cope et al. 2011). Indeed, empirical studies at the local level have shown that biomass production, often for food or heating fuel, is in fact taking place on what might be considered poor land in places like Malawi (Nalepa and Bauer 2012). In the search for 'marginal' lands on which to grow energy crops in developing economies, local populations are being alienated from their means of subsistence as the land is re-enclosed to service a market economy that requires renewable diesel rather than the subsistence economy that needed renewable heat for cooking (Baka 2014).

In order to avoid some of these issues, abandoned cropland has been proposed as a possible answer to the question of where to produce energy crops (Kang et al. 2013). Here, abandoned cropland is defined as land area that was once used for, or declared as, agricultural land, but is no longer (considered to be) performing that function. Although a significant area of land has been abandoned in North America (Zumkehr and Campbell 2013), much of this land has been reforested or urbanized. As such, and although potentially significant at regional scales, the area of abandoned cropland is not significant at national scales, as it could only help supplement a fraction of total light-duty gasoline consumption. That said, dedicated energy crops on abandoned cropland might be capable of producing enough biomass to support specific sectors, such as aviation fuel or ethylene production, and future policy might consider targeting biofuel production in these areas to find a closer match between production with consumption.

The development of biorefining and the existence of a growing market for various biofuels is leading to the appropriation of new kinds of environments, some of them environments (e.g. 'wasteland' in India) that were previously not part of the formal market economy. As a result, in some cases we are seeing the realization of environmental governance objectives through land dispossession (Baka 2014). In others, we are seeing complicated and indirect changes to regional land markets that lead to deforestation – e.g. soy taking over pasture land, and pasture land expanding into virgin areas (Lapola et al. 2010; Baird and Fox 2015) – in order to satisfy biofuel blending mandates in the name of energy sustainability.

DISCUSSION AND CONCLUSION

Much of the EEG literature focuses on the 'greening' of value chains, and refers to the emergence of a new techno-economic paradigm in which 'the environment' is integral to bottom-line thinking. Indeed, there is an increasingly strong relationship between environmental and economic governance. Investments into biorefining and biofuels are one example. In Canada, biorefining represents an attempt to move beyond Canada's legacy as a resource-based economy to be a leader in the value-added processing of biological material (see Calvert and Simandan 2015). In India, the development of green diesel is tied directly to an objective to turn 'wasteland' into productive land and to bring a 'formal' capitalist economy into rural areas.

Displacing fossil fuels across all markets is not possible owing to insufficient biomass supplies and significant economic hurdles. This constraint, combined with mounting political–economic and environmental pressures to displace fossil fuels as quickly as possible, introduces a number of strategic questions about the 'best' use of biomass. A significant part of this calculation is the economic and environmental performance of the conversion process and product. As shown above, this performance is strongly dependent on spatial dimensions in terms of the spatial distribution of resources and processing facilities. In order to maximize environmental performance and expand economic opportunities, policies and technologies are shifting investment from agriculture to forestry centres. Owing to technical challenges and few cross-sector ties, the transition is slow and highly dependent on policy support. Increasingly, policy directives are being targeted where clear technological alternatives do not exist, and where the scale of consumption more closely matches the scale of realistic production potential, e.g. aviation fuels or speciality chemicals.

What this chapter has shown is that the combination of policies and technologies used to combine economic and environmental goals through biorefining will have more profound implications than simply switching coal for wood, or petroleum for biofuel. Owing to the ways in which the materiality of the resource base participates in the fabrication of new fuel space-economies, as well as interdependencies across policies, technologies and resource materialities, the implications are much broader (see Birch and Calvert 2015). In other words, the development and implementation of biorefineries are not only a matter of identifying the proper market mechanisms or matching technologies to resources. Biorefineries are also intimately involved in social relationships to land and landscapes. In Canada (Ontario especially), for example, dubious environmental benefits from corn-to-ethanol were favoured in order to maintain what might be considered 'corn lock-in' across rural Ontario (Calvert and Simandan 2015). In India, so-called 'waste-lands' increasingly respond to the material interests of a car culture rather than those of a subsistence community through land dispossession (Baka 2014).

In light of these (unintended) impacts on social relations to land, new technologies and their implementation have stimulated new forms of environmental governance. This includes agroecological zoning to control production and harvesting, and standards and certification applied to the product, designed to minimize environmental and social costs across global supply chains. In the United States, the latest rules guiding biofuel blending limits the expansion of biomass production to the 2007 agricultural land area to avoid cropping under native or regenerated areas, and has capped the volume of corn-based ethanol that will receive government support. Clearly, the relationship between technology implementation and environmental governance is reciprocal and co-constitutive, and biorefineries offer a case-in-point (see also Kedron 2015).

As such, when considering the future of the advanced bioeconomy it is important to go beyond 'techno-economic paradigms' to also consider a new socio-ecological paradigm, i.e. to question how we integrate social systems with ecological systems through agriculture or forestry operations in the first place. Can we establish cooperative models of biofuel development that consider social justice across the entire supply chain, from harvesting and land-use to facility investment and siting? If so, how would this influence the technology, scale and product of choice? To what extent are the 'abandoned' lands that allegedly hold such promise the result of high yields from synthetically fertilized crops grown elsewhere – in other words, are abandoned lands simply a product of fossil fuel use in the first place? To what extent do our ideas about what a forest should look like, and how forested areas are currently governed, influence social acceptance of the intensive forms of forest management that would be required to achieve forest-based biorefining at scale (e.g. Pancholy et al. 2011)? Questions about how biorefineries are implicated in land tenure systems, equity and resource access, and global and regional geopolitics, are all in the queue.

These are the questions that EEG is equipped to help answer. By combining institutions, materiality, innovation and spatiality under a single analytical framework, EEG is well positioned to study the opportunities, barriers and implications of the development and implementation of biorefineries. Through EEG's conceptual framework, we better understand how biorefineries (re)connect economy–environment interactions across multiple sites and scales. Moving forward, EEG should be tied to similar epistemic projects, such as political–industrial ecology with its emphasis on 'metabolism' thinking (Newell and Cousins 2014), and recent advancements in land-change science with its emphasis on 'teleconnections' (Munroe et al. 2014). Perhaps more importantly, the application of EEG should be firmly rooted in a commitment to interdisciplinary scholarship, combining insights from ecology, forestry, agricultural/resource economics, political economy and geography, to name only a few.

REFERENCES

Asomaning, J., P. Mussone and D. Bressler. 2014. Two-stage thermal conversion of inedible lipid feedstocks to renewable chemicals and fuels. *Bioresource Technology* 158: 55–62.

Baird, I. and J. Fox. 2015. How land concessions affect places elsewhere: telecoupling, political ecology, and large-scale plantations in southern Laos and northeastern Cambodia. *Land* 4: 436–453.

Baka, J. 2014. What wastelands? A critique of biofuel policy discourse in South India. *Geoforum* 54: 315–323.

Birch, K. and K. Calvert. 2015. (Re)thinking 'drop-in' biofuels: on the political materialities of bioenergy. *Journal of Science and Technology Studies* 28(1): 52–72.

Blair, M., K. Calvert, M. Manion, S. Earley and W. Mabee. 2013. Linking analysis of market and material flow to inform Canadian forest biorefinery development. *Journal of Science and Technology for Forest Products and Processes* 3: 1–15.

Blair, M., L. Cabral and W. Mabee. 2016. Biorefinery strategies: exploring approaches to developing forest-based biorefinery activities in British Columbia and Ontario, Canada. *Technology Analysis and Strategic Management*; http://dx.doi.org/10.1080/09537325.2016.1211266.

Bridge, G. 2008. Environmental economic geography: a sympathetic critique. *Geoforum* 39: 76–81.

Bryngelsson, D. and K. Lindgren. 2013. Why large-scale bioenergy production on marginal land is unfeasible: a conceptual partial equilibrium analysis. *Energy Policy* 55: 454–466.

Cafferty, K., J. Jacobson, E. Searcy, K. Kenney, I. Bonner, G. Gresham and J. Hessl. 2014. Feedstock supply system design and economics for conversion of lignocellulosic biomass to hydrocarbon fuels. In *Conversion Pathway: Fast Pyrolysis and Hydrotreating Bio-oil Pathway*, INL/EXT-14-31211. Idaho Falls, ID: Idaho National Laboratory, US Department of Energy.

Calvert, K. and W. Mabee. 2014. Spatial analysis of biomass resources within a socio-ecologically heterogeneous region: identifying opportunities for a mixed feedstock stream. *International Journal of Geo-Information* 3(1): 209–232.

Calvert, K. and D. Simandan. 2015. A polymorphic approach to environmental policy analysis: the case of the Ethanol-in-Gasoline Regulation in Ontario, Canada. *Geografiska Annaler: Series B, Human Geography* 97(1): 31–45.

Canadian Fuels Association. 2013. The economics of petroleum refining: understanding the business of processing crude oil into fuels and other value added products, accessed from http://canadianfuels.ca/userfiles/file/Economics%20fundamentals%20of%20Refining%20Dec18-2013-Final%20ENG-1.pdf.

Cherubini, F., G. Jungmeier, M. Wellisch, T. Willke, I. Skiadas, R. Ree and E. de Jong. 2009. Toward a common classification approach for biorefinery systems. *Biofuels, Bioproducts and Biorefining* 3: 534–546.

Cope, M., S. McLafferty and B. Rhodes. 2011. Farmer attitudes toward production of perennial energy grasses in east central Illinois: implications for community-based decision making. *Annals of the Association of American Geographers* 101(4): 852–862.

Demirbas, A. 2003. Fuels recovery from municipal solid and liquid wastes (MSLW). *Energy Sources* 25(7): 713–720.

Dunnett, A., C. Adjiman and N. Shah. 2008. A spatially explicit whole-system model of the lignocellulosic bioethanol supply chain: an assessment of decentralized processing potential. *Biotechnology for Biofuels* 1: 1–13.

Eaton, W., S. Gasteyer and L. Busch. 2014. Bioenergy futures: framing sociotechnical imaginaries in local places. *Rural Sociology* 79: 227–256.

Fargione, J., J. Hill, D. Tilman, S. Polasky and P. Hawthorne. 2008. Land clearing and the biofuel carbon debt. *Science* 319: 1235–1238.

Forest Products Association of Canada (FPAC). 2010. Transforming Canada's forest products industry: summary of findings from the Future Bio-pathways Project, accessed from http://www.fpac.ca/wp-content/uploads/Biopathways-ENG.pdf.

Hines, D. 1980. Biotechnology today and tomorrow. *Enzyme and Microbial Technology* 2(4): 327–329.

Huang, H., S. Ramaswamy, U. Tschimer and B. Ramarao. 2008. A review of separation technologies in current and future biorefineries. *Separation and Purification Technology* 62(1): 1–21.

Kang, S., W. Post, J. Nichols, D. Wang, T. West, V. Bandaru and R. Izaurralde. 2013. Marginal lands: concept, assessment and management. *Journal of Agricultural Science* 5(5): 129–139.

Kedron, P. 2015. Environmental governance and shifts in Canadian biofuel production and innovation. *Professional Geographer* 67: 385–395.

Kocoloski, M., W. Griffin and H. Matthews. 2011. Impacts of facility size and location decisions on ethanol production cost. *Energy Policy* 39: 47–56.

Lamers, P., E. Tan, E. Searcy, C. Scarlata, K. Cafferty and J. Jacobson. 2015. Strategic supply system design – A holistic evaluation of operational and production cost for a biorefinery supply chain. *Biofuels, Bioproducts and Biorefining*; doi: 10.1002/bbb.1575.

Lapola, D., R. Schaldach, J. Alcamo, A. Bondeau, J. Koch, C. Koelking and J. Priess. 2010. Indirect land-use changes can overcome carbon savings from biofuels in Brazil. *Proceedings of the National Academy of Sciences of the United States of America* 107: 3388–3393.

Lynd, L., M. Laser, J. McBride, K. Podkaminer and J. Hannon. 2007. Energy myth three – high land require-ments and an unfavorable energy balance preclude biomass ethanol from playing a large role in providing energy services. In B. Sovacool and M. Brown (eds), *Energy and American Society – Thirteen Myths*. pp. 75–101. New York: Springer.

Mabee, W. 2013. Progress in the Canadian biorefining sector. *Biofuels* 4(4): 437–452.

Mesfun, S. and A. Toffolo. 2015. Integrating the processes of a Kraft pulp and paper mill and its supply chain. *Energy Conversion and Management* 103: 300–310; doi: 10.1016/j.enconman.2015.06.063.

Munroe, D., K. McSweeney, J. Olson and B. Mansfield. 2014. Using economic geography to reinvigorate land-change science. *Geoforum* 52: 12–21.

Nalepa, R. and D. Bauer. 2012. Marginal lands: the role of remote sensing in constructing landscapes for agro-fuel development. *Journal of Peasant Studies* 39(2): 403–422.

Newell, J. and J. Cousins. 2014. The boundaries of urban metabolism: toward a political–industrial ecology. *Progress in Human Geography*, in press; doi: 10.1177/0309132514558442.

Nguyen, L., K. Cafferty, E. Searcy and S. Spatari. 2014. Uncertainties in life cycle greenhouse as emissions from advanced biomass feedstock logistics supply chains in Kansas. *Energies* 7: 7125–7146.

Pancholy, B., M. Thomas, D. Solid and N. Stratis. 2011. The impact of biofuels on the propensity of land-use conversion among non-industrial private forest landowners in Florida. *Forest Policy and Economics* 13: 570–574.

Patchell, J. and R. Hayter. 2013. Environmental and evolutionary economic geography: time for EEG2? *Geografiska Annaler: Series B, Human Geography* 95(2): 1–20.

Pinzi, S., D. Leiva, I. López-García, M. Delores Redel-Macías and M. Pilar Dorado. 2014. Latest trends in feedstocks for biodiesel production. *Biofuels, Bioproducts, and Biorefining* 8: 126–143.

REN21. 2015. *Renewables 2015, Global Status Report*. Washington, DC: Renewable Energy Policy Network for the 21st Century.

Searchinger, T., R. Heimlich, R. Houghton, F. Dong, A. Elobeid, J. Fabiosa, S. Tokgoz, D. Hayes and T.-H. Yu. 2008. Use of U.S. croplands for biofuels increases greenhouse gasses through emissions from land-use change. *Science* 319: 1238–1240.

Shortall, O. 2013. Marginal land for energy crops: exploring definitions and embedded assumptions. *Energy Policy* 62: 19–27.

Sjöström, E. 2013. *Wood Chemistry: Fundamentals and Application*, 2nd edn. New York: Academic Press.

Smil, V. 2015. *Power Density: A Key to Understanding Energy Sources and Uses*. Cambridge, MA: MIT Press.

Solomon, B. 2008. Regional economic impacts of cellulosic ethanol development in the north central states. In B. Solomon and V. Luzadis (eds), *Renewable Energy from Forest Resources in the United States*. pp. 281–298. New York: Routledge.

Stephen, J., W. Mabee and J. Saddler, J. 2012. Will second-generation ethanol be able to compete with first-generation ethanol? Opportunities for cost reduction. *Biofuels, Bioproducts, and Biorefining* 6(2): 159–176.

Stephen, J., W. Mabee and J. Saddler. 2014. The ability of cellulosic ethanol to compete for feedstock and invest-ment with other forest bioenergy options. *Industrial Biotechnology* 10: 115–125.

Stuart, P. and M.M. El-Halwagi. 2012. *Integrated Biorefineries: Design, Analysis, and Optimization*. Boca Raton, FL: CRC Press.

Towers, M., T. Brown, R. Kerekes, J. Paris and H. Tran. 2007. Biorefinery opportunities for the Canadian pulp and paper industry. *Pulp and Paper Canada* 108(6): 26–29.

USDA. 2015. USDA Foreign Agricultural Service GAIN Report on Canada: Biofuels. Global Aggricultural Information Network, accessed from http://gain.fas.usda.gov/Recent%20GAIN%20Publications/Biofuels%20Annual_Ottawa_Canada_8-19-2015.pdf.

Yaman, S. 2004. Pyrolysis of biomass to produce fuels and chemical feedstocks. *Energy Conversion and Management* 45(5): 651–671.

Zhong, C.-L. and X.-M. Wei. 2004. A comparative experimental study on the liquefaction of wood. *Energy* 29(11): 1731–1741.

Zumkehr, A. and J. Campbell. 2013. Historical U.S. cropland areas and the potential for bioenergy production on abandoned croplands. *Environmental Science and Technology* 47: 3840–3847.

20. The emergence of technological hydroscapes in the Anthropocene: socio-hydrology and development paradigms of large dams

Marcus Nüsser and Ravi Baghel

The term Anthropocene has gained currency as a descriptor for the current geological epoch, characterized by human domination of the Earth. One of the most dramatic examples of human impact on planetary processes is the rapid transformation of the majority of river systems through dam building (Baghel 2014). More than half (172 out of 292) of all large river systems have been fragmented by dams, which obstruct two-thirds of all freshwater flows with their reservoirs capable of holding back more than 15% of the global runoff (Nilsson and Berggren 2000; Meybeck 2003; Nilsson et al. 2005). Apart from affecting water flows, large dams also prevent sediment transport to the sea and trap more than half (53%) of the total sediment load. If unregulated basins are included, this means that 25–30% of *all* sediments worldwide are intercepted by these artificial structures (Vörösmarty et al. 2003). This drastic alteration of the world's river systems has received much less attention than global climate change, yet it has been suggested that the 'global impact of direct human intervention in the terrestrial water cycle (through land cover change, urbanization, industrialization, and water resources development) is likely to surpass that of recent or anticipated climate change, at least over decadal time scales' (Vörösmarty et al. 2004, 513).

Large dams are among the greatest single constructions and most massive infrastructure projects built worldwide and provide water storage for hydropower generation, irrigated agriculture, industrial production and flood control. As the most significant tools employed for river basin management, large dams are considered powerful icons of modernization, economic success, national prestige and technological progress. Despite the long history of flow regulation in human civilization, such as that found within ancient China and the Middle East, the global sprawl of dam building has only emerged in the past 60–70 years. By 1950 only about 5000 of these hydraulic structures had been built, before the countries of the Global South witnessed a resurgence of dam construction soon after decolonization (Gleick 1998; McCully 2001). Notable examples include Ghana's Akosombo Dam (completed 1965), which resulted in the formation of Lake Volta, the Aswan Dam in southern Egypt (completed 1970) and the Bhakra Dam in northern India (completed 1963). Since then, China and India have become the most prominent dam-building nations. By the turn of the millennium, the world had built over 45,000 large dams with a total flooded area estimated to exceed 500,000 km² (WCD 2000). The latest synthesis of the World Register of Dams, compiled by the International Commission on Large Dams (ICOLD), includes 58,402 large dams, straddling all major river basins on earth. Together these artificially inundated reservoirs sum up to an aggregate storage capacity of about 16,100 km³ (ICOLD 2016).

Large dams are not only material artefacts of technological advancement or representative elements in the transformation of fluvial environments, but also gigantic manifestations of the social construction of nature. Over the past few decades, the erection and operation of these socio-hydrological structures have become the focus of intense debates with regard to their effectiveness for development, environmental impacts, social justice and sustainability (e.g. McCully 2001; Khagram 2004; Scudder 2005). Whereas advocates of dam construction propound their benefits as incentives for economic development, opponents emphasize a whole range of environmental, socio-economic and political costs (Nüsser 2003). Depending on their specific mode of operation, the environmental impacts of dams include fragmentation of riverine ecosystems, changes in flow patterns, modification of erosion and sedimentation processes, species extinction in freshwater and wildlife habitats, and loss of water by evaporation and contamination. Among the most commonly cited socio-economic concerns are insufficient compensation for displaced persons and the lack of their long-term development perspectives.

A huge body of literature on river control has been published and the main arguments have been repeatedly expressed in reports and case studies (e.g. Fearnside 1988; Bakker 1999; Khagram 2003; Ansar et al. 2014; Hirsch 2016; Kirchherr et al. 2016). The dominant lines of argument are rooted in the classical development paradigms of modernization, dependency and sustainability. However, large-scale physical transformation and fragmentation of rivers by dams and reservoirs are rarely framed within the broader context of geographies of technology. The term 'technological hydroscapes' (Nüsser 2014) is used here to frame the socio-hydrological nature of dam building under diverse technological and ideological settings. It captures not only the physical transformation of fluvial environments but also the implementation of new water and energy governance systems. The term combines the concepts of 'waterscapes' (Swyngedouw 2009), 'technoscapes' (Appadurai 1996), 'energyscapes' (Kaisti and Käkönen 2012) and 'cryoscapes' (Nüsser and Baghel 2014), where the suffix 'scape' emphasizes the fluid nature of these spaces. Such 'scapes' are not considered as physically delimited spaces or merely as social constructions of nature, but rather as dynamic entities constituted by complex flows of technology, funding, ideology and discourses of development and environment (Baghel and Nüsser 2010). The fundamental shift in the rationale and scale from local water use to the human domination of water systems was only made possible with the advancement of modern technology. Following a classificatory overview of large dams, this chapter retraces the origins and historical course of this technology and explores its conceptual foundations, changing development paradigms and geographical implications.

LARGE DAMS: FUNCTIONS, TYPOLOGY AND TECHNOLOGY

Dams and reservoirs serve a variety of functions. Their most important purposes are irrigation for agricultural production (49%), hydropower generation (20%), water supply for industrial and domestic use (13%) and flood control (9%). An increasing number of multi-purpose dams (17%) meet several of these objectives (ICOLD 2016). The extension of irrigated land is often portrayed as a crucial prerequisite to meet future food demands, which in turn justifies further dam construction. Their enormous economic importance is additionally demonstrated by the fact that about 675 GW of installed hydroelectric capacity produce nearly 25% of total electricity on Earth (ICOLD 2016). In order to define large dams as a distinct category, ICOLD offers a set of criteria. According to their well-established definition, a large dam is one whose height from the lowest point of foundation to the top exceeds 15 m, or whose height is between 10 and 15 m, if it meets at least one of the following five conditions: (a) the crest length of the dam is not less than 500 m; (b) the capacity of the reservoir is not less than 1 million m^3; (c) the spillway discharge potential exceeds 2000 m^3 per second; (d) the dam faces especially difficult foundation problems; and (e) the dam is of unusual design.

There are three main types of dam design, namely embankment, gravity and arch, which are primarily selected according to site topography and geological setting. Embankment dams (e.g. Nurek in Tajikistan, Tarbela in Pakistan, Tehri in India and Mohale in Lesotho), representing around 75% of global dams, are built from excavated earth and rock fill and they are usually triangular in cross-section. These most massive dams can be constructed on soft and unstable riverbeds, because their broad base distributes weight over a wide area. Gravity dams (e.g. Bhakra in India, Grande Dixence in Switzerland) are constructed from roller-compacted concrete, stone or other masonry, and entirely rely on their own weight and internal strength for stability (Figure 20.1). They are mostly built across narrow valleys with firm bedrock conditions. The third type, concrete arch dams (e.g. Katse in Lesotho), are convex upstream to transfer the force of stored water to the adjacent rock walls (Figure 20.2). They require much less concrete than gravity dams of the same length, but are limited to narrow canyons with solid rock foundation. Spillways are general structural features of dams, which are used to discharge water when the reservoir threatens to become dangerously high. The two main categories of dams are reservoir storage projects, which capture runoff and impound water for seasonal or annual storage, and run-of-river schemes. In the second type, water is diverted into head race tunnels before reappearing some distance down the valley, where the powerhouses are located. These run-of-river schemes require smaller reservoirs of inundation and operate with an increased hydraulic head. A third type of hydroelectric installation is pumping–storage facilities, using power during daily low-demand periods to pump water to a higher reservoir, from where it can be used for electricity production during peak-demand periods.

With a height of 305 m, the Jinping I Dam in China, completed in 2013, is currently the world's highest construction. At 300 m, the Nurek Dam in Tajikistan, completed in 1972, held this position for almost 40 years. The Bakhtiari Dam in Iran, under construction since 2013, will exceed the existing dams with a planned height of 315 m and the currently suspended Rogun Dam in Tajikistan may even reach an elevation of 335 m. With a surface of 8500 km^2, the Volta Reservoir behind Akosombo Dam in Ghana forms the largest artificial lake on Earth, flooding more than 5% of the country (Gleick 1998; WCD 2000; McCully 2001; ICOLD 2016; own data compilation). With 22,500 MW of installed capacity, producing 84,900 GW h of energy per year, the Sanxia project in China, also known as Three Gorges Project, located on the Yangtze River, is by far the largest hydropower plant. The gigantic project, a dream of Mao Zedong finally completed in 2010, was accompanied by a massive resettlement scheme that displaced almost 1.2 million persons (Li et al. 2001; Penz et al. 2011). Among those 20 projects causing the highest number of displaced persons, 15 are located in China (ICOLD 2016).

*Figure 20.1   With a height of 285 m, Grande Dixence is the world's highest gravity
            dam. Built between 1951 and 1965, 6 million m³ of concrete were used for
            construction. Remains of a snow avalanche are visible near the base*

Figure 20.2   Katse Dam. One of the most spectacular arch structures in the world with a height of 185 m and a crest length of 710 m, it forms a central component of the Lesotho Highland Water Project

'DAMNED DAMS': HISTORY, IDEOLOGY AND ACTORS

While extensive implementation of dam projects evolved with improvements in engineering knowledge, construction technology and progress in hydrological analyses, the global increase in technological hydroscapes is also an expression and outcome of prevailing development paradigms. The era of big dams began in the United States with the construction of Hoover Dam on the Colorado River in the 1930s. Earlier examples of extensive river control measures and dam building are known from the middle mountains of Germany, where constructions started in the early twentieth century, but the expertise remained confined domestically (Blackbourn 2006). After the Second World War, a number of large dams were built in the Soviet Union, following Stalin's concept of a 'transformation of nature into a machine for the communist state' (McCully 2001; Molle et al. 2009). Around the same time, dam building commenced in the European Alps, with Grande Dixence as one prominent example (Figure 20.1). In the Global South, large dams became icons of modernity and expressions of national prestige and emancipation from colonial rule. As the first Indian Prime Minister, Jawaharlal Nehru, stated, large dams are 'temples of resurgent India' and 'symbols of India's progress'. The modernization narrative was repeated in a different ideological setting under Mao Zedong and resulted in the construction of more than 600 large dams per year as part of the 'Great Leap Forward' (Gleick 1998; McCully 2001). Efforts at river control were not only based on development aspirations, but also upon the quest to conquer nature as a threat, or to tame, control and discipline nature as part of a 'hydraulic mission' to use every drop of water before it reaches the sea (Kaika 2006; Molle 2009; Molle et al. 2009).

Large dams are textbook examples that highlight most sensitive and contested development issues, shaped by tensions between actors, politics and economic aspirations at various scales. Key actor groups in the controversy are government agencies, river basin authorities, bi- and multilateral funding agencies, international construction associations, private sector companies, non-governmental organizations, human rights and affected people's groups (Figure 20.3). Struggles over power and influence within this constellation of place-based and non-place-based actors testify to a politicized environment, where conflicts can only be understood by considering power asymmetries and divergent interests (Baghel and Nüsser 2010). Both advocates and opponents of large hydro projects form coalitions to strengthen their position during the planning and implementation phase. The narrative of modernity, economic benefits, social progress and effective water management contrasts sharply against the counter-narrative of displacement and marginalization of local populations accompanied by the devastation of river habitats.

Nation-states and governments are definitely among the most important drivers of large dams, often making them symbols of nation-building or icons of ruling autocratic regimes. Powerful parastatal agencies and bureaucratic institutions are typically established to plan and realize these big projects and take on a life of their own, competing for influence and reasoning for their existence. The emergence of a hydraulic bureaucracy consisting of influential elites of technical engineers, hydrologists, political and financial leaders reinforces centralized planning and management of river control. One example is India's 'Ministry of Water Resources', which has a complicated history, after having variously been a 'Department of Irrigation' under ministries associated with Irrigation, Power, Mines, Scientific Research and Agriculture in different periods until finally becoming a

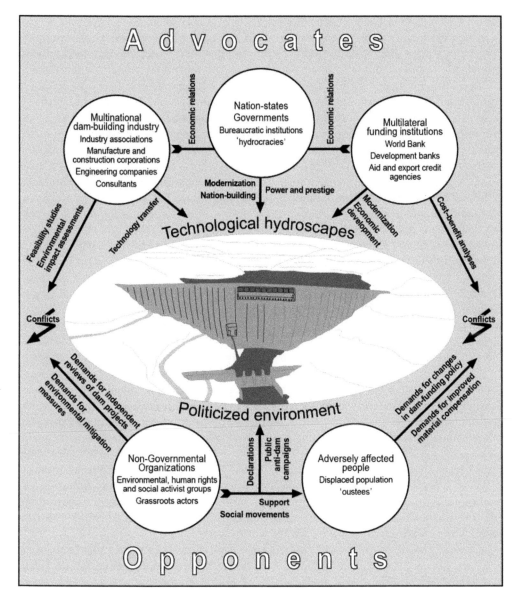

Figure 20.3   Characteristic actor constellation around large dams (modified after Nüsser 2014)

ministry of its own in 1985 (Ministry of Water Resources 2012). Pakistan's Water and Power Development Authority, established in 1958, is another case in point (Wescoat et al. 2000). This centralized government-owned organization was founded for the purpose of coordinating and giving a unified direction to the development of water and power schemes, which were previously being dealt with by respective electricity and irrigation departments at the provincial level. The emergence of such powerful and centralized parastatal agencies is by no means limited to the Asian context, as the example of the Lesotho Highlands Development Authority in southern Africa shows. These examples suggest how large dams have served changing bureaucratic structures and development priorities.

The dam-building industry, consisting of multinational engineering firms, consultants, equipment manufacturers and construction corporations, is another important driving force. These actors are organized in professional institutions such as ICOLD, which was established in 1928 to exchange experience in project design and hydraulic analysis. It consists of national committees from more than 90 countries with approximately 10,000 individual members (ICOLD 2016) and constitutes an active lobby for the propagation of dams. Advantages of technology transfer and economic progress are frequently expressed motivations for dam building in the Global South. Feasibility studies and environmental impact assessments are carried out by consultant companies, some of which are also directly involved in dam building. Dependent electricity-intensive industries (i.e. aluminium smelters) and agribusiness interests are intimately linked to the dam-building lobby as well (Fearnside 2016). The World Bank has been the most important financing agency for the dam-building industry. During recent decades, it has approved loan packages for a multitude of dams, including some of the world's most controversial megaprojects. Other major funding institutions are the multilateral development banks for Africa, Asia and Latin America as well as the bilateral aid and export credit agencies of most industrialized countries.

Since the mid-1980s, the international anti-dam movement, consisting of a network of environmental and human rights groups, has emerged. Their basic demands include independent impact assessments of projects and participation of affected people in the planning process. Working together with local groups, these 'grassroots' actors have been able to launch public opposition campaigns and declarations. Globally operating organizations like International Rivers and *Narmada Bachao Andolan* (Save the Narmada Movement) in India are prominent in this context (Khagram 2004; Nilsen 2010). The largest set of actors include the adversely affected people who suffer negative economic, social and cultural effects from construction works, impoundment and alteration of river flows. Whereas the displaced populations or 'oustees' are considered 'refugees in an unacknowledged war' (Roy 2001, 65) by activist groups, nation-states identify them as beneficiaries of prospective improvements.

However, the deep-seated differences between various interest groups cannot be reduced to a binary of modernized hydro politics versus environmental fundamentalism. Founded in 1997, the World Commission on Dams (WCD) addressed the conflicting viewpoints on large dams by including representatives of governments, industry, financial institutions, non-governmental organizations and affected people's organizations. As a forum with 68 institutions from 36 countries to reflect the diverse range of interests and to elaborate common ground in the negotiation process, it came up with five guiding principles: justice, sustainability, efficiency, participation and accountability. As a result, critiques of modernization and its neglect of broader social and environmental concerns became an integral part of the debate. The commission's final report (WCD 2000) provided the first comprehensive global and independent review of the performance and impacts of dams and presented a framework for water and energy resources development with criteria, guidelines and procedures for future decision-making. It was hoped that, with the ground-breaking work of the WCD, the controversies of the past would be buried, which they were for a period, symbolized by the drop in funding by the World Bank (Baghel and Nüsser 2010; Moore et al. 2010). However, despite on-going controversy over large dam projects, the discussion at the turn of the century was characterized by efforts to look for suitable compromises that met the requirements of different development perspectives.

THE GLOBAL PICTURE: CONTINUITIES, SHIFTS AND PARALLELS

Large dams as symbols of modernization are not solely for the internal benefit and prestige of nation-states, but also have strategic geopolitical implications (Hirsch 2016). India's Bhakra Dam, located close to the border of the Indian states Himachal Pradesh and Punjab on the Sutlej River, eventually flowing into Pakistan, is a case in point. Bhakra Dam was planned and initiated under the British administration, and was completed by Indian engineers under American supervision. Its power station on the right bank was built and later upgraded with Soviet assistance and technology during the height of the Cold War. India's control of the water flow through the dam was a major source of contestation from the Pakistan side of the Punjab owing to the need for a regular water supply for irrigation during the green revolution in both countries. This eventually led to the signing of the Indus Water Treaty in 1960 (Baghel 2014). Generally, such large hydropower projects include a flow of international expertise and financial investment, setting the blueprint for future development. In the case of China, the technological capacities needed for construction and management of hydrological infrastructure were initially provided by the Soviet Union, with thousands of engineers being trained. In the present day, China has by far the world's largest installed hydropower capacity, with aspirations for further expansion (Gleick 2012). Moreover, Chinese funding institutions and engineering companies are increasingly involved in dam projects in neighbouring and overseas countries (Brewer 2007). China's involvement in the construction of the Diamer Basha Dam in the Indus Valley of northern Pakistan, with the promise of several thousand experienced workers from the Three Gorges Dam, is just one example. This dimension of technology and knowledge transfer is in many ways similar to the expertise flowing from the Tennessee Valley Authority in the United States to India in the 1950s. Regardless of the political and ideological positions of respective countries, large dams have been considered as important foundations for national development, legitimized by the modernization paradigm (Ahlers et al. 2014).

As opposed to the Bhakra Dam, which gained widespread support as an icon for economic growth, the Narmada Dam cascade became a symbol of social inequity, resistance and an expression of failed development (Gadgil and Guha 1994; Drèze et al. 1997). Besides socio-economic and environmental concerns, the Tehri Dam (completed in 2006) in the Himalayan state of Uttarakhand, being the highest dam in India, has been criticized for the risk associated with it being located in a seismically active zone. Whereas the debate in India takes place in a democratic setting with multiple actors voicing their opinions (all too often with little real effect), the decision-making in China is primarily limited to the Communist Party with little local opposition (Dai 1998). Other prominent examples of the controversy are Pakistan's Tarbela and Diamer Basha dams, projects on the Mekong River (Bakker 1999; Matthews 2012; Merme et al. 2014; Hensengerth 2015) and Brazil's Tucurui Dam (Fearnside 1999). Earlier projects such as the Aswan High Dam in Egypt (completed 1970) or the Akosombo Dam in Ghana (completed 1965) remain contentious in terms of their long-term impacts on the environment, demography and economy of their surrounding regions.

In the Himalayan region, the governments of India, Pakistan, Nepal and Bhutan are transforming the upper reaches of the mountain drainage system into the sub-continental powerhouse of South Asia at an unprecedented pace (Baruah 2012; Huber and Joshi 2015). At an even faster rate, China taps the water resources of the Tibetan plateau to cope with growing water and energy demands in its urban agglomerations. The importance of the Himalayan region as a water tower for freshwater supply for the adjoining lowlands is now being supplemented by the additional function of a 'power tower', thereby intensifying the resource transfer from the mountains to the economic centres in the plains of the sub-continent (Erlewein and Nüsser 2011; Erlewein 2013). Prominent examples are the northern states of India, namely Himachal Pradesh, Uttarakhand, Sikkim, Arunachal Pradesh and Assam, which supply the adjoining lowlands with hydroelectric energy. Upcoming projects are also situated in territorially disputed regions, such as Arunachal Pradesh and the wider Kashmir region, where India and Pakistan are now building dams (Figure 20.4). In addition, India supports dam building in the upper

Source: Nüsser, September 2014.

Figure 20.4   *The Nimoo Bazgo Hydroelectric Plant is a run-of-the-river scheme on the Indus River, aimed at ensuring power supply to the electricity-deficient Ladakh region in northern India*

riparian countries of Bhutan and Nepal so as to fulfil the nations' energy demands as future powerhouses (Lord 2014, 2016). Major hydropower plants, which are currently under construction or in the planning phase in Nepal are primarily designed to generate power for export to India. One example is the Arun III project, which was expected to produce 404 MW. As a result of strong local and international resistance to the project it was cancelled in 1995, but is currently being redesigned as a run-of-river project. Such designs with increased hydraulic heads are widely used in the Himalayas, often in the form of a series of cascades and turbines. These constructions have been shown to have adverse hydrological consequences, such as river fragmentation and the periodic desiccation of river flow in certain parts (Erlewein 2013).

A similar situation exists in various African settings, where water and energy resources from mountain regions are transferred to downstream countries or regions. Rapid hydropower development is transforming the mountains of Ethiopia into an African power plant. Lesotho is another prominent example of an inter-basin water transfer scheme, where damming the headwaters of the highlands, which receive heavy rainfall, benefits the water-scarce metropolitan region of Gauteng in South Africa (Horta 1995; Nel and Illgner 2001; Hitchcock 2015). This has brought water revenues, hydroelectric power and major infrastructure to Lesotho; however, the local mountain dwellers, the Basotho, have lost arable and grazing land and have experienced drastic negative changes in livelihood conditions. This project was also criticized for massive corruption, similar to the case of the Turkwel Dam in Kenya (Figure 20.5), which, at more than 20 billion Kenyan shillings, ended up costing five times as much as initially proposed.

From the early enthusiasm surrounding the emergence of technological hydroscapes as economic drivers of modernization to the disillusion caused by the negative social and environmental outcome, the large dams debate has been tumultuous. Once being praised as temples of progress, they became decried as tombs of the displaced (Khagram 2003). However, with the current discussion about global warming, dam building has once again come into vogue as a climate-friendly technology, validated by global concerns for the reduction in carbon emissions. In the course of this latest disjuncture in the debate, large dams are now seen as mechanisms for mitigating the adverse effects of climate change. Not only has climate change brought a new assertiveness to the benefits of hydro-dams, it has also given birth to the emergence of climate economies, whereby funding is now relatively more easily available in the 'free market' through trade mechanisms of carbon emission certificates. This trend becomes apparent in the huge financial support whereby the international carbon trading scheme Clean Development Mechanism subsidizes dam building in China, India and Brazil as 'carbon offsetting dams' (Erlewein and Nüsser 2011; Erlewein 2014; Ahlers et al. 2015). Neoliberal ideals have taken hold, with the private sector becoming a key player in the funding and construction of large dams. However, empirical research has shown that Clean Development Mechanism projects fall short of achieving their objectives and their contribution to climate protection and sustainable development is questionable. Given the adverse impact of these technological hydroscapes on the local scale, carbon-offsetting dams are alleged to be a form of 'carbon colonialism' that exacerbates the asymmetries of problem causation and burden sharing.

Source: Nüsser, March 2016.

Figure 20.5   *With a height of 153 m and a crest length of 150 m, Turkwel Dam was constructed between 1986 and 1991. It is the highest dam in Kenya and is located on the border between the dryland counties of West Pokot and Turkana*

CONCLUSION

Various examples from different geographical regions and distinct historical, social and economic settings shed light on particularities and highlight the deep interconnections and disjunctures between different aspects of the large dam debate. The social costs of involuntary resettlement owing to large hydro projects are as dramatic as the environmental ones. The number of people flooded off their settlements, agricultural lands, forests and other resources is estimated to reach 40–80 million worldwide (WCD 2000). Besides the displaced population, other people affected by dam construction include rural dwellers residing downstream from a dam. They are often neglected in project assessments because it is assumed that they will benefit from the project; however, there are frequently significant negative downstream impacts (Scudder 2005). A huge number of case studies from the Global South provide evidence that the adverse impacts of large dams have fallen disproportionately on subsistence farmers, indigenous peoples and ethnic minorities, who often rely on common property regimes of resource utilization. Reservoirs inundate floodplain soils, woodlands, wildlife, fisheries and forests, which many local communities subsist and depend on to secure their livelihoods. Especially in mountain environments, dams force displaced inhabitants into the upper valleys, where they may cause further degradation of natural resources. Moreover, drastic natural hazards include reservoir-induced seismicity, which may lead to dam collapse and catastrophic floods (Chao 1995).

To meet the challenge of understanding the complexity of the issue, it is necessary to focus on the different development perspectives of the advocates and opponents of dam building. Framing large dams within the broader field of geographies of technology draws attention to their spatial dimension. This includes flows of technology, spatial inequalities between upstream and downstream populations, between mountains and lowlands, and between centres of decision-making and affected locations. On a national scale, power asymmetries between states lead to the transfer of benefits such as hydropower while bearing the cost of inundation. With the emergence of carbon-offsetting schemes, the costs and benefits are now traded at a global scale. Ultimately, the creation of technological hydroscapes is a political exercise carried out at the socio-hydrological interface, strongly affecting development paths at every scale (Swyngedouw 2015). Therefore large dams are not only transformative technologies that rewrite rivers and landscapes, but are also unique creations and creators of the Anthropocene.

REFERENCES

Ahlers, R., L. Brandimarte, I. Kleemans and S. Sadat. 2014. Ambitious development on fragile foundations: criticalities of current large dam construction in Afghanistan. *Geoforum* 54: 49–58.

Ahlers, R., J. Budds, D. Joshi, V. Merme and M. Zwarteveen. 2015. Framing hydropower as green energy: assessing drivers, risks and tensions in the eastern Himalayas. *Earth System Dynamics* 6: 195–204.

Ansar, A., B. Flyvbjerg, A. Budzier and D. Lunn. 2014. Should we build more large dams? The actual costs of hydropower megaproject development. *Energy Policy* 69: 43–56.

Appadurai, A. 1996. *Modernity at Large: Cultural Dimensions of Globalization*. Minneapolis, MN: University of Minnesota Press.

Baghel, R. 2014. *River Control in India. Spatial, Governmental and Subjective Dimensions*. Dordrecht: Springer.

Baghel, R. and M. Nüsser. 2010. Discussing large dams in Asia after the World Commission on Dams: is a political ecology approach the way forward? *Water Alternatives* 3(2): 231–248.

Bakker, K. 1999. The politics of hydropower: developing the Mekong. *Political Geography* 18: 209–232.

Baruah, S. 2012. Whose river is it anyway? Political economy of hydropower in the eastern Himalayas. *Economic and Political Weekly* 47(29): 41–52.

Blackbourn, D. 2006. *The Conquest of Nature: Water, Landscape and the Making of Modern Germany*. New York: W.W. Norton.

Brewer, N. 2007. Made by China: damming the world's rivers. *World Rivers Review* 22(3): 8–9.

Chao, B.F. 1995. Anthropological impact on global geodynamics due to water impoundment in major reservoirs. *Geophysical Research Letters* 22: 3533–3536.

Dai, Q. 1998. The Three Gorges Project: a symbol of uncontrolled development in the late twentieth century. In Q. Dai and J. Thibodeau (eds), *The River Dragon has Come! The Three Gorges Dam and the Fate of China's Yangtze River and its People*. pp. 3–17. London: M.E. Sharpe.

Drèze, J., M. Samson and S. Singh (eds). 1997. *The Dam and the Nation. Displacement and Resettlement in the Narmada Valley*. New Delhi: Oxford University Press.

Erlewein, A. 2013. Disappearing rivers: the limits of environmental assessment for hydropower in India. *Environmental Impact Assessment Review* 43: 135–143.

Erlewein, A. 2014. The promotion of dams through the clean development mechanism: between sustainable climate protection and carbon colonialism. In M. Nüsser (ed.), *Large Dams in Asia: Contested Environments between Technological Hydroscapes and Social Resistance*. pp. 149–168. Dordrecht: Springer.

Erlewein, A. and M. Nüsser. 2011. Offsetting greenhouse gas emissions in the Himalaya? Clean development dams in Himachal Pradesh, India. *Mountain Research and Development* 31(4): 293–304.

Fearnside, P. 1988. China's Three Gorges Dam: fatal project or step toward modernization. *World Development* 16(5): 615–630.

Fearnside, P. 1999. Social impacts of Brazil's Tucuruí dam. *Environmental Management* 24: 485–495.

Fearnside, P. 2016. Environmental and social impacts of hydroelectric dams in Brazilian Amazonia: implications for the aluminum industry. *World Development* 77: 48–65.

Gadgil, M. and R. Guha. 1994. Ecological conflicts and the environmental movement in India. *Development and Change* 25(1): 101–136.

Gleick, P. 1998. The status of large dams: the end of an era? In P. Gleick (ed.), *The World's Water 1998–1999. The Biannual Report on Freshwater Resources.* pp. 69–104. Washington, DC: Island Press.

Gleick, P. 2012. China dams. In P. Gleick (ed.), *The World's Water. Volume 7. The Biannual Report on Freshwater Resources.* pp. 127–142.Washington, DC: Island Press.

Hensengerth, O. 2015. Where is the power? Transnational networks, authority and the dispute over the Xayaburi Dam on the Lower Mekong Mainstream. *Water International* 40(5–6): 911–928.

Hirsch, P. 2016. The shifting regional geopolitics of Mekong dams. *Political Geography* 51: 63–74.

Hitchcock, R. 2015. The Lesotho Highlands Water Project: dams, development, and the World Bank. *Sociology and Anthropology* 3(10): 526–538.

Horta, K. 1995. The mountain kingdom's white oil: the Lesotho Highlands Water Project. *The Ecologist* 25(6): 227–231.

Huber, A. and D. Joshi. 2015. Hydropower, anti-politics, and the opening of new political spaces in the Eastern Himalayas. *World Development* 76: 13–25.

ICOLD. 2016. World Register of Dams, accessed 19 March 2016 from http://www.icold-cigb.org/GB/World_register/general_synthesis.asp.

Kaika, M. 2006. Dams as symbols of modernization: the urbanization of nature between geographical imagination and materiality. *Annals of the Association of American Geographers* 96(2): 276–301.

Kaisti, H. and M. Käkönen. 2012. Actors, interests and forces shaping the energyscape of the Mekong region. *Forum for Development Studies* 39(2): 147–158.

Khagram, S. 2003. Neither temples nor tombs: a global analyses of large dams. *Environment* 45(4): 28–37.

Khagram, S. 2004. *Dams and Development: Transnational Struggles for Water and Power.* Ithaca, NY: Cornell University Press.

Kirchherr, J., H. Pohlner and K. Charles. 2016. Cleaning up the big muddy: a meta-synthesis of the research on the social impact of dams. *Environmental Impact Assessment Review* 60: 115–125.

Li, H., P. Waley and P. Rees. 2001. Reservoir resettlement in China: past experience and the Three Gorges Dam. *Geographical Journal* 167(3): 195–212.

Lord, A. 2014. Making a 'hydropower nation': subjectivity, mobility, and work in the Nepalese hydroscape. *Himalaya, the Journal of the Association for Nepal and Himalayan Studies* 34(2): 111–121.

Lord, A. 2016. Citizens of a hydropower nation: territory and agency at the frontiers of hydropower development in Nepal. *Economic Anthropology* 3(1): 145–160.

Matthews, N. 2012. Water grabbing in the Mekong basin – an analysis of the winners and losers of Thailand's hydropower development in Lao PDR. *Water Alternatives* 5(2): 392–411.

McCully, P. 2001. *Silenced Rivers: The Ecology and Politics of Large Dams.* London: Zed Books.

Merme, V., R. Ahlers and J. Gupta 2014. Private equity, public affair: hydropower financing in the Mekong Basin. *Global Environmental Change* 24: 20–29.

Meybeck, M. 2003. Global analysis of river systems: from Earth system controls to Anthropocene syndromes. *Philosophical Transactions of the Royal Society B: Biological Sciences* 358(1440): 1935–1955.

Ministry of Water Resources. 2012. Brief note on organizational history of ministry of water resources. New Delhi, accessed from http://wrmin.nic.in/index2.asp?slid=283&sublinkid=542&langid=1.

Molle, F. 2009. River-basin planning and management: the social life of a concept. *Geoforum* 40(3): 484–494.

Molle, F., P. Mollinga and P. Wester. 2009. Hydraulic bureaucracies and the hydraulic mission: flows of water, flows of power. *Water Alternatives* 2(3): 328–349.

Moore, D., J. Dore and D. Gyawali. 2010. The World Commission on Dams + 10: revisiting the large dams controversy. *Water Alternatives* 3(2): 3–13.

Nel, E. and P. Illgner. 2001. Tapping Lesotho's 'White Gold': inter basin water transfer in Southern Africa. *Geography* 86(2): 163–167.

Nilsen, A. 2010. *Dispossession and Resistance in India: The River and the Rage.* London: Routledge.

Nilsson, C. and K. Berggren. 2000. Alterations of riparian ecosystems caused by river regulation. *Bioscience* 50(9): 783–792.

Nilsson, C., C. Reidy, M. Dynesius and C. Revenga. 2005. Fragmentation and flow regulation of the world's large river systems. *Science* 308(5720): 405–408.

Nüsser, M. 2003. Political ecology of large dams: a critical review. *Petermanns Geographische Mitteilungen* 147(1): 20–27.

Nüsser, M. 2014. Technological hydroscapes in Asia: the large dams debate reconsidered. In M. Nüsser (ed.),

Large Dams in Asia: Contested Environments between Technological Hydroscapes and Social Resistance. pp. 1–14. Dordrecht: Springer.

Nüsser, M. and R. Baghel. 2014. The emergence of the cryoscape: contested narratives of Himalayan glacier dynamics and climate change. In B. Schuler (ed.), *Environmental and Climate Change in South and Southeast Asia: How Are Local Cultures Coping?* pp. 138–156. Leiden: Brill.

Penz, P., J. Drydyk and P. Bose. 2011. *Displacement by Development: Ethics, Rights and Responsibilities*. Cambridge: Cambridge University Press.

Roy, A. 2001. The greater common good. In *The Algebra of Infinite Justice*. pp. 43–141. New Delhi: Penguin.

Scudder, T. 2005. *The Future of Large Dams: Dealing with Social, Environmental, Institutional and Political Costs*. London: Earthscan.

Swyngedouw, E. 2009. The political economy and political ecology of the hydro-social cycle. *Journal of Contemporary Water Research and Education* 142(1): 56–60.

Swyngedouw, E. 2015. Depoliticized environments and the promises of the Anthropocene. In R. Bryant (ed.), *The International Handbook of Political Ecology*. pp. 131–145. Cheltenham, UK and Northampton, MA, USA: Edward Elgar Publishing.

Vörösmarty, C., M. Meybeck, B. Fekete, K. Sharma, P. Green and J. Syvitski. 2003. Anthropogenic sediment retention: major global impact from registered river impoundments. *Global and Planetary Change* 39(1–2): 169–190.

Vörösmarty, C., D. Lettenmaier, C. Leveque, M. Meybeck, C. Pahl-Wostl, J. Alcamo, H. Cosgrove, H. Grassl, H. Hoff and P. Kabat. 2004. Humans transforming the global water system. *Eos Transactions American Geophysical Union* 85(48): 509–520.

WCD. 2000. *Dams and Development: A New Framework for Decision-Making*. London: Earthscan.

Wescoat, J. Jr, S. Halvorson and D. Mustafa. 2000. Water management in the Indus basin of Pakistan: a half-century perspective. *Water Resources Development* 16(3): 391–406.

21. Fracking for shale in the UK: risks, reputation and regulation

Peter Jones, Daphne Comfort and David Hillier

The identification of potentially exploitable large shale gas reserves, and plans for their subsequent commercial development by hydraulic fracturing, popularly known as fracking, within many countries of the world have generated mixed responses. In the United States, for example, the introduction of new drilling and fracturing technologies in the late 1990s saw the rapid commercial development of shale gas resources across many areas of the country, which in turn prompted the exploitation of shale gas reserves in Canada. By way of contrast, the identification and possible development of shale gas in parts of Europe have met with considerable public and political opposition. In France, which has the greatest potential shale gas resources, a moratorium on fracking has been in place since 2011, and was upheld in 2013, and Germany has not allowed any fracking since 2011. In June 2015 a majority of Members of the European Parliament voted for a moratorium on fracking which was described as 'a clear indication that public acceptance for this industry is crumbling across the EU' (Food and Water Europe 2015).

Within the UK there has recently been increasing interest in Government circles and amongst energy companies about the identification of potentially large-scale shale gas reserves and the Government 'believes that shale gas has the potential to provide the UK with greater energy security, growth and jobs' (Government of the UK 2014a). Despite this interest, exploration for shale gas is still at an early stage in the UK and there are currently no definitive or meaningful estimates of the likely shale gas reserves or of what proportion of the potential reserves *may* be practically and commercially recoverable. However the possible future commercial development of the shale gas reserves, by fracking, has also generated concerns about a wide range of environmental risks. Two linked factors seem to be important in addressing these concerns and arguably in facilitating the future development of shale gas resources within the UK. On the one hand the Government has emphasised its commitment to a regulatory regime designed to protect the environment and ensure public safety. On the other hand there is a commercial consensus that 'the industry needs to control reputation and risk' and that 'negative public opinion about environmental safety of the hydraulic fracturing process could undermine the development of this industry' (KPMG 2011, 19). With this in mind, this chapter offers a case study of the current debate surrounding the potential for fracking for shale gas in the UK. It begins with some introductory contextual thoughts on the changing and contested geographies of energy resources, describes the characteristics of shale gas and the process of fracking and outlines the scale and geography of potential shale gas reserves within the UK. The main body of the chapter provides a commentary on the environmental risks and issues associated with exploration and development of these reserves, reviews the contrasting and contested positions on the benefits and costs of shale gas development and examines the evolution of the regulatory framework with specific emphasis on planning policy and practice.

THE CHANGING AND CONTESTED GEOGRAPHIES OF ENERGY SUPPLY

In introducing the *New Geographies of Energy*, Zimmerer (2013) suggested that geography is 'crucial to addressing the multiple, interconnected dimensions of the current potpourri of global energy dilemmas and opportunities' and Bridge (2012) argued that 'the manner in which energy is captured and transformed lies at the heart of society's relationship with the natural world'. The geography of energy supply changes as new resources are discovered and old ones become depleted and/or economically unviable and as technological development makes reserves more accessible and economically recoverable. Bradshaw (2009), for example, charted the geographical dimensions of energy supply and demand and the recent global shift in the location of energy production and demand growth and argued that 'this shift is the result of increasing demand in emerging markets such as China and India' and that 'the centres of production are now focused on the Middle East, Africa and the former Soviet Union'. Changes in the geography of energy production are also occurring within countries looking to make the transition to a more sustainable energy supply based on renewable sources and in some cases to develop recently discovered fossil fuels.

In addressing the 'geographies of energy transition', Bridge et al. (2013) suggest that 'the energy challenge in the twenty-first century is to bring about a new transition towards a more sustainable energy system characterised by universal access to energy services and security and reliability of supply from efficient low-carbon sources'. Bradshaw (2010) has argued that 'we now face a global energy dilemma created by concerns about future availability of fossil fuels and the impact of their exploitation on the planetary ecosystem'. Bradshaw (2010) further suggested that within the 'developed market economies' the solution to this energy dilemma 'is being sought through increased energy efficiency, carbon trading, the development of technologies to de-carbonise fossil fuel use and electricity generation and the promotion of renewable energy and nuclear power'. That said, Bridge et al. (2013) argued that 'the geographical implications of this new energy paradigm are not well defined and a range of quite different geographical futures are currently possible'. However, within developed market economies renewable forms of energy generation cannot fully match current levels of demand and as such fossil fuels will remain a crucial element in the global energy mix into the foreseeable future. Where possible, many countries may look to exploit newly discovered indigenous fossil fuel resources while also pursuing a transition to more sustainable sources of energy supply.

At the same time in a seemingly increasingly volatile and unstable international environment concerns about the security of energy supplies loom large. Bridge (2012), for example, argued that 'the issues of energy availability and the vulnerability of fuel supplies are assuming new political prominence, so that hoary questions about depletion and security now share space on the environment and development agendas with greenhouse gas emissions and atmospheric pollution'. Energy security is a wide-ranging and complex issue and in identifying 'the key energy policy issues for energy security in the UK', Hoggett et al. (2011), for example, suggested that energy policy is 'a reflection of the sort of society that is wanted, including whether it is acceptable that the UK has large numbers of fuel poor; whether the UK should act as a responsible global nation/friend; if there is a concern about the environment; and that the balance is between environment and security'. The issue of energy security also has important geographical dimensions. Bridge et al. (2013) argued that 'the different elements of a policy to promote energy security . . . rest on assumptions about the geographical scale at which energy systems should be governed'. Further Bridge et al. (2013) suggested that 'ensuring the availability and accessibility of energy services in a carbon constrained world will require developing new ways – and new geographies – of producing , living and working with energy'.

It is important to recognise that emerging energy landscapes have become a focal point of debate within many countries and concerns are increasingly being raised about these energy landscapes and more specifically about the benefits and costs new energy developments bring to a range of stakeholders and particularly to those local communities where developments are taking place. Calvert and Mabee (2013), for example, argued that 'the unique physical properties or materialities (i.e. quality, quantity, location) of emerging energy resources are at the root of disruptive change to physical and social landscapes, and therefore of social resistance to policy efforts aimed at a sustainable energy future'. Selman (2010), for example, has argued that 'energy is likely to be a major driver of new landscapes as society seeks ways of weaning itself off fossil carbon fuels' and that 'society's increasingly earnest pursuit of sustainable development will involve landscape changes that attract protest and opposition'. More generally Jiusto (2009, 534) called for research into how society is 'contesting the next energy revolution' and Bridge (2012, 7) has emphasised the need to explore 'contemporary energy dilemmas – such as determining whether, how and for whom particular landscapes should be valued for their energy generating potential, or deciding on the geographical scale at which trade-offs between energy security and environmental impact should be made'.

SHALE GAS AND FRACKING

Shale gas is natural gas, mainly composed of methane, trapped in organic-rich shale beds often located between 1000 and 4000 m below the ground. Traditionally within the UK shale has not been seen as a reservoir rock, rather as a source rock in which gas, and oil, are stored before migrating into sandstone or limestone where they have been commercially exploited in a conventional manner. Indeed gas and oil produced from shale are often technically referred to as 'unconventional hydrocarbons'. Shale gas is accessed by fracking. The process involves drilling vertically perhaps 1500 metres or more below the surface and then drilling a number of horizontal boreholes in several directions. The horizontal drilling means that large areas of shale gas can be reached while minimising the number of surface boreholes and this facilitates drilling to less accessible locations. The fracking process involves pumping a mixture of fluids at high pressure into the shale, which creates a path for the gas to flow into the borehole and thence to the surface. Water makes up 90% of the fluids used in fracking and a large field with 1500 horizontal wells can use up to 20 million gallons of water per day. The water is mixed with gelling agents, which help to prise open the fractures, sandy materials, which hold open the fractures, chemicals, which reduce surface friction during the fracking process, and biocides, which kill bacteria.

The development of shale gas reserves includes three distinct stages, namely exploration, production and decommissioning. During the first stage two or three wells are normally drilled using a 25 m high structure known as a 'well over rig', and flow tested, to determine the incidence of shale gas reserves and this process normally takes up to 2 weeks. Production involves the commercial development of these reserves, which may, depending on the size of the reserve, continue for up to 20 years. When the shale gas reserve reaches the end of its lifespan, decommissioning involves filling the well with cement, to prevent further gas flowing into watercourses or to the surface, and capping and landscaping the well head.

The principle of fracking is not new. Explosive charges containing nitro-glycerine were first dropped down wells in the United States in the 1880s to shatter hard rock to release gas or oil. Hydraulic fracturing dates from the late 1940s, initially on an experimental basis on a gas field in Kansas in the United States, and then on a commercial basis in Oklahoma and Texas. The fracking of shale gas first took place on a demonstration basis in the 1970s but it was early in the twenty-first century before the technique began to be employed on a large-scale commercial basis. Since then developments in drilling and exploitation technology have seen dramatic growth in the fracking of shale gas within the United States. By 2013 shale gas was estimated to account for the largest share of total US natural gas production (US Energy Information Administration 2013) and to have transformed the energy landscape within the United States. Shale gas resources are now being exploited in West Virginia, Pennsylvania and New York State in the east across to Colorado and New Mexico and from Michigan in the north and as far south as Texas. In summarising trends within the United States, KPMG (2013, 2) suggested that the commercial development of shale gas reserves will continue 'for the foreseeable future'. At the same time KPMG (2013, 8) reports that 'inconsistent environmental regulations' have 'led investors to shun certain states, such as New York, in favour of those which are more supportive of development, such as Texas, North Dakota, Pennsylvania and West Virginia'.

Globally the Institute for Energy Research (2015) estimated that the total technically recoverable shale gas reserves are some 7299 trillion m^3 with China, Argentina, Algeria, the United States and Canada accounting for 45% of this total. While the term technically recoverable reserves is used to describe the volume of shale gas that could be produced with current technology, three factors, namely the cost of drilling and establishing wells, the volume of gas produced from a well during its lifetime and the price received for the gas, shape the economics of recovery. China has the largest shale gas resources in the world but many of these are located deep below the surface in mountainous rocky desert areas. The installation of production equipment and the construction of pipeline connections to the existing gas network seem likely to impede the commercial exploitation of these resources. In Australia there are sizeable shale gas reserves in both South and West Australia and in the Northern Territories. In Southern Australia, for example, production drilling began in 2012 in the Cooper Basin and here optimistic estimates suggest that up to 25,000 wells may be in production by the late 2020s (UCL International Energy Policy Institute 2013). That said, here and elsewhere in Australia, the need for the development of new pipelines to transport gas to existing networks and thence to centres of market demand and regulatory problems in allowing access to existing pipelines by new contractors may well slow the pace of development. Within Western Europe shale gas reserves have been identified in the Netherlands, Ireland, France, Germany, Poland, Romania, Bulgaria, Denmark, Sweden and Norway, as well as in the UK, but KPMG (2011, 12) suggested that, as reserves in a number of these countries 'tend to be close to populated areas and as European environmental laws tend to be quite strict, the potential for significant shale gas production there in the near future seems unlikely'.

POTENTIAL SHALE GAS RESERVES IN THE UK

Within the UK there are several areas where Carboniferous and Jurassic shale beds have the potential to produce shale gas, including sizeable areas of north-west, central and eastern England, smaller parts of south and north-east England, central Scotland and Northern Ireland. Although the commercial development of shale gas has been under-way in the United States for over 20 years, exploration for shale gas reserves within the UK is still very much in its infancy. There are currently no national estimates of how much shale gas will be technically and economically recoverable. The geological condi-tions are complex in that many of the shale basins are not large continuous structures, such as those found in many North American shale regions, but more typically comprise small fault-bounded sub-basins (Advanced Resources International 2013). At the same time the exploratory process is costly with some estimates suggesting that the average cost of drilling an exploratory well in the UK is some £6 million compared with £2.4 million in the United States (Ratcliffe 2014).

The British Geological Survey, in association with the UK Government's Department for Energy and Climate Change, has undertaken a number of shale resource estimates for some areas of the UK. In 2013, for example, the British Geological Survey published their estimate of shale gas resources in the Bowland–Hodder Shale Gas Resources under-lying an area stretching from north Wales and Blackpool in the east to Scarborough and Nottingham in the west (British Geological Survey 2015a). Given geological uncertainty, this estimate ranged from 822 trillion cubic feet (tcf) to 2281 tcf with the central estimate being 1329 tcf. That said the British Geological Survey stressed that 'not enough is yet known to estimate a recovery factor' nor to estimate 'how much gas may be ultimately produced' (British Geological Survey/Department of Energy and Climate Change 2013, 3). Estimates of the Carboniferous shales in the Midland Valley of Scotland ranged from 49 to 135 tcf with the central estimate being 80 tcf, but the British Geological Survey suggested that 'the relatively complex geology and limited amount of good quality con-straining data result in a higher degree of uncertainty to the Midland Valley of Scotland shale gas estimate than the Bowland–Hodder' study (British Geological Survey 2015b).

A number of small energy companies, including Cuadrilla, IGas, Third Energy and Celtique Energie, have undertaken test drilling wells, principally in West Lancashire, Cheshire, Manchester, Somerset, East Yorkshire, South Wales and Northern Ireland. Cuadrilla, for example, began drilling in 2010 at Preese Hall in Lancashire, but following some seismic activity associated with the hydraulic fracturing, the company suspended exploration activity and plugged the well. In response the UK Government announced a moratorium on fracking in July 2011 but following further investigations and consul-tations permission was given to resume exploratory drilling in December 2012. More recently Cuadrilla recommenced exploratory drilling, and obtained planning permis-sion for such drilling, elsewhere in Lancashire. Dart Energy have acquired planning permission for exploratory fracking in Dumfries and Galloway and submitted planning applications for exploration in the Falkirk and Stirling area of central Scotland. The Scottish Government announced a moratorium on all consents for fracking for shale gas in January 2015 and the Welsh Government imposed a similar moratorium the following month and in the light of these developments the main body of this chapter focuses on fracking for shale gas in England.

ENVIRONMENTAL ISSUES AND RISKS

The momentum behind shale gas development within the UK has been accompanied by growing and increasingly vocal concerns about the environmental impact of fracking. A wide range of environmental issues and risks have been identified. These include climate change; fugitive carbon dioxide and methane emissions; water use, waste water treatment and water pollution; seismic activity; air pollution; noise; visual intrusion; damage to valued and heritage landscapes; and the fragmentation and loss of habitats, damage to species and reductions in bio-diversity. The potential environmental risks are manifest at a variety of, often partly interlinked, spatial and temporal scales. Concerns about carbon dioxide emissions and climate change might, for example, be seen to be global, although they have implications for the UK Government's national targets on the reduction of greenhouse gas emissions.

Shale gas, like other natural gases, is not a low-carbon source of fuel and the large-scale development of shale gas would certainly not be consistent with a transition towards a more sustainable energy supply system. Methane can be emitted at a number of stages within the fracking process and such fugitive emissions are a particular concern in that methane has high global warming potential. Research on potential climate change impacts of shale gas (Tyndall Centre for Climate Change Research 2011, 110) concluded that 'without a meaningful cap on global carbon emissions, any emissions associated with shale gas are likely to be additional, exacerbating the problem of climate change'. More pointedly, Friends of the Earth (2013a) claimed that 'burning shale gas could set the world on course for catastrophic climate change' and 'have a major impact on investment in renewable energy needed to decarbonise the energy sector'.

The initial drilling process and the fracking of shale gas require large volumes of water. Meeting these demands in areas where other users are already finding it difficult to meet their water needs and that are vulnerable to water shortages may generate increasing stress on resources across wide geographical areas. Following the drilling of a well perhaps as much as 80% of the fracturing fluid, which may be saline and contain naturally occurring radioactive materials, returns to the surface and requires treatment before being returned to natural watercourses. That said, although the fracking fluid may be pumped into boreholes at discrete locations, once deep underground it is often difficult to predict its migration and concerns may arise about the contamination of drinking water over a wide area. Groundwater can also be contaminated by fugitive methane.

During the shale gas exploration and production stages a range of gaseous emissions can pass into the air not only from the wells themselves but also from the diesel-powered machinery at the drilling site. These emissions can lead to the formation of ozone, photochemical oxidants and particulate matter which can be damaging to human health. While earthquakes can be induced by fracking, shale rock is inherently weak and any resultant seismic activity is normally too small to be noticed at ground level. During the initial drilling phase the delivery of equipment, materials and water and the increase in vehicle movements can cause environmental disruption and there is also noise pollution from the drilling process. Fracking also has a significant footprint on the landscape. Land clearance is required, with up to 2 hectares required for each well head plus any land required for improved road access, and this can damage or destroy amenity, landscapes and habitats, reduce bio-diversity and lead to soil erosion.

There are also social concerns about the disruption fracking could bring to small communities, and to their traditional ways of living and working and of the possible impact on property prices and land values. There are concerns, for example, about the capacity of local infrastructure to cope with the attendant increase in traffic, employees and drilling equipment and worries that the chemicals used in the fracking process could pose health risks. In some rural areas there are fears that fracking operations may lead to a reduction in the number of tourists and of the income tourism has traditionally generated, while proposed fracking operations may have an effect on house prices, on potential purchasers' perceptions, on the availability of mortgages and on property insurance in the immediate vicinity of such operations. The employment of horizontal drilling could also have adverse property impacts across a much wider geographical area.

More general concerns have been expressed about the cumulative impact of a number of environmental (and social) risks outlined above in areas such as south-west Lancashire in the north of England, for example, where much of the initial fracking activity in the UK has been concentrated. In a wide-ranging report on the potential environmental risks arising from fracking operations in Europe for the European Commission, AEA, for example, suggested that the development of shale gas reserves may span a wide geographical area and argued that 'cumulative risks need to be taken into account in risk assessment' (AEA 2012, 24). The AEA report classed the cumulative impacts associated with water resources; ground and surface water contamination; gas emissions; land take; risks to bio-diversity; noise impacts; and traffic as all being 'high' (AEA 2012, vi). More specifically research on the large Marcellus shale gas reserves in the United States (Evans and Kiesecker 2014) concluded 'our analysis reveals it will be the cumulative impacts that pose the greatest challenge for landscape level conservation'.

REPUTATION

Public concern about many of the potential environmental risks associated with the fracking of shale gas reserves is generally seen to pose a significant threat to the successful commercial development of these reserves. In taking 'a global perspective' on the 'risks that could dim the future of shale gas', KPMG (2011, 18), for example, suggested that 'the industry needs to control reputation risk and turn public opinion round' and that 'negative public opinion about environmental safety of the hydraulic fracturing process could undermine the development of this industry, particularly where the process is used in – or directly under – populated areas' (KPMG 2011, 19). More specifically within the UK in identifying 'reputation' as one of the main barriers to enabling commercial production to go ahead, the Institute of Directors (2013a, 137) suggested that 'without a social licence to operate the industry will find it more difficult and more time consuming to obtain the necessary approvals to undertake exploration, and subsequent production activities'. In a similar vein KPMG (2013, 25) argued that 'If the UK is to meet the government's goals and extract shale gas on a commercially viable basis, the sector needs to overcome regulatory and market barriers and manage negative public views on exploration' (KPMG 2013). A battle has certainly been underway within the UK to win the public's hearts, minds and confidence particularly, although certainly not entirely, within local communities where exploratory fracking for shale is underway or planned. While it would be an oversimplification to suggest that those who wish to pursue, encourage and support the commercial development of shale gas and those who oppose its development sing from the same, if very contrasting, hymn sheets, two simple illustrative examples provide some basic insights into the case for and against shale gas development and how the battle for reputation is currently being played out.

First, a number of national organisations and local groups have been mobilising against shale gas exploration and production. These groups are generally well organised at the grassroots level, their case draws on a wide range of research evidence and they also tap into powerful community emotions. They have been harnessing information and communication technologies and social media to good effect and some have taken direct action to blockade sites in an attempt to stop exploratory drilling activity. At the local level a large number of opposition groups have emerged and are linked under the umbrella of 'Frack Off: Extreme Energy Action Network'. In July 2013, 21 local groups were listed on the pressure group's website (Frack Off: Extreme Energy Action Network 2013) but by November 2015 the number of local groups had risen to 202, spread throughout much of the UK (Frack Off: Extreme Energy Action Network 2015a). Local group Trowbridge Area Frack Free, for example, 'is for anyone in the Trowbridge area who is concerned about the impact fracking will have' and claims 'we want to share information with the public and raise awareness of what fracking means for the environment, wildlife, house prices and the increase in heavy traffic on our roads' (Frack Off Extreme Energy Action Network 2015a). In a similar vein Frack Free York is 'a grassroots group set up to raise awareness about and to prevent new forms of gas extraction in our local community. We provide a channel for action and work together with local and national groups' (Frack Off Extreme Energy Action Network 2015a).

Nationally Frack Off outlined 'The Fracking Threat to the UK' in graphic terms, namely 'Fracking is a nightmare! Toxic and radioactive water contamination. Severe air pollution. Tens of thousands of wells, pipelines and compressor stations devastating our countryside and blighting communities. All while accelerating climate change. And to produce expensive gas that will soon run out' (Frack Off: Extreme Energy Action Network 2015b). More widely, but equally graphically, Frack Off argued that 'fracking is just a symptom of a much wider problem. As easier to extract energy resources are exhausted by the unsustainable energy consumption of the present system, we are resorting to ever more extreme methods of energy extraction' (Frack Off: Extreme Energy Action Network 2015b). Frack Off also argued 'at present we are on a course which leads towards a world dominated by energy extraction and where most of the energy produced is used to run the extraction process while people live and die in its toxic shadow' and that 'the present system's addiction to massive amounts of energy is driving this headlong rush towards oblivion and unless something is done to stop it we will all be dragged down into hell with it' (Frack Off: Extreme Energy Action Network 2015b). Extreme Energy

Action Network provides a range of 'Campaign Resources' on its website including impact image resources, films, flyers and fact sheets and workshops.

Secondly energy companies, the business community and the UK Government have stressed the benefits that shale gas development will bring and have looked to assuage environmental and social concerns. The energy company, Cuadrilla, for example, argues that such development 'has been shown to have significant benefits for the communities in which operations take place, the regions that host them and for the rest of the country as well' (Cuadrilla 2015a). These benefits are described as 'jobs and investment', 'energy security', 'community benefit' and 'tax revenue' (Cuadrilla 2015a). Cuadrilla claims to be 'part of the community it operates within' and to be 'keen to make a contribution to community life' (Cuadrilla 2015b). Cuadrilla also claims that throughout its operations 'robust safety measures are in place to protect the environment' (Cuadrilla 2015c). Cuadrilla has also undertaken a number of other public engagement activities designed 'to provide residents and representatives with factual information about what is involved in the exploration for natural gas in shale rock' (Cuadrilla 2015d). These activities included the distribution of newsletters to residents living near to current and proposed drilling sites, site visits, presentations to community groups and a free phone community helpline.

More widely some sections of the UK business community have been keen to emphasise the economic benefits that the development of shale gas could generate. The UK's Institute of Directors argued that 'shale gas could represent a multi-billion pound investment, create tens of thousands of jobs, reduce imports, generate significant tax revenue and support British manufacturing' (Institute of Directors 2013b, 2). More specifically the UK's Institute of Directors claimed that 'cement and steel manufacturers, equipment manufacturers, drilling service companies and water treatment specialists would form important parts of the supply chain' and that 'spending by employees of the industry and its supply chain would benefit local businesses including restaurants, shops, pubs, theatres and hotels' (Institute of Directors 2013b, 2). The UK Government has clearly sought to make a strong economic case for the development of shale gas reserves. In 2014, David Cameron, the then-UK Prime Minister, for example, claimed that 'we're going all out for shale. It will mean more jobs and opportunities for people and economic security for our country' (Government of the UK 2014b). Edward Davey, the then Secretary of State for Energy and Climate Change, argued that shale gas is 'a national opportunity' and more specifically 'an opportunity for investment, jobs and tax revenues' (Government of the UK 2013b). At the same time the Government has also looked to answer many of the environmental concerns outlined earlier. A study of the potential greenhouse gas emissions from the production of shale gas in the UK, for example, commissioned in 2012 by the UK Government's Department of Energy and Climate Change (MacKay and Stone 2013, 37), concluded that 'with the right safeguards in place, the net effect on UK greenhouse gas emissions from shale gas production in the UK will be relatively small'. The Government has also looked to present shale gas as the 'cleanest fossil fuel' (Department of Energy and Climate Change 2013, 10), which would help, as part of a diverse energy mix, to act as a bridge in the transition to a low-carbon future.

The Government has also stressed that shale gas development 'must be done in partnership with local people' and that it wants 'to encourage a shale industry that is safe and doesn't damage the environment' (Government of the UK 2013a). In March 2013 the Government announced the creation of the new Office of Unconventional Gas and

Oil within the Department of Energy and Climate Change. This Office plans, inter alia, to 'bring forward proposals to ensure people benefit from shale gas production if there are future developments in their area' (Government of the UK 2013c). In 2014 the Government introduced a package of benefits, including financial support, for communities located close to exploratory wells and local councils in such areas will be able to retain 100%, as opposed to the existing 50%, of business rates from any shale gas developments (Government of the UK 2014b).

REGULATION AND THE PLANNING FRAMEWORK

Shale gas within the UK is owned by the state, under the Petroleum Act of 1988, and a Petroleum and Exploration and Development Licence is required for the development of shale gas reserves. As of November 2015, the UK government had issued licences to a range of energy companies for 203 blocks, each about 4 miles square, and these licences confer exclusive rights to undertake exploratory drilling and production of shale gas (White et al. 2015). Licences in themselves do not give consent for fracking and a number of other permissions are required before a company can begin exploratory or production drilling for shale gas. More specifically companies must gain access rights from the landowners, obtain both the relevant environmental permits to drill from the UK's Department for Energy and Climate Change, meet the UK's Health and Safety Executive's health and safety regulations and obtain local authority planning permission. Where fluids used in the fracking process contain pollutants, for example, then an environmental permit must be obtained from the UK's Environment Agency. The Environment Agency will also take account of the potential impacts of fracking on groundwater levels and the appropriate consents may be required before drilling can commence. The Health and Safety Executive is responsible for monitoring safe working practices and the integrity of borehole operations.

Within the UK it is the local minerals planning authority that is responsible for determining if shale gas exploration and production by fracking is acceptable at specific sites. Given the scale of recent estimates of shale gas reserves local minerals planning authorities in many parts of the UK seem likely to face a growing number of applications for shale gas exploration and production. Some national planning guidelines have recently been published which might be seen to help local minerals planning authorities in determining such applications. The National Planning Policy Framework (NPPF) for England and Wales published in 2012, for example, did not explicitly mention fracking and thus it offered nothing by way of specific guidance for local planning authorities. That said, potentially contradictorily, the NPPF stressed the need 'to help increase the use and supply of renewable and low carbon energy, local authorities should recognise the responsibility on all communities to contribute to energy generation from renewable and low carbon sources' (Department for Communities and Local Government 2012, 22). The NPPF also emphasised the need 'to respond to the changes that new technologies offer us', to 'accommodate the new ways by which we will earn our living in a competitive world' (Department for Communities and Local Government 2012, 1) and to 'give great weight to the benefits of mineral extraction, including to the economy' (Department for Communities and Local Government 2012, 34).

However, in 2013 the Government published planning practice guidance for onshore oil and gas exploration and production for England. This guidance provides advice on 'how shale gas development should proceed through England's planning system' (Department for Communities and Local Government 2013) and included advice on development management procedures, environmental impact assessment, determination of planning applications and decommissioning and land restoration. This guidance on the need to conduct an environmental impact assessment, for example, suggested that such an assessment would only be required 'if the project is likely to have significant environmental effects' and that 'it is unlikely that an Environmental Impact assessment will be required for exploratory drilling operations' (Department for Communities and Local Government 2013, 13). Planning authorities are also advised to take account of the possible cumulative effects of one or more applications for shale gas development within an area, but here again the advice is that such cumulative effects are unlikely at the exploration phase. The guidance lists some 16 environmental issues including noise, landscape character, land contamination and flood risk, which should be addressed by planning authorities. More generally local planning authorities are advised that they must ensure that shale gas development is appropriate to its location and that it does not have an unacceptable adverse impact on the natural or historic environment or human health (Department for Communities and Local Government 2013). In determining planning applications for shale gas exploration and production, local authorities were advised that, while they should not consider the demand for, or the alternatives to shale gas, they should 'give great weight to the benefits of mineral extraction' (Department for Communities and Local Government 2013, 15).

While the guidance sought to provide greater clarity about the planning process for shale gas exploration and extraction, it was not universally well received. Within the planning profession some critics have argued that this guidance was weighted in favour of granting permission. A principal planner at Savills, the UK's leading estate agency, for example, was reported as arguing that the guidance was akin to a presumption in favour of the development of shale gas resources and more specifically that 'rather than just introducing controls over how decisions would be made, the guidance implies that government wants to see them go through' (Planning Resource 2013).

Pinsent Masons (2013, 2), a UK-based law firm with specific expertise in energy and natural resources and real estate, for example, suggested that the guidance was not comprehensive. More specifically Pinsent Masons argued 'there are areas where some in the industry may find that guidance is lacking: for example, in its failure to tackle key questions such as how planning boundaries should be drawn for directional and horizontal drilling once the appropriate rock formation is reached, how to deal with issues where the surface and subsurface are in different ownership and the way in which the guidance deals with the consideration of alternatives in the context of need and demand'. More critically Friends of the Earth (2013b) has criticised this guidance, arguing that it 'will ride roughshod over local concerns about shale gas exploration and development with little regard for the impact on the wellbeing of local people or the environment' and that it is 'little more than a carte blanche to dispatch dirty energy companies into the British countryside to start sinking thousands of new fracking wells'.

In June 2015 the first, and currently the only, planning applications to permit shale gas production by fracking in the UK, on two sites, at Roseacre Wood in Preston and at

Little Plumpton, between Blackpool and Preston in Fylde, West Lancashire, submitted by the energy company Cuadrilla, were rejected by Lancashire County Council planning authority. The application at the Plumpton site, for example, was rejected for two reasons: firstly, 'The development would cause an unacceptable adverse impact on the landscape, arising from the drilling equipment, noise mitigation equipment, storage plant, flare stacks and other associated development. The combined effect would result in an adverse urbanising effect on the open and rural character of the landscape and visual amenity of the residents contrary to policy DM2 Lancashire Minerals and Waste local Plan and Policy EP11 of the Fylde Local Plan'; secondly, 'The development would cause unacceptable noise impact resulting in a detrimental impact on the amenity of local residents which could not be adequately controlled by condition contrary to Policy DM2 of the Lancashire Minerals and Waste Local Plan and Policy EP27 of the Fylde Local Plan' (Lancashire County Council 2015). In July 2015 Cuadrilla announced their intention to formally appeal against Lancashire County Council's refusal of planning permission for fracking at the two sites and the appeals were subsequently submitted in September 2015.

Seemingly, although not explicitly, in response to Lancashire County Council's rejection of these two applications and perhaps because of the signal it might be seen to send to other local planning authorities, in August 2015, the UK Government announced that 'shale gas planning applications will be fast tracked through a new dedicated planning process' (Government of the UK 2015). The objective was 'to ensure shale applications can't be frustrated by slow and confused decision making amongst councils (local planning authorities), which benefits no one' and a number of specific measures were included in the announcement. The Secretary of State for the Department for Communities and Local Government can call in shale gas planning applications on a case by case basis, thus removing the decision-making process from the local planning authority, and can also call in shale gas applications that have not been determined by local planning authorities within the 16 week statutory time frame. More pointedly, where local authorities repeatedly fail to determine shale gas applications within the statutory time frame they could lose their right to determine any such future applications. At the same time the emphasis will ensure that any applications called in and all appeals are prioritised by the Government's Planning Inspectorate.

While it remains to be seen how these new measures will play out in reality, they attracted widespread criticism when announced. Local authority politicians in Lancashire, for example, expressed concerns about proposals that may effectively take decisions about the fracking of shale gas away from the locally elected representatives. At the same time there are also concerns that, in submitting planning applications for fracking, shale energy companies may include large amounts of detailed technical data and documentation in support of their application that the local planning authority may find very difficult to assimilate and evaluate within the statutory 16 week time frame. Where local community groups and environmental organisations also look to make detailed and wide-ranging representations to the local planning authority, this may further exacerbate delays and effectively play into the hands of the applicants. Friends of the Earth (2015) argued that 'bulldozing fracking applications through the planning system, against the wishes of local people and councils, will simply fan the flames of mistrust and opposition. Local authorities have been following the rules. These changes are being made because the Government doesn't agree with the democratic decisions councils have been making'. More generally a report on the potential environmental impacts of fracking for shale gas undertaken for a range of UK nature conservation organisations concluded that 'the current regulatory regime is not fit for purpose and therefore unable to adequately manage serious environmental risks that may arise from individual projects and cumulative development' (Moore et al. 2014, 26).

CONCLUSION

The commercial exploitation of shale gas reserves is very much at the exploratory stage in the UK, but the pressures for the commercial development of these reserves by fracking have gained momentum in a number of areas. Opinion is sharply divided about the potential economic benefits and environmental risks of such development. While the UK Government and the business community have generally been keen to stress the economic benefits, the development of shale gas could bring nationally and locally, a range of environmental pressure groups are energetically and vociferously opposed to such development. Within the shale gas industry there certainly is a broad consensus that promoting positive messages about shale gas development and managing and countering many of the negative public views about such developments are essential if shale gas resources are to be successfully exploited commercially. To this end a number of the energy companies have engaged public relations companies to develop comprehensive, coherent and co-ordinated media relations campaigns in an attempt to win hearts and minds at both the local and national levels. However the scale of the challenges should not be underestimated. The independent global risks consultancy, Control Risks (2013, 1), for example has argued that 'the oil and gas industry has largely failed to appreciate social and political risks and has repeatedly been caught off guard by the sophistication, speed and influence of anti-fracking activists'.

Given current Government thinking, local minerals planning authorities in many parts of the UK may receive a growing number of planning applications for shale gas exploration and development and they seem likely to have the primary regulatory responsibility for determining whether initial exploration for, and subsequent production of, shale gas reserves goes ahead. As such in looking to reconcile competing interests at the local level, planning authorities may have to balance the potential inward investment and job creation benefits claimed for such exploration and development and their commitments to sustainability and the transition to a low-carbon future and deeply held local environmental and community concerns. That notwithstanding, there is a body of opinion that suggests that the direction of thinking adopted by the UK Government is, at best, flawed and at worst, weighted in favour of the development of shale gas reserves. More generally the potential economic benefits and environmental risks associated with fracking for shale gas can be seen in terms of a local and national framework. Thus while major national economic and energy benefits are claimed for the development of shale gas, the environmental risks are concentrated at the local level.

REFERENCES

Advanced Resources International. 2013. World shale gas and shale oil resource assessment, accessed 25 March 2014 from http://www.adv-res.com/pdf/A_EIA_ARI_2013%20World%20Shale%20Gas%20and%20 Shale%20Oil%20Resource%20Assessment.pdff.

AEA. 2012. Support to the identification of potential risks for the environment and human health arising from hydrocarbons operations involving hydraulic fracturing in Europe, accessed 25 March 2014 from http://ec.europa.eu/environment/integration/energy/pdf/fracking%20study.pdff.

Bradshaw, M. 2009. The geopolitics of global energy security. *Geography Compass* 3(5): 1920–1937.

Bradshaw, M. 2010. Geographical energy dilemmas: a geographical perspective. *Geographical Journal* 176(4): 275–290.

Bridge, G. 2012. Teaching energy issues in geography, accessed 25 March 2014 from http://www.heacademy.ac.uk/assets/documents/STEM/Teaching-Energy-Issues-in-Geography-GEES_2012.pdff.

Bridge, G., S. Bouzarovski, M. Bradshaw and N. Eyre. 2013. Geographies of energy transition: space, place and the low carbon economy. *Energy Policy* 53: 331–340.

British Geological Survey. 2015a. Bowland shale gas, accessed 9 November 2015 from http://www.bgs.ac.uk/research/energy/shaleGas/bowlandShaleGas.html.

British Geological Survey. 2015b. The Carboniferous shales of the Midland Valley of Scotland: geology and resource estimation, accessed 9 November 2015 from http://www.bgs.ac.uk/research/energy/shaleGas/mid landValley.html.

British Geological Survey/Department of Energy and Climate Change. 2013. The Carboniferous Bowland shale gas study; geology and resource estimates, accessed 24 February 2014 from https://www.gov.uk/government/uploads/system/uploads/attachment_data/file/226874/BGS_DECC_BowlandShaleGasReport_MAIN_REPORT.pdff.

Calvert, K. and W. Mabee. 2013. Energy transition management as a spatial strategy: geographical implications of the transition towards renewable energy, accessed 9 November 2015 from http://www.geog.psu.edu/sites/default/files/Calvert_Geogs%20of%20RE_working%20paper.pdf.

Control Risks. 2013. The global anti-fracking movement, accessed 25 March 2014 from http://www.control risks.com/Oversized%20assets/shale_gas_whitepaper.pdf.

Cuadrilla. 2015a. Benefits, accessed 9 November 2015 from http://www.cuadrillaresources.com/benefits/.

Cuadrilla. 2015b. Community benefits, accessed 9 November 2015 from http://www.cuadrillaresources.com/benefits/community-benefit/.

Cuadrilla. 2015c. Protecting our environment, accessed 9 November 2015 from http://www.cuadrillaresources.com/protecting-our-environment/.

Cuadrilla. 2015d. Part of the community, accessed 9 November 2015 from http://www.cuadrillaresources.com/about-us/part-of-the-community/.

Department for Communities and Local Government. 2012. National Planning Policy Framework, accessed 25 March 2014 from https://www.gov.uk/government/uploads/system/uploads/attachment_data/file/6077/2116950.pdf.

Department for Communities and Local Government. 2013. Planning practice guidance for onshore oil and gas, accessed 24 March 2014 from https://www.gov.uk/government/uploads/system/uploads/attachment_data/file/224238/Planning_practice_guidance_for_onshore_oil_and_gas.pdf.

Department for Energy and Climate Change. 2013. Developing onshore shale gas and oil – facts about fracking, accessed 24 March 2014 from https://www.gov.uk/government/uploads/system/uploads/attachment_data/file/265972/Developing_Onshore_Shale_Gas_and_Oil__Facts_about_Fracking_131213.pdf.

Evans, J. and J. Kiesecker. 2014. Shale gas, wind and water: assessing the potential cumulative impacts of energy development on ecosystem services within the Marcellus Play. *PLoS One* 9(2), accessed 20 March 2014 from http://www.plosone.org/article/info%3Adoi%2F10.1371%2Fjournal.pone.0089210.

Food and Water Europe. 2015. For first time, a majority of MEPs vote for an immediate moratorium on fracking, accessed 9 November 2015 from http://www.foodandwatereurope.org/pressreleases/majority-meps-vote-for-immediate-fracking-moratorium/.

Frack Off: Extreme Energy Action Network. 2013. Local groups, accessed 30 July from http://frack-off.org.uk/local-group-specific-pages/.

Frack Off: Extreme Energy Action Network. 2015a. Local groups, accessed 3 November 2015 from http://frack-off.org.uk/local-group-specific-pages/.

Frack Off: Extreme Energy Action Network. 2015b. Fracking threat to the UK, accessed 3 November 2015 from http://frack-off.org.uk/fracking-hell/.

Friends of the Earth. 2013a. Briefing: unconventional, unnecessary and unwanted, accessed 25 March 2014 from http://www.foe.co.uk/resource/briefings/shale_gas.pdf.

Friends of the Earth. 2013b. Govt. planning guidance on fracking slammed, accessed 25 March 2014 from https://www.foe.co.uk/resource/press_releases/govt_planning_guidence_on_fracking_slammed_19072013.

Friends of the Earth. 2015. UK government announces new fast-track fracking policy, accessed 3 November 2015 from https://www.foe.co.uk/resource/press_releases/govt-fast-track-fracking-through-planning-system_13082015.

Government of the UK. 2013a. Estimates of shale gas resource in North of England published, alongside a package of community benefits, accessed 24 March 2014 from https://www.gov.uk/government/news/estimates-of-shale-gas-resource-in-north-of-england-published-alongside-a-package-of-community-benefits.

Government of the UK. 2013b. The myths and realities of shale gas exploration, accessed 24 March 2014 from https://www.gov.uk/government/speeches/the-myths-and-realities-of-shale-gas-exploration.

Government of the UK. 2013c. New office to look at community benefits for shale gas projects, accessed 25 March 2014 from https://www.gov.uk/government/news/new-office-to-look-at-community-benefits-for-shale-gas-projects.

Government of the UK. 2014a. Providing regulation and licencing of energy industries and infrastructure, accessed 24 March 2014 from https://www.gov.uk/government/policies/providing-regulation-and-licensing-of-energy-industries-and-infrastructure/supporting-pages/developing-shale-gas-and-oil-in-the-ukk.

Government of the UK. 2014b. Local councils to receive millions in business rates from shale gas development, accessed 25 March 2014 from https://www.gov.uk/government/news/local-councils-to-receive-millions-in-business-rates-from-shale-gas-developments.

Government of the UK. 2015. Faster decision making on shale gas for economic growth and energy security, accessed 5 November 2015 from https://www.gov.uk/government/news/faster-decision-making-on-shale-gas-for-economic-growth-and-energy-security.

Hoggett, R., C. Mitchell, B. Woodman and P. Baker. 2011. The key energy policy issues for energy security in the UK, accessed 25 March 2015 from http://geography.exeter.ac.uk/catherinemitchell/The_Key_Energy_Policy_Issues_for_Energy_Security_in_the_UK_-_Summary_Report.pdf.

Institute of Directors. 2013a. Getting shale gas working, accessed 17 November 2015 from http://www.iod.com/influencing/policy-papers/infrastructure/infrastructure-for-business-getting-shale-gas-working.

Institute of Directors. 2013b. Getting shale gas working: summary report, accessed 17 November 2015 from http://www.iod.com/influencing/policy-papers/infrastructure/infrastructure-for-business-getting-shale-gas-working.

Institute for Energy Research. 2015. US miles ahead in global shale race, accessed 20 July 2015 from http://instituteforenergyresearch.org/analysis/only-four-countries-produce-shale-oilgas/.

Jiusto, S. 2009. Energy transformations and geographic research. In N. Castree, D. Demeritt, D. Rhoads and D. Liverman (eds), *A Companion to Environmental Geography.* pp. 533–551. London: Blackwell.

KPMG. 2011. Shale gas – a global perspective, accessed 25 March 2014 from http://www.google.co.uk/#bav=on.2,or.r_qf.&fp=881ed480936369b9&q=KPMG+2011.+%E2%80%98Shale+Gas-A+Global+Perspective%E2%80%99%2c.

KPMG. 2013. Shale development: global update, accessed 5 July 2015 from http://www.kpmg.com/Global/en/IssuesAndInsights/ArticlesPublications/shale-gas/Documents/shale-development-global-update.pdf.

Lancashire County Council. 2015. Development control committee minutes of the meeting held on 23, 24, 25 and 29 June 2015 at 10.00 am in Council Chamber, County Hall, Preston, accessed 29 October 2015 from http://council.lancashire.gov.uk/documents/s67167/FinalDCminsJunewithannexs.pdf.

MacKay, D. and T. Stone. 2013. Potential greenhouse gas emissions associated with shale gas extraction and use, accessed 24 February 2014 from https://www.gov.uk/government/uploads/system/uploads/attachment_data/file/237330/MacKay_Stone_shale_study_report_09092013.pdf.

Moore, V., A. Beresford and B. Gove. 2014. Hydraulic fracturing for shale gas in the UK. RSPB, Sandy, Bedfordshire, accessed 25 March 2014 from https://www.google.co.uk/#q=Moore%2C+V.%2C+Beresford%2C+A.+and+Gove%2C+B.+2014.+%E2%80%98Hydraulic+fracturing+for+shale+gas+in+the+UK%E2%80%99.

Pinsent Masons. 2013. Shale gas & fracking toolkit: part 3, accessed 25 March 2014 from http://www.pinsentmasons.com/PDF/ShaleGasFrackingPart3.pdf.

Planning Resource. 2013. Warning of fracking guidance bias, accessed 14 February 2014 from http://www.planningresource.co.uk/article/1192695/warning-fracking-guidance-bias.

Ratcliffe, V. 2014. Brakes put on UK shale gas revolution, accessed 25 March 2014 from http://www.ft.com/cms/s/0/99d6a16e-a3ba-11e3-88b0-00144feab7de.html#axzz2wxUlbB2rr.

Selman, P. 2010. Learning to love the landscapes of carbon-neutrality. *Landscape Research* 35(2): 157–171.

Tyndall Centre for Climate Change Research. 2011. Shale gas: an updated assessment of environmental and climate change impacts, accessed 25 March 2014 from http://www.mylocalfuneraldirector.co.uk/Corporate/Fracking/1/Shale%20gas%20update%20-%20full%20report.pdf.

UCL International Energy Policy Institute. 2013, Shale gas in Australia: the policy options, accessed 9 November 2015 from https://www.ucl.ac.uk/australia/files/shale-gas-in-australia-green-paper-final.

US Energy Information Administration. 2013. Technically recoverable shale oil and shale gas resources, accessed 20 February 2014 from http://www.eia.gov/analysis/studies/worldshalegas/pdf/overview.pdf?zscb=599773466.

White, E., M. Fell, L. Smith and M. Keep. 2015. Shale gas and fracking. House of Commons Library; briefing paper number SN06073, accessed 9 November 2015 from http://researchbriefings.files.parliament.uk/documents/SN06073/SN06073.pdf.

Zimmerer, K. 2013. Approaching the new geographies of global energy: analytics and assessment of current energy landscapes and alternatives. In K. Zimmerer (ed.), *The New Geographies of Energy*. pp. 1–10. London: Routledge.

22. Geography of geothermal energy technologies
Edward Louie and Barry Solomon

Geothermal energy is released from within the Earth, and the term is often associated with its use for electric power generation. The Earth's heat comes primarily from the radioactive decay of uranium, thorium and potassium, as well as residual heat from the kinetic energy of the Earth's formation. The temperature increases with depth from the surface because of a natural geothermal gradient, which can range from 15 to >50°C/km and averages around 25°C (Blackwell et al. 2006). This temperature difference between the Earth's interior and the surface creates an upward heat flow. As a result, geothermal energy is potentially available worldwide. However, its practical use is site-specific, since its spatial concentration and vertical accessibility vary geographically owing to geological and other factors. Because of this uneven and in some cases quite limited accessibility, only a modest amount of the total geothermal energy resource can be used by people.

There are many ways to use geothermal energy directly without generating electricity, and it thus has a long history. The first use of geothermal energy was undoubtedly for hot springs, perhaps in Babylonia and ancient Greece, although the exact origins are uncertain and date back to prehistoric times. The use of geothermal energy at hot springs or other 'direct use' applications is possible with water temperatures below 90°C, and more likely 40°C or less for baths and swimming pools. Geothermal energy use in hot springs, spas and public baths became common in the Roman Empire, Iceland and England as well, and in many other parts of Europe, the Middle East and Asia (Cataldi et al. 1999). Such direct use of geothermal energy requires little or no technology to effectively capture and contain the heat, and thus can be very attractive.

Geologists as well as geographers have made important contributions to the study of geothermal energy, and there is overlap. Geologists have been mainly concerned with categorizing, assessing and quantifying the geothermal energy resource base. They are also interested in studying the spatial distribution, characteristics and geological controls on site-specific geothermal resources. These steps are required before we can determine how much of the energy can be used by people, and where. After this is done, geothermal energy use is then based on current and emerging technologies, which often require engineering. Geographers, in contrast, place more emphasis on the nexus between geothermal energy and the environment, human needs and impacts, and sustainability.

This chapter begins by reviewing the geography literature on geothermal energy technologies. While not a major research subject in geography, important work on geothermal energy has been done since the late 1970s. We can divide this literature into five main foci: ecological and environmental effects; land use planning; public opinion, institutional and social barriers; sustainability; and place-based case studies. After the literature review, we will discuss several conceptual issues and debates surrounding geothermal energy use. These include its status as a renewable energy source, and where it should or should not be used. The next section will address the main categories of geothermal energy technologies, and the associated geographical and spatial issues. This discussion will be divided into electricity generation and various types of direct use applications. The chapter will close with some conclusions.

THE LITERATURE ON GEOTHERMAL ENERGY

Ecological and Environmental Effects

Mike Pasqualetti has been the most prolific geographer to research geothermal energy technologies. However, Pryde (1977) was actually the first to study geothermal energy, and provided a review of the resources in the western United States based on satellite data. Two of Pasqualetti's earliest contributions focused on better understanding of the environmental quality issues at geothermal electric power plants, and comparing them with conventional electricity generating stations. He conducted an early review of these issues for six countries (Pasqualetti 1980). He found that the environmental impacts and constraints are much less serious in the case of geothermal plants, although they vary by country. All geothermal sites have challenges and conflicts and competition in land use, although most environmental concerns are site-specific. In the case of the United States, for example, he found hydrogen sulfide (H_2S) emissions, which can damage human health and cause a foul odor, to be the major concern. In other countries land use, subsidence, water pollution and wastewater disposal were more salient. Another environmental problem, first recognized in the 1980s, is hazardous waste generation from drilling activities and power generation (Pasqualetti and Dellinger 1989). Hazardous material generated at geothermal power plants includes H_2S sludge, hydrogen peroxide and sodium hydroxide, among others. Several serious accidents and spills involving hazardous waste haulage from The Geysers geothermal power plant in northern California occurred from 1984 to 1987, although such problems since then have been greatly reduced (Pasqualetti and Dellinger 1989).

Land Use Planning

The site-specific nature of geothermal energy, especially for direct use, has received attention. This is because it is more efficient and cost-effective to use geothermal heat close to the resource. As noted by Pasqualetti (1983, 1990), this has important implications for land use planning. In a study of eight potential development sites in the western United States, he found that potential energy density, commercial energy demand, land use characteristics and zoning restrictions were most important in determining project viability (Pasqualetti 1983). In another study, Pasqualetti (1986) developed a simple, low-cost site suitability technique to select sites for direct use of geothermal energy for district heating or cooling. In a precursor to later geographic information system (GIS) analyses, this study used map overlays for Scottsdale, Arizona to determine the most compatible sites. The characteristics analyzed included user energy density, zoning, land parcel size, vacancy and ownership for sites with geothermal resources. Geothermal energy studies in the 1970s and 1980s gave only passing consideration to the impacts on terrestrial ecosystems and wildlife, since these effects are minor. Copeland et al. (2011), in a recent study of energy development in the western United States, showed that over three-quarters of the geothermal leases are on shrublands. However, geothermal energy's footprint is much smaller than that of other types of energy development, especially oil and natural gas.

Social and Institutional Barriers

Turning to the social side of geothermal energy, there has been significant research on public opinion and social barriers to facility development. However, most of this work is broader in scope than geothermal, covering a range of renewable energy technologies, and much of the research has been conducted outside of geography by sociologists, political scientists, psychologists and others. In the next two paragraphs, we will discuss three important papers written by geographers on these topics.

Walker (1995) considered public opinion towards geothermal energy development, along with hydroelectric, tidal and especially wind power. While the public generally favors geothermal energy use and other renewable sources of energy, a major exception occurred in Hawai'i. On Hawai'i's Big Island, a geothermal power plant proposal of up to 500 MWe was protested and blocked by Native Hawaiians, on and off from 1972 to 1990. Reasons for local opposition included: the large scale of the project; lack of local community benefits; lack of sensitivity to environmental and cultural impacts (e.g. the site was on sacred grounds near the highly active Kilauea Volcano); skepticism over predictions of impacts; and locational factors of high landscape sensitivity. Thus, the common 'green' characteristics of renewable energy technologies do not guarantee public support at all times and places.

Brown and Chandler (2008) catalogued over 30 institutional barriers that impede the market uptake of some clean energy technologies. Geothermal energy use in particular has grown at a much slower rate than wind and solar energies. Many of the barriers favor incumbent energy technologies, while in other cases laws trail behind and inhibit technological progress, or are simply poorly designed. These barriers are divided into fiscal, regulatory and statutory categories. While many have been addressed through state or federal public policies, others continue to be ignored. The authors refer to some of these government interventions as 'public failures' that should be addressed, along with market failures such as pollution externalities. Pasqualetti (2011) considered the social barriers to greater use of renewable energy sources. A review of the two major geothermal energy power plants in California, at The Geysers and in the arid Imperial Valley, showed that social barriers were raised in the form of conflicting land uses. In one case, H_2S odor led to complaints from nearby lakeshore resort users and proprietors, which led to stricter air pollution control but hazardous waste problems for many years. In the other case, fear of adverse effects on the region's farm sector slowed energy development, until better planning protocols and guidelines were established.

Sustainability

With growing interest in sustainable development, and use of sustainable energy resources, renewable energy technologies have received greater scrutiny. A semi-quantitative case study to evaluate the sustainable development of a small geothermal power plant in Iran was reported by Phillips (2010), and compared with a second plant in Turkey. A Rapid Impact Assessment Matrix method was used based on 25 criteria. The calculated level of sustainable development for the Iranian plant was found to be positive, but weak. This is because while there were positive socioeconomic effects, the negative environmental effects were nearly as strong. Similar results were found for the Turkish case. To help overcome institutional and geographic barriers to geothermal and other types of renewable energy development, GIS and remote sensing techniques can play an important role. Calvert et al. (2013) reviewed the role that these techniques and related concepts can play in decision support and information management. Traditionally, the focus has been on top-down site suitability analysis. The authors addressed four research needs for improving renewable energy mapping.

Place-based Case Studies of Geothermal Energy Development

Numerous place-based studies of geothermal energy development have been completed. We will focus here on cases in the United States, the world's leading geothermal power producer. One of the most comprehensive and earliest was for Southern California's Imperial Valley, as reported by Pasqualetti et al. (1979). While a public opinion survey showed that geothermal electricity development was favored by almost 90% of Imperial County residents, many environmental problems were uncovered, all of which seemed controllable. Additionally, while the socioeconomic effects would be mostly positive, the export of the majority of the electricity would limit local benefits. A case study of geothermal electricity development in Northern California was discussed earlier, although it focused solely on hazardous waste problems (Pasqualetti and Dellinger 1989). Sommer et al. (2003) reported a third California case, in rural, mountainous Mammoth Lakes (near the Nevada border). This study investigated the use of district energy and the local spatial economics in the low-density town, with a software program called HEATMAP©. The authors concluded that geothermal district energy was feasible across six geographic scenarios.

A Hawaiian case study focused on the Kilauea region of the Big Island, and used the tools of GIS and landscape ecology (Griffith et al. 2002). Two development scenarios were incorporated into GIS, reflecting condensed versus dispersed development, along with landscape metrics. Human-induced disturbance versus natural landscape change from lava flow were also compared. The authors found that landscape patch shape and fragmentation would not appreciably change with the geothermal development, although the dispersed development would result in much more open edge forest. In addition, human disturbance would have a greater simplifying effect on vegetation patch size than natural disturbance, as well as increased fragmentation (Griffith et al. 2002).

CONCEPTUAL ISSUES AND DEBATES

Is Geothermal Energy Renewable?

Geothermal energy is commonly considered renewable. However, this assumption depends on the rate of heat extraction relative to the rate of heat replenishment via conduction (Sanyal 2005). Since the Earth's heat flow rate varies (Figure 22.1), geothermal energy is a diffuse energy source with an average continental heat flow rate of 0.06 W/m^2. Still, energy extraction is often concentrated to a small geographic point as in the case of

SMU Geothermal Laboratory Heat Flow Map of the Conterminous United States, 2011

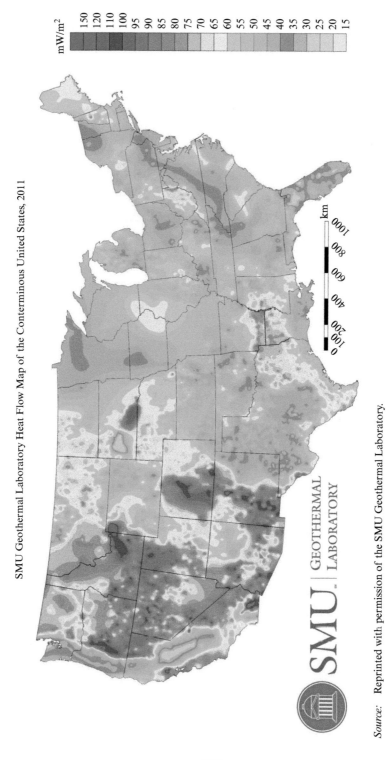

mW/m²
150 120 110 100 95 90 85 80 75 70 65 60 55 50 45 40 35 30 25 20 15

km
0 1000 200 400 600 800 1000

SMU | GEOTHERMAL LABORATORY

Source: Reprinted with permission of the SMU Geothermal Laboratory.

Figure 22.1 Heat flow map of the contiguous United States

electricity generation. In order to meet demand, we utilize energy transfer by convection. Convection only occurs where there are void spaces and the presence of a thermal fluid. Areas with porous and/or fractured rocks with naturally occurring groundwater are ideal. Water can also be injected into hot porous and/or fractured rock, and ongoing research into enhanced geothermal systems is attempting to cost-effectively harness and enhance the geothermal energy transfer from hot rocks that lack sufficient fractures and/or porosity. Convection allows geothermal energy to be harnessed volumetrically, although the heat supply is delivered based on area. Like a hot water tank, it is possible to use hot water faster than the heater can replenish it. As a result, in most if not all instances geothermal energy is harnessed faster than the renewable heat flow rate (Sanyal 2005).

In electricity generation, exhaustion of the thermal fluid, groundwater or steam is often encountered well before encountering significant reductions in the reservoir temperature. This is because we are only able to access a fraction of the stored heat in the geology (Iregui et al. 1978). The Geysers geothermal field is a good example of this phenomenon, where the rated capacity has been lowered over time. The dashed line in Figure 22.2 shows the decline in generation predicted to have occurred had make-up water not been injected. Sanyal and Enedy (2011) predicted that, with continued injection and monitoring, The Geysers will continue to produce power for the next century with moderate declines in generation capacity because of pressure declines caused by cooling.

Less intensive thermal energy technologies such as ground source heat pumps (GSHPs) drawing heat from vertical boreholes are often considered a renewable source of heat; however, even this is incorrect. For example, 10^5 m² of land with a heat flow of 0.06

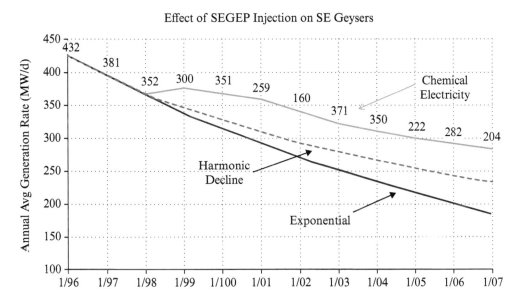

Effect of SEGEP Injection on SE Geysers

Source: Grande et al. (2004). Reprinted with permission of Murray Grande and the Geothermal Resources Council.

Figure 22.2   *The Geysers power generation capacity maintained by effluent injection*

W/m^2 would be needed to supply 6 kW of thermal energy, an amount capable of being extracted by a single 120 m deep vertical borehole heat exchanger (BHE). Extracting heat at 50 W/m^2 at the borehole site far exceeds the natural heat flow of the area, and will cause thermal draw-down similar to the cone of depression from groundwater extraction.

Is Geothermal Energy Clean?

Electricity produced by geothermal energy emits far lower quantities of air pollution than carbon-based sources, although they are not zero like solar, wind and tidal energy. Open-loop geothermal power plants like The Geysers can emit small amounts of nitrogen, H$_2$S, carbon dioxide (CO$_2$) and radon into the air depending on the geology. CO$_2$ emissions from geothermal power plants are ~20 times less and sulfur oxides are ~100 times less than from coal-fired power plants (Holm et al. 2012). Emissions of H$_2$S can be removed with control equipment. Hot water and condensed steam from geothermal sources can also contain mercury, arsenic, boron and other heavy metals, but since the liquid is re-injected into the ground, the environmental risk is minimal. Closed-loop power plants and low-temperature direct uses such as ground exchange heat pump systems are able to harness geothermal energy with nearly zero emissions (Fridleifsson 2001).

Is Geothermal Energy Sustainable?

If humanity's utilization of geothermal energy is not renewable, can it be sustainably harnessed, that is, can the utilization of geothermal energy be managed to last a long time – 30, 50 or 100 years or more, enough time for future generations to find other suitable energy sources? Sanyal (2005) defined the sustainable geothermal extraction level as the production level that can sustain a power plant through its amortized life of 30 years, taking management strategies such as make-up water injection into account. He determined that the sustainable heat extraction rate is '5 to 45 times the renewable capacity, with ten times being most likely' (Sanyal 2005, 1). Geothermal field recovery after 30 years of extraction at this level is predicted to take 100–300 years according to Sanyal (2005), or closer to 1000 years according to Pritchett (1998). Draining the hot water tank faster than the heating element can resupply it can be a sustainable strategy for geothermal energy extraction. Sanyal (2005) argued that geothermal energy could reduce pollution today and save fossil fuels for future generations to use in more purposeful ways than generating heat and electricity.

In the case of GSHPs drawing heat from vertical boreholes, temperature measurements made by Eugster and Rybach (1999) found that temperature decreases caused by thermal extraction tend to stabilize. These findings informed the numeric model shown in Figure 22.3. Thus, a BHE is not renewable during use, although it is sustainable and renewable in the long term. By using a GSHP in both heating and cooling modes, the sustainability of a system can be increased, since the ground acts partly as a seasonal storage medium for thermal energy and partly as energy source (Paksoy 2007).

Should Geothermal Energy be Harnessed in/near National Parks and Volcanoes?

Geothermal energy's power and abundance is visible at sites such as Yellowstone National Park in the western United States. Hot springs and geysers, such as the famous Old Faithful,

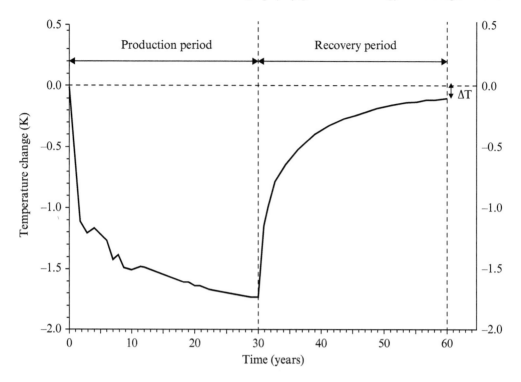

Source: Eugster and Rybach (1999). Reprinted with permission of Walter J. Eugster.

Figure 22.3   *Calculated ground temperature change at a depth of 50 m and at a distance of 1 m from a 105 m long BHE over a production period and a recuperation period of 30 years each*

have inspired many to speculate on the feasibility of harnessing geothermal energy at Yellowstone and other national parks with geothermal features. Owing to the intersecting regulations from the Mineral Leasing Act of 1920 and the Geothermal Steam Act of 1970 codified in CFR 3201, 'lands within all National Park Systems in the United States are off limits to federal geothermal leasing' (US Bureau of Land Management 2005). Furthermore, the 'BLM cannot issue leases on lands where the Secretary has determined that geothermal development could reasonably likely result in a significant adverse effect on a significant thermal feature in a park' (US Bureau of Land Management 2005). Because of these regulations, the immense geothermal energy that powers Old Faithful will remain for the enjoyment of people who visit it rather than harnessed for power production.

Spectacular displays of the Earth's geothermal energy often form other national parks, and the continuously erupting volcanoes of Kilauea and Mauna Loa that make up Hawaii's Volcanoes National Park are no exception. Thirty-five miles outside the park though, neither the dangers of an active volcano nor opposition from local Native people were enough to thwart the development of a small geothermal power plant in Kilauea. Ormat Technologies purchased the Puna Geothermal Venture in 1989, an experimental 3

MWe portable geothermal power plant, and further developed wellheads and generation capacity (Teplow et al. 2008). By 1993, Ormat began producing 27 MWe of electricity and in 2011 added another 8 MWe for a total of 35 MWe (Kaleikini et al. 2011). Similarly in southwestern Iceland, on the Reykjanes peninsula there are three even larger power plants and the famous Blue Lagoon geothermal spa resort, which draw their energy from a volcanic system (Kritmannsdóttir and Ármannsson 2003).

Despite the ban on geothermal development inside US national parks, there are still plenty of geothermal features available for development across the country that total 3442 MWe of power generation capacity (Matek 2014). An additional 50% of geothermal power generating capacity is in various stages of planning and development, scheduled to be completed by the early 2030s. Globally, there is currently 12,000 MWe of geothermal power generation capacity (Matek 2014). However, use of geothermal energy via heat pumps and directly for heating buildings, greenhouses, aquaculture ponds, swimming pools, baths and snow melting, as well as heating and cooling industrial processes dwarfs that of power generation. Over 17,000 MWt of geothermal energy have been harnessed for the above purposes in the United States; globally over 70,000 MWt have been harnessed (Lund and Boyd 2015). The next section of this chapter discusses the spatial and geographic implications of the different applications of geothermal energy.

SPATIAL AND GEOGRAPHIC IMPLICATIONS OF GEOTHERMAL TECHNOLOGIES

Technology has expanded humanity's ability to harness geothermal energy. Prior to geophysical survey technologies including seismic, seismo-electric, electrical, gravimetric, magnetic and electromagnetic, our ability to harness geothermal energy was limited to surface features such as hot springs. Water at hot springs was used for bathing and swimming and pumped or channeled to heat nearby buildings. Geophysical survey technologies have enabled us to survey underground for the size and location of areas with desirable geothermal conditions (high temperatures, fractures, porous geology and groundwater). The US Department of Energy (DOE) funds research on: (a) enhanced geothermal systems (EGS), which are human made reservoirs created where there is hot rock but insufficient or little natural permeability or fluid saturation – an EGS can greatly expand the amount of harnessable geothermal energy; (b) improving the accuracy of geophysical survey techniques for pinpointing geothermal resources; and (c) recovering minerals from geothermal brines (Stephens and Jiusto 2010).

Electric Power Plants

There are three types of geothermal electric power plants: dry steam, single/double-flash and binary cycle. The leading producers are the United States (16,600 GWh in 2015), the Philippines (9646), Indonesia (9600), New Zealand (7000), Mexico (6071) and Italy (5660) (Bertani 2015). The choice of plant depends on the temperature and pressure of the geothermal resource. Dry steam power plants are the oldest and harness geothermal sites with temperatures from 240 to 300°C to produce superheated steam. The steam is piped directly to a turbine that drives a generator to produce electricity. The cooled,

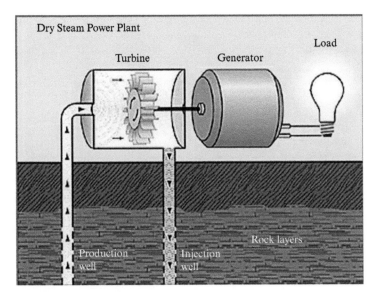

Dry Steam Power Plant

Source: US Department of Energy (2015).

Figure 22.4   Dry steam power plant

condensed steam is then reinjected into the reservoir (Figure 22.4). There are very few dry steam geothermal power plants because of the low number of dry steam reservoirs.

Single- and double-flash steam power plants are more common. High pressures below ground allow water in geothermal reservoirs to remain liquid at temperatures above 100°C. Single- and double-flash steam power plants harness liquid-dominated geothermal reservoirs that are 180°C or more. When the water reaches the surface it is depressurized in a tank that causes it to flash to steam, which drives a turbine-generator (Figure 22.5). The remaining liquid in the tank can be pumped to a second tank and further depressurized to produce more steam. This is called a double-flash geothermal power plant.

Lower-temperature liquid-dominated reservoirs with temperatures of 120–200°C require the hot water to be pumped to the surface. The water is passed through a heat exchanger where it boils a secondary fluid that has a much lower boiling point than water. The vaporized secondary fluid then drives a turbine (Figure 22.6). The lower required reservoir temperatures of a binary cycle power plant greatly increase the amount of harnessable geothermal energy and the number of suitable sites. A small power plant at Chena Hot Springs in Alaska operates on a reservoir with water that is only 74°C. It is the lowest-temperature geothermal resource worldwide that produces commercial power (Bertani 2015, 6).

The DOE hosts an interactive map with heat flow and existing plants, and plants in development overlaid (Figure 22.7). The location of the existing and proposed power plants gives one an idea of the locations with not only favorable heat resources, but also favorable hydrothermal and geological conditions.

The intersection of favorable hydrothermal and geological conditions on a global heat flow map is shown in Figure 22.8. The areas enclosed by dashed lines have favorable hydrothermal conditions. From this map it is clear why Iceland is able to meet 25% of its

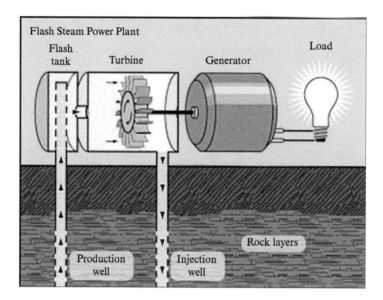

Source: US Department of Energy (2015).

Figure 22.5 Flash steam power plant

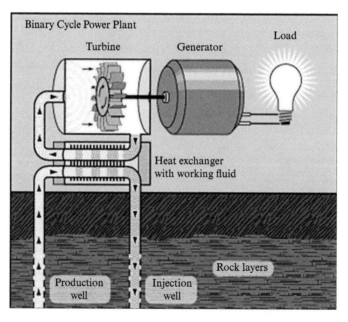

Source: US Department of Energy (2015).

Figure 22.6 Binary cycle power plant

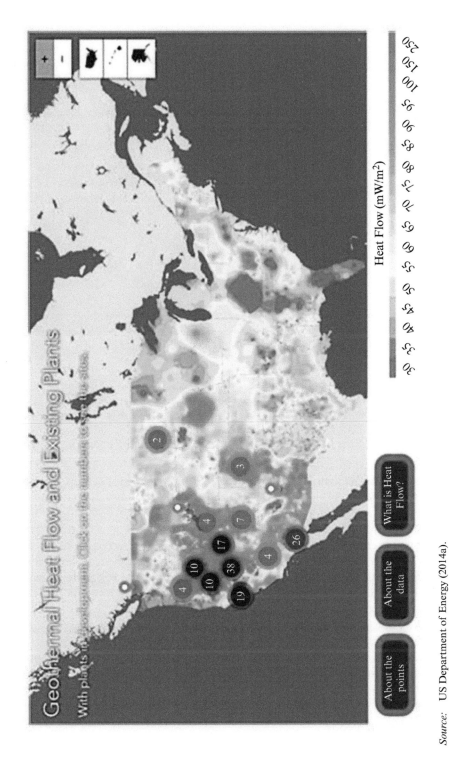

Source: US Department of Energy (2014a).

Figure 22.7 Geothermal heat flow of the contiguous United States

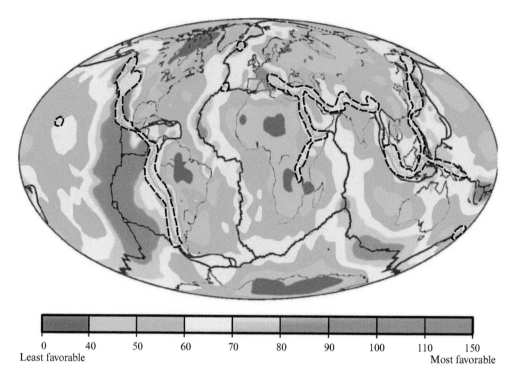

0   40   50   60   70   80   90   100   110   150
Least favorable   Most favorable

Source: IPCC (2012). Reprinted with permission.

Figure 22.8   Global comparison of favorable hydrothermal reservoirs with heat flow

total electricity demand with geothermal energy; it is located in an area with favorable
heat flow and hydrothermal geology. Other countries that produce high fractions of their
electricity from geothermal energy include El Salvador and Kenya (24%), Costa Rica
(15%) and the Philippines (14%) (Bertani 2015). At this map's resolution, many of these
countries are located in areas with only moderate heat flow, which further illustrates the
importance of the nexus of heat, thermal fluid and porosity in making a location favorable to geothermal energy. Since these three factors are highly geographically heterogeneous, higher-resolution maps are needed to precisely locate a favorable area. Furthermore,
heat flow measurements are point data and the sampling interval is not homogenous.
Areas with low population and/or low economic development have less data resolution
and thus more prone to interpolation errors.

Areas with high heat flow but no favorable hydrothermal conditions may one day be
economically harnessed using an EGS. Major technological and economical barriers still
need to be overcome. Should this occur, EGS can double the global geothermal power
generation forecast from 70 to 140 GWe (Bertani 2015). As a fledgling energy technology,
EGS is in a crowded space competing with other technologies that are better funded and
have more support and public attention (Stephens and Jiusto 2010).

Direct Use

Direct uses of geothermal energy are much more geographically diverse as they can take advantage of lower temperatures and smaller heat resources (US Department of Energy 2014b). Direct use of geothermal energy is expanding faster than electric power use (Lund and Boyd 2015). The most widespread direct use of geothermal energy is with GSHPs, followed by spas, baths and swimming pools. The next largest direct user is district heating. Minor applications include greenhouses, aquaculture, agriculture heating, industrial heating and cooling, snow melting, frost protection irrigation, thermal regulation of animal shelters, spirulina cultivation, desalination and sterilization of bottles. By the end of 2014, these direct uses amounted to an estimated 70,239 MWt and supplied an estimated 587,786 TJ/year of thermal energy. This translates to an equivalent of 350 million barrels of oil that would have emitted 148 million tonnes of CO_2 to the atmosphere (Lund and Boyd 2015).

Heat Pumps

GSHPs enable the geothermal energy at any location to be harnessed. The geothermal resource does not even need to be hot enough to heat a building since a GSHP can concentrate the thermal energy to a temperature suitable for building conditioning. As a result of its flexibility, GSHPs account for 71% (325,028 MWt) of the worldwide direct geothermal use capacity (Lund and Boyd 2015). Used at an average capacity factor of 0.2, GSHPs supplied 325,028 TJ/year of thermal energy by the end of 2014 (Lund and Boyd 2015). The leading countries in terms of installed capacity are the United States, China, Sweden, Germany and France in descending order (World Geothermal Congress 2015). Even with almost 3 million GSHP units installed by 2009 worldwide (International Energy Agency 2010), the geographic spread of these units has been diffuse enough that thermal interference between systems has not been a problem. However, as the number of installations increases, regulations will need to be implemented to address thermal interference, especially in metropolitan areas since the thermal extraction area can extend well beyond property boundaries (Williams and Preene 2015).

Spas, Baths and Swimming Pools

Baths, swimming pools, developed hot springs and hot tubs heated by geothermal energy account for an estimated capacity of 9140 MWt, and used 119,381 TJ in 2014 (Lund and Boyd 2015). The largest geothermal energy use for baths and swimming pools occurs in China, Japan, Turkey, Brazil and Mexico. The quantities are conservative, however, since there are many natural hot springs that are not included in these data.

District (Space) Heating

Geothermal systems tend to benefit from economies of scale. The startup survey and engineering cost and efforts to develop a reservoir are relatively fixed, thus it is often most cost-effective to heat multiple buildings and even a whole community in a district heating system. As a result, 88% of the total direct geothermal space heating capacity is used for district heating (Lund and Boyd 2015). The largest geothermal energy user for district heating is China (33,710 TJ/year) followed by Iceland (29,400 TJ/year) and Turkey (8,885 TJ/year) (Lund and Boyd 2015). Expanding direct use of geothermal energy for space heating has been made possible by improved drilling machines, reservoir analysis and management. These factors coupled with available geothermal capacity have enabled the city of Reykjavík, Iceland and surrounding communities, with a population over 200,000, to be almost completely heated by geothermal energy (Ragnarsson 2013).

Underutilized Sources of Geothermal Energy

Here we will explore underutilized sources of geothermal energy such as closed flooded underground mines and heat co-products (steam and hot water) from oil and gas drilling. When underground mines close they often fill up with groundwater and surface water, as dewatering pumps cease. Convection currents in flooded underground mines can bring higher-temperature geothermal energy from depth closer to the surface, where it can be more economically accessed (Marot and Harfst 2012). The large volume and depth of underground mines also allow for large thermal loads to be met using only a few wells. Since 2008, the town of Heerlen, the Netherlands has been heated and cooled using geothermal energy from an abandoned coalmine (Verhoeven et al. 2014). The system has been gradually expanded to heat additional buildings and incorporate additional heat sources; in 2015 Heerlen's geothermal district heating system reached 500,000 m² of building space (Verhoeven et al. 2014). Around 20 flooded underground mines have been harnessed for geothermal energy internationally, yet there are thousands of closed/abandoned mines in the United States and over a million worldwide (Wolkersdorfer 2008). Although the water temperature of most flooded underground mines is too low for power generation, they can be harnessed by GSHPs to condition buildings, greenhouses, aquacultures, baths and swimming pools.

The 823,000 US oil and gas wells produce around 25 billion barrels of hot water per year yet the electricity used at these wells come from refined diesel, natural gas or transmission lines (US Department of Energy 2012). Historically this hot water was an inconvenience and costly to treat, and disposed of as waste. Research by the DOE and industry partners into a small, modular, scalable, binary cycle geothermal power plant is underway with several demonstration systems deployed at a few sites (US Department of Energy 2010). In the near future all oil and gas wells may be equipped with a small geothermal power plant to supply necessary electricity. Furthermore, computer analysis has demonstrated the feasibility of re-using closed, abandoned oil and gas wells by injecting water to extract geothermal energy (Bu et al. 2012). However, the energy recovery from an average well may only be 54 kW. Nonetheless, with approximately 2.5 million abandoned oil and gas wells in the United States alone, hundreds of GW of dispatchable electricity could be harnessed. The challenge lies in developing a small, inexpensive, geothermal power plant and factoring in transmission line accessibility.

CONCLUSIONS

Geographers have complemented the research of geologists on geothermal energy, and have sought to better understand the ecological and environmental effects of consuming geothermal resources. With such basic research largely completed, scholars have turned to land use conflicts, sustainability and place-based studies of geothermal energy. While geothermal is often considered renewable, sustainable and clean, the implications of its development are more complex. Much of this research has been based in the United States, where geothermal energy use is greatest. However, other nations such as the Philippines, Indonesia, New Zealand, Iceland, Italy and China are also major consumers of geothermal energy, both for electricity generation and for direct use.

A diversity of technologies and applications has been used with geothermal energy. In the case of electric power generation, single- and double-flash steam plants have dominated. Dry steam plants are more efficient and cost-effective, but are limited by a lack of suitable, near-surface reservoirs. Flash plants require the conversion of water from high-pressure reservoirs to steam to drive a turbine-generator. Direct use of geothermal energy is dominated by a few applications: heat pumps, spas, baths, pools and district heating. Given the widespread spatial availability of geothermal energy and its attractive features, its adoption rate can be expected to expand worldwide.

REFERENCES

Bertani, R. 2015. Geothermal power generation in the world 2005–2015 update report. In *Proceedings World Geothermal Congress 2015*, Melbourne, Australia, 19–25 April 2015, accessed from https://pangea.stanford.edu/ERE/db/WGC/papers/WGC/2015/01001.pdf.

Blackwell, D., P. Negraru and M. Richards. 2006. Assessment of the enhanced geothermal system resource base of the United States. *Natural Resources Research* 15: 283–308.

Brown, M. and S. Chandler. 2008. Governing confusion: how statutes, fiscal policy, and regulations impede clean energy technologies. *Stanford Law and Policy Review* 19: 472–509.

Bu, X., W. Ma and H. Li. 2012. Geothermal energy production utilizing abandoned oil and gas wells. *Renewable Energy* 41: 80–85.

Calvert, K., J. Pearce and W. Mabee. 2013. Toward renewable energy geo-information infrastructures: applications of GIScience and remote sensing that build institutional capacity. *Renewable and Sustainable Energy Reviews* 18: 416–429.

Cataldi, R., S. Hodgson and J. Lund. 1999. *Stories from a Heated Earth: Our Geothermal Heritage*. Davis, CA: Geothermal Resources Council.

Copeland, H., A. Pocewicz and J. Kiesecker. 2011. Geography of energy development in Western North America: potential impacts on terrestrial ecosystems. In D. Naugle (ed.), *Energy Development and Wildlife Conservation in Western North America*. pp. 7–22. Washington, DC: Island Press. pp. 7–22.

Eugster, W. and L. Rybach. 1999. How renewable are borehole heat exchanger systems? *Bulletin d'Hydrogeologie* 17, accessed from https://pangea.stanford.edu/ERE/pdf/IGAstandard/EGC/1999/Eugster.pdf.

Fridleifsson, I. 2001. Geothermal energy for the benefit of the people. *Renewable and Sustainable Energy Reviews* 5: 299–312.

Grande, M., S. Enedy, W. Smith, J. Counsil and S. Jones. 2004. NCPA at the geysers. *GRC Bulletin* July–August: 155–161.

Griffith, J., C. Trettin and R. O'Neill. 2002. A landscape ecology approach to assessing development impacts in the tropics: a geothermal energy example in Hawaii. *Singapore Journal of Tropical Geography* 23: 1–22.

Holm, A., D. Jennejohn and L. Blodgett. 2012. Geothermal energy and greenhouse gas emissions. Washington, DC, Geothermal Energy Association, accessed from http:// geo-energy.org/reports/Geothermal GreenhouseEmissionsNov2012GEA _web. pdf.

International Energy Agency. 2010. Renewable energy essentials: geothermal, accessed from https://www.iea.org/publications/freepublications/publication/Geothermal_Essentials.pdf.

IPCC. 2012. *IPCC Special Report on Renewable Energy Sources and Climate Change Mitigation. Prepared by Working Group III of the Intergovernmental Panel on Climate Change.* New York: Cambridge University Press.

Iregui, R., A. Hunsbedt, A. Kruger and A. London. 1978. *Analysis of Heat Transfer and Energy Recovery in Fractured Geothermal Reservoirs,* SGP-TR-31. Stanford, CA: Stanford Geothermal Program, Stanford University.

Kaleikini, M., P. Spielman and T. Buchanan. 2011. Puna geothermal venture 8MW expansion. *GRC Transactions* 35: 1313–1314, accessed from http://pubs.geothermal-library.org/lib/grc/1029420.pdf.

Kritmannsdóttir, H. and H. Ármannsson. 2003. Environmental aspects of geothermal energy utilization. *Geothermics* 32: 451–461.

Lund, J. and T. Boyd. 2015. Direct utilization of geothermal energy 2015 worldwide review. In *Proceedings, World Geothermal Congress 2015,* Melbourne, 19–25 April 2015.

Marot, N. and J. Harfst. 2012. Post-mining potentials and redevelopment of former mining regions in Central Europe – case studies from German and Slovenia. *Acta Geographica Slovenica* 52: 99–110.

Matek, B. 2014. Annual U.S. and global geothermal power production report. Geothermal Energy Association, accessed from http://geo-energy.org/events/2014%20Annual%20US%20&%20Global%20Geothermal%20 Power%20Production%20Report%20Final.pdf.

Paksoy, H.Ö. (ed.). 2007. *Thermal Energy Storage for Sustainable Energy Consumption: Fundamentals, Case Studies and Design.* Dordrecht: Springer.

Pasqualetti, M. 1980. Geothermal energy and the environment: the global experience. *Energy* 5: 111–165.

Pasqualetti, M. 1983. The site specific nature of geothermal energy: the primary role of land use planning in nonelectric development. *Natural Resources Journal* 23: 795–814.

Pasqualetti, M. 1986. Planning for the development of site-specific resources: the example of geothermal energy. *Professional Geographer* 38: 82–87.

Pasqualetti, M. 1990. The land use focus of energy impacts. In J. Cullingworth (ed.), *Energy, Land, and Public Policy.* pp. 99–136. New Brunswick, NJ: Transaction.

Pasqualetti, M. 2011. Social barriers to renewable energy landscapes. *Geographical Review* 101: 201–223.

Pasqualetti, M. and M. Dellinger. 1989. Hazardous waste from geothermal energy: a case study. *Journal of Energy and Development* 13: 275–295.

Pasqualetti, M., J. Pick and E. Butler. 1979. Geothermal energy in Imperial County, California: environmental, socio-economic, demographic, and public opinion research conclusions and policy recommendations. *Energy* 4: 67–80.

Phillips, J. 2010. Evaluating the level and nature of sustainable development for a geothermal power plant. *Renewable and Sustainable Energy Reviews* 14: 2414–2425.

Pritchett, J. 1998. Modeling post-abandonment electrical capacity recovery for a two-phase geothermal reservoir. *GRC Transactions* 22: 521–528.

Pryde, P. 1977. Geothermal energy potential in western United States. *Journal of Geography* 76: 170–174.

Ragnarsson, A. 2013. Geothermal energy use, country update for Iceland. In *European Geothermal Congress,* Pisa, Italy, 3–7 June 2013. pp. 1–11.

Sanyal, S. 2005. Sustainability and renewability of geothermal power capacity. In *Proceedings World Geothermal Congress,* 2005, Antalya, Turkey. pp. 1–13.

Sanyal, S. and S. Enedy. 2011. Fifty years of power generation at the Geysers geothermal field, California – the lessons learned. In *Proceedings, Thirty-sixth Workshop on Geothermal Reservoir Engineering,* 31 January to 2 February 2011, SGP-TR-191, Stanford University, Stanford, CA.

Sommer, C., M. Kuby and G. Bloomquist. 2003. The spatial economics of geothermal district energy in a small, low-density town: a case study of Mammoth Lakes, California. *Geothermics* 32: 3–19.

Stephens, J. and S. Jiusto. 2010. Assessing innovation in emerging energy technologies: socio-technical dynamics of carbon capture and storage (CCS) and enhanced geothermal systems (EGS) in the US. *Energy Policy* 38: 2020–2031.

Teplow, W., B. Marsh, J. Hulen, P. Spielman, M. Kaleikini, D. Fitch and W. Rickard. 2008. Dacite melt at the Puna geothermal venture wellfield, Big Island of Hawaii. *GRC Transactions* 33: 989–994, accessed from http://pubs.geothermal-library.org/lib/grc/1028595.pdf.

US Bureau of Land Management. 2005. Title 30. Mineral lands and mining. Geothermal resources, accessed from http://www.blm.gov/style/medialib/blm/wo/MINERALS__REALTY__AND_RESOURCE_ PROTECTION_/energy/geothermal_eis.Par.23503.File.dat/steam.pdf.

US Department of Energy. 2010. Geothermal energy production with co-produced and geopressured resources. DOE/GO-102010-3004, accessed from http://www.nrel.gov/docs/fy10osti/47523.pdf.

US Department of Energy. 2012. Geothermal power/oil and gas coproduction opportunity. DOE/EE-0699, accessed from http://energy.gov/sites/prod/files/2014/02/f7/gtp_coproduction_factsheet.pdf.

US Department of Energy. 2014a. Mapping geothermal heat flow and existing plants, accessed from http:// energy.gov/articles/mapping-geothermal-heat-flow-and-existing-plants.

US Department of Energy. 2014b. Geothermal technologies program, direct use, accessed from http://www. nrel.gov/docs/fy04osti/36316.pdf.

US Department of Energy. 2015. Electricity generation, accessed from http://energy.gov/eere/geothermal/ electricity-generation.

Verhoeven, R., E. Willems, V. Harcouët-Menou, E. De Boever, L. Hiddes, P. Op't Veld and E. Demollin. 2014. Minewater 2.0 project in Heerlen the Netherlands: transformation of a geothermal mine water pilot project into a full scale hybrid sustainable energy infrastructure for heating and cooling. *Energy Procedia* 46: 58–67.

Walker, G. 1995. Renewable energy and the public. *Land Use Policy* 12: 49–59.

Williams, H. and M. Preene. 2015. Addressing sustainability in ground source heat pump projects. In *Proceedings World Geothermal Congress 2015*, Melbourne, Australia, 19–25 April 2015. pp. 1–6.

Wolkersdorfer, C. 2008. *Water Management at Abandoned Flooded Underground Mines: Fundamentals, Tracer Tests, Modelling, Water Treatment*. Berlin: Springer.

World Geothermal Congress. 2015. WGC 2015 celebrates industry success story, accessed from http://www. geothermalpress.com/release-wgc-2015-celebrates-industry-success-story/.

23. LEED buildings
Julie Cidell

LEED refers to the Leadership in Energy and Environmental Design rating system of the non-profit United States Green Building Council (USGBC). LEED is not a set of standards, which are more technical, specific guidelines, but a rating system that produces an overall score. The ratings produced by LEED comprise one example of designating green buildings, or structures that are designed to have less of an environmental impact than a traditional building of the same type.

There are obviously many ways to define and implement the concept of a green building, which has been considered both a strength and a weakness (O'Neill and Gibbs 2014; Stenberg and Räisänen 2006). In this article, I concentrate on certified buildings, highlighting in particular the LEED system. LEED was developed by the USGBC in the late 1990s to provide an explicit, attainable and flexible rating system for green buildings. Since its inception, LEED has gone through four major revisions and many smaller adjustments. It is the most widespread rating system in terms of the number of countries where it is used and the number of buildings certified. Other rating systems such as BREEAM, CASBEE and Green Star are more common in the UK, Japan and Australia, respectively, and they will be referenced in this chapter as well. While there are many other rating systems in the world besides LEED, they are broadly comparable in terms of their requirements and outcomes. For example, a study of the ratings systems used in the United States, UK, Japan, Hong Kong, and China found that although LEED was the most stringent of the group, the outcomes in terms of building performance were roughly equal across systems (Lee 2012).

Using the LEED rating system as an example, two of the most important components of a green building are energy and water conservation. Both of these are significant sources of credits, some referring to a percentage reduction in energy or water use compared with a 'normal' building of the same type, others referring to the use of green power, xeriscaping or other specific technologies designed to reduce environmental impact. However, there are other categories within these rating systems that mean green buildings are about more than saving energy and/or water. These include indoor environmental quality (e.g. low-emission paints and carpets), materials and resources (e.g. local materials or rapidly renewable materials such as bamboo), and the site of the building (e.g. close to transit or reusing an existing building). This holistic, flexible perspective means that certified green buildings can be very different from each other, based on which credits they earn. While there is certainly criticism that the term 'green building' is therefore too amorphous to be useful, others argue that it is this flexibility that has enabled the relatively broad uptake of the concept among the construction and design sectors.

This chapter first gives a short history of green building rating systems, concentrating on LEED. It then describes four aspects of green buildings and the research that has been done about them: spatial (e.g. the distribution of green buildings within a country or city); economic (e.g. if there is an extra cost to building green); social (e.g. considering indoor environments as worthy of protection); and environmental (are green buildings really 'green'?). The conclusion addresses some of the criticisms and pushback against LEED and what the implications are for sustainable building more broadly.

THE HISTORY OF LEED

A main goal of many green building rating systems has been to not only produce greener structures, but also to change the way the building industry works. One of the founders of the USGBC, Rick Fedrizzi, has noted that one of the goals of LEED is to make itself obsolete, to the point that more sustainable ways of building no longer need to be rated or certified but become part of business-as-usual (e.g. Kriss 2014c). In order to accomplish this goal, the requirements for certification have to be regularly reviewed and raised so that only the highest-quality buildings are certified, particularly at higher rating levels.

The general setup of LEED and similar rating systems is that a few initial credits are mandatory, such as a percentage reduction in energy and water usage compared with a traditional building. Beyond that, project owners and designers can choose which credits and how many they want to pursue, making it a very flexible system. There are multiple levels of certification; certified, silver, gold and platinum for LEED; four-star, five-star and six-star for Green Star; pass, good, very good, excellent and outstanding for BREEAM, etc. A building is *registered* once its owner announces their desire to seek LEED certification, and *certified* once the building is complete and has been verified. While some systems like Green Star or BREEAM rely on in-person inspections, LEED is based on paperwork: documenting the work of producing the building in order to verify that credits have been achieved.

The pilot projects for LEED were developed in 1998, with 19 initial projects (Kriss 2014a). The rating system was officially launched in 2000 as LEED 2.0 with 51 projects, followed shortly by LEED 2.1 in 2003. By 2004, there were 100 certified buildings (Kriss 2014b). Uptake of the system was greatly enhanced by its adoption by various levels of government: the General Services Administration required that all new federal projects be LEED-certified starting in 2003, and Chicago was the first of many local jurisdictions to require certification for public buildings, starting in 2004. The virtuous circle of government mandates and increasing private sector capabilities meant that LEED started to grow more rapidly, so that by 2007 there were 1000 certified projects (Cidell 2009a). As of 2015, there are over 70,000 buildings with LEED certification, about 50,000 of which are housing and 20,000 commercial (Kriss 2014a). This totals about 5% of the total buildings in the United States, but about 20% of the floor space, indicating the tendency for LEED to be used for larger projects (CBRE 2015).

As LEED has expanded, it has gone from two systems – New Construction and Existing Buildings – to separate rating systems for specialized uses such as health care, schools and housing, as well as operations and maintenance to certify the performance of the building over time. (The concept of a green building has since been extended to a sustainable neighborhood, as in LEED for Neighborhood Design, but that is not considered here.) Individuals can also be certified or accredited as LEED Green Associates after taking a certifying exam, or as Accredited Professionals after working on a LEED certification project. As of 2014, there were 32,000 Green Associates and 59,000 APs (Kriss 2014a).

The flexibility of LEED is one of the reasons for its success, but this same flexibility leads to variations in where and how it is applied. The following section summarizes work that has been done on the spatial distribution of LEED, followed by discussions of its economic, social and environmental implications.

SPATIAL STUDIES OF LEED

Most of the academic work on green buildings, unsurprisingly, has come from the fields most directly concerned with their production: architecture, design, engineering, construction management, etc. However, there are a number of ways in which a spatial perspective can be brought to bear on green buildings. For example, some of the most basic work on LEED has been on analyzing the spatial patterns of its use. Within the United States, such studies include research on the spatial distribution of LEED buildings (Cidell 2009a). The general pattern included a concentration of green buildings in some expected areas such as the Pacific Northwest. However, there were other clusters that were not expected, including Grand Rapids, Michigan, which at the time had the highest number of green buildings per capita in the United States. Even so, it is important to remember that, overall, only 5% of total US commercial space is LEED-certified (CBRE 2015). .

The spatial distribution of green buildings can be explained by multiple factors, including the presence of a mandatory green building policy, large population size, a wealthier and more educated population, the presence of a champion within city government, support from local universities and more individuals certified as green building professionals (Cidell and Cope 2014; Fuerst et al. 2014; Mason et al. 2011; Van Schaack and BenDor 2011). Political affiliation (of a mayor or the population at large) does not seem to influence the number of green buildings in a metropolitan area and economic incentives are also not generally significant. Neighborhood effects at the regional scale are inconsistent between studies that looked at commercial properties (Fuerst et al. 2014) and those that looked at all certified buildings (Mason et al. 2011; Cidell and Cope 2014; Simcoe and Toffel 2014). In other words, it is not clear how much of an influence cities with green buildings have on their neighbors. At a metropolitan scale, however, green buildings tend to cluster in CBDs and in outlying major commercial districts (Kaza et al. 2013), suggesting that there is an agglomeration economy operating at a neighborhood or sub-regional scale.

Since 'green buildings' are flexible in definition and practice, a spatial perspective also needs to consider if the concept of a green building is implemented differently in different locations. Some components of a green building rating system are location independent, such as the use of low-emitting paints and sealants or rapidly renewable materials. However, many credits vary in either their feasibility or importance depending on the region and site of the building, including a reduction of water usage in landscaping and the use of local materials in construction ('local' here meaning from within a 500 mile radius). Indeed, a revised version of the LEED criteria awarded 'extra credit' for pursuing regionally appropriate credits, such as a greater reduction in water usage in a project in Arizona. For the most part, there does seem to be a relationship between the local environment and the pursuit of certain credits (Cidell and Beata 2009). In particular, regional differences are more pronounced for the spatial credits such as use of local materials than the location-independent credits such as use of rapidly renewable materials. Similarly, a comparison of climatic conditions to LEED credits earned found that credits with a stronger spatial component, such as public transportation availability, urban heat island mitigation and water-efficient landscaping, were influenced by local climatic factors such as diurnal temperature range, precipitation amount and/or days with sun (Cheng and Ma 2015). Both of these studies suggest that project teams are, in fact, taking into account local and regional conditions in their building design.

The large size of the US market combined with the early activity of the USGBC here means that few studies have been done in other countries on the spatial patterns of green buildings. Nevertheless, as green buildings are a worldwide phenomenon, so too is their study. For example, South Africa's 50 Green Star-rated buildings are concentrated in the commercial capital of Johannesburg and the political capital of Pretoria, indicating the dual roles of the commercial and government sectors in promoting and demonstrating green building (Rogerson 2014). Chegut et al. (2014) mapped green buildings in the UK according to BREEAM rating, although only for informational purposes and not for analysis. In Canada, the credits most likely to be influenced by climate and regional location were *less* likely to be earned by Canadian projects than they were on US projects, even with Canada's version of LEED being adapted to the climate (Da Silva and Ruwanpura 2009). China's attempts to develop its own rating system rather than borrowing from a Western model have resulted in a number of different systems in place at local and regional scales, making it difficult to analyze the situation across the entire country (Geng et al. 2012). Case studies in Malaysia, India, Turkey, Sweden, Australia and other locations indicate the worldwide nature of the phenomenon, and the need to study it in different physical environments and political contexts.

ECONOMIC ASPECTS

One of the most commonly cited reasons for pursuing green building certification is to save money through lower energy costs. However, one of the most common perceptions about green buildings is that they cost more than conventional buildings to produce. The USGBC has tackled this question from its inception, arguing that a certified building should return any extra costs within a few years owing to reduced energy costs and increased employee productivity. Nevertheless, this question continues to be debated in the literature and in practice, especially when it comes to speculative projects. There is also the question of spillover effects: if the stated goal of the USGBC and similar organizations is to change business as usual so that every architect and contractor is using green building methods, can changes in those economic sectors be identified based on green building activity? Although the number of certified buildings is small in most regions, a case can nevertheless be made that such spillovers do exist.

The perception that green buildings cost more to build than conventional buildings has existed from the start of green building rating systems and continues to the present day. In fact, higher upfront costs are often perceived as the largest barrier to green buildings (Chan et al. 2009). At the same time, there is a lasting perception that there are good business reasons to pursue certification (although this perception is not universal; see Wong and Abe 2014 on Japan). These reasons include lower long-term costs of owning and operating the building and a market advantage from being able to advertise as 'green' or 'sustainable' (Chan et al. 2009; Eichholtz et al. 2010). Another reason to pursue green certification may be to increase employee productivity, as buildings with more natural sunlight and dispersed temperature controls have been shown to reduce absenteeism and increase productivity (Edwards 2006); this will be considered in more detail in the following section. At the same time, increased property values are not always a good thing; there is some evidence for green gentrification through the certification process (Chegut et al. 2014). These perceptions have been verified empirically by studies in the UK, which found a huge 24% premium for BREEAM-certified buildings within London (Chegut et al. 2014), although they note that green buildings may be produced by organizations with generally higher standards (e.g. Class A office space).

The economic impacts of green buildings can be larger than the structures themselves. If one of the goals of LEED and other rating systems is to transform the building industry, it is worth considering the extent to which that has happened. One study of the architects and contractors behind each of the certified green buildings within four US metropolitan areas found that most firms were locally based or at least had a local branch, suggesting that knowledge of how to produce green buildings has diffused spatially (Cidell 2014). At the same time, there was very little overlap in team members from one project to the next. This suggests that green building experience is *not* being centralized within a few expert firms but is spreading widely throughout architecture and contracting firms, validating the hope that the transformation of the industry is underway. On a related note, a study in California found a clear connection between municipalities mandating certification for public buildings and the increased use of green building practices among architects, contractors and realtors (Simcoe and Toffel 2014), also suggesting that the industry is, in fact, being transformed. Within the industry, there is also a strong perception that the role of government in the form of subsidies or changes in the building code is key to changing the usual methods of building (Chan et al. 2009).

Green buildings can also be used to rethink or expand upon geographic theory. For example, the 'new economic geography' emphasizes the importance of including the discursive and the material in geographic studies of the economy, both of which are clearly a key part of green buildings (Boyle and McGuirk 2012). At the same time, the increased desire for flexibility within the workplace and for visible corporate social responsibility also drives the inclusion of 'sustainable' elements into building design and operation. In other words, the economic benefits of green buildings are not only in terms of energy savings or increased productivity, but also in enhancing a company's own brand through connecting to LEED, Green Star or a similarly well-known rating system. Similarly, work within economic geography that considers the production of knowledge can also be applied here, in particular how knowledge is formed through a process of bricolage, cobbling bits and pieces together from multiple sources into a new concept or phenomenon (Faulconbridge 2013). Drawing on communities of practice, global networks and urban policy mobilities, Faulconbridge demonstrates that the production of green buildings requires a combination of mobile and local knowledges and institutions (see also Cidell 2014). Using a different theoretical framework, that of socio-technical studies (more on this below), O'Neill and Gibbs (2014) emphasize the multiple scales or levels at which actors shape decisions about what and how to build. They argue that, instead of thinking of individuals as 'green entrepreneurs' or atomized actors, we should consider the networks and institutions within which they operate and how those networks and institutions might be strengthened. Taken together, these three studies demonstrate that firms can, in fact, be encouraged to engage in more sustainable behaviors and that this engagement happens through knowing local context as well as understanding mobile concepts like green building rating systems. They also suggest that considerable work remains to be done on applying the theories of economic geography to environmental issues.

SOCIAL ASPECTS

One of the potentially significant contributions of green buildings to the social sciences is that they encourage us to think about the hybrid nature of the socio-environment. In other words, since most people in developed countries spend nearly all of their time indoors (ALA 1996), perhaps indoor air quality, lighting and atmosphere should be considered as part of 'the environment'. This dimension dovetails with a call to consider workplaces as sites of environmental action and justice, not only homes and residential neighborhoods (Edwards 2006; Pellow 2002). At the same time, socio-technical studies with their insistence on the intertwined nature of technology and social practices emphasize that building a structure to meet 'green' qualifications is not enough; we must also consider the users of the structure and how their practices and habits will affect that structure's environmental impact (Edwards 2006; O'Neill and Gibbs 2014). In other words, the workplace should be considered as an important site of livability, and a number of elements of green building rating systems are designed to deal with that.

There is certainly some overlap with the previous section in terms of the effect of green buildings on productivity. There can be a fine line between the underlying goals of improving the workplace to increase productivity and thus the bottom line, or providing a healthier environment for one's employees. Allowing more lighting and fresh air, reducing exposure to mold and volatile organic compounds, and increasing visual access to the outdoors have all been shown to increase employee satisfaction and health, and therefore to increase productivity. Each of these elements may be used to obtain credits in a green building rating system. However, while multiple studies show a higher rate of occupant satisfaction in buildings that are certified as green (e.g. Edwards 2006; Pei et al. 2015; Singh et al. 2010), other studies question those findings when all aspects of satisfaction are considered (Altomonte and Schiavon 2013; Liang et al. 2014). For example, increased daylight and air quality are positives, but since many new commercial buildings tend to use an open floor plan which results in less visual and aural privacy (Edwards 2006), the overall effect may be that there is no significant improvement (Altomonte and Schiavon 2013). Alternatively, while some factors such as improved indoor air quality are positive, thermal comfort may be lower as compared with a conventional building (Liang et al. 2014).

These studies highlight in various ways that designing a building for maximum sustainability needs to consider how it will be used. This is not only for practical reasons – employees will open windows, adjust thermostats, bring in space heaters, etc. – but also because of larger social concerns: 'Green offices in particular raise issues of the democratic control of space and access to healthy working conditions' (Edwards

2006, 193). A tightly sealed building or centralized temperature controls may heighten overall building efficiency but leave workers feeling too warm, too cold or without fresh air (e.g. Liang et al. 2014; Pei et al. 2015). In other words, if we are to consider the workplace as a 'liveable' environment, we need to also consider how a green building is not just a physical structure, but a place of dwelling, where individuals' practices and habits shape or even perform how the building is used and functions (Kraftl 2006).

Such a perspective might be achieved by drawing on socio-technical studies. In brief, this literature attempts to explain how technological innovations lead to broader social changes, with an eye towards understanding how to go about effecting such change (Bulkeley et al. 2015; Gibbs and O'Neill 2015). 'Niches' or experiments are where ideas are tried out and tested at a small scale, such as an individual building or city. At a broader scale, 'regimes' are the networks of institutions, cultures, rules and practices that condition whether niches develop into something larger within the industry or remain one-off experiments. Finally, 'landscapes' are the wider socio-technical realms, including average users who might not know or care about green building elements but nevertheless are key to ensuring their successful implementation. This multi-level perspective can be useful for considering how to make green building technologies and practices mainstream because of its movement from small, individual projects to wider changes in society. It also enables us to see how difficult this can be, given both active and passive resistance to change from regimes within the building and construction industries and from landscapes of consumers (Gibbs and O'Neill 2015; Lovell 2005). For example, homeowners are not passive recipients of green housing technologies like solar hot water heaters or rainwater tanks (Bulkeley et al. 2015). They maintain and live with those technologies, adapting them to their own preferences and uses (or not, as the case may be) (Lovell 2005).

However, more work remains to be done on exactly *how* ideas or concepts move from the niche to the regime stage, including the role of power and politics and the contradictions inherent within regimes (Gibbs and O'Neill 2014). While there is work to suggest the broader take-up of green buildings after implementation of a municipal policy, for example (Cidell and Cope 2014), it remains to be determined how the politics of 'greening' influence this on a broader scale (see the Conclusion). Furthermore, there may be too much focus on niches because they are relatively easy to identify and support, when larger-scale changes of regimes and landscapes are what is really necessary (Lovell 2007). Nevertheless, the consideration of green buildings as social as well as technological products is an important one, especially if the goal is to transform the broader landscape of the built environment.

ENVIRONMENTAL ASPECTS

Finally, and perhaps most importantly, there is the environmental component of a green building. If a green building is designed to reduce environmental impact, is there evidence to show that this does, in fact, happen? One of the early criticisms of LEED was that certification could be made based on a checklist alone, and that once certification was registered and the plaque was on the wall, there was no way to ensure that the building lived up to its estimated potential. A newer version of LEED for Operations and Maintenance takes this into account, requiring the recording and analysis of post-occupancy use for energy and water. Similar programs exist in China (Pei et al. 2015) and under Green Star and BREEAM.

(This also makes it possible for the same building to earn certification two times, once for being built and later for being operated.) There are also broader theoretical frameworks or approaches that can be applied to understand the relationship between green buildings and the environment, namely ecological modernization and political ecology.

For the most part, empirical post-occupancy studies have found that certified green buildings do, in fact, have a reduced environmental impact compared with standard buildings. However, many buildings only do so in some areas but not others (e.g. energy versus water), while some buildings use more energy or water than their standard counterparts (Liang et al. 2014; Moschandreas and Nuanual 2008). For example, an early study found that, as compared with similar buildings that were not certified, the LEED buildings performed better only 39% of the time, and were worse in some measurements 10% of the time (Moschandreas and Nuanual 2008). This study was based on a survey of building occupants as well as energy bills and interviews with building designers for the first 10 LEED-certified buildings in Chicago, and it noted that none of the buildings were evaluated by their owners for their performance within two years of their construction. A larger study of 100 buildings across the United States found that *on average*, LEED-certified buildings use less energy than their counterparts, even though about a third of the buildings studied actually used *more* (Newsham et al. 2009). Furthermore, there was no correlation between the level of certification (gold, silver, etc.) and the amount of energy saved. More recently, a Chinese study focusing on indoor environmental quality compared survey data with air quality measurements and found that green buildings achieve the same indoor temperature with less energy use, resulting in greater comfort for occupants (Pei et al. 2015). There has been less work on the other elements of a green building: the use of non-car commuting options, the environmental impact of using local and/or renewable materials, etc. In short, research is cautiously optimistic about whether or not LEED-certified buildings have measurably better environmental performance compared with standard buildings – although this is the ostensible reason they exist.

This cautious optimism is one reason why green buildings have been taken up as a prime example of ecological modernization, 'a policy concept characterizing the types of processes toward sustainability rather than judging the results of these processes' (Jensen and Gram-Hanssen 2008, 148). Ecological modernization argues that capitalist systems can incorporate ecological concerns in a modified version of business as usual, or that the same systems that created our current environmental problems can also be used to solve them. For example, green buildings in Denmark have been promoted not as an 'alternative lifestyle', but as an improvement upon standard ways of building (Jensen and Gram-Hanssen 2008). By keeping green buildings from being visibly different, builders might appeal to more consumers, but they also run the risk of failing to activate consumer support for broader changes. Similarly, China's top-down approach demonstrates that, while the number of certified green buildings may be growing rapidly, wider societal changes such as coalition building have not occurred to make this system sustainable (Zhou 2015). In this case, more *political* modernization is needed in order to produce lasting environmental effects. Green building rating systems might be attractive because they do not offer too drastic a change from business as usual, but that also limits their ability to effect larger change.

Of course, LEED certification may serve other environmental functions besides reducing emissions or water usage. For one, it may signal a company or institution's interest in or commitment to being environmentally friendly. For example, local governments that incorporate the LEED rating system in their policies do so in part so that they can be known for being green – and therefore can be known as leaders, at least at this point in time (Cidell 2014). Another possible function of the LEED rating system is to promote a certain type of city. Using an urban political ecology approach to analyzing the LEED standards suggests that the standards themselves and the ways they have been interpreted and refined lead to four aspects of a desirable city (Cidell 2009b). The first is flexibility and reflexivity: flexible in terms of what it means to be 'green', thanks to the multi-point rating system, and reflexive in terms of being regularly reviewed and updated. The second is a highly documented city: since LEED certification is based not on physical inspection of the finished building but documentation of the building process, every item and method that goes into the building must be tracked and recorded along the way. The third is that, although there might be a stated desire for social sustainability, the outcomes are neutral at best. Finally, a city of green buildings is planned as a socio-natural hybrid, since the available credits include biophysical and social criteria in order to improve both human and non-human habitats. Taken together, these four elements suggest that there are subtle changes underway in how urban environments are governed and constructed because of LEED and similar programs, even if their measured empirical impact is not what was expected.

CONCLUSION

The spatial, social, economic and environmental aspects of LEED-certified or green buildings can be thought of in both empirical and theoretical terms, as outlined above. Spatial clustering of buildings exists, as do regional differences in which credits are chosen in the certification process. In economic terms, while the perception exists that it costs more to build green, the values of certified properties are generally higher, indicating the perception that the higher costs are worth it. While the evidence is equivocal as to whether green buildings have the positive environmental impact they are expected to – both indoors and out – there is a strong suggestion that the desired transformation of the building industry is happening. From a theoretical perspective, the frameworks of urban political ecology, socio-technical studies, practice theory and ecological modernization have all been brought to bear on green buildings. There is great potential for additional contributions to policy mobilities and environmental justice, as well as considering the workplace as an environment alongside home and the outdoors.

Unsurprisingly, there are political elements to LEED and other rating systems as well. In particular, there has been some pushback against LEED from conservative politicians and from some traditional elements of the building industry within the United States. For example, the third major version of the LEED rating system was delayed for a year because of conflict over whether only materials designated by the Forest Stewardship Council could count as sustainably harvested, or if an alternative designation such as that of the Sustainable Forestry Initiative (an industry-led group) could be used instead. Three states with a significant timber industry – Alabama, Georgia and Maine – subsequently banned LEED certification for public buildings as a result, even though FSC-certified wood accounts for only one possible credit and is in no way mandatory. On a broader scale, there has been concern over local green building legislation that requires the use of the LEED rating system, which essentially means that government is mandating the use of a non-profit organization's services. The usual workaround has been to required 'LEED or its equivalent' (Cidell and Cope 2014). However, in some jurisdictions, the ideological backlash has been significant enough that the use of LEED has been banned.

In addition to outright opposition, while the flexibility of green building rating systems has been one of the reasons behind their rapid uptake, there is some danger in this flexibility. It may seem like an advantage to be able to meet other agendas under the guise of a green building, such as projecting a green image or reducing unemployment. However, the term 'green building' may become so flexible as to be vacated of any meaning, leading to broader suspicions about the environmental movement (Stenberg and Räisänen 2006). One of the strengths of LEED has been its constant revision over time, both in terms of becoming more specialized and by increasing its requirements. As long as the USGBC can continue this kind of revision, it stands a good chance of achieving its goals of transforming the building industry – at least where it is allowed to.

REFERENCES

ALA. 1996. Indoor air pollution: an introduction for health professionals. American Lung Association: New York, accessed 18 January 2016 from http://www.epa.gov/indoor-air-quality-iaq/indoor-air-pollution-introduction-health-professionals.

Altomonte, S. and S. Schiavon. 2013. Occupant satisfaction in LEED and non-LEED certified buildings. *Building and Environment* 68: 66–76.

Boyle, T. and P. McGuirk. 2012. The decentred firm and the adoption of sustainable office space in Sydney, Australia. *Australian Geographer* 43(4): 393–410.

Bulkeley, H., V. Castán Broto and G. Edwards. 2015. *An Urban Politics of Climate Change: Experimentation and the Governing of Socio-Technical Transitions*. London: Routledge.

CBRE. 2015. National Green Building Adoption Index, accessed from http://www.cbre.com/~/media/files/corporate%20responsibility/green-building-adoption-index-2015.pdf.

Chan, E.H.W., Q.K. Qian and P.T.I. Lam. 2009. The market for green building in developed Asian cities – the perspectives of building designers. *Energy Policy* 37: 3061–3070.

Chegut, A., P. Eichholtz and N. Kok. 2014. Supply, demand and the value of green buildings. *Urban Studies* 51(1): 22–43.

Cheng, J.C.P. and L.J. Ma. 2015. A data-driven study of important climate factors on the achievement of LEED-EB credits. *Building and Environment* 90: 232–244.

Cidell, J. 2009a. Building green: the geography of LEED-certified buildings and professionals. *Professional Geographer* 61(2): 1–16.

Cidell, J. 2009b. A political ecology of the built environment: LEED certification for green buildings. *Local Environment* 14(7): 621–633.

Cidell, J. 2014. Mapping the green building industry: how local are architects and general contractors? *Tijdschrift voor Economische en Sociale Geografie* 105(1): 79–90.

Cidell, J. and A. Beata. 2009. Spatial variation among green building certification categories: does place matter? *Landscape and Urban Planning* 91(3): 142–151.

Cidell, J. and M. Cope. 2014. The effect of municipal policy on U.S. green building activity. *Journal of Environmental Planning and Management* 57: 1763–1781.

Da Silva, L. and J. Ruwanpura. 2009. Review of the LEED points obtained by Canadian building projects. *Journal of Architectural Engineering* 15(2): 38–54.

Edwards, B. 2006. Benefits of green offices in the UK: analysis from examples built in the 1990s. *Sustainable Development* 14: 190–204.

Eichholtz, P., N. Kok and J. Quigley. 2010. Doing well by doing good: green office buildings, *American Economic Review* 100: 2494–2511.

Faulconbridge, J. 2013. Mobile 'green' design knowledge: institutions, bricolage and the relational production of embedded sustainable building designs. *Transactions of the Institute of British Geographers* 38: 339–353.

Fuerst, F., C. Kontokosta and P. McAllister. 2014. Determinants of green building adoption. *Environment and Planning B* 41: 551–570.

Geng, Y., H. Dong, B. Xue and J. Fu. 2012. An overview of Chinese green building standards. *Sustainable Development* 20(3): 211–221.

Gibbs, D. and K. O'Neill. 2014. Rethinking sociotechnical transitions and green entrepreneurship: the potential for transformative change in the green building sector. *Environment and Planning A* 46: 1088–1107.

Gibbs, D. and K. O'Neill. 2015. Building a green economy? Sustainability transitions in the UK building sector. *Geoforum* 59: 133–141.

Jensen, J. and K. Gram-Hanssen. 2008. Ecological modernization of sustainable buildings: a Danish perspective. *Building Research and Information* 36(2): 146–158.

Kaza, N., T. Lester and D. Rodriguez. 2013. The spatio-temporal clustering of green buildings in the United States. *Urban Studies* 50(16): 3262–3282.

Kraftl, P. 2006. Ecological architecture as performed art: Nant-y-Cwm Steiner School, Pembrokeshire. *Social and Cultural Geography* 7(6): 927–948.

Kriss, J. 2014a. Part 1: from a simple idea to a several-hundred-billion-dollar industry, accessed 2 February 2016 from www.usgbc.org/articles/simple-idea-several-hundred-billion-dollar-industry.

Kriss, J. 2014b. Part 2: a green building explosion, accessed 2 February 2016 from http://www.usgbc.org/articles/part-2-green-building-explosion-2003-2009.

Kriss, J. 2014c. Part 3: challenges and opportunities (2010–present), accessed 2 February 2016 from http://www.usgbc.org/articles/part-3-challenges-and-opportunities-2010-present.

Lee, W.L. 2012. Benchmarking energy use of building environmental assessment schemes. *Energy and Buildings* 45: 326–334.

Liang, H.-H., C.-P. Chen, R.-L. Hwang, W.-M. Shih, S.-C. Lo and H.-Y. Liao. 2014. Satisfaction of occupants toward indoor environment quality of certified green office buildings in Taiwan. *Building and Environment* 72: 232–242.

Lovell, H. 2005. Supply and demand for low energy housing in the UK: insights from a science and technology studies approach. *Housing Studies* 20(5): 815–829.

Lovell, H. 2007. The governance of innovation in socio-technical systems: the difficulties of strategic niche management in practice. *Science and Public Policy* 34(1): 35–44.

Mason, S., T. Marker and R. Mirsky. 2011. Primary factors influencing green building in cities in the Pacific Northwest. *Public Works Management and Policy* 16(2): 157–185.

Moschandreas, D. and R. Nuanual. 2008. Do certified sustainable buildings perform better than similar conventional buildings? *International Journal of Environment and Sustainable Development* 7(3): 276–292.

Newsham, G., S. Mancini and B. Birt. 2009. Do LEED-certified buildings save energy? Yes, but . . . *Energy and Buildings* 41: 897–905.

O'Neill, K. and D. Gibbs. 2014. Towards a sustainable economy? Socio-technical transitions in the green building sector. *Local Environment* 19(6): 572–590.

Pei, Z., B. Lin, Y. Liu and Y. Zhu. 2015. Comparative study on the indoor environment quality of a green office building in China with a long-term field measurement and investigation. *Building and Environment* 84: 80–88.

Pellow, D. 2002. *Garbage Wars: The Struggle for Environmental Justice in Chicago*. Cambridge, MA: MIT Press.

Rogerson, J. 2014. Green commercial property development in urban South Africa: emerging trends, emerging geographies. *Bulletin of Geography, Socioeconomic Series* 26: 233–246.

Simcoe, T. and M. Toffel. 2014. Government green procurement spillovers: evidence from municipal building policies in California. *Journal of Environmental Economics and Management* 68: 411–434.

Singh, A., M. Syal, S. Grady and S. Korkmaz. 2010. Effects of green buildings on employee health and productivity. *American Journal of Public Health* 100(9): 1665–1668.

Stenberg, A.-C. and C. Räisänen. 2006. The interpretive flexibility of 'green' in the building sector: diachronic and synchronic perspectives. *International Studies of Management and Organization* 36(2): 31–53.

Van Schaack, C. and T. BenDor. 2011. A comparative study of green building in urban and transitioning rural North Carolina. *Journal of Environmental Planning and Management* 54(8): 1125–1147.

Wong, S.-C. and N. Abe. 2014. Stakeholders' perspectives of a building environmental assessment method: the case of CASBEE. *Building and Environment* 82: 502–516.

Zhou, Y. 2015. State power and environmental initiatives in China: analyzing China's green building program through an ecological modernization perspective. *Geoforum* 61: 1–12.

24. The interaction of pipelines and geography in support of fuel markets

Jeff D. Makholm

In 2013, Ernest Moniz – now celebrated for his role in the Iran nuclear deal but then Director of the MIT Energy Initiative – spoke about shale gas as a 'game changer' and a 'bridge to a lower carbon planet'.[1] However, shale gas production is as yet only an American phenomenon despite a world that seems to have abundant shale gas resources (Makholm 2016). Why, in a world full of geological shale resources and multinational oil and gas companies, is shale gas strictly an American phenomenon?

The answer lies largely with pipelines and the way that they work with or against geography to govern the spatial allocation of economic activity. The world's major pipelines represent a pretty unexceptional industry of century-old technology. Pipelines ship fuel over vast inland spaces much more efficiently than other modes of energy transport. More than four-fifths of the world's pipelines ship natural gas – an inconvenient fuel in that it requires pipelines to serve even the smallest customers.

Gas pipelines present the best way to study the geography of this particular technology. The utter dependency of the gas industry on pipelines shows how the treatment of pipelines by governments, and their resulting role in supplying fuel to cities and industrial areas from geographically dispersed sources, has such profound effects on fuel decisions and international trade. Table 24.1 shows both the age of the industry and its only recent penetration to some areas of the world. It also shows that some continents are veritably covered with this prosaic technology and others are not. Why? Some of the differences are due to sources too obvious to challenge economists, e.g. gas as a space heating fuel – North America needs much more of that than Africa. Yet other differences are more challenging to study. Pipelines are highly capital intensive, directional and immobile, and thus risky for investors given the possibilities for 'hold-up' and other forms of economic opportunism.[2]

The essential problem with the world's pipelines is that they are difficult to study at anything but a comparatively trivial level.[3] While all pipelines perform essentially the same physical functions (shipping fuel from one spot to another), the institutions that surround them are stubbornly dissimilar from country to country. Those institutions, experiences and political exigencies from one region do not translate to others, and such differences are more often than not buried in documents and experiences that are difficult for outsiders to study.

Studying the economic geography of pipelines is impossible without taking account of such institutions (Makholm 2012). This paper is about where harmony or dissonance exists between today's gas pipelines and the geographies they serve. My purpose is to try to demystify this critical industry – so weighted-down by obscure institutions, politics and history – so as to help this prosaic transport technology to become a more integral part of the study of geography and international trade.

Table 24.1   The reach and historical origin of gas pipelines, 2013

Continent	Gas pipelines (miles)	Percentage of world gas pipelines	Pipeline miles per 100 square miles	First major gas pipeline
North America	1,245,250	72.4%	16.33	1904
Europe	255,901	14.9%	3.97	Early 1970s
Asia	91,364	5.3%	1.49	1969
South America	42,654	2.5%	1.07	1949
Middle East	38,408	2.2%	2.77	1997
Africa	25,634	1.5%	0.51	1964
Australia/New Zealand	19,878	1.2%	0.73	1969
1.1.6. Total	1,719,088	100.0%		

Note: Gas pipelines account for 80.2 percent of the world's pipeline mileage, oil and refined products pipelines 19.8 percent.

Source: CIA World Factbook – Field Listing: Pipelines, accessed from https://www.cia.gov/library/publications/the-world-factbook/fields/2117.html.

PIPELINE COSTS AND RELATED FUEL MARKETS

Economists have studied pipelines (both gas and oil), particularly Makholm (2012). Most of the earlier economic analysis of pipelines is much older. Troxel (1936, 1937a,b) made important contributions to the economic analysis of gas pipelines and their regulation in the United States. Cookenboo (1955) wrote about the economics of oil pipelines, but to modern eyes seems unduly preoccupied with neoclassical concepts of natural monopoly – to the extent that he recommended giant mandatory joint ventures from oil-producing areas to maximize the size, and minimize the unit cost, of shipping oil. Kahn (1971) devotes a size-able section of *The Economics of Regulation* to pipelines, having himself been involved in many of the disputes involving the regulation of pipelines and gas prices in the 1950s and 1960s. Kahn weighs the advantages of joint planning (which preoccupied Cookenboo) against the advantages of whatever genuine competitive rivalry the pipeline industry could support.

Some existing economic studies of pipelines were a by-product of the study of oil or gas commodity markets. For example, MacAvoy and Breyer (1974), Adelman (1972) and Tussing and Barlow (1984) studied pipelines as methods of serving the gas industry. Mabro and Wybrew-Bond (1999) have studied pipelines, somewhat tangentially, as a part of the sources of gas that now supply Europe. Business historians have focussed more intently on the cost structures and practical markets for pipelines, including Johnson (1956, 1967), Castaneda (1993, 1999) and Castaneda and Smith (1996). Legal scholars have also con-tributed to the literature on pipelines, including Rostow (1948), Rostow and Sachs (1952), Wolbert (1951, 1979) and Pierce (1988). The evident scholarly debate between Rostow and Wolbert is both somewhat dated and takes on more of the appearance of opposing legal briefs than an economic analysis of pipelines. However, Rostow was decades ahead of his time in outlining the basis for contract-based transport with pipelines. One political scientist, Sanders (1981), combined an economic view with a pluralist political perspective to analyze how Congress chose among competing constituencies to write the 1938 US Natural Gas Act that ultimately supported competitive US interstate pipeline transport.

Most of these economic analyses of pipelines basically emerged from the conflicts between pipeline owners and the customers they served that were heard either before regulators or in the US courts. Outside of the United States and Canada, there were no major private pipelines until the end of the twentieth century and no legal disputes to chronicle. Subsequent to the privatization of many pipelines in Europe and elsewhere, the definition of private regulated pipeline property has remained somewhat vague, and disputes have been less likely to be heard in the courts. The newness of the privately owned pipelines, the less clearly defined limits on government involvement in their practices and the uncertain split between EU and national jurisdictions for regulatory disputes in Europe have contributed to a shaky foundation for economic analysis there. Various writers (Vasquez et al. 2012; Glachant et al. 2013) have written about European gas markets, but only treating pipelines as 'notional' abstractions in gas markets – not as technologies.

THE TECHNOLOGY OF PIPELINE TRANSPORT

Pipelines themselves appear to be the quintessential natural monopolies driven by seemingly limitless economies of scale. The technology of pipeline costs has been well known for a very long time. Part of the technology is driven by the simple ratio of the volume of a circle to its circumference – or some function of π. The rest is driven by the interaction of pipeline size, distances and pressure differentials (perhaps boosted by compression along the way).

Archimedes: Declining Unit Pipeline Capacity Costs

The capacity of a pipe is a squared function of its radius ($\pi r2$) while the cost is a function of its surface area ($2\pi r$) – hence ever-declining unit capacity costs with larger diameters. It is a relationship between size and cost that Archimedes of Syracuse, who mastered an iterative derivation of π in the third century BC, would have understood. That means that a single pipeline would appear to be the least expensive way to serve the market for any conceivable size of pipeline – the definition of a natural monopoly for modern economists (Figure 24.1).

Troxel (1936) made the first study of the cost structure of gas pipelines. He used Federal Trade Commission data in the mid-1930s to perform his own economic cost calculations regarding pipeline construction – as there was no federal regulation of pipelines, nor data collection, at the federal level. Using his own assumptions regarding the contributions of capital and operating costs for pipelines, he found a roughly constant relationship between the diameter of gas pipelines and the cost-per-mile per inch of diameter, with a mile of 3.5 inch pipelines costing an average of $1320 per inch of diameter, much the same as the $1216 cost per inch of diameter of an 18 inch line.

Such a relationship between diameter and cost has been confirmed repeatedly. Trends in gas pipeline construction costs for 1954, 1959 and 1960, collected by pipeline filings

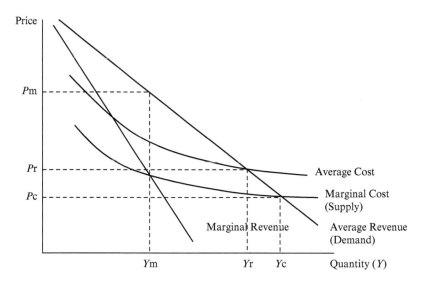

Figure 24.1   The cost structure of natural monopoly

before the Federal Power Commission, showed the same roughly constant cost in cost/mile/inch over a wide range of pipeline sizes (Leeston et al. 1963). The same is true for data based on Federal Energy Regulatory Commission gas pipeline permit filings from 1980 and 1994 (McAllister 1998). Figure 24.2 plots these data for gas pipeline construction costs in 1980 dollars, from 1935 to 1980, with a simple nonlinear trend line for each data series. The data collected at various times on gas pipeline construction costs confirm the rule of thumb: pipeline costs are generally linear in diameter, with capacity increasing with diameter at a factor of the ratio of area to radius, or $2\pi r$. There is nothing in these data to indicate that the average unit capacity cost for gas pipeline capacity is anything other than downward sloping over the entire range of pipeline sizes – a natural monopoly.

Weymouth: Pressure and Geography

Pipelines combine with compression to create the capacity to ship gas from one location to another. Once in transmission lines, the flow of gas is caused by pressure differential, which becomes lower through friction as the gas moves from the inlet towards the outlet of the pipeline. The greater the pressure differential, the greater will be the capacity of the line, following Weymouth's formula:[4]

$$Q = 670.8d^{8/3}\sqrt{\frac{P_1^2 - P_2^2}{GL}}$$

Here, Q is the cubic feet per hour, d is the internal diameter of the pipe, $P1$ and $P2$ are the inlet and outlet pressures, respectively, G is the specific gravity of the gas and L is the length of the pipe. While this formula was modified over time as pipeline companies gathered new

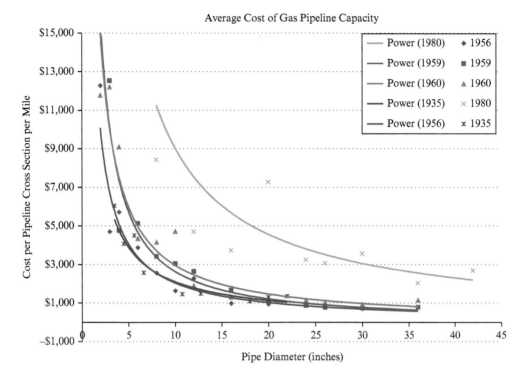

Figure 24.2   Average cost of gas pipeline capacity, 1935–1980

information on construction and operating costs, those modifications are not material in the search for economies of scale or the natural monopoly features of single pipelines traversing the countryside. The capacity of the lines is an exponential function (8/3, or 2.66) of the diameter, and it drops by the square root of the inverse of the pipeline length.

Thus, the longer the pipeline, the greater is the need to enhance the pressure along the way in order to deliver acceptable gas pressures at the outlet. Capacity expansions in this respect can be achieved in three ways: (a) greater pipeline diameter; (b) insertion of compression along the line to step-up the pressure; and (c) 'looping' (i.e. installing additional sections of pipeline alongside the original line) at various strategic places on the pipeline to increase the effective diameter of the line, in particular just downstream of compression stations. Compression is also highly capital intensive. Across a wide range of companies in the United States, the capital cost of compression capacity is on the order of 10 times the annual operating costs of the compressor stations, in both labor and compression fuel (Leeston et al. 1963, 84–85).

MARKET PROBLEMS WITH PIPELINES

Pipelines cross a lot of public and private land and political boundaries with capital that is sunk both literally and metaphorically. It should thus be expected that the placement and funding of pipelines in democracies involves considerable government and regulatory intervention. When governments intervene, they tend to want to deal with problematic aspects of pipelines that can frustrate competitive fuel markets or the construction of any pipelines at all.

Pipeline size and compression contribute to the total cost of transporting gas. They imply cost structures that lead to natural monopoly. In some industries, like local electric, gas and water public utilities, natural monopoly cost structures have long convinced governments that competition is impractical or impossible and that public regulation or ownership is a necessity. However, it is a mistake to apply local public utility concepts to pipelines as they serve diverse fuel markets. This is because fuel markets expand incrementally over time to serve growing demand, and the pipelines that serve them exist in a diverse spatial landscape, where there are multiple ways of supplying particular markets from existing gas fields or import points. Both in theory and practice, when put into such contexts, pipelines are rather poor natural monopolies.

Pipeline capital is also land-bound and immobile, providing difficulties for those who finance them. Uncertainty or commercial opportunism at either end of a pipeline can strand the facilities and undermine the value of invested capital – either for pipelines or the producers or consumers they serve. If investors build pipelines, they make interlocking alliances with fuel suppliers and users. The closest of such alliances is formal vertical integration. However, vertical integration is problematic: oil and gas producers tend to use vertically integrated pipelines as weapons against non-integrated competitors. Alliances can also come through contracts – which also have the ability to be used to harm competition in fuel markets. Governments have taken various approaches to facilitating the flow of capital to pipeline transport – to efficiently serve the public's need for fuel – while trying to prevent oil- and gas-producing companies from using pipeline access to harm competition. The United States and Canada sought a solution through the regulation of independent investor-owned pipelines by appointed commissions. Most of the rest of the world turned to various forms of direct state investment and control for vertically integrated monopolies.

When pipelines are regulated as independent rival suppliers, it supports both the efficient geographic allocation of fuel supply, the application of new technology (as in unconventional gas production) and the proliferation of competitive pipeline routes and construction. However, if they are regulated like natural monopolies (as in Europe), it undermines such a role for pipelines – serving effectively to bar competitive entry and slow the entry of new pipelines. In such circumstances, the tie between pipelines and geography can effectively disappear.

The two economic problems burden every major pipeline system in the world. The approach taken to solving them in the various global regulatory jurisdictions determines whether and how this technology is applied to serve fuel markets. Thus, a review of the nature of – and solutions to – these problems in the United States and Europe (the two largest gas markets, which account for more than 87 percent of the world's gas pipelines) is useful.

Natural Monopoly Cost Structures versus Time and Space

Multi-period economic investigations of natural monopolies deal with whether monopolies are sustainable in production environments where firms expand over time. In other words, a firm is an inter-temporal natural monopoly if its product cannot be produced at a lower cost by any combination of two or more firms as demand grows over time (Baumol et al. 1982, 408–413). Using a simple two-period model, Baumol and his colleagues were able to show that an incumbent natural monopoly with lumpy capital additions could be unsustainable when it came to the point of expanding its system even in the presence of declining costs. Baumol's model and others like it, however, are bound by highly restrictive assumptions. In the real world, incumbents have many practical and political options, including contracts and methods of price regulation, to bar entry.[5]

However, pipeline incumbents in the world find it hard to deter entry anyway. While between any two points there are manifest economies of scale in employing the largest diameter pipe available, the sources and markets – as they grow and shift – provide ample possibilities for competition between existing and prospective competing lines, for three reasons:

(1) Shifting resources – pipelines costs are geography and the locationally fixed and irretrievably sunk (figuratively and literally). Natural gas comes from geological formations that inexorably decline – a 'wasting asset'. Recoverable gas fields deplete while gas pipelines, representing non re-deployable capital, do not. The application of new technology to unconventional gas production through 'fracking' represents a highly sped-up shifting of the location of economically recoverable reserves.
(2) Administrative politics – new pipeline entry involves considerations beyond cost alone. The land required by pipelines is the ultimate scarce commodity – major gas pipelines need lots of it, often requiring government approvals for rights-of-way and permissions to cross private land. The natural monopoly view that a single regulated gas pipeline company could serve at lower cost has often failed to sway policy-makers and regulators.[6]
(3) Regulated pricing – Baumol's theories of natural monopoly assume uniform pricing, which is not a universal practice. The federal regulator of US interstate pipelines now requires the segregation of costs and the separation of prices for different vintages of capacity. Such 'incremental pricing' prevents incumbent pipelines from drawing upon the market value of old capacity to price incremental capacity below its cost.

Given these geological, technological, political and administrative problems with barring entry into pipeline markets, pipeline companies are unreliable natural monopolies – unless they enjoy the protection of governing institutions designed to frustrate competitive pipeline entry. Such is common enough around the world – the EU is the clearest example.

Transacting with Pipelines: 'Coasian Bargaining' and Competition

Pipelines are pipelines, in technology and operation, no matter where they are. However, the way governments treat them is highly dissimilar in different parts of the world. In particular, the role and regulation of gas pipelines, and the competitiveness of the gas markets they serve, are very different in the two largest gas markets of North America and Europe.

What distinguishes the US gas pipelines stems from its market in pipeline capacity rights that in turn depends on specific regulatory actions that remove the ability of pipeline interests to obstruct gas markets. Such actions allowed for deregulated markets in US pipeline capacity rights and the vigorous, technology-driven competition in the gas markets they serve. Unregulated markets built on such regulatory actions have formed in radio bandwidth and pollution control, among others, and exhibit 'Coasian bargaining', named after the 1991 Nobel laureate in economics, Ronald Coase (1959). Such markets are defined by the open trade in intangible property rights created when regulation defines such rights, tells the market who possesses them and facilitates a 'frictionless' exchange between willing sellers and buyers. Regulation in such markets merely defines the rights and informs the market – the re-sale prices in those rights are left to the forces of supply and demand.

The trading of US interstate gas pipeline capacity is a successful industrial application of Coasian bargaining where gas pipeline capacity rights have become a freely tradable shipper property. The Coasian market in US gas pipeline capacity ensures competitive use and competitive entry (in pipelines, storage and new shale gas production). It has made the pipeline market competitive and has caused gas producers and consumers to be sensitive to the competitive cost of shipping – as with any other transport technology.

WHETHER PIPELINE REGULATIONS ACKNOWLEDGE GEOGRAPHY – OR NOT

The two market problems with pipelines invite inevitable regulatory control over pipeline systems. Whether such regulations acknowledge geography into pipeline relationships with their users depends heavily on history and the institutional and political endowments of the places where pipelines operate. Different histories and endowments led the United States to embrace geography and Europe to resist.

Inherited Institutions

As described in Makholm (2015), several institutional factors were key to breaking down the barriers to a competitive US gas market: (a) unified federal regulation and the effective structural unbundling of US gas pipelines (i.e. breaking the corporate link between interstate pipelines and local distributors); (b) effective and fully transparent Federal Energy Regulatory Commission control of capacity licensing and cost-based pipeline tariffs; (c) the total functional unbundling of interstate pipelines (i.e. unbundling of the gas from the pipe); and (d) formation of effective markets for access to licensed pipeline transport capacity. There are key institutional features of Europe's gas industry that inhibit transparency in the nature and cost of gas supply and impede the formation of EU-wide regulation that would permit competitive pipeline entry on a continental scale as in North America (Makholm 2010; Correljé et al. 2013). These institutions include: (a) internal sovereign borders and a proliferation of member-state regulators that guard their jurisdictions and prohibit an effective regulatory body for trans-EU gas shipments; (b) secrecy regarding pipeline company financial and operating data; (c) structural vertical integration between cross-border gas pipelines and member-state local gas distributors; and (d) functional vertical integration where pipeline companies effectively transport their own gas supplies. Such institutions mean that vertically integrated EU member-state gas companies and their gas suppliers are effectively protected from any meaningful threat of competitive entry for either gas supply or pipeline transport.

The Technological Model for Regulation

Transacting by contract between points on the US interstate gas pipeline system is relatively straightforward, as the investor-owned pipelines were engineered and licensed to handle the distinct contractual needs of suppliers and consumers, despite the commingling of the gas supplies themselves. It is a simple accounting task to track how much gas flows in and out of the pipeline at each location – the basis for operational restrictions and tariffs. Overall, the knowledge of how fast gas actually flows – no faster than the average galloping horse – is the basis for licenses, contracts, tariffs and the normal operational restrictions needed to make sure that the whole group of users is reliably served with the minimum of investor capital. The resulting pipeline transactions have been transparently fair among the various users and relatively easy to regulate (Makholm 2012).

Such forthright point-to-point transacting is impossible under current technology for AC electricity grids. It is common knowledge that electricity travels around the AC grid at the speed of light, and that such speed (in addition to technical features of power grids, such as 'loop flow' and 'reactive power') makes it impossible to predict where a particular power plant's output will go at a particular point in time.[7] Therefore, technology itself prevents the incorporation of any realistic notion of geography into AC transmission tariffs.[8] Given current technology, competitive power markets connected through AC grids have no other choice than to make electricity grid pricing and regulatory rules geography-free.

A consequence of Europe's institutional and political endowments is that the EU prohibits the use of point-to-point pipeline contracting in its Third Legislative Package in favor of a regulatory regime that treats pipelines as grid-like natural monopolies. The regime effectively obscures the distance-based price signals that would otherwise emerge naturally on such comparatively low-technology gas pipeline systems.[9]

Geographic Gas Markets with Competitive Pipeline Transport

In North America, investor-owned gas pipelines and geographies work together to link fuel resources and cities in a larger competitive equilibrium, pricing transport at the margin by consuming part of the gas resource in route literally – as compression fuel – consistent with 'iceberg' transport in spatial market models first proposed by Samuelson (1952).[10] The exploitation of new technology in shale gas production – the competitive

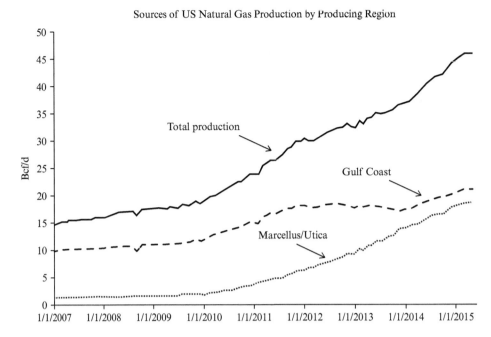

Figure 24.3   Shift in the geography of US gas production

entrant in a deregulated American gas market – has had a large effect on the continent of North America. It has prompted a boom in competitive entry of long-distance private pipeline construction to access the shale gas fields, driven down the gas price to the lowest level in decades, tilted the competitive electricity generating mix away from coal and led to a sharp interest in American exports of liquefied natural gas. Throughout North America, the technological shale revolution has upended the geography of gas supply (Figure 24.3). The major cities in eastern Canada now obtain their gas from the Marcellus shale fields of Pennsylvania rather than Alberta (their traditional source since the 1950s that is five times farther away). Pipelines from the Gulf Coast that used to take gas north now bring gas south. Planned liquefied natural gas import terminals in the United States have been converted for export – just a few years later. One of the largest gas pipelines on the continent – the 1679 mile Rockies Express Pipeline – commissioned in 2009 to bring Rocky Mountain gas to eastern markets, just switched directions.[11] These are profound changes in the spatial allocation of North American economic activity with the Marcellus/Utica shale gas in Pennsylvania and West Virginia substantially altering the source of gas in the United States and its direction of flow in regional pipelines.

There are various ways to show how geography and pipeline markets interact in the United States. Perhaps the best way is to chart how different regions of the country use the deregulated market in pipeline capacity to capture market price differences for gas at different locations – referred to as basis differentials. Prices at various trading locations are often significantly different, reflecting the availability of gas and the pipeline transport options between those points. An example is the great shock to the US gas market in

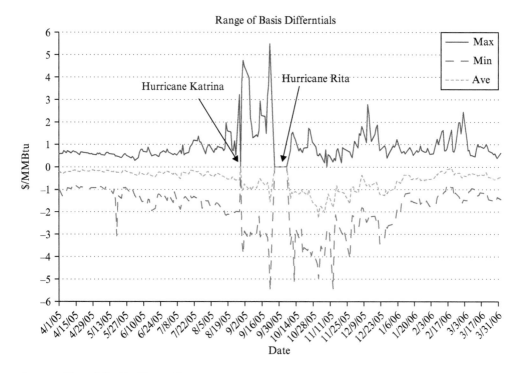

Range of Basis Differntials

Source: Natural Gas Intelligence Press.

Figure 24.4   The hurricane season of 2005

the late summer of 2005, during hurricane season in the Gulf of Mexico. Two hurricanes disrupted a large portion of the US gas supply and production. In addition to completely shutting down the Henry Hub in Louisiana (the main trading point for gas futures) for a day and week, respectively, hurricanes Katrina and Rita led to different and larger than normal supply–demand imbalances across the country. Figure 24.4 shows the range and average of the 84 basis differentials relative to the price at the Henry Hub between April 2005 and April 2006. Despite the shocks, pipeline capacity was re-sold and redirected according to changing market conditions in the aftermath of the hurricanes and spot gas markets cleared.[12]

Figure 24.5 shows how the basis differential for the New York region grew over the years, as local distribution gas companies used their long-term contracts for their own winter heating purposes, leading to lower 'resale' capacity for gas-fired electricity genera-tors over time. Such a growing price differential prompted pipeline companies to offer long-term contracts to build capacity in both New York and New England.[13]

Non-geographic Gas Markets with Barriers to Pipeline Entry

Europe's gas pipelines, for their part, are stubbornly disconnected from geography. EU regulations ban the pricing of gas pipeline service within member-states based on

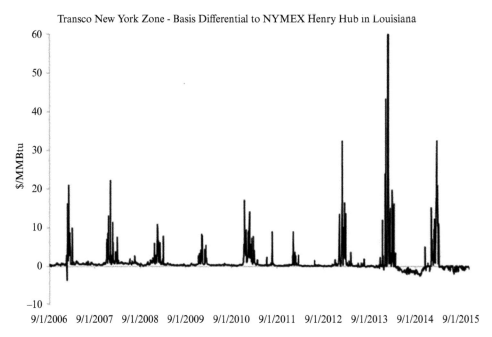

Figure 24.5   The values of delivery capacity to New York (2006–2015)

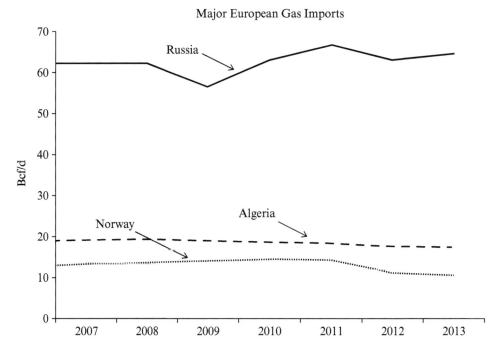

Figure 24.6   Consistency in the geography of European gas imports

geographical distances through the existing lines. Major additions to Europe's gas transport capacity come from EU-taxpayer-funded, 'projects of common interest' rather than investors. The major historical sources of natural gas to Europe in the 1980s (Russia, Norway, Algeria) remain (Figure 24.6). Shale gas remains in the ground, coal displaces gas for power generation and oil-linked gas prices have cost EU consumers more than half a trillion dollars from 2009 to 2015 vis-à-vis their US counterparts (Makholm 2015). Poland, which imports about 60 percent of its gas from western Siberian Russian fields more than 2000 miles distant, faces institutional and political barriers to shortening the source of gas to shale resources literally underfoot.[14]

If European pipelines and geography worked together in a larger equilibrium as in North America, then the gas market would build and re-direct pipelines to change the location of the fuel and drive down its cost, as in North America. Russia would be in the same boat as gas producers in the US Gulf Coast – bypassed by unconventional extraction technology closer to gas consumers. Yet as of 2016, Europe seems lost, without location-based pipeline transport, in what Isard (1956) once called a 'wonderland of no spatial dimension' where, despite the high cost, the geography of supply for the continent does not change.

CONCLUSION

Pipelines have perhaps a unique place in the study of the geography of technology. Pipelines sit in place for decades – literally buried far upstream from consumers. The cost of the fuels they transport is figuratively buried in many millions of heating bills, in electricity prices, in manufactured goods and in motor fuel. There is little novel technology – and surely no romance – in pipelines as they have crossed the countryside transporting fuel from one point to another for over a century. In some ways they are similar to other inland transport technologies, such as canals and railroads. Yet in important ways in modern energy economies around the world, they are unique in the economic problems they pose and the way governments address them.

The organization of the industry and whether geography plays any part in the spatial location of fuel supply is necessarily about institutions of economic governance: public or private ownership, legal systems, political boundaries and regulatory legislation. Any analysis of the industry that abstracts from such institutions has no chance of making sense of the organization of major pipeline systems or when they work with geography or fight it.

NOTES

1. https://www.washingtonpost.com/news/wonk/wp/2013/03/04/is-fracking-a-bridge-to-a-clean-energy-futu re-ernest-moniz-thinks-so/ (accessed 14 December 2015).
2. By opportunism I use Oliver Williamson's definition: '[s]elf-interest seeking with guile, to include calculated efforts to mislead, deceive, obfuscate, and otherwise confuse. Opportunism should be distinguished from simple self-interest seeking, according to which individuals play a game with fixed rules that they reliably obey' (Williamson 1996, 378). The hold-up problem with pipelines is well described in Klein et al. (1978, pp. 297–326).

3. By trivial, I mean comparing costs, capacities and mileages over time as if it were a study of productivity in making and laying pipelines rather than the nature of how pipeline systems serve regional fuel markets.
4. See Leeston et al. (1963, 69, 78). The formula is attributed to Mr R. Weymouth (from a paper read in 1912 before the Society of Mechanical Engineers). Other, more empirical but generally equivalent approximations to Weymouth's formula appeared later, known as the *Panhandle* and *Modified Panhandle* equations.
5. See Makholm (2012), *The Political Economy of Pipelines*, pp. 35–37, for a long discussion of the theoretical weakness of Baumol's natural monopoly framework given contracting and price regulation.
6. For example, US regulators during the boom in pipeline construction after the Second World War regularly favored more than one pipeline applicant in expansions of the interstate system to the east and west coasts of the continent. See Makholm (2012), *The Political Economy of Pipelines*, pp. 42–43.
7. Loop flow refers to power flow along an unintended path that loops away from the most direct geographic path or contract path. Reactive power describe the background energy movement in an AC system arising from the production of electric and magnetic fields.
8. Sally Hunt (2002, 399–400) recognized the technical distinction between gas pipeline system and power transmission networks; 'the pipeline contracts are not complicated – they are for point-to-point transport on a given system – a contract path. In electricity . . . the laws of physics mean that electricity does not flow over the designated contract path'.
9. Third Legislative Package, Explanatory Memorandum, Section 1.2. Proposal of the European Parliament and of the Council amending Directive 2003/55/EC concerning common rules for the internal market for natural gas.
10. Interstate natural gas pipelines frequently require that customers contribute a small percentage of the volumes of natural gas tendered for transportation service to provide fuel for compressors. See FERC docket no. RM07-20-000, Fuel Retention Practices of Natural Gas Companies, Notice of Inquiry (20 September 2007).
11. Rockies Express Pipeline LLC, docket no. CP14-498-000, Order Issuing Certificate (issued 27 February 2015).
12. Energy Information Agency, Natural Gas Weekly Update, 29 September 2005.
13. Federal Energy Regulatory Commission, Final Environmental Impact Statement on Algonquin Gas Transmission, LLC's Algonquin Incremental Market Project (docket no. CP14-96-000), issued 23 January 2015.
14. So much so that Poland had a contract to import liquefied natural gas from Qatar, which contains some of the most onerous liquefied natural gas price and take-or-pay terms in the world, accessed 14 December 2015 from http://www.reuters.com/article/conoco-poland-shalegas-idUSL5N0YR2R320150605.

REFERENCES

Adelman, M. 1972. *The World Petroleum Market*. Baltimore, MD: Johns Hopkins University Press.
Baumol, W., J. Panzar and R. Willig. 1982. *Contestable Markets and the Theory of Industrial Structure*. New York: Harcourt Brace Jovanovich.
Castaneda, C. 1993. *Regulated Enterprise: Natural Gas Pipelines and Northeastern Markets, 1938–1954*. Columbus, OH: Ohio State University Press.
Castaneda, C. 1999. *Invisible Fuel: Manufactured and Natural Gas in America, 1800–2000*. New York: Twayne.
Castaneda, C. and C. Smith. 1996. *Gas Pipelines and the Emergence of America's Regulatory State: A History of Panhandle Eastern Corporation, 1928–1993*. Cambridge: Cambridge University Press.
Coase, R. 1959. The Federal Communications Commission. *Journal of Law and Economics* 2: 1–40.
Cookenboo, L. Jr. 1955. *Crude Oil Pipe Lines and Competition in the Oil Industry*. Cambridge, MA: Harvard University Press.
Correljé, A., M. Groenleer and J. Veldman. 2013. Understanding institutional changes: the development of institutions for the regulation of natural gas transportation systems in the US and the EU. European University Institute Working Paper RSCAX 2013/07.
Glachant, J., M. Hallack and M. Vazquez. 2013. *Building Competitive Gas Markets in the EU*. Cheltenham, UK and Northampton, MA, USA: Edward Elgar Publishing.
Hunt, S. 2002. *Making Competition Work in Electricity*. New York: Wiley.
Isard, W. 1956. *Location and Space-Economy*. New York: Wiley.
Johnson, A. 1956. *The Development of American Petroleum Pipeline: A Study in Private Enterprise and Public Policy, 1862–1906*. Ithaca, NY: Cornell University Press.
Johnson, A. 1967. *Petroleum Pipelines and Public Policy: 1906–1959*. Cambridge, MA: Harvard University Press.

Kahn, A. 1971. *The Economics of Regulation: Principles and Institutions* (two volumes). New York: John Wiley.
Klein, B., R. Crawford and A. Alchian. 1978. Vertical integration, appropriable rents, and the competitive contracting process. *Journal of Law and Economics* 21(2): 297–326.
Leeston, A., J. Crichton and J. Jacobs. 1963. *The Dynamic Natural Gas Industry*. Norman, OK: University of Oklahoma Press.
Mabro, R. and I. Wybrew-Bond (eds). 1999. *Gas to Europe: The Strategies of Four Major Suppliers*. Oxford: Oxford University Press.
MacAvoy, P. and S. Breyer. 1974. *Energy Regulation by the Federal Power Commission*. Washington, DC: Brookings Institution.
Makholm, J. 2010. Seeking competition and supply security in natural gas: the US experience and the European Challenge. In F. Lévêque, J.-M. Glachant, J. Barquín, C. von Hirschhausen, F. Holz and W. Nuttal (eds), *Security of Energy Supply in Europe*. pp. 21–55. Cheltenham, UK and Northampton, MA, USA: Edward Elgar Publishing.
Makholm, J. 2012. *The Political Economy of Pipelines: A Century of Comparative Institutional Development*. Chicago, IL: University of Chicago Press.
Makholm, J. 2015. Regulation of natural gas in the United States, Canada, and Europe: prospects for a low carbon fuel. *Review of Environmental Economics and Policy* 9(1): 107–127.
Makholm, J. 2016. Why does most shale gas worldwide remain in the ground? *Natural Gas & Electricity* 33(1): 29–32.
McAllister, E. (ed.). 1998. *Pipeline Rules of Thumb Handbook*, 4th edn. Houston, TX: Gulf.
Pierce, R. 1988. Reconstituting natural gas, from the wellhead to the burnertip. *Energy Law Journal* 9(1): 1–57.
Rostow, E. 1948. *A National Policy for the Oil Industry*. New Haven, CT: Yale University Press.
Rostow, E. and A. Sachs. 1952. Entry into the oil refining business: vertical integration re-examined. *Yale Law Journal* 61: 856–914.
Samuelson, P. 1952. The transfer problem and transport costs. *Economic Journal* 62: 278–304.
Sanders, M. 1981. *The Regulation of Natural Gas: Policy and Politics, 1938–1978*. Philadelphia, PA: Temple University Press.
Troxel, E. 1936. Long-distance natural gas pipe lines. *Journal of Land & Public Utility Economics* 12(4): 344–354.
Troxel, E. 1937a. Regulation of interstate movements of natural gas. *Journal of Land & Public Utility Economics* 13(1): 20–30.
Troxel, E. 1937b. Some problems in state regulation of natural gas utilities. *Journal of Land & Public Utility Economics* 13(2): 188–203.
Tussing, A. and C. Barlow. 1984. *The Natural Gas Industry: Evolution, Structure and Economics*. Cambridge, MA: Ballinger.
Vazquez, M., M. Hallack and J. Glachant. 2012. Designing the European gas market, more liquid and less natural? *Economics of Energy & Environmental Policy* 3(1): 25–38.
Williamson, O. 1996. *The Mechanisms of Governance*. New York: Oxford University Press.
Wolbert, G. Jr. 1951. *American Pipe Lines: Their Industrial Structure, Economic Status and Legal Implications*. Norman, OK: University of Oklahoma Press.
Wolbert, G. Jr. 1979. *U.S. Oil Pipe Lines*. Washington, DC: American Petroleum Institute.

25. The evolution of solar energy technologies and supporting policies

Govinda Timilsina and Lado Kurdgelashvili

Solar energy has been used by humanity since pre-historic times. The ancient Greeks used building orientation, different materials and tree shading to utilize passive solar architecture for heating and cooling buildings (Brower 1992; Perlin 2004). When more efficient forms of energy, mainly fossil energy sources, became available, progress on the development of solar energy technologies did not occur for centuries. In the late 1990s, solar and other renewable energy technologies received attention as the supply of fossil fuel energy, especially oil, became volatile and expensive owing to wars and political maneuvering. These renewable forms of energy gained momentum in the early 21st century mainly owing to the climate change debate and also the concerns of countries that are dependent on imports for their energy supply. The spirit and speed of solar energy development being observed today has not been witnessed in the history of any other energy technology development.

Solar energy has expanded to include diverse sets of technologies ranging from simple solar box driers to complex technologies that produce electricity. Technologies utilizing solar energy resources can be divided into two broad categories: first, solar thermal technologies that convert solar radiation to thermal energy, which can be directly used either for heating purposes (e.g. solar hot water systems) or converted further into electricity (e.g. Concentrating Solar Power, CSP). A second type of technology is solar photovoltaics, which generates electricity from sunlight using the photovoltaic effect. Both solar thermal and photovoltaic technologies have experienced rapid growth over the last decade. The global capacity of solar photovoltaic technologies expanded more than 52-fold from 3.4 GW in 2004 to 177 GW in 2014 and the global capacity of concentrated solar thermal power generation increased by 11-fold from 0.4 to 4.4 GW in the same period (REN21 2015).

The growth of solar and other renewable technologies is driven by government policies. A series of sustained fiscal and regulatory incentives were adopted in almost all major economies. Incentives such as feed-in-tariffs (FIT), renewable energy portfolio standards, dispatching priority and transmission access have created an environment that has substantially boosted private sector financing for solar energy. The facilitating environment caused solar energy to expand several hundred-fold in only two decades. In addition, the costs of solar energy technologies have dropped rapidly over the last four decades. However, the economics of solar and other renewable technologies, such as wind, is still not favorable compared with conventional sources of energy such as coal, oil and natural gas. Existing policies are expected to further facilitate development of solar energy technologies and markets.

This chapter provides an overview of the evolution of solar energy technologies and policies to accelerate their adoption. It is organized as follows: the next section provides a brief history of development of solar energy technologies, followed by the discussion of the evolution of solar energy markets. The policy frameworks that contributed to the rapid expansion of the solar energy market are briefly discussed next followed by the cost trends of solar energy technologies. The Conclusions and Final Remarks wrap up the chapter.

THE EVOLUTION OF SOLAR ENERGY TECHNOLOGY

Humanity's use of solar energy spans thousands of years. Ancient Greeks used building orientation, different materials and tree shading to utilize passive solar architecture for heating and cooling buildings (Brower 1992, Perlin 2004). Beginning with Archimedes' story, where 'burning mirrors' were used against an attacking Roman fleet in the 3rd century BC, several references were made to solar energy use for lighting fires and water desalination. Interest in 'burning mirrors' rekindled in the 17th and 18th centuries, and by the late 18th and 19th centuries, solar energy use for cooking, water heating and industrial processes was also being considered (Cheremisinoff and Regino 1978).

Today, harnessing the sun's energy includes a diverse set of technologies that range from simply using the sun to dry crops to direct generation of electricity using photovoltaic cells. Solar energy technologies can be divided into two broad categories: solar thermal applications that convert solar radiation to thermal energy which can be directly used (e.g. solar hot water systems) or converted further into electricity (e.g. CSP), and applications that directly generate electricity from sunlight using the photovoltaic effect. The following section examines the modern history of these technologies.

Solar Thermal Technology (Heating)

The modern application of solar energy started with cooking. Swiss scientist Horace de Saussure created the first solar collector in 1767 when he built a small-scale greenhouse with five layers of glass boxes turned upside down on a black table to cook fruits (Cuce and Mert Cuce 2013; Saxena et al. 2011). The work on solar cookers was continued by English astronomer Sir John Herschel in the 1830s. During his expedition in South Africa, he attempted to cook with a similarly insulated box (Nahar 2001; Arenas 2007).

An important step towards large-scale utilization of solar energy came when American inventor Clarence Kemp patented the first commercial solar water heater called 'the Climax' in 1891. The solar water heater had a simple design with a water tank and a glass-covered wooden box (a hot box) integrated in one enclosure (Robles et al. 2014; Shukla et al. 2009). It could easily be integrated with the existing household plumbing, plus glazing over the tank could help to reduce heat losses during the night (Hren and Hren 2010). High levels of solar radiation combined with high fuel prices in California and neighboring Arizona made these markets primary targets for 'Climax' solar water heaters. From 1895 to 1900, around 1600 units were installed in southern California alone (Smyth et al. 2006; Perlin 2004).

In 1908, William J. Bailey significantly improved the water heater design when he separated the solar heat collector from the insulated water storage tank, which was placed above the collector. The separation of the collector from the storage tank decreased the amount of water present in the collector, thereby allowing faster heating during the day and reducing thermal losses during the night. This design allowed water to be stored overnight and used the next day (Fortuin and Stryi-Hipp 2012). The system also relied on the thermosiphon principle aiding the circulation of water in the collector and storage tank without a pump (Shukla et al. 2009; Seddegh et al. 2015). The introduction of Bailey's 'Day and Night' systems drove Kemp's 'Climax' out of major solar heater markets in California, Arizona and Hawaii (Seddegh et al. 2015). The 'Day and Night' system also solved the problem of water freezing in 1913, when the company introduced antifreeze solution containing alcohol in tubes located in the collector, which passed through the water tank (Hren and Hren 2010). The heat from the tubes was extracted through a coil of pipe within the storage tank, which acted as a heat exchanger. Nowadays, this type of system is referred to as a closed-loop water heating system (1001 Home Efficiency Tips n.d.).

In its first 10 years of operation, Bailey sold more than 4000 of his 'Day and Night' systems. However, in the 1920s, the discovery of large quantities of natural gas in southern California ushered in an era of low-cost fossil fuels and made solar heaters uneconomical. Bailey himself shifted to gas-powered hot water heater manufacturing (Jones and Bouamane 2012) and sold the patent for the 'Day and Night' system to a firm in Florida, where high fuel prices still made the system very attractive. From 1935 to 1941, propelled by a housing boom, up to 60,000 systems were installed in Florida (1001 Home Efficiency Tips n.d.). By 1941, more than half of the State's houses were using solar energy to heat water. During the Second World War, owing to restrictions on copper, a major component of solar water heaters, the industry had a serious setback. After the war, solar water heater sales briefly recovered. The largest US electric utility company, Florida Power and Light, however, aggressively marketed electric water heaters at discounted prices to increase electricity sales, consequently discouraging the use of solar water heaters (Perlin 2004).

While solar water heater production stalled in the United States, solar thermal applications were still developing worldwide. In Japan, significant fuel prices facilitated the demand for solar hot water systems. In 1947, the Japanese inventor Yamamoto designed the first widely used commercial solar water heater in Japan. The heater consisted of a basin with its top covered by glass. The later models comprised stainless steel pipes enclosed in a glass-covered box, a design upon which many modern Japanese units are still based. Around 4 million of these units had been sold in Japan by 1969 (Smyth et al. 2006; Perlin 2004). Faced with energy shortages in Israel, Levi Yissar manufactured solar water heaters similar to Bailey's design and, by 1967, 5% of all households in the country had solar water heaters (Perlin 2004).

Following the 1974 energy crisis, in the United States, Freeman A. Ford developed low-cost, easy-to-replace polymer absorbers, replacing more expensive copper in solar collectors (Perlin 2004). The collector did not require glazing and was ideal for heating a swimming pool, where the pool itself would serve as a storage unit. The outdoor swimming season perfectly coincided with the availability of solar energy, making the system ideal for its application. Ford's company FAFCO became the dominant solar water heater company in the United States. During its 40 years of operation it produced 66 million square feet of swimming pool heating collectors (equivalent to 4.3 GW_{th}) (FAFCO 2009).

Another significant development in solar thermal technology was the introduction of evacuated tube collectors. During the 1980s, collectors developed at the University of Sydney comprised two concentric glass tubes separated by a vacuum. The 'Sydney tube' design presented minimal problems for mass production (Garg 1987; Morrison 2013) and played a crucial rule in the solar water heating market.

Beginning in the 1970s, China became interested in solar water heaters. Initially China relied on flat plate collectors (similar to Bailey's design), but they underperformed during winter seasons and evacuated tube collectors were deemed more suitable (Qiu et al. 2015). During the 1990s, as China shifted towards evacuated tubes, their share in the newly installed solar water heaters gradually grew, reaching 30% by 1998 and 85% by 2002 (Zhiqiang 2005). In 2003, Himin, the leading solar water heater manufacturer in China, obtained the license for the 'Sydney tube' (Morrison 2013). As a result, in China the cumulative installed capacity of evacuated tube collectors grew from 30.8 GW_{th} in 2003 to 242.8 GW_{th} by 2013 (Weiss et al. 2005; Mauthner et al. 2015), making the country the largest consumer of solar thermal energy in the world.

Solar Thermal (Electricity)

The origins of solar thermal electricity, also referred to as CSP, can be traced back to the 19th century. In the 1860s a French mathematician, Augustin Mouchot, combined the box heat trapping idea with the burning mirror concept and built a solar oven (Cuce and Mert Cuce 2013; Saxena et al. 2011). Mouchot and his assistant Abel Pifre were also known for constructing a conical mirror for focusing solar radiation on a copper boiler placed along the axis of the cone. Several engines were demonstrated by Mouchot from 1866 to 1878, while Pifre began using parabolic reflectors, which were predecessors of the parabolic dish collectors used today (Pytlinski 1978; Delyannis 2003). The French government, however, concluded that the solar insolation was too weak for industrial applications in France (Perlin 2004). About the same time, John Ericsson, a Swedish-American scientist, was developing a parabolic solar collector (a forerunner of the parabolic trough reflectors) connected to a small steam engine (Cheremisinoff and Regino 1978).

In California and Egypt, work on solar powered engines was continued by British entrepreneur Aubrey Eneas, and American inventor Frank Shuman, respectively. Both locations had high solar insolation and expensive coal prices. Eneas began producing large conical collectors adopted from Mouchot's design in South California. The collectors, however, could not fare well against strong winds and hail (Jones and Bouamane 2012). In 1912, Shuman and his partners constructed a steam engine powered by solar energy for irrigation in Egypt, using parabolic trough collectors (Perlin 2004; Kalogirou 2014). The project was commercially successful and Shuman was asked to build a number of other plants. But the outbreak of the First World War and the subsequent decrease in fossil fuel costs halted commercial viability of solar energy (Spencer 1989; Jones and Bouamane 2012). For the next 50 years, while there were some attempts to use solar energy for mechanical or electric power,[1] low fossil fuel prices made power generation from solar energy uneconomical.

In the 1960s Italian scientist Giovanni Francia developed the first linear Fresnel reflector. It comprised a number of long strips of flat reflectors (mirrors) above a receiver running in parallel with rotational axis of reflectors (Mills 2004; Delyannis 2003). Owing to technical problems, however, Fresnel reflectors did not last long and there were no significant power production projects based on this technology until the end of the 1990s (Mills 2004; Haberle 2012). During the 1970s parabolic trough reflectors, parabolic dishes and central receiver systems showed much more promise. In 1979 the world's first commercial solar power plant with 5 MW capacity was built in Albuquerque, New Mexico at the Sandia National Laboratories test facility. The plant was based on central receiver systems, also referred as a power tower, comprising 220 heliostats (polished mirrors with a two-axis tracking system) focusing on a solar tower serving as a central receiver (Delyannis 2003; Kalogirou 2014). It was followed by Solar One, a 10 MW system located in Barstow, California, which operated from 1982 to 1988. Solar One was later replaced by another 10 MW system (Solar Two) integrated with a 24 hour thermal energy storage, which operated from 1996 to 1999. Several other demonstration systems were built around the world (Alexopoulos and Hoffschmidt 2012).

From 1984 to 1990 LUZ International Ltd built nine Solar Electric Generating Systems (SEGS) based on parabolic trough reflectors. The plants were installed in the Mojave Desert in California with a total capacity of 354 MW. In the early 1990s, SEGS generated around 95% of the world's solar thermal electricity (Brower 1992). Even though LUZ was bankrupted in 1991, owing to political and regulatory hurdles, SEGS was considered to be a successful project. For the next 15 years there were no significant developments in the United States and until recently SEGS remained the largest commercial CSP plant in the world (CSP Today 2012).

The new wave of CSP development in the United States started in 2006 with a 1.2 MW parabolic trough Saguaro Power Plant in Tucson, Arizona, followed by a much larger (75 MW) plant in Nevada (Nevada Solar One) in 2007. Since then several large-scale CSP power plants have been built or are under construction. The largest of them, the 500 MW Palen Solar Electric Generating System in California using a power tower, is due to come online in 2016 (National Renewable Energy Laboratory 2015). Interest in CSP was rekindled not only in the United States but also in other countries. Encouraged by favorable feed-in policies, 50 large-scale projects came into operation in Spain between 2007 and 2013. The majority of these projects (45) were based on a parabolic trough and all except one had 50 MW capacity. The remaining three were based on power towers and two were based on Fresnel reflectors (National Renewable Energy Laboratory 2015).

Photovoltaics

French scientist Edmond Becquerel discovered the photovoltaic effect in 1839 using solid electrodes placed in an electrolyte solution. In 1870s, English scientists Willoughby Smith, William Adams and Richard Day discovered the photovoltaic effect of selenium. As a solid semiconductor material, selenium exhibits increases in electricity conductivity after it is exposed to sunlight and it also produces an electrical current. In 1883, American scientist, Charles Fritts made the first photovoltaic (PV) cell from selenium with only 1–2% efficiency. The photovoltaic effect remained unexplained until 1905, when Albert Einstein published a paper on this phenomenon, for which he was awarded the Nobel Prize in 1921 (Lamont 2012).

The next major breakthrough in PV technology development came by the accidental discovery of strong photovoltaic properties in silicon by Daryl Chapin, Calvin Fuller and

Gerald Pear at Bell Labs in 1953. A year later these researchers created the silicon PV cell with 4.5% efficiency, and after a few months its efficiency was increased to 6% (Perlin 2004; Lamont 2012). Commercialization of PV cells started when Hoffman Electronics acquired the license from Bell Labs in 1956. Soon after, Hoffman Electronics silicon cell efficiency reached 8% in 1957, and 14% in 1960 (Jones and Bouamane 2012; Lamont 2012).

PV cells found their first practical application in space, where they proved to be significantly superior to electrochemical batteries. Whereas batteries lasted only for weeks, PV cells could operate for years (Perlin 2004). In 1958, the United States launched Vanguard I, the first satellite with PV cells, and other satellites soon followed (Jones and Bouamane 2012; Lamont 2012). Over time, the size of PV cells launched with satellites increased. In 1958, Vanguard had less than 1 W of power; by 1968 NASA's astronomical observatory was powered by 1000 W PV cells (Lamont 2012). In the early 1960s the cost of PV was US$40,000/W (Schaeffer 2005). Even at this high price, PV cells were viable for space applications but they were prohibitively expensive for everyday use.

In the late 1960s, the price of solar cells went down to US$200/W, but it was still too expensive. American scientist Elliot Berman wanted to reduce PV costs to US$20/W (Perlin 2004). In 1969, he joined the Exxon Corporation and started looking for cost-reduction options. After his initial efforts failed, in 1973 he established the 'Solar Power Corporation', a subsidiary of Exxon, and began to explore cheap silicon wafers that had been rejected by the semiconductor industry (Jones and Bouamane 2012).

During the 1970s, PV cells started providing power to: (a) offshore oil rigs to replace less reliable electrochemical batteries; (b) oil pipelines to keep them corrosion free in remote locations; and (c) railroad crossing signals (Perlin 2004). Other common applications, such as watches and calculators, soon followed (Jones and Bouamane 2012). Another important market for PV was powering telecommunication services for remote rural areas. This was particularly visible where Telecom Australia considered PV as the preferred power source for remote locations during the 1970s. In developing countries PV also powered water pumps and provided essential services for locations with no access to grid-supplied electricity (Perlin 2004).

In 1972, the Institute of Energy Conversion (IEC) was established at the University of Delaware, focusing on thin-film PV cell development. IEC was the world's first laboratory dedicated to PV research and development. Through IEC efforts, thin-film cell efficiency improved from around 3–5% in 1972 to 10% in 1980. In 1973 the University of Delaware built one of the world's first residential houses with PV, named 'Solar One'. The system served as a demonstration project for showing the value of grid-connected PV (University of Delaware 2015). Then, in 1982 ARCO Solar built the first utility-scale 1 MW PV system connected into the distribution grid of the Southern California Edison utility (Jones and Bouamane 2012). It was followed by a 6 MW PV system installed in central California in 1983 (Blazev 2012). By then the annual cell and module production volume in the United States exceeded 10 MW (US Energy Information Administration 2012).

During the 1980s collapsing oil prices and the elimination of incentives for renewable energy in the United States created severe challenges for the PV industry. The 1980s and 1990s were referred to as the 'Valley of Death' for PV companies (Yergin 2011, 534). Several leading PV manufacturing companies went bankrupt or had to shift to overseas markets (Jones and Bouamane 2012). In 1983, out of the 12.6 MW cells and module produced by the United States PV industry only 15% was for exports, whereas by 1990, out of the 13.8 MW produced, 55% was for exports (US Energy Information Administration 2012). During the 1980s and 1990s, Germany and Japan became the primary markets for PV systems.

In 1983, the first 4 kW grid-connected PV system in Europe was installed on the roof of a residential building in Munich. Yet by 1990, the German PV market was still in its infancy, accounting for only 1.5 MW in cumulative installed power (Erge et al. 2001). In 1990, the German government initiated the 1000 Roofs Measurement and Analysis Program and the world's first feed-in-tariff to spur interest in the technology (Byrne and Kurdgelashvili 2011). The Japanese government started similar programs promoting PV installation in public and residential buildings, which proved to be very successful. By 1998, Japan, with a cumulative capacity of 133 MW, surpassed both the United States with 100 MW and Germany with 54 MW in PV installations (International Energy Agency 2012).

During the 1980s, significant progress was made in efficiency improvements and cost reductions. The laboratory-obtained cell efficiency of single-crystal silicon increased from 17% in the early 1980s to 23% in 1990. Efficiency for thin-film cells increased from 10 to 15% (US Department of Energy 1991) and PV module prices decreased from US$44/W in 1976 (in 1986 dollars) to US$4–5/W (Brower 1992). By 1992, the University of South Florida had developed a 15.9% efficient thin-film cell made from cadmium telluride. During the 1990s, significant progress was made in developing next-generation PV cells. In 1994, Jerry Olson, working at the National Renewable Energy Laboratory (NREL), combined gallium indium phosphide and gallium arsenide, producing a two-junction cell with 30% efficiency (National Renewable Energy Laboratory 2012). In 1999, NREL set a new record for two-junction cells reaching 32.3% efficiency under sunlight concentrated to 50 times normal strength. In the same year NREL, using copper indium gallium diselenide, set the record for thin-film solar cells at 18.8% (US Department of Energy 2000).

By 2001, cumulative global PV installation reached 1.0 GW (International Energy Agency 2012). By then, Japan, led by companies like Sharp, Kyocera and Sanyo, surpassed the United States, and became a leading PV cell manufacturing country in the world as well as a leader in PV deployment (International Energy Agency 2002). In 2005, Japan was overtaken by Germany in terms of the highest cumulative PV capacity (International Energy Agency 2012). Likewise, PV cell manufacturers, led by Germany's Q-Cell, were catching up with Japanese companies, and in 2007 it became the number one PV cell manufacturer in the world (International Energy Agency 2008; Yergin 2011, 578). Competition soon emerged in China from Suntech Power, which focused on low-cost manufacturing, including 'de-automated' parts of production using low-cost labor. With lower costs and government policies promoting manufacturing and exports, China started dominating PV cell production (Yergin 2011). By 2011, China was producing more than half of the world's PV cells (International Energy Agency 2012) and since then it has maintained its leading position (International Energy Agency 2015).

In 2008, cumulative global PV installation surpassed 10 GW. Four years later, cumulative PV installation reached 100 GW, and by 2014 cumulative PV installation stood at around 180 GW (International Energy Agency 2015). Within just 20 years, global PV capacity increased 1000-fold. PV already provides more than 1% of global electricity and in Italy, Greece and Germany it generates more than 5% (International Energy Agency 2015).

Table 25.1   Key milestones in the history of solar energy development

Time	Milestones
1800–1900	➤ French scientist Edmond Becquerel discovered the PV effect in 1839. ● French mathematician August Mouchet and his assistant, Abel Pifre, constructed the first solar-powered engines in 1860s which were the predecessors of modern dish collectors. ➤ In 1870s, English scientists Willoughby Smith, William Adams and Richard Day discovered PV effects for selenium. ● A Swedish-American scientist John Ericsson introduced parabolic cylindrical solar collectors during 1868–1886. ➤ In 1883, American inventor Charles Fritts made the first PV cell from selenium with 1–2% efficiency. o Baltimore inventor Clarence Kemp developed the first commercial solar water heater in 1891.
1900–1950	➤ In 1905 Albert Einstein explained the PV effect. o William J. Bailey significantly improved solar water heater design by separating the collector from the insulated storage tank in 1909. ● In 1912, Frank Shuman constructed a steam engine powered by solar energy using parabolic trough collector for irrigation in Egypt.
1950–1960	➤ In 1954 Daryl Chapin, Calvin Fuller and Gerald Pearson developed the silicon PV cell at Bell Labs with 4.5% efficiency. ➤ Hoffman Electronics improved silicon cell efficiency to 8% in 1957, and 14% in 1960. ● In 1960s, Italian scientist, Giovanni Francia developed the first linear Fresnel reflector.
1960–1970	o Freeman A. Ford developed low-cost plastic to act as the solar collector for warming swimming pools.
1970–1980	➤ In 1970s Elliot Berman established the 'Solar Power Corporation' a fully owned subsidiary of Exxon, bringing cost of PV down to US$20/W. ➤ The University of Delaware built 'Solar One,' one of the world's first PV powered residences in 1973. ● In 1979 the world's first commercial solar power plant with 5 MW capacity was built in Albuquerque, New Mexico.
1980–1990	➤ In 1980 thin-film solar cell exceeded 10% efficiency. ➤ In 1982 the first megawatt-scale PV power station developed by ARCO Solar was commissioned in Hisperia, California. It was followed by 6 MW PV system in central California in 1983. ● From 1984 through 1990 LUZ International Ltd built nine Solar Electric Generating Systems in the Mojave Desert in California. The plant used parabolic trough reflectors and with 354 MW of total installed capacity was the largest solar power plant in the world at that time. ➤ In 1983, the US annual shipments in PV cells and modules exceeded 10 MW. ➤ Laboratory-obtained cell efficiency of single-crystal silicon increased from 17% in the early 1980s to 23% in 1990. In the same time frame, efficiency for thin-film cells increased from 10 to 15%.
1990–2000	➤ In 1994 Jerry Olson, working at NREL, combined gallium indium phosphide and gallium arsenide, producing a two-junction cell with 30% efficiency. ➤ By 1998 Japan and the United States had surpassed 100 MW in cumulative installation. Germany, with 54 MW installations, closely followed.

Table 25.1   (continued)

Time	Milestones
1990–2000	➤ In 1999, NREL set a new record for two-junction concentrator cell efficiency at 32.3%. In the same year, a new record was set by NREL for thin-film PV cells at 18.8%.
2000–2010	➤ In 2001, cumulative global PV capacity reached 1 GW.
	➤ In 2008, cumulative global PV capacity exceeds 10 GW.
2010–2015	➤ In 2012, cumulative global PV capacity reached 100 GW.
	➤ In 2014, 1% of global electricity came from PV.

Notes:
o   Thermal (heating)
●   Thermal (electric)
➤   PV

Thus, the PV industry, which originally focused on niche applications such as the space program and providing power to remote locations, has become a mainstream power source.

THE EVOLUTION OF SOLAR ENERGY MARKETS

Solar Thermal for Heating

Solar thermal systems for water and space heating have experienced significant growth. In 2000, the cumulative installed capacity of solar thermal systems was around 50 GW_{th} and it was more than 350 GW_{th} by the end of 2013. The dominant market for solar heating was China with 70% of total existing global capacity. The European Union (EU) was the next largest market with 9% of the global capacity. By 2013, within the EU, Germany had around 40% of the cumulative capacity, more than Austria, Greece, Italy and Spain combined. The United States accounted for 4% of the global capacity. Besides China, among the developing economies, Turkey, Brazil and India with 3%, 2% and 1% of the installed capacity respectively, boasted relatively a large share of the market (Mauthner et al. 2015).

Solar thermal heating technologies vary greatly in scale as well as type. The technologies include glazed flat-plate collectors, evacuated tube collectors and unglazed collectors that use water as the energy carrier, as well as glazed and unglazed air collectors.

Figure 25.2 shows the cumulative global capacity of the solar thermal systems for water heating by type. In 2013, 70.5% of the global installed collector capacity was based on evacuated tube collectors. This type of collector is dominant in China, where it accounts for more than 90% of all installed collectors. In India, share of evacuated tube collectors is nearly 50%. In the United States and Australia, the predominant type of solar collector is unglazed water collectors used for pool heating; both account for 75% in their total cumulative capacities. In the other countries, flat-plate water collectors dominate the market. For example, more than 80% of installed capacity in Europe was of this type in

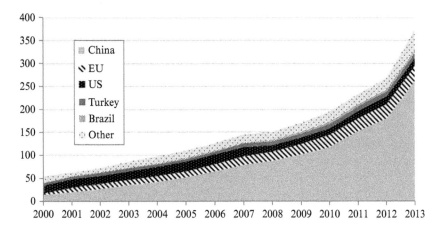

Source: Mauthner et al. (2015).

Figure 25.1 Installed capacity (GW_{th}) of solar thermal systems by location

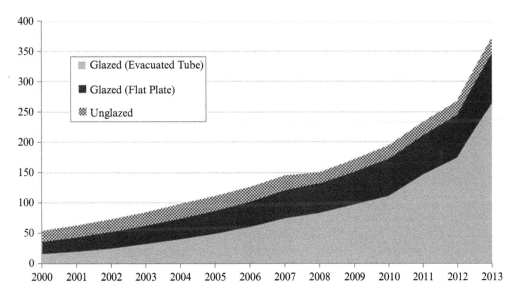

Source: Mauthner et al. (2015).

Figure 25.2 Installed capacity (GW_{th}) of solar thermal systems by type

2013. The use of air collectors is nearly negligible. Air collectors represent only 0.1% of the cumulative global capacity (Mauthner et al. 2015).

In 2013, there were around 111 million solar water heating systems in operation in the world. Eighty percent of these systems were used for domestic water heating in single-family houses; 9% of them were larger domestic hot water systems for multi-family houses, hotels, hospitals and schools; 6% were used for swimming pool heating. The remaining 4% of systems provided space heating combined with water heating or delivered heat to district heating networks (Mauthner et al. 2015). In terms of solar water heater penetration, Cyprus and Israel, with 93 and 85% of households using solar water heaters, respectively, boasted leadership (REN21 2015). Not surprisingly these countries were also among the leaders in terms of per capita installations. In 2013, other countries with high per capita install capacity were Austria, Barbados, Greece and Palestine, followed by China and Germany, which were also the leading countries in terms of the total cumulative installations (Mauthner et al. 2015).

As noted above, the majority of solar thermal systems were used for domestic water heating. These systems typically provided 40–80% of hot water demand (REN21 2015). Interest in larger domestic water heating systems for hotels, multi-family houses, hospitals, schools and other large complexes was also increasing. Their use was noticeable in Middle East and North Africa (MENA),[2] where nearly half of solar thermal systems were used for these purposes. The use of solar collectors for space heating was growing in Europe. In Austria and Germany, space heating systems accounted for 40% of the systems (REN21 2015). Space heating systems could typically displace between 20–30% of conventional energy used for building heating. In Europe, solar thermal systems combined with various backup heat sources (hybrid systems) were gaining popularity, and they accounted for 18% of cumulative solar thermal capacity (Mauthner et al. 2015).

Solar Thermal Energy for Electricity

Electricity production from solar thermal energy is based on CSP technologies. Three major CSP technologies are currently in use: parabolic trough collectors; central receivers (also referred to as power towers); and linear Fresnel reflectors. Among these technologies, parabolic trough collectors are the most widely used. As noted above, in the United States from 1984 to 1990, nine power plants in the Mojave Desert of California were built based on parabolic trough collectors. They were the first commercially operated CSP power plants. No other commercially operated CSP power plant was built until 2006–2007.

In 2006–2007, three CSP plants were completed, one MW parabolic trough plant in Arizona, a 75 MW parabolic trough plant in Nevada and an 11 MW central receiver plant in Spain. Since then, several CSP projects have been built in the United States and Spain, and these two countries have dominated the CSP market in the last 10 years. By 2015, Spain was leading with 2.3 GW cumulative installed capacity, and the United States had 1.9 GW capacity. Combined, they represented nearly 90% of global cumulative CSP capacity (Figure 25.3). Starting from 2011, within four years, India installed five CSP systems with a combined capacity of 228 MW, and became the third largest market for CSP. Several other countries are actively developing CSP projects. Among them, Chile, China, Israel, Morocco, South Africa and United Arab Emirates already have or are planning to install at least 100 MW of CSP capacity in their countries within the next two years (National Renewable Energy Laboratory 2015).

Before 2013, nearly all CSP projects were based on parabolic trough technology. By the end of 2015, this technology still dominated the CSP market, occupying 84% of cumulative CSP capacity. In 2013–2014, the 392 MW Ivanpah Solar Electric Generating System using power tower technology was built in California's Mojave Desert (National

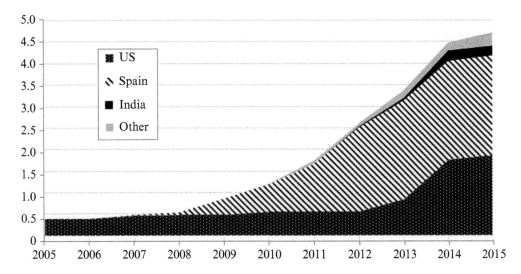

Source: National Renewable Energy Laboratory (2015).

Figure 25.3 Installed CSP capacity (GW) by location

Renewable Energy Laboratory 2015). Several other smaller Power Tower systems were built in Australia, China, India, Spain and Turkey. Combined, these plants represent 12% of global cumulative CSP capacity. The remaining 4% of global capacity was represented by linear Fresnel reflectors (Figure 25.4).

The leading countries for linear Fresnel reflectors were France and Spain. All three existing CSP plants in France use linear Fresnel reflectors. Spain hosts a Puerto Errado 2 thermosolar power plant with 30 MW capacity, which is the largest linear Fresnel reflector in the world.

Photovoltaics

In the last 15 years, the global photovoltaics market has experienced phenomenal growth. The global cumulative PV capacity increased from less than 1 GW in 2000 to 177 GW in 2014 (Figure 25.5).

In 2014, nearly half of the cumulative PV installations were in Europe, where Germany and Italy had 38 and 19 GW of capacity, respectively. Spain, France and the UK each had more than 5 GW of installed PV capacity. In Asia, China and Japan were leaders with 28 and 23 GW installed capacity, respectively. In the United States, cumulative installed PV capacity stood at 18 GW. From the remaining countries, Australia and India had 4 and 3 GW of capacity, respectively, followed by Korea and Canada, with 2 GW each (International Energy Agency 2015).

The driving force for the ongoing deployment of PV power plants has been the emergence of the grid-connected PV market. During the 1990s, at the initial phase of PV commercialization, most of the PV applications were off-grid. Over the years, the share

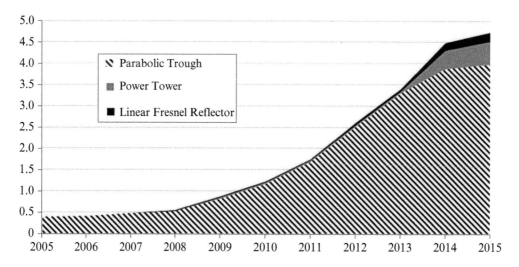

Source: National Renewable Energy Laboratory (2015).

Figure 25.4 Installed CSP capacity (GW) by type

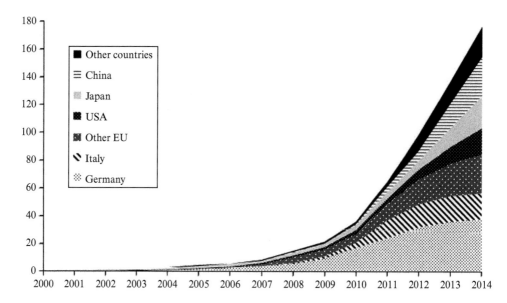

Source: International Energy Agency (2015).

Figure 25.5 Installed CSP capacity (GW) by location

of grid-connected PV has steadily increased. By 2014, the grid-connected installations represented more than 99% of new installed PV capacity (Figure 25.6).

Nevertheless, the off-grid PV remains a viable solution for electrification in developing countries such as India, Bangladesh and Peru, where governments have introduced

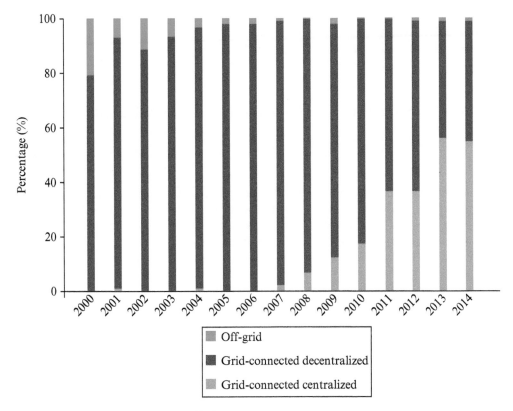

Source: International Energy Agency (2015).

Figure 25.6   PV deployment by installation type

programs to help rural electrification through off-grid PV deployment (International
Energy Agency 2015).

In the early 2000s, PV deployment was dominated by grid-connected distributed
systems, serving residential and commercial customers. The late 2000s witnessed rapid
growth of large centralized utility-scale PV systems. By 2014, more than half of the new
PV installations were grid-connected centralized systems (International Energy Agency
2015). Some of these systems were very large scale, with over 50 MW capacity. At least
70 of these systems were operational by early 2015. The largest of them were Topaz Solar
and Desert Sunlight, deployed in California with 550 MW capacity each (REN21 2015).
Both projects were built by First Solar, the leading thin-film PV manufacturer in the
United States (First Solar 2015).

In 2014, PV cell production was estimated to reach over 46.7 GW, including about
3.6 GW of thin-film capacity. Thus, thin-film production represented only 8% of the
market. Nearly half of thin-film capacity was produced by First Solar based on cadmium
telluride. Such a small share of thin-film PV was due to strong competition from crys-
talline silicon cells, which have experienced significant cost reductions in recent years.
Sixty-one percent of PV cells were produced in China, 16% in Taiwan, 6% each in Japan

and Malaysia and 3% in Korea. The former leaders Germany and the United States produced just 2% each. Although PV cell manufacturing was at its highest level, with 70 GW of production capacity, the industry still had significant spare capacity (International Energy Agency 2015).

EVOLUTION OF POLICIES TO STIMULATE EXPANSION OF SOLAR ENERGY

United States

The major policy initiative before 1980 that opened the door for the promotion of solar energy in the United States was the introduction of the Public Utilities Regulatory Policy Act (PURPA) in 1978. This Act obliged electric utilities to purchase power from small renewable generators and co-generators at their avoided cost.[3] However, PURPA could not be implemented until 1981 owing to legal challenges. The Energy Tax Act of 1978, for the first time, created a provision of credits in both personal and business income taxes to promote renewable energy. According to this law, private residents who used solar, wind or geothermal sources of energy could get, as a tax credit, 30% of the cost of the equipment up to US$2000 and 20% of costs greater than US$2000, up to a maximum of US$10,000. Businesses were allowed to receive tax credits up to 25% of the initial cost of renewable energy equipment (Plante 2014). In the same year, President Carter signed the Solar Photovoltaic Energy Research, Development, and Demonstration Act of 1978 (H.R. 12874), which authorized an aggressive program of research, development and demonstration of solar PV energy technologies in the United States. The act made electricity generated from PV systems economically competitive with electricity produced from conventional sources (ProCon.Org 2013).

PURPA and the 1978 Energy Tax Act helped promote renewable energy during the 1980s and the total installed capacity of renewable energy during the decade was 12,000 MW[4] (i.e. geothermal, small hydro, biomass, solar thermal and wind). However, the share of solar in the total installed renewable capacity (excluding hydro) was almost negligible (approximately 0.1%). The noticeable expansion of solar energy in the United States did not happen until the late 1990s and early 2000s, when various policies including Renewable Energy Portfolio Standards (RPS), Public Benefit Funds and Net Metering were introduced in the second half of the 1990s in some states. These policies expanded in other states in the following decade. Massachusetts was the first state to introduce RPS in 1997 with a target of 9% by 2009 and increasing by 1% per year thereafter. Connecticut and Wisconsin followed Massachusetts in 1998 and Maine in 1999 with 10, 2.2 and 30% targets, respectively. The RPS expanded into 30 states and Washington, DC by 2010. Not only the number of states introducing RPS increased but also the targets (i.e. percentage renewables in the total electricity supply in a state) set by RPS in most states were increased. For example, New York introduced in 2004 an RPS of 24% to be achieved by 2013; this target increased to 29% to be met two years later in 2015. Similarly, California introduced in 2002 an RPS of 20% to be met by 2017; it was expanded to 33% to be met by three years later in 2020 (Martinot et al. 2005; REN21 2011).

European Union

EU policies to support solar energy started in the early 1990s or late 1980s with an exception of government support for research and development funding for renewable energy technologies.[5] The EU introduced series of Directives to promote renewable energy in its member states. In 1997, the EU adopted a Directive on Renewable Energy to increase the share of renewable energy to 22% of the total electricity generation mix or 12% of the total final energy consumption of EU-15 by 2010. This directive was replaced with a new one in 2001 to accommodate the enlargement of the EU and also to highlight the role of renewables in contributing to its Kyoto obligations on climate change mitigation. The overall target of 22% of electricity mix set forth in the 1997 Directive was not changed. In 2009, EU further replaced the 2001 Directive. The 2009 Directive set an overall EU target of at least 20% of its total energy needs with renewables by 2020. To achieve the overall 20% EU targets, individual EU member states set their national renewable energy targets from 10% in Malta to 49% in Sweden. Each of the EU member countries also disclosed its action plans to meet the national targets.

While the EU set a renewable energy target for each member state and the EU as a whole, member states introduced different policies to promote renewable energy, including solar energy. In fact, several EU member states started their policies before the EU set the targets. While it is not feasible to discuss different policies in each member state, we briefly present here the evolution of the most common policy instrument, feed-in-tariff (FiT) or similar market instrument, in the EU countries which led the promotion of solar energy globally (e.g. Germany and Spain).

Germany introduced the FiT through the 1990 Electricity Feed Law (Lauber and Mez 2004). The law required electric utilities to provide grid access to renewable electricity producers and to purchase electricity generated from renewable sources at rates of 65–90% of the average tariff (rates for solar and wind were set at 90% of the average tariff).[6] The 1990 law was the Renewable Energy Sources Act in 2000, which specifically differentiated FiT by type of renewable energy technology. The FiT for solar energy in Germany varied across project size, type (standalone versus mounted in buildings; Brown 2013).[7] Various European countries followed Germany to introduce either the FiT or similar fiscal incentives to promote renewable energy, including solar. They included Switzerland in 1991, Italy in 1992, Denmark in 1993, Spain and Greece in 1994, Sweden in 1998, Portugal, Norway and Slovenia in 1999, France and Latvia in 2001, Austria and the Czech Republic in 2002, Cyprus, Estonia, Hungary and Slovakia in 2003, and Ireland in 2004. By 2010, all EU member states except Belgium, Malta, the Netherlands and Romania had introduced the feed-in policy (REN21 2011). These four countries used renewable tradable certificates instead of feed-in policies.

Feed-in policy was the major driver for the expansion of solar energy in Germany, Italy Spain and the Czech Republic. For example, Germany installed more PV in 2010 than the entire world did the previous year; in the Czech Republic, installation of PV increased by more than 30-fold in three years (from 65 MW in 2008 to 2000 MW by the end of 2010); France more than tripled its new PV additions in 2010 compared with the previous year; and Greece more than quadrupled its PV installation in 2010 compared with the previous year (REN21 2011). Spain was the global leader on CSP installation before 2010 owing to one of the most generous FiT schemes in the world. Italy introduced a special FiT scheme for PV, known as the Conto Energia, in 2005 which helped solar capacity to expand by more than 1000-fold in the next 7 years (REN21 2012).

Massive changes have occurred in solar energy policies in Europe since 2012. The rapid expansion of solar energy and the policies supporting this growth attracted consumer criticism. Consumers were concerned over the rising bills for the electricity owing to the higher penetration of costly solar energy. At the same time, technological innovation and market development also caused the cost of solar technologies to decrease. As a result, governments started to revise the financial incentives provided to solar energy. For example, the Czech Republic passed new legislation in 2010 to slow the rate of PV installations, which had increased more than 30-fold in three years (from 65 MW in 2008 to 2000 MW by the end of 2010) and influenced (i.e. increased) the average price of electricity in the country. Similarly, Italy announced in 2011 that it would cut its FiTs for solar PV by up to 45% by 2013. Spain completely suspended the FiT and market premium incentives for new renewables as of January 2012 (Brown 2013). Bulgaria put an annual cap on new projects receiving the FIT prices by applying a quota in 2011 (REN21 2012).

China

China has become the world leader in annual PV capacity additions over the last few years (REN21 2015). It also supplies more than two-thirds of global solar PV modules (International Renewable Energy Agency 2014). China's significant renewable energy policy surfaced only after 2005. In 2006, China enacted the Renewable Energy Law, which required electric utilities to pay full price for electricity generated by renewable energy sources while offering consumers of renewables-generated electricity discounted rates. In 2009, the law was amended to require utilities to purchase all renewable power generated in China. In 2007, the National Development Research Council released the Medium and Long-Term Development Plan for Renewable Energy, which required power companies with installed capacity of over 5 GW to have 3% of non-hydro renewable by 2010 and 8% by 2020. These plans set 10% renewable targets in China's total energy supply by 2010 and 15% by 2020. During the financial crisis in 2008, China provided a US$586 billion Stimulus Package, allocating a major portion to renewable energy industries (Howell et al. 2010). This support caused China to become the leader of manufacturing and export of solar PV modules. In 2009, China also introduced the FiT. In the same year, the Golden Sun Demonstration Program was launched to subsidize 50% of the investment cost for grid-connected solar power systems (Worldwatch Institute 2010). In November 2014, Chinese President Xi Jinping, in a joint announcement with US President Barack Obama, declared that China would reach a peak in its CO_2 emissions by 2030 along with an increase in the share of non-fossil fuels to 20% of its total primary energy consumption. These policies led China to increase solar power by more than 400-fold in the 10 years between 2005 and 2015.

India

In India, policies and strategies to promote renewable energy started with the formation of a government department for non-conventional energy resources in 1982, which was transformed to the Ministry of New and Renewable Energy Resources in 1992 (Ministry

Table 25.2   *Policy milestones to promote solar energy in countries with higher solar capacity installed*

Time period	United States	Europe	India	China
Before 1980	● Public Utilities Regulatory Policy Act in 1978 that allows electricity utilities to purchase electricity from third-party suppliers ● 1978 Energy Tax Act to provide personal and corporate tax credits for renewables ● The Solar Photovoltaic Energy Research, Development, and Demonstration Act of 1978 (H.R. 12874)			
1980–1990	● Federal R&D for renewable stood at US$1.3 billion in 1980 ● 1980 Windfall Profits Tax Act gave tax credits for alternative fuels		Establishment of a government department for non-conventional energy resources in 1982	
1990–2000	● Federal production tax credit took effect as part of the Energy Policy Act of 1992 ● Renewables portfolio standards (RPS) were introduced in four states (Massachusetts 9% by 2009 and +1%/year thereafter; Connecticut 10% by 2010; Wisconsin 2.2% by 2011; and Maine 30% ongoing) ● Public benefit funds (PBF) were established in 14 states through a small surcharge to promote RE ● Net metering policies were enacted in 33 states starting from 1996; these policies allowed individual homes and businesses to sell	EU's 1997 Directive on renewable energy to increase its share of total electricity generation mix by 12% by 2010 Feed-in policies Germany, Denmark, Greece, Italy, Spain and Switzerland	A separate ministry – Ministry of New and Renewable Energy Resources was created in 1992	

Table 25.2   (continued)

Time period	United States	Europe	India	China
	their surplus electricity generation produced from renewable sources, particularly solar PVs			
2000–2010	New RPS in 26 states and DC as well as expansions of the RPS in four states were introduced (Massachusetts 9% by 2009 and +1%/year thereafter; Connecticut 10% by 2010; Wisconsin 2.2% by 2011; and Maine 30% ongoing)	2009 EU Directive to supply 20% of final energy demand through renewable sources by 2020. Individual EU member countries' national renewable energy targets and corresponding action plans to meet the national targets		Renewable Energy Law enacted in 2006 requiring utilities to pay full price for renewable electricity while offering it to consumers at discounted rates The Medium- and Long-Term Development Plan of 2007 set renewable energy targets (10% by 2010 and 15% by 2020) In 2009 FiT and the Golden Sun Demonstration Program (GSDP) were introduced. The GSDP provided up to 50% investment subsidy for grid-connected solar power systems
After 2010		Substantial modification in feed-in policy including reduction or elimination of feed-in-tariff	In 2011, a renewable energy target of 18% of the total installed electricity generation capacity by 2022 was set, of which solar power would supply 20,000 MW through grid and another 2000 MW through off-grid The solar target was increased to 100 GW in 2015	Chinese President declared in 2014 that China would peak its CO_2 emissions by 2030 and increase the share of non-fossil fuels to 20% of its total primary energy consumption

Table 25.2   (continued)

Time period	United States	Europe	India	China
			Jawaharlal Nehru National Solar Mission was launched to help accomplish the solar energy target and 5% of the 100 GW target was met by the end of 2015	

Source: Ministry of New and Renewable Energy Resources (2011); REN21 (n.d.).

of New and Renewable Energy Resources 2011). The Ministry set a target of 20,000 MW grid-connected solar power, 2000 MW of off-grid solar power and 20 million square meters of solar water heating system by 2022. Within the off-grid component, there was a separate target of providing solar lights for 20 million rural households in 1000 villages by 2022 (Ministry of New and Renewable Energy Resources 2011). The Ministry also set a target of installing at least 1000 solar cooking systems by 2022. In 2015, the new government increased the solar target to 100,000 MW by 2022. However, these targets were aspirational and realization of even the first target of 20,000 MW was challenging.

The government launched a national mission named after the country's first Prime Minister, Jawaharlal Nehru (Jawaharlal Nehru National Solar Mission) to accomplish the targets. The mission has three phases: Phase 1, 2011–2013; Phase 2, 2013–2017; and Phase 3, 2017–2022. The first phase aimed to add 1000 MW through bundling with thermal power supplied by the state-owned national thermal power corporation and 100 MW was to be implemented through financial support of IREDA, a financing wing under the Ministry of New and Renewable Sources. At the end of the first phase, bidding of 950 MW was completed. The bidding price varied between Rs.10.49 and Rs.12.76 per kW h (Ministry of New and Renewable Energy Resources 2013). As of November 2015, 4681 MW (or 4.7% of the 100 GW by 2022 target) has been commissioned (Ministry of New and Renewable Energy Resources 2015).

Other Countries

Besides the major developers of solar energy (United States, EU, China, India), many other countries have implemented policies to promote solar energy along with other renewable sources for energy. Other developed countries besides the United States and EU (e.g. Australia, Canada, Japan, New Zealand) introduced renewable energy policies in the 1990s. By 2000, there were only a few developing countries with policies to support renewable energy. Eight more developing countries besides China and India (Brazil, the Dominican Republic, Egypt, Malaysia, Mali, the Philippines, South Africa and Thailand) had introduced renewable energy policies by 2005 (REN21 2005). Another 23 developing countries had introduced renewable policies by 2010 (REN21 2011). As of

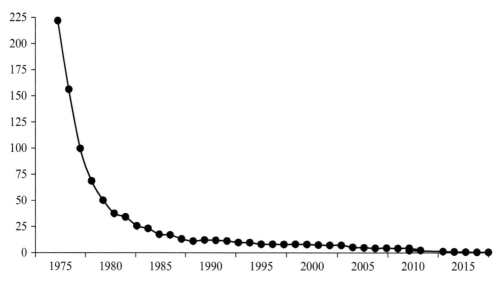

Figure 25.7   Trend of the average price of solar PV module (2010 US$/W)

early 2015, 164 countries had set renewable energy targets and 145 countries had introduced renewable energy supporting policies (REN21 2015). While feed-in incentives have
been overhauled or reduced in many developed countries, especially in the EU member
states, they are being increasingly introduced in the developing countries.

SOLAR ENERGY COST TRENDS

Solar energy technologies have demonstrated a dramatic decline over the last few decades.
The average price of a PV module was reported to be US$76.67/W in 1977.[8] It dropped
more than 100-fold by 2014 to approximately US$0.5/W.[9] In real terms, the price dropped
by 475 times over the same period.[10] Figure 25.7 presents the trend of the average price of
solar PV expressed in the 2010 constant price.

One of the main factors causing the rapid drop of solar technology costs over the last
few decades is market development, although the market was mainly driven by government policies. The development of the market can be portrayed by the total installed
capacity of solar power generation. Figure 25.8 shows that the average price of PV
modules dropped by 22% for every doubling of installed capacity.

Costs of Solar Thermal Technologies

Development of solar thermal technologies for electricity generation is more recent.
Therefore, it does not have a long history of cost reduction. Figure 25.9 presents the

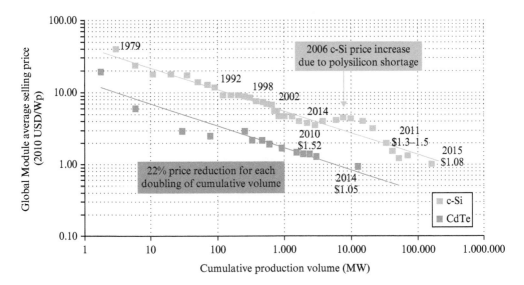

Source: International Renewable Energy Agency (2012a).

Figure 25.8   *Average PV module price along with installed capacity*

trend of unit electricity generation from two types of concentrated solar power technologies, trough and tower technologies in the United States. Unlike solar PV technologies, the cost trend of CSP is not very clear although it is decreasing. The unit cost of
electricity generation from trough technologies has decreased from US$0.65 per kW h
in 1980s to US$0.35/kW h in 2005 and is expected to further drop to US$0.2/kW h in
2015 (International Renewable Energy Agency 2012b). The cost of tower technology also
shows a similar trend of cost reduction over the same time horizon (1980–2015).

CONCLUSIONS AND FINAL REMARKS

Solar energy is the most common and abundant source of energy for humanity. However,
it was not utilized to supply basic energy services (e.g. light, heat and motive power) until
recently because of its high costs and lack of proper technologies to store and convert to
useful energy services. Over the last two decades, solar energy has experienced unprecedented advances in terms of technological innovation and its deployment to produce
major energy services/commodities (i.e. heat and electricity). Global capacity of energy
generation using solar heat and radiation expanded by a 100-fold over two decades.
Technological innovations have caused the costs of solar energy production to drop by
1000% in four decades. No other source of energy exhibited this type of development
momentum in a very short period of time.

   The growth of solar and also other renewable technologies is driven mainly by government policies. Almost every country in the world is interested in developing solar energy
and all major economies around the world have not only introduced policy instruments

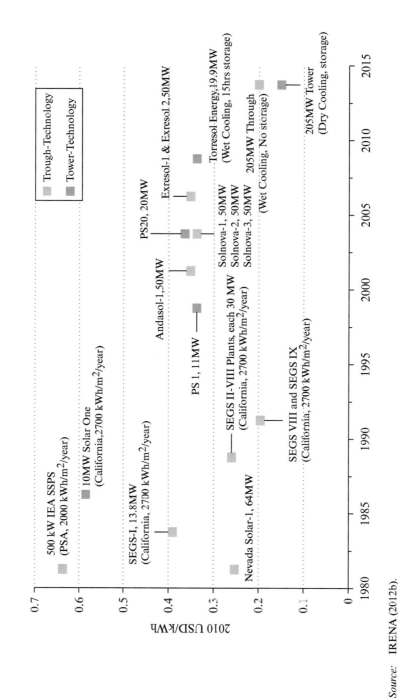

Source: IRENA (2012b).

Figure 25.9   Unit costs of electricity generation from solar thermal technologies

to stimulate the expansion of solar energy but are also sustaining its use for a long time. However, solar energy development is still facing challenges because it is still not favored economically compared with conventional sources of energy. Intermittency is another big hurdle for solar energy to compete with conventional sources. For a sustained future of solar energy, the costs of technologies need to be cut further and innovations are required in storage technologies so that the limitations caused by intermittency can be resolved.

NOTES

1. For example, in 1929 American space engineer R.H. Goddard designed a solar power plant, using a large parabolic mirror focusing solar energy on a steam turbine housed at the focal point (Hagemann 1962). In 1935, another American scientist, Charles G. Abbot used a parabolic trough collector to convert solar energy to mechanical power (Moya 2012).
2. The MENA region includes Israel, Jordan, Lebanon, Morocco, Palestinian Territories and Tunisia (Mauthner et al. 2015).
3. The cost of power generation the utility avoids from its system if it purchased from a third party. Normally, this cost is smaller than the purchasing cost of renewable energy. The renewable energy supplying the third party is then compensated through the payment of the gap (i.e. difference between cost of renewable supplied by the third party and the avoided cost of the utility) by the government.
4. More than half of this capacity (6,100 MW) was installed in California alone (Martinot et al. 2005).
5. The cumulative International Energy Agency government funding on renewable energy R&D during 1977–1989 was estimated to be US$11 billon (1989 price), which was 7% of the total R&D budget (Laughton 1990).
6. Owing to the high cost, solar energy, however, could not benefit much from the 1990 Electricity Feed Law. Rather municipal government direct financial support, for example that started in Aachen (also known as the Aachen model), Länder market introduction programmes, was more helpful in promoting PV programs in Germany in the early 1990s (Lauber and Mez 2004).
7. In 2012 Germany replaced FiT with a market premium scheme, which was designed so that the premium would decline over time based on both pre-determined and market responsive schedules.
8. http://cleantechnica.com/2014/09/04/solar-panel-cost-trends-10-charts/.
9. PV Magazine Global, accessed 25 January from http://www.pv-magazine.com/investors/module-price-index/#axzz3yGCDTvxD.
10. The real costs of solar PV, when expressed in 2010 constant price (converted using US GDP deflator) would be US$222/W and US$0.47/W in 1977 and 2014, respectively.

REFERENCES

1001 Home Efficiency Tips. n.d. *A Brief History of Solar Energy*, accessed 26 December 2015 from http://www.1001-home-efficiency-tips.com/solar_energy_history1.html.

Alexopoulos, S. and B. Hoffschmidt. 2012. Concentrating receiver systems (solar power tower). In C. Richter, D. Lincot and C. Gueymard (eds), *Solar Energy*. pp. 29–71. New York: Springer.

Arenas, J. 2007. Design, development and testing of a portable parabolic solar kitchen. *Renewable Energy* 32: 257–266.

Blazev, A. 2012. *Photovoltaics for Commercial and Utilities Power Generation*. Lilburn, GA: Fairmont Press.

Brower, M. 1992. *Cool Energy: Renewable Solutions to Environmental Problems*. Cambridge, MA: MIT Press.

Brown, P. 2013. *European Union Wind and Solar Electricity Policies: Overview and Considerations*. CRS Report: R43176. Washington, DC: Congressional Research Service (CRS), accessed 28 January 2016 from https://www.fas.org/sgp/crs/row/R43176.pdf.

Byrne, J. and L. Kurdgelashvili. 2011. The role of policy in PV industry growth: past, present and future. In A. Luque and S. Hegedus (eds), *Handbook of Photovoltaic Science and Engineering*, 2nd edn. pp. 39–81. New York: Wiley.

Cheremisinoff, P. and T. Regino. 1978. *Principles and Applications of Solar Energy*. Ann Arbor, MI: Ann Arbor Science.

CSP Today. 2012. *The History of American CSP: A Timeline*, accessed from http://www.csptoday.com/usa/pdf/USAtimeline.pdf.

Cuce, E. and P. Mert Cuce. 2013. A comprehensive review on solar cookers. *Applied Energy* 102: 1399–1421.

Delyannis, E. 2003. Historic background of desalination and renewable energies. *Solar Energy* 75: 357–366.

Erge, T., V. Hoffmann and K. Kiefer. 2001. The German experience with grid-connected PV-systems. *Solar Energy* 70: 479–487.

FAFCO. 2009. *FAFCO Celebrates 40-Year Anniversary in the Solar Water Heating Industry*, 28 October, accessed 27 December 2015 from http://www.fafco.com/news/default.aspx.

First Solar. 2015. *Projects: Proven Results*, accessed 25 January 2016 from http://www.firstsolar.com/en/About-Us/Projects.aspx.

Fortuin, S. and G. Stryi-Hipp. 2012. Solar collectors, non-concentrating. In C. Richter, D. Lincot and C. Gueymard (eds), *Solar Energy*. pp. 378–398. New York: Springer.

Garg, H. 1987. *Advances in Solar Energy Technology. Vol. 1, Collection and Storage Systems*. Berlin: Springer.

Haberle, A. 2012. Linear fresnel collectors. In C. Richter, D. Lincot and C. Gueymard (eds), *Solar Energy*. pp. 72–78. New York: Springer.

Hagemann, E. 1962. R. H. Goddard and solar power 1924–1934. *Solar Energy* 6: 47–54.

Howell, T., W. Noellert, G. Hume and A. Wolf. 2010. *China's Promotion of the Renewable Electric Power Equipment Industry: Hydro, Wind, Solar, Biomass*. Washington, DC: National Foreign Trade Council, accessed 28 January 2016 from http://www.nftc.org/default/Press%20Release/2010/China%20Renewable%20Energy.pdf.

Hren, S. and R. Hren. 2010. *A Solar Buyer's Guide for the Home and Office: Navigating the Maze of Solar Options, Incentives and Installers*. White River Junction, VT: Chelsea Green.

International Energy Agency. 2002. *Trends in Photovoltaic Applications in Selected IEA Countries between 1992 and 2001*. IEA-PVPS.

International Energy Agency. 2008. *Trends in Photovoltaic Applications in Selected IEA Countries between 1992 and 2007*. IEA-PVPS.

International Energy Agency. 2012. *Trends in Photovoltaic Applications in Selected IEA Countries between 1992 and 2011*. IEA-PVPS.

International Energy Agency. 2015. *Trends in Photovoltaic Applications in Selected IEA Countries between 1992 and 2014*. IEA-PVPS.

International Renewable Energy Agency. 2012a. *Renewable Energy Technologies: Cost Analysis Series*. Volume 1 Power Sector, Issue 4/5 Solar Photovoltaic. Abu Dhabi: IRENA.

International Renewable Energy Agency. 2012b. *Renewable Energy Technologies: Cost Analysis Series*. Volume 1 Power Sector, Issue 2/5 Concentrating Solar Power. Abu Dhabi: IRENA.

International Renewable Energy Agency. 2014. *Renewable Energy Prospects: China, REmap 2030 Analysis*. Abu Dhabi: IRENA, accessed 28 January 2016 from http://irena.org/remap/IRENA_REmap_China_report_2014.pdf.

Jones, G. and L. Bouamane. 2012. *Power from Sunshine: A Business History of Solar Energy*. Working Paper: 12–105. Cambridge, MA: Harvard Business School.

Kalogirou, S. 2014. *Solar Energy Engineering Processes and Systems*, 2nd edn. Boston, MA: Elsevier.

Lamont, L. 2012. History of photovoltaics. *Comprehensive Renewable Energy* 1: 31–45.

Lauber, V. and L. Mez. 2004. Three decades of renewable electricity policies in Germany. *Energy & Environment* 15(4): 599–623.

Laughton, M. 1990. *Renewable Energy Sources*. Published on behalf of the Watt Committee on Energy: Report Number 22. New York: Elsevier Applied Science.

Martinot, E., R. Wiser and J. Hamrin. 2005. Renewable energy markets and policies in the United States. *Renewable Energy Information on Markets, Policy, Investment and Future Pathways*, accessed 28 January 2016 from http://www.martinot.info/Martinot_et_al_CRS.pdf.

Mauthner, F., W. Weiss and M. Spörk-Dür. 2015. *Solar Heat Worldwide: Markets and Contribution to the Energy Supply 2013*. IEA Solar Heating & Cooling Programme, accessed from http://www.iea-shc.org/Data/Sites/1/publications/Solar-Heat-Worldwide-2015.pdf.

Mills, D. 2004. Advances in solar thermal electricity technology. *Solar Energy* 76: 19–31.

Ministry of New and Renewable Energy Resources. 2011. *Strategic Plan for New and Renewable Energy Sector for the Period 2011–17*. New Delhi: Government of India.

Ministry of New and Renewable Energy Resources. 2013. *Further Revised Draft Guidelines for Selection of 750 MW Grid Connected Solar Power Projects Under Batch-I of JNNSM Phase-II*. New Delhi: Government of India, accessed 28 January 2016 from http://mnre.gov.in/file-manager/UserFiles/further-revised-VGF_750MW_Guidelines_for-grid-solar-power-projects.pdf.

Ministry of New and Renewable Energy Resources. 2015. *Visibility Plan for Status of Implementation of*

Various Schemes to Achieve 1,00,000 MW Solar Plan. New Delhi: Government of India, accessed 28 January 2016 from http://mnre.gov.in/file-manager/UserFiles/GW-Solar-Plan.pdf.

Morrison, G. 2013. Solar water heating in Australia, accessed 28 December 2015 from http://users.tpg.com.au/t_design/papers/Morrison%20AuSES2013.pdf.

Moya, E. 2012. Parabolic-trough concentrating solar power (CSP) systems. In K. Lovegrove and W. Stein (eds), *Concentrating Solar Power Technology: Principles, Developments and Applications*. pp. 197–239. Oxford: Woodhead.

Nahar, N. 2001. Design, development and testing of a double reflector hot box solar cooker with a transparent insulation material. *Renewable Energy* 23: 167–179.

National Renewable Energy Laboratory. 2012. NREL scientists spurred the success of multijunction solar cells, accessed 5 January 2016 from http://www.nrel.gov/docs/fy12osti/53604.pdf.

National Renewable Energy Laboratory. 2015. *Concentrating Solar Power Projects*, accessed 31 December 2015 from http://www.nrel.gov/csp/solarpaces/.

Perlin, J. 2004. Solar energy, history of. In C. Cleveland (ed.), *Encyclopedia of Energy*, Vol. 5, pp. 607–622. Boston, MA: Elsevier.

Plante, R. 2014. *Solar Energy – Photovoltaics and Domestic Hot Water: A Technical and Economic Guide for Project Planners, Builders and Property Owners*. San Diego, CA: Academic Press.

ProCon.Org. 2013. *Explore Pros & Cons of Controversial Issues: Historical Timeline: History of Alternative Energy and Fossil Fuels*, accessed 28 January 2016 from http://alternativeenergy.procon.org/view.timeline.php?timelineID=000015.

Pytlinski, J. 1978. Solar energy installations for pumping irrigation water. *Solar Energy* 21: 255–262.

Qiu, S., M. Ruth and S. Ghosh. 2015. Evacuated tube collectors: a notable driver behind the solar water heater industry in China. *Renewable and Sustainable Energy Reviews* 47: 580–588.

REN21. n.d. *Renewable Energy Policy Network for 21st Century*, accessed 28 January 2016 from http://www.ren21.net/.

REN21. 2005. *Renewables 2005: Global Status Report*. Renewable Energy Policy Network for the 21st Century.

REN21. 2011. *Renewables 2011: Global Status Report*. Renewable Energy Policy Network for the 21st Century.

REN21. 2012. *Renewables 2012: Global Status Report*. Renewable Energy Policy Network for the 21st Century.

REN21. 2015. *Renewables 2015: Global Status Report*. Renewable Energy Policy Network for the 21st Century.

Robles, A., V. Duong, A. Martin, J. Guadarrama and G. Diaz. 2014. Aluminum minichannel solar water heater performance under year-round weather conditions. *Solar Energy* 110: 356–364.

Saxena, A., S. Pandey and G. Srivastav. 2011. A thermodynamic review on solar box type cookers. *Renewable and Sustainable Energy Reviews* 15: 3301–3318.

Schaeffer, J. 2005. *Solar Living Source Book: 30th Anniversary Edition*. Hopland, CA: Gaiam Real Goods.

Seddegh, S., X. Wang, A. Henderson and Z. Xing. 2015. Solar domestic hotwater systems using latent heat energy storage medium: a review. *Renewable and Sustainable Energy Reviews* 49: 517–533.

Shukla, A., D. Buddhi and R. Sawhney. 2009. Solar water heaters with phase change material thermal energy storage medium: a review. *Renewable and Sustainable Energy Reviews* 13: 2119–2125.

Smyth, M., P. Eames and B. Norton. 2006. Integrated collector storage solar water heaters. *Renewable and Sustainable Energy Reviews* 10: 503–538.

Spencer, L. 1989. A comprehensive review of small solar-powered heat engines: part I. A history of solar-powered devices up to 1950. *Solar Energy* 43: 191–196.

University of Delaware. 2015. *IEC History*, accessed 4 January 2016 from http://www.udel.edu/iec/history.html.

US Department of Energy. 1991. *Photovoltaics Program Plan: FY 1991–FY 1995*. Golden, CO: Department of Energy.

US Department of Energy. 2000. *Photovoltaic Energy Program Overview: Fiscal Year 1999*. Washington, DC: DOE/GO-102000-0963.

US Energy Information Administration. 2012. *Annual Energy Review 2011*. Washington, DC: US Government Printing Office.

Weiss, W., I. Bergmann and G. Faninge. 2005. *Solar Heat Worldwide: Markets and Contribution to the Energy Supply 2003*. IEA Solar Heating & Cooling Programme, accessed from http://www.iea-shc.org/data/sites/1/publications/Solar_Heat_Worldwide-2005.pdf.

Worldwatch Institute. 2010. *Renewable Energy and Energy Efficiency in China: Current Status and Prospects for 2020*. Report 182. Washington, DC: Worldwatch, accessed 28 January 2016 from http://www.worldwatch.org/system/files/182%20China%20Energy.pdf.

Yergin, D. 2011. *The Quest: Energy, Security, and the Remaking of the Modern World*. New York: Penguin Press.

Zhiqiang, Y. 2005. Development of solar thermal systems in China. *Solar Energy Materials and Solar Cells* 86: 427–442.

PART VI

MANUFACTURING
TECHNOLOGIES

26. Just-in-time and space
Ruth Rama and Adelheid Holl

Just-in-time (JIT) is a manufacturing and inventory management technique originally developed by the Toyota Motor Corporation in the 1950s. Initially meant to produce goods for consumer demand in terms of time, volume and quality, the JIT concept currently also implies production with the minimum waste of time and resources. According to a former Toyota executive, JIT is a complex method that primarily includes a multi-skilled workforce[1] whose members can each be allocated to various types of jobs (Kaneko and Nojiri 2008). Secondly, Total Quality Control is guaranteed by the full participation of the entire workforce. Third, JIT implies the elimination of inventories and the efficient utilization of equipment and labor; both measures are meant to shorten the production process, speed capital turnover and reduce expenditures. Although the management literature points to many other practices encompassed by JIT, consensus has yet to be reached as to what exactly constitutes JIT practices (for a list of practices proposed in the management literature, see Mackelprang and Nair 2010). The economic geography literature uses terms such as 'true JIT' and 'full JIT' to signal whether certain practices have been adopted; for instance, multiple deliveries per day would constitute a 'true JIT' system according to certain interpretations (Egeraat and Jacobson 2005). Kaneko and Nojiri (2008) argue that the efficient implementation of JIT also requires the use of an electronic information network. The latter is crucial since parts are manufactured on the reception of orders, and not in advance. The order, or *kanban* (in the Japanese terminology), specifies the types of parts needed by the client, their reference numbers and quantities, the deadline for delivery and the destination of output (Kaneko and Nojiri 2008).[2]

As described by certain authors, rather than hardware technology, JIT is a 'soft' technology comprising procedures, processes, know-how and skill (Swamidass and Winch 2002). However, the adoption of JIT is often associated to the use of advanced manufacturing technology, such as robotics, computer numerically controlled machines, computer-aided design and manufacturing, and automated materials handling systems (Bergman et al. 1999). The employment of this type of equipment enables firms to produce customized items and to switch inexpensively and rapidly from one task to another.

Just-in-time has been known under, or is related to, several terms, such as Toyota Production System, World Class Manufacturing and Lean Production (LP).[3] Lean Production, for instance, is a concept coined by researchers of the Massachusetts Institute of Technology for a philosophy of production that emphasizes the minimization of resources used in manufacturing activities (Bhamu and Singh Sangwan 2014). Mackelprang and Nair (2010) opine that the relationship between JIT and LP has not been clearly defined in the extant literature. They observe that, for numerous authors, JIT and LP are highly similar, while for others, LP encompasses more than does JIT. Nevertheless, Mackelprang and Nair (2010) concede that JIT is at least an important component of LP. The term *lean logistic* refers, specifically, to logistic systems that monitor the geographic location of raw materials, parts and finished products with the lowest costs

(Moyano-Fuentes and Sacristán-Díaz 2012). Other authors use the term 'time-based competition' which, in their view, encompasses JIT (Egeraat and Jacobson 2005).

In the West, the interest drawn by JIT in academia and in industry was fueled by the successful performance of Japanese car makers, as compared with the less successful performance of their Western rivals. The JIT concept was introduced to the US audience at the beginning of the 1980s (Bhamu and Singh Sangwan 2014). Egeraat and Jacobson (2005) report that studies published in the 1990s viewed the Fordist mass-production system as an 'old' system that would soon be replaced. In this context, JIT was viewed as an alternative to the traditional Fordist (or Taylorist) system. Traditional production systems are associated to the utilization of dedicated one-process manufacturing equipment and to mass production channeled to large homogeneous markets for standardized items. It was at that time that Milgrom and Roberts (1990, 511) wrote 'the mass production model is being replaced by a vision of flexible multiproduct firm that emphasizes quality and speedy response to market conditions while utilizing technologically advanced equipment and new forms of organization'.

Since the 1980s, JIT has been implemented by many non-Japanese firms, and although it was initially developed in the automobile industry, it has also become increasingly adopted in other sectors, such as the electronics industry (McCann and Fingleton 1996), the textile and apparel industry (Abernathy et al. 1999; Bruce et al. 2004) and the food sector (Bourlakis and Bourlakis 2004). As pointed out later, there is, however, a lack of quantitative studies on the degree of JIT adoption in various countries and sectors that would enable a precise assessment of the geographic diffusion of JIT. Nevertheless, as argued in Amosse and Coutrot (2011), the rise of the 'Toyotist model' has been paralleled with a resistance of traditional production systems. Those authors do not support the view of a generalized implementation of JIT.

Most of the analyses of JIT originate in various strands of the management literature, such as that on operations management, engineering management, purchasing management and logistics management. These studies have mainly looked at the relationship between JIT adoption and firm performance, and supply chain management issues. However, these strands of research work have displayed few inroads into spatial aspects of JIT (Moyano-Fuentes and Sacristán-Díaz 2012). Nevertheless, a meta-analytic investigation of quantitative management studies observed that the geographic location of the industrial plant may, positively or negatively, influence the effects of the adoption of JIT on the performance of the firm (Mackelprang and Nair 2010). Even in economic geography and regional studies, relatively few analyses have been devoted to spatial aspects of JIT, of which most are case studies, often based on small samples of companies; quantitative studies remain relatively rare.

MAIN ISSUES

This section discusses several of the main issues concerning the geography of JIT that have arisen in the literature.

Is JIT Applicable in Countries other than Japan?

Studies published in the 1980s and 1990s intensely debated the applicability of JIT in the West (Mair 1992; Moyano-Fuentes and Sacristán-Díaz 2012; Rutherford 2000). One reason for this debate was that the achievements of this system seemed strongly dependent on specific features of Japanese culture, values and work organization. An early study cited, for instance, the lifetime employment system, work as the center of individuals' lives, and the opportunities of promotion to managerial responsibilities offered to blue-collar workers in Japan (Sugimori et al. 1977). There was a view that differences in the use of technology between countries could be culture-induced (Swamidass and Winch 2002). Therefore, one of the first issues inquired by Western researchers was whether JIT could be successfully implemented in countries other than Japan. Furthermore, the international expansion of Japanese multinational corporations stimulated the debate about the possible transfer or re-embedding of the JIT system abroad (Rutherford 2000). Certain authors believed that the adoption of practices, such as autonomous teams, multi-skilling and employee incentives, would guarantee, on their own, a better performance of any firm whatever its location, while others reasoned that 'optimal practices *per se* do not exist independently of a firm's environment context and strategic choices' (Amosse and Coutrot 2011, 787). The literature now provides evidence of the worldwide geographic diffusion of JIT. This system has been adopted all over the world, even in emerging economies. Multinational corporations, especially the foreign subsidiaries of Japanese firms, have acted as the major vehicles of its international diffusion (Florida and Kenney 1992; Ó Huallacháin and Wasserman 1999; Rutherford 2000).

Is the Degree of JIT Adoption Homogeneous across Countries?

The empirical literature has corroborated the view that JIT can be transplanted to other countries. However, as shown below, it also suggests that JIT international diffusion is uneven across countries, since it depends on national specificities, such as physical geography, culture and cost structure. After reviewing the management literature, Moyano-Fuentes and Sacristán-Díaz (2012) note that still very little is known about the *degree* to which JIT is adopted in different countries.

One of the main limitations in the literature is that there is a general lack of comparable statistical data. In general, very few quantitative studies on the degree of JIT adoption at country level have been published (Bergman et al. 1999). Most analysis in the field of JIT is based on case studies. Those surveys regarding JIT that have been carried out have often targeted small samples and specifically firms that have implemented JIT or sectors and industries that are more likely to use JIT. This makes it difficult to derive information on the general degree of JIT implementation in industries and countries.

Gale (1999) studies JIT adoption among US manufacturing firms. In that study, 48% of surveyed establishments reported using JIT and about 30% used JIT and supplied to a JIT company at the same time. Approximately one-third of establishments neither used JIT in their own plant nor supplied to a customer that used JIT. Waterson et al. (1999) provide survey data for large UK manufacturing firms. In their survey, about 40% reported using JIT entirely or a lot, while another 40% reported not using JIT at all or only very little. Clegg et al. (2002) extended the research by Waterson et al. (1999) and surveyed large companies in the UK, Australia, Japan and Switzerland using a common questionnaire. Except for Switzerland, they reported a similar uptake of JIT, with around 40% reporting that they used JIT entirely or a lot. In Switzerland, the reported uptake of JIT among the surveyed firms was considerably higher, with about 60% stating that they used JIT entirely or a lot. Swamidass and Winch (2002) conducted surveys of manufacturers in the UK and the United States by again using a common questionnaire in order to compare the degree of adoption of JIT and other technologies in these two countries. They found that British manufacturers were more prone to use JIT than US manufacturers. They reasoned that, given the cultural proximity of the two countries, this difference could not be culture-induced and, instead, proposed that the greater degree of adoption in the UK may be explained by the shorter physical distances, which facilitate frequent deliveries of inputs and components, and by the heavy inflow of Japanese investment.

Moreira and Alves (2008) surveyed electronic, metal parts and paint manufacturers in Portugal. They state that one-third of the surveyed firms reported using JIT. This figure is similar to that reported in Holl et al. (2013) for medium-sized and large Spanish manufacturing firms. However, Moreira and Alves (2008) also stress that not all firms that report using JIT have actually implemented consistent JIT practices on a plant-wide basis and many only fulfill certain conditions of JIT or have implemented only limited JIT practices. This raises important issues for any comparison of JIT implementation. Without a clear consensus on what practices have to be fulfilled in order to be regarded as a JIT system, any comparative study will face important difficulties that cannot be fully remedied by using Likert scale measures regarding the amount of JIT use in production and delivery activities.

Surveys conducted in Argentina, Brazil and Romania to study the predisposition of firms towards the implementation of JIT with some guarantee of success (e.g. acceptance of teamwork) suggest that actual levels of adoption may be relatively low in certain emerging economies (Glaser-Segura et al. 2011). The possible exceptions to this low level of adoption within these countries include specific industries controlled by foreign multinational corporations, such as automobiles (Ó Huallacháin and Wasserman 1999).

There are even fewer longitudinal studies that have analyzed the evolution of the spread of JIT at country level and over time. This lack of data makes it difficult to systematically assess the extent of the use of JIT practices and their evolution over time. A recent exception is Amosse and Coutrot (2011). On analyzing a very large panel of data spanning from 1992 to 2004, these authors observed 'a Toyotization of France's productive base' (p. 802). However, they also found that the so-called 'Toyotist model', which features JIT, organizational methods, innovative human resource management, high employee participation and low rates of unionization, co-existed in France with traditional production models. This was notably the case of the 'neo-Taylorist model', featuring low innovation, low investment in human resource management, strict management control and minimal employee participation, and an average trade union presence. We take the above-mentioned research results as symptomatic of the fact that the JIT system has not replaced traditional production systems, as proposed by early studies, but rather has worked side by side with them, even in advanced countries.

Is the Degree of JIT Adoption Homogeneous across Regions?

As documented below, the literature suggests that, within a country, the degree of JIT adoption may vary across regions. According to Amosse and Coutrot (2011), certain French industries (e.g. automobile industries) tend to adopt JIT while others (e.g. production of intermediary goods) tend to adopt traditional production systems. Although these authors do not deal with the regional aspects of the question, we infer from their analysis that the industrial structure of regions may contribute towards explaining the uneven geographic spread of JIT across localities. As outlined by Mackelprang and Nair (2010), geographic location may influence the successful implementation of JIT. Different spatial conditions, as well as differences in industry-relationship climate and strategies of firms, may also promote an uneven spread of JIT practices within countries. As shown by a comparison between two large Japanese companies, Toyota and Nissan, the various spatial conditions faced by different factories may affect the degree of JIT adoption, even within the same national industry (Mair 1992). For instance, the unreliability of transportation infrastructures in certain Japanese regions has limited the degree of adoption of the *kanban* system, a typical JIT practice; this circumstance obliged factories located there to maintain inventories larger than usually recommended by the JIT philosophy. Holl et al. (2013) study the spatial pattern of JIT adoption for a sample of medium-sized and large Spanish manufacturing firms. Their results show that JIT adoption rates are indeed related to transport accessibility. Adoption rates are higher in smaller cities with higher transport accessibility compared with the largest urban areas, thereby indicating that possibly congestion in those areas reduces the benefits that firms may obtain from JIT implementation. Other studies that highlight the importance of the transportation system for the successful implementation of JIT to ensure timely delivery include Smith and Florida (1994) and Klier (1999 and 2000). Gale (1999) studied adoption of JIT among rural and metropolitan US manufacturing firms and observed no significant impact of urbanization on adoption probabilities.

Within a large multi-plant company, the type of manufactured products and the industrial climate may encourage certain individual industrial plants to adopt JIT, while other plants may continue employing traditional production systems (Herod 2000). This circumstance may influence JIT adoption in the different regions in which these plants are located. Several studies show that substantial labor availability and non-unionized workers are regional attractors for JIT industrial plants, both in the home country and abroad; the targeted areas may be rural areas (Herod 2000; Kaneko and Nojiri 2008; Lee 2003; Linge 1991; Mair 1992; Ó Huallacháin and Wasserman 1999). Certain authors believe that workers and trade unions have been insufficiently taken into consideration as determinants of the geography of JIT (Lee 2003; Rutherford and Gertler 2002).

DOES JIT ADOPTION PROMOTE THE CLUSTERING OF FIRMS?

At the beginning of the 1990s, Mair (1992) noted that, in theory, close proximity with its suppliers may help a factory to solve the logistical problems of JIT implementation. The notion of a reduction of transaction costs was at the root of the alleged advantages of proximity. However, with time, the development of information and communications technology and the fall of transportation costs induced a generalized perception of the 'death of distance'. Consequently, proximity between plants was perceived as less important. However, McCann and Sheppard (2003) suggested that this consideration could be true for industries producing raw materials, agricultural products, minerals or standardized manufactured items, but not for other industries. They observed that, in industries in which the demand-lead-times had fallen or in industries that marketed varied or complex items, spatial transaction costs appeared not to have decreased over the previous decades and, in some cases, even to have increased. In their opinion, the problem was that previous researchers had taken into account only the reduction of transportation costs but not other major logistic costs, such as those of inventory holding. When these costs are considered, McCann and Sheppard (2003) reasoned, it can be concluded that the costs of geographic distance in these industries are higher than previously thought. This consideration, they argue, shines a new critical light on the perceived wisdom that distance does not matter for firms that produce items characterized by high value per weight.

In the context of this debate, the empirical literature asked whether JIT adoption was likely to change the spatial structure of an existing industry. As the burden of JIT implementation tends to fall on supplier firms (Helper and Sako 1995) much of the literature has focused on how JIT adoption stimulates the geographical concentration of suppliers. The empirical evidence regarding this question is divided. On the one hand, there are studies that argue that the implementation of JIT has little impact on the geography of an industry. Mair's (1992, 91) study of JIT manufacturing in Japan reaches the conclusion that 'existing factory locations are not quickly changed'. Pallarès-Barberà (1998) studied the Spanish automobile industry and argues that JIT adoption did not cause a substantial change in the regional distribution of the sector. She argues that JIT does not necessarily induce changes in the geography of an existing industry, although 'new' areas may emerge. Sadler (1994) studied the automobile component industry in Western Europe and concluded that there was no clear association between JIT and spatial clustering.

This literature reported that difficulties derived from the physical dispersion of industrial plants have sometimes been solved, in Japan and Europe, by building warehouses in intermediate localities; by setting up truck circuits that collect parts and inputs from suppliers dispersed across a wide area; or by using the services of independent logistic agents (Egeraat and Jacobson 2005; Kaneko and Nojiri 2008; Pallarès-Barberà 1998).

On the other hand, there is also extensive literature that argues that the adoption of JIT has indeed contributed towards alteration of the geography of previously existing national industries. Klier (1999) and Klier and McMillen (2008) find evidence of increased clustering at the regional level (rather than the local level) in the US auto supplier industry ever since the 1980s that coincides with the implementation of JIT production in the industry. Klier and Rubenstein (2013) also argue that JIT has played an important role in the restructuring of the US automobile industry. Ó Huallacháin and Wasserman (1999) provide similar evidence for the automobile components industry in Brazil.

Several frequently cited issues are related to the value, volume, weight and variety of components, and to costs of transportation (Egeraat and Jacobson 2005). The position of suppliers along the value-chain may also determine whether they cluster near their client (Kaneko and Nojiri 2008). According to the aforementioned study, certain

Japanese companies that employ the JIT system prefer geographically close relationships with a small number of first-tier suppliers. In contrast, they purchase parts requiring labor-intensive manufacturing processes from suppliers located in other Asian countries or in Latin America owing to lower manufacturing costs. In other cases, it is not high-value-added components but rather bulky components which are sourced locally even if their unit value is low, the reason being that bulkiness increases inventory costs (Egeraat and Jacobson 2005). Frigant and Zumpe (2014) also stress that JIT does not require an absolute need for proximity, but the issue of proximity rather depends on the character-istics of the component to be transported. Proximity constraints are largest for bulky or fragile components, which exert higher transportation costs. Suppliers delivering such inputs directly to the final assembly plant show a greater tendency for clustering. The same applies for suppliers delivering pre-assembled modules or systems that, increas-ingly, not only deliver just-in-time but also just-in-sequence. However, in the case of components that offer important economies of scale in production and that are shared by several clients, the parts supplier may deliver over greater distances from one single production site to several dispersed assembly plants and optimize flows through sophis-ticated logistics. Component producers that operate further upstream in the value-chain are the least constrained by proximity.

Holl et al. (2010) explore the issue of the alleged propensity of JIT towards the pro-motion of the clustering of firms from a different perspective. They ask whether the use of JIT is more likely to encourage proximity in inter-firm relationships than the use of traditional manufacturing systems. In a sample of Spanish automobile and electronics firms, they test whether contractors that employ JIT display the same geographical pat-terns as contractors that employ a traditional approach to manufacturing. They find that JIT adopters are more prone to prefer intra-regional subcontracting of production, than companies using traditional manufacturing methods. Even among firms using a similar form of governance (i.e. subcontracting of production), proximity is more important for JIT adopters than for users of traditional manufacturing systems (size of plant, ownership, product innovation and other variables checked in the econometric model). The explanation offered for this finding is that JIT requires more flexibility, and entails more time constraints and uncertainty than traditional production systems. These characteristics of JIT, it is argued, rely more heavily on the geographical proximity of contractors and suppliers in order to reduce transaction costs.

Florida and Kenney (1992) argue that JIT adoption led to a different locational logic in the Japanese foreign direct investment in the US steel industry by redirecting the industry spatially towards the automobile centers in order to deliver steel on a just-in-time basis. Lampón et al. (2015) study relocation processes in the Spanish components industry. Their findings indicate that lean supply requirements act as a restriction for plant mobility. Companies that had implemented JIT show a reduced probability for international relocation. This is interpreted as a higher proximity requirement of those firms. As they are characterized by strong spatial links to the local, regional and national market, relocation would imply greater potential risk.

The geography of JIT may be affected not only, as mentioned above, by the industrial relationships prevailing within a company, but also by the nature of inter-firm rela-tionships. Much seems to depend on the exercise of power within the value-chain: for instance, certain monopsonies are able to persuade their suppliers to move near to their plant (Mair 1992). Other JIT adopters operating in the same national industry are unable to do so, since their suppliers channel their components and inputs to several factories and prefer, thus, to serve all their customers from a centric location.

Nevertheless, the role of geographic proximity for JIT cannot be viewed without considering the quality of the transportation infrastructure, advancements in information technology, and the rise of specialist freight handling and logistics firms including 4PL (Fourth-Party Logistics) firms that make highly sophisticated configurations of supply networks feasible. When analyzing issues of proximity in relation to JIT, both space and time are intrinsic dimensions that need to be considered together. In fact, JIT can be viewed as part of broader corporate strategies where time becomes the benchmark and firms compete on time compression and on what has been termed as 'time-based competition' (Kasarda 1999; Demeter 2013).

JIT AND OTHER TECHNOLOGIES

A less-frequently tackled topic is that of the comparison of the respective spatial patterns of JIT and other technologies. For a sample of medium-sized and large Spanish manufacturing firms, Holl et al. (2013) found that patterns of the adoption of JIT differ from patterns of the adoption of CAD/CAM: that of CAD/CAM is higher in larger cities, while adoption of JIT is higher in areas with better transport accessibility and in smaller cities (size and other characteristics of the industrial plant checked in the econometric model). The above-mentioned study concludes that JIT differs from other technologies since it relates directly to the spatial co-ordination of the internal production organization of firms with the external productive environment and depends on the quality of the transport system. Just-in-time depends notably on rapid, frequent and reliable deliveries and communication.

CONCLUSIONS

This chapter reviewed the literature on the spatial aspects of the JIT technique. Rather than hardware technology, JIT has been described as soft technology encompassing manufacturing and inventory techniques meant to prevent the waste of time and resources. In the literature, the JIT concept is related to concepts such as Lean Production, Toyota Production System and 'time-based competition'. Most of the analyses of JIT come from various strands of the management literature, which displays little interest in spatial aspects. Consideration of such aspects in relation to JIT is also relatively infrequent even in the literature on economic geography and regional studies, and most available analyses are case studies often based on small samples of companies.

Thirty years ago, the success of Japanese car makers compared with that of less successfully performing rivals brought with it a practical and an academic interest in JIT to the West. The possibility of implementing JIT in countries other than Japan stimulated much controversy in early literature since the success of this system seemed to be strongly associated with specificities of Japanese culture. When JIT started to be diffused internationally, many believed that the old Fordist manufacturing system was at a turning-point. This interpretation was in consonance with the parallel theoretic formulations of the literature on industrial districts and flexible specialization. In fact, the lack of quantitative studies limits the systematic assessment of the current degree of JIT adoption across space and its evolution over time. Another shortcoming of the literature is the lack of consensus on what exactly constitutes JIT practices, as the precise meaning of this term often varies. There is a need for a clear definition of JIT practices and for more quantitative studies. With these limitations in mind, the picture that emerges from the studies available is that of an uneven degree of adoption of JIT across countries. Moreover, in contrast to early expectations, there was no sharp rupture with the Fordist system, and JIT currently co-exists with traditional manufacturing systems, even in advanced countries. The degree of adoption of JIT also seems to be markedly uneven *within* countries, depending on regional factors, such as physical characteristics, transportation systems, industrial structure and industrial relationships.

Much of the literature has focused on the impact of the adoption of JIT on the geographic concentration of suppliers. Transaction costs economics provided a theoretical framework of analysis to study the alleged advantages of physical proximity. The empirical evidence remained inconclusive, since trends towards the geographic concentration of suppliers were by no means generalized. Later, the widespread perception that proximity among plants was becoming less important was a consequence of the rapid development of information and communication technology, and of the fall in transportation costs. Some authors, nevertheless, contested this view. They argued that inventory costs and other logistic costs had not been taken into account in previous research work; in certain industries, notably those that depend on a rapid response to customers' demand, spatial transaction costs may even have increased. Currently, the latest tendencies in the automobile industry swing towards the increased use of synchronized sequential JIT deliveries, which may even heighten the role of space.

In the context of this debate, another theme that came to the fore was concerned with the possible impact of the adoption of JIT on the location of existing industries. The empirical evidence was divided. Analyses of Japan and Europe tended to demonstrate that the adoption of JIT produced no substantial geographic changes, while analyses of the Americas provided evidence that the geography of existing national industries, especially that of the automobile industry, was indeed altered.

In the literature, other, less-frequently tackled topics concern quantitative comparisons of the spatial patterns of JIT adopters and those of non-JIT adopters, that is, industrial plants that use traditional manufacturing systems. Another example is found in the studies on spatial patterns of the diffusion of JIT, as compared with patterns of other technologies. Other spatial implications of JIT that have not been directly addressed in the chapter refer to its impact on trade patterns. Keane and Feinberg (2007), for example, argue that JIT constituted a major contribution towards the dramatic increase in intra-firm trade. The literature suggests several avenues for future research work. One segment of the field claims that greater consideration should be given to workers and trade unions as determinants of JIT geography. Another segment highlights the need to take accessibility and connectivity into account when studying JIT adoption rates across space.

NOTES

1. Certain authors contest the idea that the adoption of JIT enables workers to acquire new marketable skills (Herod 2000). This point seems to be supported by empirical evidence (Holl et al. 2013).
2. Note, however, that not all companies which adopt JIT use the *kanban* system owing, among other reasons, to the difficulties of road transportation (Mair 1992).
3. For other conceptualizations and terms used in the JIT literature, see Cowton and Vail (1994).

REFERENCES

Abernathy, F., J. Dunlop, J. Hammond and D. Weil 1999. *A Stitch in Time*. New York: Oxford University Press.
Amosse, T. and T. Coutrot. 2011. Socio-productive models in France: an empirical dynamic overview, 1992–2004. *Industrial and Labor Relations Review* 64(4): 786–817.
Bergman, E., E. Feser and A. Kaufmann. 1999. Lean production systems in regions: conceptual and measurement requirements. *Annals of Regional Science* 33(4): 389–423.
Bhamu, J. and K. Singh Sangwan. 2014. Lean manufacturing: literature review and research issues. *International Journal of Operations and Production Management* 34(7): 876–940.
Bourlakis, C. and M. Bourlakis. 2004. The future of food supply chain management. In M. Bourlakis and P. Weightman (eds), *Food Supply Chain Management*. pp. 221–230. Oxford: Blackwell.
Bruce, M., L. Daly and N. Towers. 2004. Lean or agile: a solution for supply chain management in the textiles and clothing industry? *International Journal of Operations and Production Management* 24(2): 151–170.
Clegg, C., T. Wall, K. Pepper, C. Stride, D. Woods, D. Morrison, J. Cordery, P. Couchman, R. Badham, C. Kuenzler, G. Grote, W. Ide, M. Takahashi and K. Kogi. 2002. An international survey of the use and effectiveness of modern manufacturing practice. *Human Factors and Ergonomics in Manufacturing* 12(2): 171–191.
Cowton, C.J. and R.L. Vail. 1994. Making sense of just-in-time production: a resource-based perspective. *Omega* 22(5): 427–441.
Demeter, K. 2013. Time-based competition – the aspect of partner proximity, *Decision Support Systems* 54: 1533–1540.
Egeraat, C.v. and D. Jacobson. 2005. Geography of production linkages in the Irish and Scottish microcomputer industry: the role of logistics. *Economic Geography* 81(3): 283–303.
Florida, R. and M. Kenney. 1992. Restructuring in place: Japanese investment, production organization, and the geography of steel. *Economic Geography* 68(2): 146–173.
Frigant, V. and M. Zumpe. 2014. Are automotive global production networks becoming more global? Comparison of regional and global integration processes based on auto parts trade data. MPRA working paper 55727.
Gale, H.F. 1999. Adoption of just-in-time manufacturing by rural and urban plants. *Review of Regional Studies* 29(2): 157–174.
Glaser-Segura, D.A., J. Peinado and A. Reis Graeml. 2011. Fatores influenciadores do sucesso da adoção da produção enxuta: uma análise da indústria de três países de economia emergente. *RAUSP – Revista de Administração da Universidade de São Paulo* 46(4): 423–436.
Helper, S. and M. Sako. 1995. Supplier relations in Japan and the United States: are they converging? *Sloan Management Review* 36: 77–84.
Herod, A. 2000. Implications of just-in-time production for union strategy: lessons from the 1998 General Motors–United Auto Workers dispute. *Annals of the Association of American Geographers* 90(3): 521–547.
Holl, A., R. Pardo and R. Rama. 2010. Just-in-time manufacturing systems, subcontracting and geographic proximity. *Regional Studies* 44(5): 519–533.
Holl, A., R. Pardo and R. Rama. 2013. Spatial patterns of adoption of just-in-time manufacturing. *Papers in Regional Science* 92(1): 51–67.
Kaneko, J. and W. Nojiri. 2008. The logistics of just-in-time between parts suppliers and car assemblers in Japan. *Journal of Transport Geography* 16(3): 155–173.
Kasarda, J. 1999. Time-based competition and industrial location in the fast century. *Real Estate Issues* 14(2): 24–29.
Keane, M.P. and S. Feinberg. 2007. Advances in logistics and the growth of intra-firm trade: the case of Canadian affiliates of U.S. multinationals, 1984–1995. *Journal of Industrial Economics* 55(4): 571–632.
Klier, T. 1999. Agglomeration in the U.S auto supplier industry. *Economic Perspectives* 23(1): 18–34.

Klier, T. 2000. Does 'just-in-time' mean 'right-next-door'? Evidence from the auto industry on the spatial concentration of supplier networks. *Journal of Regional Analysis and Policy* 30(1): 41–57.

Klier, T. and D. McMillen. 2008. Clustering of auto supplier plants in the United States: generalized method of moments spatial logit for large samples. *Journal of Business and Economic Statistics* 26(4): 460–471.

Klier, T. and J. Rubenstein. 2013. The evolving geography of the US motor vehicle industry. In F. Giarratani, G. Hewings and P. McCann (eds), *Handbook of Industry Studies and Economic Geography*. Cheltenham, UK and Northampton, MA, USA: Edward Elgar Publishing.

Lampón, J., S. Lago-Peñas and J. González-Benito. 2015. International relocation and production geography in the European automobile components sector: the case of Spain. *International Journal of Production Research* 53(5): 1409–1424.

Lee, Y.-S. 2003. Lean production systems, labor unions, and greenfield locations of the Korean new auto assembly plants and their suppliers. *Economic Geography* 79(3): 321–339.

Linge, G. 1991. Just-in-time: more or less flexible? *Economic Geography* 67(4): 316–332.

Mackelprang, A. and A. Nair. 2010. Relationship between just-in-time manufacturing practices and performance: a meta-analytic investigation. *Journal of Operations Management* 28: 283–302.

Mair, A. 1992. Just-in-time manufacturing and the spatial structure of the automobile industry: lessons from Japan. *Tijdschrift voor Economische en Sociale Geografie* 83(2): 82–92.

McCann, P. and B. Fingleton. 1996. The regional agglomeration impact of just-in-time input linkages: evidence from the Scottish electronics industry. *Scottish Journal of Political Economy* 43(5): 493–518.

McCann, P. and S. Sheppard. 2003. The rise, fall and rise again of industrial location theory. *Regional Studies* 37(6–7): 649–663.

Milgrom, P. and J. Roberts. 1990. The economics of modern manufacturing: technology, strategy and organization. *American Economic Review* 80(3): 511–528.

Moreira, M. and R. Alves. 2008. A study on just-in-time implementation in Portugal: some empirical evidence. *Brazilian Journal of Operations and Production Management* 5(1): 5–22.

Moyano-Fuentes, J. and M. Sacristán-Díaz. 2012. Learning on lean: a review of thinking and research. *International Journal of Operations and Production Management* 32(5): 551–582.

Ó Huallacháin, B. and D. Wasserman. 1999. Vertical integration in a lean supply chain: Brazilian automobile component parts. *Economic Geography* 75(1): 21–42.

Pallarès-Barberà, M. 1998. Changing production systems: the automobile industry in Spain. *Economic Geography* 74(4): 344–359.

Rutherford, T.D. 2000. Re-embedding, Japanese investment and the restructuring buyer–supplier relations in the Canadian automotive components industry during the 1990s. *Regional Studies* 34(8): 739–751.

Rutherford, T.D. and M.S. Gertler. 2002. Labour in 'lean' times: geography, scale and the national trajectories of workplace change. *Transactions of the Institute of British Geographers* 27(2): 195–212.

Sadler, D. 1994. The geographies of just-in-time: Japanese investment and the automotive components industry in Western Europe. *Economic Geography* 70(1): 41–59.

Smith, D. and R. Florida. 1994. Agglomeration and industrial location: an econometric analysis of Japanese-affiliated manufacturing establishments in automotive-related industries. *Journal of Urban Economics* 36(1): 23–41.

Sugimori, Y., K. Kusunoki, F. Cho and S. Uchikawa. 1977. Toyota production system and kanban system materialization of just-in-time and respect-for-human system. *International Journal of Production Research* 15(6): 553–564.

Swamidass, P. and G. Winch. 2002. Exploratory study of the adoption of manufacturing technology innovations in the USA and the UK. *International Journal of Production Research* 40(12): 2677–2703.

Waterson, P., C. Clegg, R. Bolden, K. Pepper, P. Warr and T. Wall. 1999. The use and effectiveness of modern manufacturing practices: a survey of UK industry. *International Journal of Production Research* 37(10): 2271–2292.

27. Robotics

Antonio López Peláez

Robotics is a mature technology in the industrial sector with impressive growth rates: sales increased from the 65,000 units sold in 1995 to 229,000 in 2014 (IFR 2015a). From a geographical perspective, Asia accounted for two-thirds of all industrial robots installed in 2014. The countries with the largest stock of active industrial robots are also the most industrialized: China, Japan, the United States, Germany and South Korea. As in other key technologies, the concentration of industrial robots in industrialized countries should not obscure the fact that the stock of industrial robots is also growing substantially in many other countries, making investment in robots a general feature of economic growth worldwide.

Collaboration between human workers and robots is on the rise. New production models based on this collaboration and the so-called 'cobots' (robots that work cooperatively) are redefining industry, and can be analyzed in reindustrialization schemes such as China's Made in China 2025, the US reindustrialization, Japan's rejuvenation strategy and the EU's Industrial 4.0 program.

Service robotics is also expanding rapidly (IFR 2015b). New generations of service robots in fields such as medicine or in the home permit the automation of more tasks, and at the same time enable a more intense interaction with users. In a technological context characterized by what we call 'the Internet of things', interconnected service robots will be able to perform new tasks and become a key element in households, leisure and dependent care. For example, the International Federation of Robotics (IFR) estimates that 35 million units of service robots will be sold for personal use over the period 2015–2018.

We are therefore in the midst of a new cycle of technological transformations that are changing both industry and services. In the field of robotics, the main features of this new cycle are: first, connectivity (the Internet of things, cloud computing and big data); second, flexibility and adaptability to the environment (the new generations of robots are more flexible, mobile and cooperative, thus increasing productivity in diverse contexts); and third, the ability to work side-by-side with humans.

ROBOTICS: AN HISTORICAL CONTEXT

The considerations of thinkers and writers about the possibility of making machines to perform human activities and replace human labor (Aristotle 1962, 26), or any other type of activity, have been accompanied by numerous attempts to create human-like automata capable of performing sundry movements. For example, the mechanical dolls created by J. de Vaucanson in the eighteenth century, which were of a similar size to humans, could perform various musical pieces. Beyond the dreams of creating a mechanical human being, the term 'robot' was first used by Karel Capek in 1917 (Capek 1996) in his play *R.U.R. (Rossum's Universal Robots)* and later popularized in the fiction of Isaac Asimov. Asimov used the word 'robot' for the first time in a short story titled 'Runaround' published in 1942. In a story published in the magazine *Galaxy Science Fiction* in 1947, Asimov formulated the three laws of robotics, to which he later added a previous or zeroth law. The implications of these laws for the development of future technology have been analyzed by R. Clarke (1993, 1994). Asimov's laws are as follows:

Law 0: a robot may not harm humanity or, by inaction, allow humanity to come to harm.

Law 1: a robot may not injure a human being or, through inaction, allow a human being to come to harm except where such orders would conflict with the previous Law.

Law 2: a robot must obey the orders given to it by human beings, except where such orders would conflict with the First Law.

Law 3: a robot must protect its own existence as long as such protection does not conflict with the First or Second Law.

It is important to distinguish two large areas in robotics: industrial robots and service robots. In industry, robotics is widely implemented because the automation of the largest possible number of processes has become a priority. In the service sector, robotics is expanding at a great speed, and will no doubt entail a revolution in coming years in terms of both automating tasks and modifying human–machine relationships. In this section, we analyze some relevant data regarding robotics to then examine in depth the debates on robotics technology in the following section.

INDUSTRIAL ROBOTICS: ALWAYS AND EVERYWHERE

As a scientific discipline, robotics and industrial robots date back to the second half of the twentieth century. In an industrial context characterized by increasing demands for productivity, lower costs and better quality of life, a series of technologies were developed and patented in the early 1950s that enabled the creation of the first industrial robots: numerical control (technology that allows the actions of a machine tool to be controlled by means of numbers) and telechirics (technology that enables the use of a human-controlled remote manipulator).

Cyril Walter Kenward applied for the first British patent for a robot in 1954, which was granted in 1957. George C. Devol laid the foundations for the development of modern industrial robots by inventing a device for magnetically recording and reproducing electrical signals to control a machine and jointly developed his patent on Programmed Article Transfer in 1961. The first commercially available robot, marketed by Planet Corporation, was introduced in 1959. The installation of industrial robots began in the United States thanks to the collaboration between Joseph F. Engelberger and George C. Devol, who developed the UNIMATE universal robot and formed Unimation in 1962. General Motors installed the first UNIMATE robot in 1961 to perform spot welding and extract die castings.

Industrial robotics was developed to achieve higher levels of automation through the incorporation of robots, which adds a key feature to the capabilities of automatic machines: the capacity to perform various tasks using sensors and artificial intelligence

Table 27.1   Development phases of industrial robotics

First Phase	1950: Development of master-slave manipulators by the Argonne National Laboratory to handle radioactive materials.
Second Phase	1958–1969: Development and installation of the first industrial robots by Unimation in the United States.
Third Phase	1970: Laboratories at Stanford University and MIT developed computer-controlled robots.
Fourth Phase	1975–1980: Application of a microprocessor to industrial robots to transform the image, capabilities and features of robots.
Fifth Phase	1980–2010: Developments in applied computing, sensors and artificial intelligence make it possible to develop robots that can adapt to the environment and make real-time decisions in any situation.
Sixth Phase	2010–onwards: Development of new collaborative robots (cobots) capable of interacting with humans in different interconnected contexts (the Internet of things, cloud computing, big data).

to achieve a given learning ability. In 1979, the US Robotics Industries Association defined an industrial robot as a 'reprogrammable, multifunctional manipulator designed to move materials, tools or specialized devices through variable movements programmed to execute tasks automatically'.

In 1982, this definition was widely accepted and manual manipulators were no longer considered 'robots'. For those known as 'fixed-sequence robots', four general classifications were established: variable-sequence robots, playback robots, numerically controlled robots and intelligent robots. Owing to the rapid development of robotics technology and the emergence of new service robots, these definitions have been reformulated. Thus, ISO 8373 defines an industrial robot as an 'automatically controlled, reprogrammable, multipurpose manipulator, programmable in three or more axes, which can be either fixed in place or mobile for use in industrial automation applications'.

The development of industrial robotics can be grouped into six large phases, culminating today with the development of cooperative robots capable of working side-by-side with humans (Table 27.1).

A proper analysis of the industrial robots market must take into account the evolution of the current operational stock of robots and industrial robot sales, their distribution by industries and applications, as well as robot density (number of robots per 10,000 workers employed in the industry), which indicate the level of automation. In this line, according to data provided by the IFR, 1,480,778 robots were installed in 2014. Over the next four years (2015–2018), the IFR estimates a double-digit growth in the stock of robots installed, which is expected to reach 2,327,000 million units at the end of 2018 (Table 27.2).

The main sectors that will lead this growth are the automotive industry and the electrical/electronics industry. This increase in number of robots entails a higher robot density:

The average global robot density is about 66 industrial robots installed per 10,000 employees in the manufacturing industry. The most automated markets are the Republic of Korea, Japan and Germany. In 2014, the Republic of Korea had again the highest robot density in the world by far

Table 27.2   Estimated operational stock of multipurpose industrial robots in selected countries (number of units)

Country	2013	2014	2015[a]	2018[a]
Americas	**226,071**	**248,430**	**272,000**	**343,000**
Brazil	8,564	9,557	10,300	18,300
North America (Canada, Mexico, United States)	215,817	236,891	259,200	323,300
Other America	1,690	1,982	2,500	1,700
Asia/Australia	**689,349**	**785,028**	**914,000**	**1,417,000**
China	132,784	189,358	269,900	614,200
India	9,677	11,760	14,300	27,100
Japan	304,001	295,829	297,200	291,800
South Korea	156,110	176,833	201,200	279,000
Taiwan	37,252	43,484	50,500	67,000
Thailand	20,337	23,893	27,900	41,600
Other Asia/Australia	29,188	43,871	60,000	96,300
Europe	**392,227**	**411,062**	**433,000**	**519,000**
Czech Republic	8,097	9,543	11,000	18,200
France	32,301	32,233	32,300	33,700
Germany	167,579	175,768	183,700	216,800
Italy	59,078	59,823	61,200	67,000
Spain	28,091	27,983	28,700	29,500
UK	15,591	16,935	18,200	23,800
Other Europe	81,490	88,777	97,900	130,000
Africa	**3,501**	**3,874**	**4,500**	**6,500**
Not specified by country	**21,070**	**32,384**	**40,500**	**41,500**
Total	**1,332,218**	**1,480,778**	**1,664,000**	**2,327,000**

Note: [a]Forecast.

Source: IFR 2015a.

due to continued installation of a large volume of robots in recent years. 478 industrial robots were in operation in 2014 per 10,000 employees. The robot density in Japan further decreased to 314 units, and in Germany it continued to increase to 292 units. The United States, which is one of the five biggest robot markets regarding annual supply, has a robot density of 164 units in 2014. The robot density in China, the biggest robot market since 2013, reached 36 units in 2014 unveiling the huge potential for robot installations in this market. (IFR 2015a, 17)

It is important to highlight that robot density is much higher in the automotive industry, which serves to indicate the extent of automation. 'Japan had by far the highest robot density in the automotive industry. 1,414 industrial robots were installed per 10,000 employees in the automotive industry. It is followed by Germany with 1,149, the United States with 1,141 and the Republic of Korea with 1,129 units' (IFR 2015a, 17).

Table 27.3   Estimated operational stock of service robots at year-end (number of units)

Type of service robot	2013	2014	2015–2018[a]
Service robots in defense	9,800	11,000	58,800
Milking robots	4,790	5,180	28,000
Medical robots	1,412	1,224	7,800
Logistic systems	2,100	2,644	14,500
Service robots for personal and domestic use	3,500,000	4,700,000	35,000,000

Note: [a]Forecast.

Source: IFR (2015b).

Service Robotics: A New and Versatile Companion

In their article 'Robots in Service Industry', Nostrand and Sampson (1990) proposed the following definition of a service robot: a service robot is a mobile or stationary electro-mechanical device, fitted with one or more independent arms, controlled by a computer program and that performs non-industrial service tasks (Nostrand and Sampson 1990). However, this definition does not take into account some fundamental aspects of service robotics. Taking as a starting point the exclusion of manufacturing, service robots can be defined as robots that provide services to people or facilities, but do not 'manufacture' products. Service robots should operate autonomously within certain limits depending on their design and solve navigation-related problems in a more or less defined open environment. In 1998, the IFR proposed the following tentative definition of service robots: 'A service robot is a robot which operates semi- or fully autonomously to perform services useful to the well-being of humans and equipment, excluding manufacturing operations' (IFR 1999, 23).

   The IFR differentiates between two types of service robots: professional service robots and service robots for personal and domestic use. The main areas in the first category are defense applications, milking robots, medical robots and logistic systems, while the second category includes robots for personal and domestic use (vacuum and floor cleaning, lawn-mowing robots and entertainment and leisure robots; Table 27.3).

   The growth in the stock of both types of service robots will entail a profound change in our uses and customs not only in professional fields, such as medicine and the military, but also in the domestic environment. In this regard, service robotics is already a key strategy to address aging and elder care in some countries. For example, the Singapore government has introduced the RoboCoach to keep older people physically fit (Holmes 2015).

THEORETICAL PERSPECTIVES

The interaction of humans and machines, and especially the possibility of developing robots that can reproduce human behavior and intelligence, is a fascinating field of research. Not only do we transfer our capabilities to robots, but we are increasingly inclined to implant technology and machinery in our own bodies, to the point of making robots more like humans than machines, and transforming humans into machines that are more like robots, such as cyborgs. From the perspective of the so-called NBIC convergence (Roco and Bainbridge 2002), new possibilities have arisen for transforming the human species by way of a post-modified human life, in which artificial intelligence, genetics and robotics will make way for a new stage in the history of humankind. The debate on the emergence of a new species has sparked heated discussions in leading forecasting journals (*Technological Forecasting and Social Change* 2006, 95–127).

Moreover, the new models of social relationships that will be established within this context, in which the frontiers between humans and machines will be blurred, highlight the extent to which the current technological trajectories and social models in which we design and implement robots will shape the immediate future in an increasingly irreversible way (as has happened in the past with other crucial technologies). Not only do we need to consider the emergence of a new species, but also the development of a new social subject (as robots are also designed to work as a team, learn together and interact socially with humans and other robots) in an ontological way. Recent research shows the emergence of a new companion for purposes of leisure and entertainment, or a partner for romantic and sexual relationships (Kalla 2015). Given that technology is developed and implemented according to market demands, it will come as no surprise that, while we are debating the rights and obligations of robots, they will quickly colonize the sphere of our private lives, and become much more to us than a videogame or a pet – perhaps even a romantic partner.

Four main areas of debate have arisen around robotics: the now classic debate dating back to Aristotle on the emergence of an automaton that could be a new human being, a notion that was redefined by Moravec (1999) and Kurzweil (2005); the debate on the impact of robotics on employment in the industrial and service sector and the complex relationship between technology and employment (Brynjolfsson and McAfee 2014; Ford 2015); the changes in human–machine interactions brought about by advanced robotics (Levy 2007); and, finally, the effects of this technology on power. For example, in the defense sector, military robots could represent a crucial competitive advantage in the coming years. Will we see a new *robotics divide*, which will redefine and expand the impact of the already widely studied *digital divide* (López Peláez 2014a)?

In this context, it is important to highlight three characteristics of technological developments in the field of robotics: the convergence of technologies (which augurs profound changes as it will enable integrating intelligence, emotions and machines into a single entity, industrial or service robots); the speed at which computer and microelectronics technologies are being developed; and the lower costs of automated and robotic systems. These three features allow us to reasonably assume that the new automatic and robotic systems, in both industry and services, will be increasingly competitive and reliable and perform a larger number of tasks.

The increasing computational power of machines in the twentieth century has resulted in a growth curve that has surpassed exponential growth and driven down their cost. In this regard, Moravec (1999), in a classic text in the field of robotics, analyzed the exponential growth of the computing power of a $1000 machine in the twentieth century (Figure 27.1) using what he called 'MIPS' (the calculation capacity of $1000 of computer). As the author explains:

Steady improvements in mechanical and electromechanical calculators before World War II had increased the speed of calculation a thousandfold over manual methods from 1900 to 1940. The pace quickened with the appearance of electronic computers during the war, and 1940–1980 saw a millionfold increase. Since then the pace has been even quicker, a pace that would make humanlike robots possible before the middle of the next century. (Moravec 1999, 61)

In Figure 27.1, the vertical axis is logarithmic and the divisions represent thousand-fold increases in computer power over manual methods. The upward curve indicates faster than exponential growth and an accelerating rate of innovation (Moravec 1999).

CONCEPTUAL AND POLITICAL DEBATES

There are two major conceptual and political debates regarding robotics. The first is the classic debate on employment and robots, which in turn brings us to a broader debate on technology and labor. The second is the debate on how to reorganize society around the expansion of service robotics, especially robotics in key areas such as defense, outer space or the Internet. As concerns the second question, the issue of power and the effect of technology on the geopolitical balance is becoming increasingly important.

Robotics and Labor

As regards robotics and labor, the main debates focus on the impact of robotics on the volume of employment (i.e. do robots create or destroy jobs?), the transformation of types of labor and organizational changes to tap into the potential of automatic and robotic systems. The growing stock of industrial robots, and now of service robots, has redefined the classic debate on technology and employment, a debate that has accompanied us since the first industrial revolution.

The impact of robotics and advanced automation occupies a prominent position in the sociological reflections of Beck on the 'risk society' (Beck 1992), Schaff on the 'occupations society' (Schaff 1985), Gorz on the 'free-time society' (Gorz 1997), Tezanos on the 'post-labor society' (Tezanos Tortajada 2001), Coriat on the 'technological revolution' (Coriat 1994) and Manuel Castells (1996) on the 'network society'. As for the direct impact on the volume of employment, the controversy is ongoing. Suffice it to cite the now classic research of Leontief and Duchin (1985), and the criticism they received for assuming that the final demand remains fixed (Castells 1996). Today, the debate on robotics and labor is a topical one (Rotman 2015; Brynjolfsson and McAfee 2014; Ford 2015).

When a global vision is adopted, the reference point is the growth or decline of industrial employment worldwide. From this perspective, the crisis of industrial employment in Europe and the United States in the last two decades of the twentieth century was accompanied by an increase in industrial employment in other parts of the world such as China. However, although China was the largest market for industrial robots in 2015, it should be noted that a remarkable process of reindustrialization is occurring in the United States, partly driven by lower labor and energy costs. In addition to the volume of employment, it is important to consider that the jobs which are created do not have the

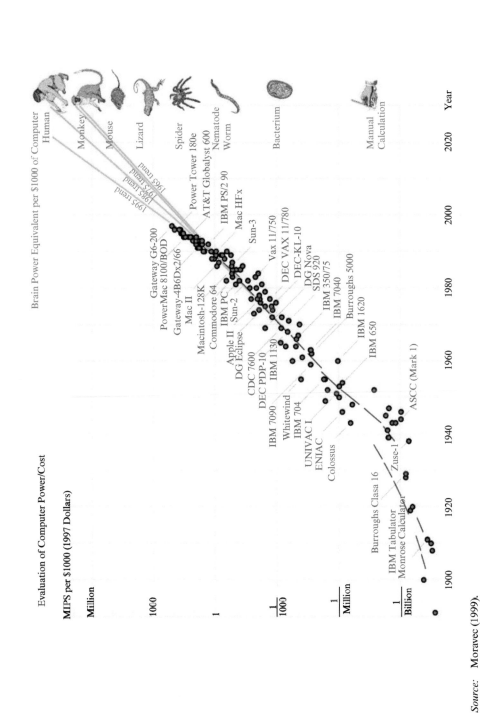

Source: Moravec (1999).

Figure 27.1 Evolution of computer power/cost

same characteristics as jobs that are automated, and redesigning organizations is key to unlocking the potential of advanced automation technologies.

The trend to automate the maximum possible number of activities based on the available technology presents us with a horizon in which industrial labor (and progressively certain areas of the service sector) is no longer defined in terms of the capacity and pace of human labor. It is increasingly defined from the viewpoint of the potential of robots. The increasing intensity and pace of work owing to the implementation of automated systems leads to the 'intensification' of labor and overload for workers operating with robots. Investment in robotics appears to be a central element in the fight for competitiveness, as it enables higher levels of quality and efficiency, while reducing labor costs.

The need for high-skilled workers, increasing labor intensification and the new demands arising from cooperative work between humans and robots requires new skills and specific training. This transformation of labor in the automated society means that education must also change: advanced technologies require workers to go the extra mile and do initial and continuing training, thus redefining the skills needed to find a 'job' and pursue a 'professional career' as understood in industrial societies until now.

Robotics and Power

From a historical perspective, 'the ability or inability of societies to master technology, and particularly technologies that are strategically decisive in each historical period, largely shapes their destiny, to the point where we could say that while technology *per se* does not determine historical evolution and social change, technology (or the lack of it) embodies the capacity of societies to transform themselves, as well as the uses to which societies, always in a conflictive process, decide to put their technological potential' (Castells 1996, 33).

Science and technology play a key role in structuring our societies. Indeed, from the moment the first tools were developed by *Homo sapiens*, technology has always played a key role in human existence. However, it is also true that the importance of technology today, and our own self-knowledge acquired via such innovations as the cell phone, computers, television and the Internet, make one wonder about the consequences of the way we design, develop and implement new technologies, and how they contribute to strengthening hierarchical structures or asymmetric distributions of power.

It is precisely this kind of question that inspires fantasy literature to explore the conflict between humans and machines, people and androids, or cyborgs, sentient beings with both organic and cybernetic parts, and robots and humans. If technologies, both old and new, play a key role in the economy, in the production of goods and services, in behavior patterns, in forms of social and political organization, in leisure and in forms of participation, to speak of a newly emerging technologically advanced society, we should ask ourselves who is included and who is excluded from technological dynamics. That question has been raised in the last three decades in relation to the Internet (the 'digital divide'), and now we need to ask that same question in relation to the technology that seeks to replicate intelligent behavior in autonomous machines: robotics.

Will robotics, which is characterized by converging technologies, rapid progress and a progressive reduction in costs, be incorporated into our societies under the current market model? Will it create a new technological, social and political divide? Will it transform power relations between individuals, groups, communities and states in the same way as crucial military technology? Will it change our everyday social interactions by forming part of our daily lives in the same way it has changed industry where industrial robotics is already mature and fully established? These are crucial questions, because, as we have seen in the past with other key technologies in Western societies, technology is not neutral, and its foreseeable impacts arise from the socioeconomic models in which we are immersed.

All technology is implemented in an unequal way, owing both to its cost and usage requirements and the fact that new technologies are almost exclusively used first by what is referred to as 'early adopters' to progressively expand into ever-wider layers of the population. In a parallel manner, the type of technology designed depends on the needs it aims to satisfy, and the logic that lies behind every technology application, thus causing a set of frequently unanticipated, and often unwanted, side effects.

To analyze the impact of robotics on power from this point of view, and to what extent there exists a divide between those who have access to such technology and those who do not (both individuals and companies or countries), the following issues should be considered (López Peláez 2014b):

- First, the economic as well as science and technology resources that are needed to develop robotics technology in all areas.
- Second, the ability of companies, users and civil society to reorganize in order to increase economic productivity and incorporate industrial and service robotics into a wider number of spheres.
- Third, the market economy model and distribution of existing resources in advanced societies.
- Fourth, areas in which robotics technology entails redefining power in relation to military and space programs, as well as in the Internet (in which so-called 'intelligent agents' play a crucial role in accessing goods and services, which are increasingly only available online).

Given these dimensions, we can define the robotics divide as follows: the distance or separation between those individuals, companies and states that possess the economic, as well as scientific and technological capacity and resources to develop robotics technology, have redefined their spheres of production and leisure in order to incorporate robots, can make the necessary investments in those spheres, have developed and have at their disposal advanced robots in the military and aerospace field and the Internet, and those individuals, companies and states that do not have these resources (López Peláez 2014b, 22).

This distance or separation implies higher levels of economic, military and technological power for those individuals, companies and states that possess robotics technology, and especially in critical areas such as aerospace programs or military combat robots that could gain a competitive advantage which would significantly alter the balance of power between one country and another. As a result, military and aerospace robotics, along with developments in robots for the Internet, will become a strategic issue affecting competition between countries, especially between those that are set to play a leading role in the twenty-first century such as the United States, China, India and Russia.

In leisure and domestic life, as well as healthcare and the care for disabled or dependent individuals, new service robots will provide a competitive advantage, and could have a significant impact by reducing the need for immigrant workers (spheres of activity which are currently very labor intensive). To a certain extent, the robotics divide could introduce a new actor into our social life, the robot, which could become an assistant, a romantic partner (Levy 2007) or even a warrior, and while robots may not alter job creation in absolute terms, they could have a significant impact on some labor intensive sectors (i.e. the immigrant workforce in Western countries). New technologies, in particular robotics, could become a new factor to take into account when analyzing immigration flows (López Peláez and Krux 2003). In a more personal context, and to some extent one that runs parallel to the so-called digital divide as noted in some of the chapters in this book, service robotics could become a crucial technology in terms of both access and user requirements. The difference between having and not having a robot may become not only a visible sign of socioeconomic status, but also a symbol of power and wealth, thus creating a gap between the technology 'haves' and 'have nots' in the sense of Bourdieu. Limited access to robots could also be a clear predictor of social exclusion as they could perform many tasks that would greatly ease the burdens of daily life (not only in terms of mobility or cleaning, but also in terms of emotions and relationships). Table 27.4 summarizes some of the features of the emerging robotics divide at three different levels: states, companies and individuals.

Table 27.4   Consequences of the robotics divide in the twenty-first century

States		Companies		Individuals	
Access to advanced robotics technology	No access to advanced robotics technology	Access to advanced robotics technology	No access to advanced robotics technology	Access to advanced robotics technology	No access to advanced robotics technology
Economic growth and enhanced productivity. Greater military power. Border control. Technological innovation. Conquered space.	Lower economic growth and productivity. Less military power. Less border control. Lower level of technological innovation Increasingly distanced from the space race.	Higher productivity. Higher levels of automation. New business niches.	Lower productivity. Lower levels of automation. Lower competitiveness in new business niches such as aerospace.	Automation of domestic chores. New forms of leisure and services. Better employment opportunities. More educational resources associated with robots in the classroom and at home.	Greater difficulties for the disabled. More activities related to domestic chores. Less educational resources.

Source: López Peláez (2014a).

TRENDS

The current trends in industrial and service robotics have revealed a concentration of robots in just a few countries. Given that robotics is a critical technology, the asymmetry between the five countries with over 70% of robots installed in 2014 (China, Japan, USA, Germany and South Korea) and the rest of countries points to an increasing distance or divide between some countries and others. Perhaps it is also a sign of the asymmetric distribution of robots in other key and less developed sectors such as industry or defense. The IFR foresees a significant expansion in the field of industrial robotics in the period 2015–2018, in terms of both the number of installed robots and areas of activity. In the service sector, the growth forecasts are even higher (IFR 2015b).

Experts in robotics (Figure 27.2) predict that automated tasks will account for 80% of all activities by 2050. It should be noted that this sector, the main consumer of robots and advanced automation systems, serves as a model of reference. The experts also predict that 70% of all tasks in other sectors, such as chemicals, petroleum, coal, rubber and plastic products, metal products and footwear and textiles will be automated by the year 2050. By the year 2020, 31.5% of the activities in the chemicals, petroleum, coal, rubber and plastics sectors will be automated, while 50% of activities in the automotive sector will be automated. Two characteristics can be observed in the medium term (2025): first, the increased automation of tasks in all areas of activity; and second, an expansion that will tend to homogenize the levels of automation reached in various areas of industrial activity – 65% of all jobs in the automotive sector and 42.5% of all jobs in the footwear and textiles sector

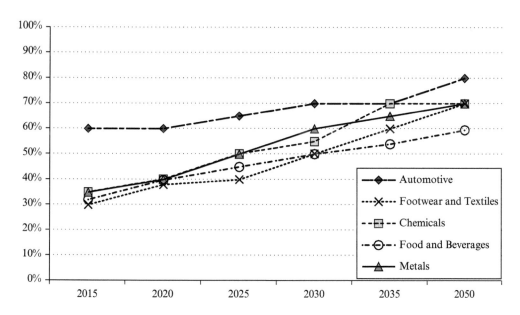

Source: López Peláez and Segado Sánchez-Cabezudo (2014).

Figure 27.2 Predictions of the evolution of the percentage of activities that will be automated in various industrial sectors

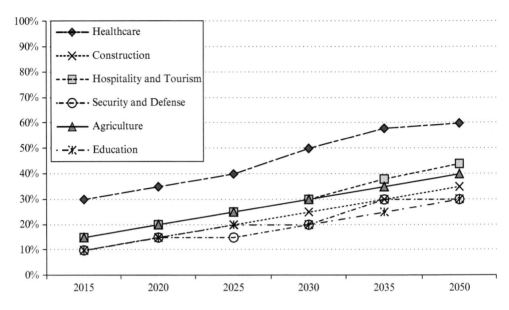

Source: López Peláez and Segado Sánchez-Cabezudo (2014).

Figure 27.3   Predictions on the percentage of activities that will be automated in the agricultural and service sectors

will be automated. By 2050, this convergence will be higher, with 60% of all jobs in the food and beverage sector and 80% in the automotive sector being automated.

In the service sector, experts predict robotics to develop significantly in four main areas: security and defense, the healthcare sector, the tourist industry and education and research robots (Figure 27.3).

From the perspective of the geography of technology, development trends in the field of robotics show, first, the increasing automation of an ever larger number of areas of activity. Secondly, they highlight the strategic impact of robotics in terms of power, welfare, social relations and health. Companies, firms and individuals who have access to these technologies will have a crucial competitive advantage, both in terms of economic competitiveness and military power, and, of course, in the field of welfare, elder care or health.

Robot intelligence and capabilities will introduce an alter ego in our social environment, in the same way that they already play a central role in performing tasks in industrial sectors. All of this is going to materialize in cosmopolitan but unequal and stratified societies, which will possibly create new forms of inequality that will superimpose themselves on older forms of inequality. At the same time, new opportunities will arise, and, to some extent, the worlds of work and entertainment will be redefined, and possibly even the world of intimate relationships (Levy 2007).

Finally from a technological standpoint, the convergence of biology, robotics and artificial intelligence suggests a three-fold horizon of possibilities: human genetic enhancement in order to improve our capabilities and make way for a new stage in the evolution of the species; the development of enhanced artificial intelligence until finally producing autonomous machines and robots capable of improving and repairing themselves, which will give rise to sentient beings that are more intelligent than humans; and a mixed future, in which machines such as nanorobots will coexist with biological improvements to the human brain, producing a mixture of information and communication robot-human technologies, which will also permit us to take a new leap in the history of life on earth and beyond.

REFERENCES

Aristotle. 1962. *Politics*. Madrid: Espasa-Calpe.

Beck, U. 1992. *Risk Society: Towards a New Modernity*. London: Sage.

Brynjolfsson, E. and A. McAfee. 2014. *The Second Machine Age: Work, Progress and Prosperity in a Time of Brilliant Technologies*. New York: W.W. Norton.

Capek, K. 1996. *R.U.R. Robots Universales Rossum*. Madrid: Alianza Editorial.

Castells, M. 1996. *The Rise of the Network Society. The Information Age: Economy, Society and Culture, Vol. I*. Cambridge, MA: Blackwell.

Clarke, R. 1993. Asimov's laws for robotics: implications for information technology. Part 1. *Computer* December: 53–61.

Clarke, R. 1994. Asimov's laws for robotics: implications for information technology. Part 2. *Computer* January: 57–65.

Coriat, B. 1994. *L'Atelier et le chronomètre*. C. Bourgois (ed.), Paris: Coriat.

Ford, M. 2015. *Rise of the Robots. Technology and the Threat of a Jobless Future*. New York: Basic Books.

Gorz, A. 1997. *Misères du présent, richesse du possible*. Paris: Galilée.

Holmes, O. 2015. Singapore introduces Robocoach to keep older citizens in shape. *The Guardian*, 14 October 2015, accessed from http://www.theguardian.com/technology/2015/oct/14/singapore-introduces-robo coach-to-keep-older-citizens-in-shape.

IFR. 1999. *World Robots 1999*. Geneva: UN/International Federation of Robotics.

IFR. 2015a. *World Robotics. Industrial Robots 2015*. Geneva: UN/International Federation of Robotics.

IFR. 2015b. *World Robotics. Service Robots 2015*. Geneva: UN/International Federation of Robotics.

Kalla, S. 2015. Human-like robots are getting jobs. *Forber/Techn*, 20 January 2015, accessed from http://www.forbes.com/sites/susankalla/2015/01/20/human-like-robots-are-getting-jobs/#7022de03220f.

Kurzweil, R. 2005. *The Singularity is Near: When Humans Transcend Biology*. New York: Viking Penguin.

Leontief, W. and F. Duchin. 1985. *The Future Impact of Automation on Workers*. New York: Oxford University Press.

Levy, D. 2007. *Love and Sex with Robots: The Evolution of Human–Robot Relationships*. London: Harper.

López Peláez, A. (ed.). 2014a. *The Robotics Divide. A New Frontier in the 21st Century?* New York: Springer.

López Peláez, A. 2014b. From the digital divide to the robotics divide? Reflections on technology, power and social change. In A. López Peláez (ed.), *The Robotics Divide. A New Frontier in the 21st Century?* pp. 5–24. New York: Springer.

López Peláez, A. and M. Krux. 2003. New technologies and new migrations: strategies to enhance social cohesion in tomorrow's Europe. *IPTS Reports* 80: 11–17.

López Peláez, A. and S. Segado Sánchez-Cabezudo. 2014. From 'Singularity' to inequality: perspectives on the emerging robotics divide. In A. López Peláez (ed.) *The Robotics Divide. A New Frontier in the 21st Century?* pp. 195–218. New York: Springer.

Moravec, H. 1999. *Robot: Mere Machine to Transcendent Mind*. New York: Oxford University Press.

Nostrand, J. and E. Sampson. 1990. Robots in service industry. In R. Dorf (ed.), *Concise International Encyclopedia of Robotics*. New York: John Wiley & Sons.

Roco, M. and W. Bainbridge. 2002. *Converging Technologies for Improving Human Performance. Nanotechnology, Biotechnology, Information Technology and Cognitive Science*. Arlington, VA: National Science Foundation.

Rotman, D. 2015. Who will own the robots? *MIT Technological Review*, accessed from http://www.technology review.com/featuredstory/538401/who-will-own-the-robots/.

Schaff, A. 1985. *¿Qué futuro nos aguarda?. Las consecuencias sociales de la segunda revolución industrial*. Barcelona: Crítica.

Technological Forecasting and Social Change. 2006. 73(2): 95–127.

Tezanos Tortajada, J.F. 2001. *El trabajo perdido. ¿Hacia una sociedad postlaboral?* Madrid: Biblioteca Nueva.

28. The geography of nanotechnology
Scott W. Cunningham

This chapter explores the geography of nanotechnology. Nanotechnology is a novel, interdisciplinary study of the material properties of objects at very small scales (Porter and Youtie 2009). The foundations of the science were set over 30 years ago with the invention of scanning probe microscopy (Binnig et al. 1982). Now there is considerable hope that nanotechnology will lead to a range of new products and businesses. As a result policy-makers are today assessing the geographical incidence and economic consequences of this new technology.

The purpose of this chapter is to review a small but significant literature on nanotechnology policy. The review will synthesize a range of geographical theories from regional studies, economic geography and economic sociology. The chapter begins with a brief definition of nanotechnology, and a discussion of the funding and planning that promotes this new technology.

DEFINITION AND SIGNIFICANCE OF NANOTECHNOLOGY

Nanotechnology operates at a scale of one to hundreds of nanometers. This is the scale of molecules and collections of atoms. At this scale there are simultaneous processes of conventional chemistry as well as quantum physics. At these very small scales the surface becomes extremely important and dominates material properties. Fundamental properties of materials are often size-dependent.

Using nanotechnology techniques it becomes possible to tune matter to achieve a range of different properties, including melting point, fluorescence, electrical conductivity, magnetic permeability and chemical reactivity. Because of these unconventional features of nanotechnology, there is a wealth of opportunity to exploit these new material properties for the human betterment (National Nanotechnology Initiative 2016).

Nanotechnology emerged out of a substrate of existing knowledge in materials science, microscopy and applied physics in the early 1980s. While the promise of the field was recognized relatively early (Feynman 1959), it was ultimately a work of popular science that catalyzed policy attention for the field (Drexler 1986). This stimulated considerable new funding and initiatives for the new technology.

The impetus for nanotechnology is strongly driven by the physical capability to manipulate atoms. The same instruments designed to inspect and visualize atoms are also highly suitable for moving and arranging atoms into new configurations. This 'top down' approach to nanotechnology is increasingly met with a set of 'bottom up' or industrial chemistry approaches to finely configure molecules.

Porter and Youtie (2009) note that in its ideal form nanotechnology should be a broad and interdisciplinary field of study. Yet critics have suggested that nanotechnology is merely a loose collection of knowledge with little substantive integration between fields of knowledge (Anderson et al. 2007). The field is loosely organized around the field of material science, yet physical and multidisciplinary chemical knowledge is an increasingly important component of research.

There are high ambitions for nanotechnology. The European Commission (2010) anticipates that nanotechnology will contribute to a number of grand societal challenges, including aging, climate change, optimal resource use and the creation of sustainable production systems. More than 30 US states have invested public monies to capture and grow new nanotechnology initiatives in their region. The ideal is for nanotechnology to become a fundamental source of high-technology earnings, entrepreneurship and employment (Shapira and Youtie 2008).

Nanotechnologies are already present today. Nanotechnology advances are mediated across a wide range of intermediate products including catalysts, coatings, paints and raw materials such as rubber. Furthermore nanotechnology is stimulating significant amounts of innovation across a wide range of industries including microelectronics, heavy equipment and aerospace (Roco and Bainbridge 2005).

Over US$250 million have been invested in nanotechnology research and development over the past 11 years. Nearly a quarter of this funding has been from public sources (US$67.5 million). The European Commission and European member states have been out-funding the United States in nanotechnology since 2007, with French and German national spending being exceptionally strong (Hullman 2006; Harper 2015). The extent of this funding can be hard to evaluate, since nanotechnology is often hidden in programmatic funding – especially for health care research. Similarly the purchasing power and effectiveness of funding vary widely by region (Harper 2015).

An existing literature on nanotechnology development asks how universities, governments and industry can set forth better policies and strategies to promote the growth of nanotechnology innovation. This question of how to make better technology policy leads naturally to a deeper inquiry into why certain regions are more active and more successful in nanotechnology.

A further set of questions asks why these activities are conducted in certain locations but not others. These are especially important questions given the magnitude of funding that is invested in nanotechnology. The answers to such a question require a deeper examination of a range of literatures from economics, geography and regional science. Given the diversity of research and the diversity of component technologies in nanotechnology, it is unreasonable then to expect there to be a single set of policies for promoting nanotechnology.

Like the policy impacts of nanotechnology, the human impacts of nanotechnology are likely to be highly multi-faceted (Roco and Bainbridge 2005). Some European member states are extremely cautious about the health and safety impacts of nanotechnology (Hullman 2006). Furthermore there is a fear in some industries about a consumer backlash against nanotechnology components in products. A body of new research in nanotechnology provides an integrative assessment of health and safety impacts (Goud 2006; Youtie et al. 2011).

This material provides an important impetus to questions concerning the distribution of nanotechnology production capabilities. It remains unclear whether these capabilities will be spatially highly distributed, or instead concentrated in comparatively few institutions and regions (Huang et al. 2011). This section described emerging work in the field of nanoscience and nanotechnology, and also some of the aspirations that policy-makers place on nanotechnology in developing prosperous new regions and renewed prosperity for nations.

The second section synthesizes current research on nanotechnology policy, outlining in particular the geographical determinants of these policies and why these policies are deemed to be effective. The previous material is presented without critique. Therefore in the third section a critique of the selection and use of geographical ideas is presented, and new opportunities for research are suggested. A summary and conclusion to the chapter are presented at the end.

RESEARCH IN NANOTECHNOLOGY POLICY

The section reviews much of the research on nanotechnology policy, paying particular attention to the methods which are used. Geographical reasoning in nanotechnology policy is quite abundant. Nonetheless this reasoning is rarely made very explicit. The purpose of this section is therefore to clearly set forth the geographical underpinning of the current literature on nanotechnology policy, utilizing the selected geographical concepts discussed in the previous section. As an organizing device the section considers a series of macro-, meso- and micro-scale explanations for the geographical incidence of nanotechnology innovation. Finally, I draw some conclusions about the locations and incidence of nanotechnology research and development.

Research evaluation attempts to measure the inputs and outputs of the research and development process. In the field of nanotechnology policy, research evaluation helps to quantify the impacts of public and private policies on the national and the local level. Inputs to research and development include measures such as skilled personnel and funding. Output indicators ideally include quantities such as new product offerings and new jobs created.

Quantifying the outputs of research and development is often quite difficult however, so measures of accomplishment such as research papers published, or patents issued, are often used as surrogates. Throughput indicators, which attempt to open the sometimes mysterious processes of research and development activity, are also sometimes used. In the nanotechnology policy literature many varied measures are used including the interdisciplinary character of research, spatial concentration, institutional quality or characteristics, degree of patenting dominance, and local and international research collaborations.

Macro-scale Explanations

Most nanotechnology research today is funded by private organizations (Harper 2015). Roco (2011) offers a range of development indicators which are up to date to 2008. These indicators include employment, value to market and venture capitalist deals. The indicators include values for the United States and the rest of the world.

The remainder of nanotechnology research and development, or roughly one-quarter of the funding, is financed by public organizations. These organizations are positioned at multiple levels in the research system. In Europe there is nanotechnology funding from both the European Commission, for all of Europe, as well as funding for individual member states. The United States provides funding for nanotechnology at both state and national levels. As noted earlier, there is also very extensive national funding from nations in Asia, including China, Japan, and South Korea. The policy literature is predominantly focused on public funding at the national level.

At a macro-scale nanotechnology is heavily dependent on national or supra-national funding. This funding is, in turn, often dependent on specific regional development or integration goals, which are made for specific regions at the national level. A final set of precursors hinge on the quality of national institutions – whether there is adequate regulation and governance and adequate provision of infrastructure, and whether there is a strong tradition of scientific research and development.

The research differs in emphasis, with some researchers focusing on the quality of non-material inputs such as institutions and policy, and others focusing more on funding initiatives. The national systems of innovation approach often attempts to wrap these factors into a single, integrated perspective of national prospects (Islam and Miyazaki 2006; Miyazaki and Islam 2007; Shapira et al. 2011). Some proponents of this school merge previous scholarship about systems of innovation (Carlsson et al. 2002) with actor-network theory (Bell and Callon 1994). The resultant approach especially emphasizes networks of actors, and their strategic choices, focusing particularly on economic, technical and market decision-making.

Although nanotechnology systems of innovation attempt a description of both the macro-level and the micro-level, they lend themselves to a more aggregated and more meso-level of analysis. The approach offers a constructive means of incorporating strategic choice and decision-making. Nonetheless the result is presented in a simple, and sometimes even coarse, manner.

Much of the work at the national level focuses on research outputs in the form of scientific papers (Islam and Miyazaki 2006; Porter and Youtie 2009; Takeda et al. 2009; Huang et al. 2011). Publicly funded research is naturally upstream of the market, and publications may be the best indicators of basic research accomplishment. Exemplary research giving funding indicators includes Hullman (2006) and Calignano (2013). National funding is therefore also of strong interest.

Comparatively fewer examples of patent analyses are given at the national level (Shapira et al. 2011). Nor do many papers in this strand of research give concrete examples of commercial products or applications (Miyazaki and Islam 2007). Coccia et al. (2012) offer some measures of research talent and the workforce. It is also worth noting research that provides a more institutional description (Jia et al. 2011). Many of these papers also mix indicators, providing for instance a mix of patent and publication indicators (Shapira et al. 2011), or a robust set of measures across a single nation (Coccia et al. 2012).

Macro-level analyses of nanotechnology production are largely descriptive in character. The work provides tables of primary sources (such as compilations of patents and publications). Other work re-uses or re-incorporates secondary sources (such as national databases of funding, or industrial databases of product sources). There is comparatively little modeling work, perhaps given the challenges of creating simple linear models of research input and research output. However the reasoning on derivative indicators is often sophisticated in character (Miyazaki and Islam 2007; Porter and Youtie 2009).

Given this research it is difficult to say who is 'winning' the global nanotechnology race. Certainly the United States is the incumbent, with historically the highest funding, a large number of research institutions which are on the cutting edge, and the largest numbers of publications and patents. However on many of these indicators the United States is now lagging. Europe has outpaced US nanotechnology funding for the past nine years. Japan and South Korea are leading in product announcements and commercial developments. China is increasingly taking the lead in publication outputs. Purchasing power considerations also mean that China can purchase relatively more given the same monetary input.

Meso-scale Explanations

National or supra-national choices for nanotechnology work themselves out at the regional level. This occurs in a number of different ways. Funding direct to universities creates a skilled workforce with new industrial knowledge and capabilities. Given a sufficient agglomeration of high-tech firms, research funding to universities can often create spillovers to the local industrial district (Varga 2000).

The industrial choice to locate in a nanodistrict is also strongly determined by the quality of the local infrastructure. New firms add their own high-tech plants or capital-intensive equipment to their district. Such firms may anchor the growth of an industrial ecology, as smaller companies and start-ups contribute to the district. Capital equipment and instrumentation is of particular importance for the growth of the nanotechnology district. Meyer (2007) provides indicators suggesting that instrumentation provides a common platform for a range of nanotechnology fields of interest. Furthermore instrumentation nanoscience is the most frequently cited, and therefore industrially utilized, field of nanotechnology research. For instance, the leading patenting sector across a range of countries is instrumentation. Only Japan, with extensive patenting in nanotechnology and cosmetics, counters this pattern. We do not have adequate data for new nanotechnology enterprises, but their founding may depend on boundary-spanning star scientists able to move fluidly from academic to industrial settings.

The existing literature on nanotechnology policy discusses both institutional and firm-level decision-making. Rothaermel and Thursby (2007) focus upon mechanisms for knowledge acquisition in the firm. These acquisition efforts are clearly quite different from prior evidence previously collected on biotechnology innovation (Powell et al. 2005). Quantitative methods, based on patent counts and general linear models, are the most commonly used analytical method here.

Heinze and Kuhlman (2008) describe the often fraught relationships between national laboratories, industry and universities. As an alternative these authors use a much more qualitative approach based on interviews and case study methodologies. Other authors attempt a more spatial or geographical explanation of districts and their evolution.

Frenken et al. (2010) examined a number of science-based industries in order to better clarify processes of scientific collaboration and institutional coordination. Spatial determinants vary widely by industry. Cunningham and Werker (2012) studied a range of different proximity measures when investigating nanotechnology collaboration in Europe. Both knowledge and physical proximity play roles in determining collaborative outcomes. Both sets of authors use publication counts, coupled with an econometric modeling approach.

Shapira and Youtie (2008) offer an exploratory analysis of the nanodistricts in the United States where nanotechnology publication and patenting occur. They utilize a cluster analysis methodology. The authors derive multiple clusters, with a correspondingly complex spatial distribution over the United States. The authors do not directly cite earlier foundational work on clusters or district (Feldman and Florida 1994; Markusen 1996), instead offering their own original interpretation of the nanodistrict. Compare for instance the work of Clark et al. (2009), who are much more explicit in the use of these concepts.

Nonetheless the work replicates the fundaments fairly well. Most of the research and development activity in the United States is occurring in a few technology-leading regions. These include Chicago, the Bay Area of California, Boston, the District of Columbia, Baltimore and New York City. These regions show strong involvement of corporations in publication and patenting. These regions may represent a new form of satellite platform district, with a high concentration of international and global players.

Nanobio research has its own distinctive profile. Existing biotechnology sectors in Los Angeles and San Diego are involved. There is strong involvement by both corporations and universities. There has been a strong push to publication in recent years, and the work is also very well cited. These regions may be a modern exemplar of flexible production, or the Marshallian industrial district. Nonetheless the districts constitute a relatively small proportion of the whole research output – some 10–15% of the totals.

Micro-scale Explanations

The whole success of the policy argument depends in large part at what happens at the micro-level. State and national funding is anchored at the macro- and meso-levels, and the choices of firms to create industrial districts is anchored at the meso-level. Nonetheless these policies will not be successful if the researchers and developers themselves do not take advantage of the locational opportunities which are provided to them.

Two papers in particular explore these concepts of local networks and professional ties in nanotechnology. Avenel et al. (2007) examined the knowledge base of the firm, using both publication and patenting indicators. Derived indicators illustrate the diversity of the knowledge used by the firm. Smaller firms are involved in a wide variety of science and technology topics. Larger firms, in contrast, focus on vertically integrated silos of knowledge, perhaps organized around specific product or process offerings.

Zucker and Darby (2007) examined scientists in the biotechnology sector. The authors demonstrate that entrepreneurial scientists limit knowledge flow in scientific research, but actively promote the production and commercialization of new technologies. The appropriation of new scientific knowledge means that local ties are needed for up-and-coming scientists to receive training. The work is largely archival in character.

A second paper by two of the lead authors more specifically explores the creation of knowledge in nanotechnology (Zucker et al. 2007). This is an econometric model based on publication and patents a public knowledge base of nanotechnology – the NanoBank. Their model incorporates growth and heterogeneity in knowledge production, by US region. The model convincingly shows how research funding (in this case the US National Science Foundation) anchors growth in nanotechnology publication.

The geographical prognosis for this work is quite clear. Nanotechnology knowledge production already seems quite locked in. There are several different processes of knowledge production and utilization at work, and these different processes may well be highly specialized by region. Whether the corporate strategies or government policies evidenced here are productive ones is yet another question.

It is much less clear that a good balance of national funding by region is offered, and whether this balance of funding matches regional capabilities. The public research funding is very concentrated; a considerable amount of funding is going to relatively few universities. The data and modeling work of Zucker et al. (2007) is performed for the United States; outcomes for other nations or regions are even less clear. Small company policy, which is absorbing a varied body of knowledge, seems like an effective response. Larger companies, however, are only selectively absorbing nanotechnology research, perhaps in support of specific and incremental research along existing product lines.

Summary

Figure 28.1 summarizes the arguments of this section. The argument was structured in layers – macro-, meso- and micro-level explanations of nanotechnology innovation. A variety of different institutional and geographical factors are considered by the literature. The policy measures available to decision-makers vary widely by the scale under consideration.

In order to integrate the discussion it is also helpful to consider several distinct chains of causality. These causal chains cut across the scales of analysis. The brief discussion to follow will emphasize three separate chains of action. Nonetheless all three chains are hypothesized to operate congruently and concurrently.

The first chain, which I call the 'institutions of production', involves the long-term creation of institutions that are favorable for high-tech development. These institutions encourage firms and other organizations to locate close to high-technology production facilities, including expensive banks of instrumentation. The co-location of these facilities with other resources leads to a rapid diffusion of innovation to those who are most likely to use the new knowledge or technology.

The second chain, which I call the 'cluster design problem', begins with the desire of national policy-makers to elevate the economic importance of particular districts or regions. Specific policy instruments are deployed to attract organizations with complementary assets, thereby building a vibrant ecology of new technology designers and developers. The cluster then benefits from knowledge spillovers, and becomes an autonomous and self-sustaining economic district.

The third chain, which I call 'public institutions of science', is dependent on public funding for research and development. The rationale for public funding hinges on the argument that individual organizations cannot uniquely appropriate the benefits of basic research, resulting in an under-investment by all. Thus national funding, and even regional funding, is often spent to create or anchor public research laboratories. These laboratories create new workforces, deliver new capabilities, and meaningfully embed knowledge in local social networks. The arguments presented here are the 'received wisdom' of the literature on nanotechnology policy. This review synthesizes arguments provided in a piece-meal fashion across multiple sources. This also, by necessity, elevates some arguments while neglecting others.

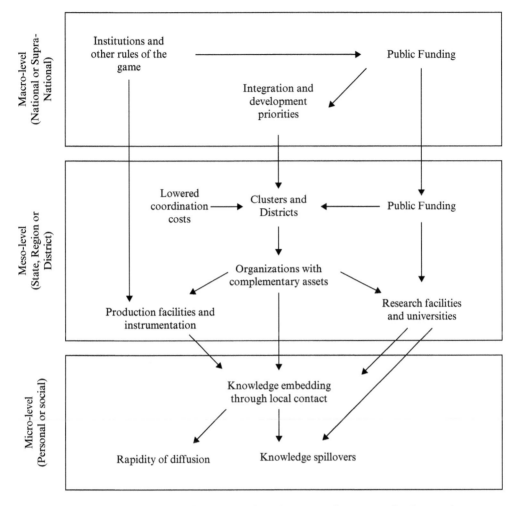

Figure 28.1   Macro-, meso- and micro-level explanations for nanotechnology policy

DISCUSSION AND CRITIQUE

The next section examines current debates concerning this tacit model of policy. General implications for policy and strategy are discussed. The state of knowledge concerning nanotechnology policy is broken down into scientific, technological and geographical factors. The state of the debate is thereby revealed. Furthermore in this section a critique is also offered concerning the state of research in nanotechnology policy. A conceptual and theoretical critique is offered. Then, a more general methodological critique is offered, most especially in light of the geographical synthesis provided in the previous sections.

There is no one nation clearly winning the nanotechnology race. In fact there are many distinct measures of nanotechnology research and development performance. Indeed there are equally varied possible and desirable policy outcomes as well. The winners may be determined in part by what sorts of economic and development outcomes are being sought. Research in the field appears very path-dependent, and with 30 years of basic research already conducted, progress may be locked in to a comparatively small number of regions.

In the United States at least, nanotechnology research and development is located in a number of high-tech corridors and is being performed by large companies in pursuit of incremental innovation in long chains of production and supply. The much lauded regions of Boston and the Bay Area are performing this research, but then so are a number of regions which are less traditionally credited with high-tech developments. These include New York, Chicago and Baltimore.

One prominent debate in the field concerns the very definition of nanotechnology. Nanotechnology is not really a single and coherent field of innovation (Anderson et al. 2007; Meyer 2007). Instead it is a loose bundle of related knowledges, which are slowly growing more coherent over time (Islam and Miyazaki 2006; Rafols et al. 2011). Because of this policies for nanotechnology must in the future be much more specific about which nanotechnologies are being addressed. Only then will it be fruitful to seek out specific spatial, institutional or sectoral correlates.

The transition process from research to development to commercialization may be underway (Coccia et al. 2012). Recent published research, which is undoubtedly a lagging indicator of actual commercialization progress, reveals major new efforts in materials (Rao and Cheetham 2001), medicine (Bertrand et al. 2014), the environment (Pendergast and Hoek 2011), manufacturing (Rogers and Nuzzo 2005) and energy (Howard and Spotnitz 2007). Despite this progress in commercialization, the actual character of nanotechnology commercialization remains to be seen.

Nanomaterials are not consumer products; these new materials are incorporated into other products to increase quality and performance. The resultant value chains are long and distributed, and demonstrate great flexibility in the kinds of products which can be produced. This presents fundamental challenges to directing and governing the innovation process (Rafols et al. 2011). Furthermore nanotechnology innovation is largely incremental, not radical. Nanotechnology is incorporated into existing products, rather than generating entirely new products (Meyer 2007).

The first view concerns whether nanotechnology will become a general purpose technology, and therefore become strongly dispersed throughout the economy (Roco 2011). The alternative perspective holds that nanotechnology expertise is highly knowledge-dependent, resulting in a path dependency and a strong spatial concentration of nanotechnology activities (Huang et al. 2011).

Nanotechnology offers new features not seen in previous sectors, including high-tech manufacturing or biotechnology. Nonetheless the literature offers limited evidence, and very mixed conclusions concerning the potential regional incidence of nanotechnology. One possible scenario could see nanotechnology broadly incorporated into a range of material goods and technologies.

The original conception of the general purpose technology originated with Breshnahan and Trajtenberg (1995). These authors note that major technological changes come in cascades, often led by growth in relatively few sets of technologies. These so-called general purpose technologies generally open up new opportunities rather than presenting final solutions. Nonetheless the governance of these technologies is problematic because the technologies are produced by a variety of actors, dispersed throughout the economy with a wide range of economic incentives. Traditional measures of industrial policy including strong intellectual property protections, the limitation of foreign competition and a relaxed anti-trust policy may do more harm than good because these instruments may harm investments in complementary areas of application.

If nanotechnology is primarily a general purpose technology then it should broadly diffuse throughout the economy. Practical training for designers, technicians and engineers would remain significant, as would ready access to this new workforce. The resultant industrial districts might look like a mix of Marshallian and state-anchored districts. Pavitt's supply-dominated sectors might be highly relevant here (Pavitt 1984).

Another vision might see nanotechnology stimulating new research inside the boundaries of large firms and their suppliers. The resultant district structure might look much like the satellite platform districts as described by Markusen (1996). Malerba and Orsenigo (1996) identify a range of sectors where research is concentrated in a few, large firms. Both sets of geographical impacts might be true.

Nanotechnology is highly heterogeneous in knowledge as well as in sector. While nanotechnology is increasingly founded in materials science and chemistry, nanobio is rapidly growing to be a discipline in its own right. Material science may see more, distributed industrial districts, while medical and biotechnological applications may see a high spatial concentration of activities.

In the final part of this section we turn to a theoretical and methodological critique of nanotechnology policy research. The processes at work in nanotechnology policy operate at multiple scales. While it is challenging to create multi-level statistical models, it is a well-established field with many well-developed applications. These multi-level statistical models may aid in generalization, while partitioning (if not fully explaining) observed heterogeneities by sector or by region.

The field of regional studies is making increasing used of spatially explicit models (Autant-Bernard 2009). This work has not diffused into the field of nanotechnology policy. This deficiency may stem from the use of concepts like cluster and district, where space, distance and proximity are implicit in the theory. However as nanotechnology knowledge diffuses more widely, and as there become a wider variety of research participants and innovation outputs, the need for spatial statistics of observed phenomena will only grow in urgency.

The specialty makes major use of publication and patent indicators. These are effective if partial measures of part of the research and development process, but it is not entirely clear that these are suitable measures of the theoretical constructs that are being used. As Meyer (2007) writes, there is only intermittent contact between the science and technology of nanotechnology. Individual research articles provide a fragmented view over the whole of the policy arena. This stems in part from the comprehensive character of the nanotechnology innovation process. Nonetheless it would be beneficial to knowledge in the field to have more end-to-end studies, even if they incorporated more qualitative or case-based research designs.

Particularly under-theorized are the micro-scale processes that anchor the theories of nanoclusters or nanodistricts. Finally, researchers seem to have adopted earlier geographical and economic theories in a loose and analogous manner. The whole of the theory should be used, including the more normative recommendations which emerge from the foundations. Alternatively nanotechnology policy researchers should reach from findings given a great variety of empirical measures to develop a new body of theory.

CONCLUSIONS

As noted, there are unresolved issues in nanotechnology policy and policy research. These issues invite a program of new research in geography on a few outstanding questions. The purpose of this concluding section is to outline three geographical questions where more research is needed. The final point is to summarize and conclude the chapter.

One area of concern and interest is the geography of the modern university. What are the types and varieties of universities, and why are some universities located where they are located? The brief review noted that the university is a construct of the medieval guild system. Therefore part of the answer has to do with the same geographical patterns of growth that lead to the creation of cities. However, the modern technology university emerged out of distinct forces, and was funded by novel mechanisms. The geographical incidence of the technology university is likely to be very different, and the regional impact of these choices likely to be very profound.

A second question concerns geographical thinking within national systems of innovation. A broad outline of theory concerning high-tech development is emerging. However current institutional and geographical research remains in an unnecessary opposition. Both kinds of research will be needed to better underpin pressing questions concerning the pattern of development in the so-called 'BRICK' nations – Brazil, Russia, India, China and South Korea. Existing work in geography, institutionalism and technology policy is not sufficient to fully explain and analyze these nations.

Nonetheless there is a strong foundation with which to begin such a work. Nelson (1993) provides a theoretical framework and extensive case studies including nations of China, South Korea and Brazil. Storper (1995, 1997) provides additional welcome perspective on the role of regions in the world economy. Haggard (1990) describes the politics of growth in formerly marginalized, newly industrialized countries. There are also a number of relevant and high-profile case studies examining the regional prerequisites for success in a modern economy (Woolcock 1998; Yeung 2009).

Bueno de Mesquita et al. (2004) expand the geographical and institutional account of high-tech development. Perhaps most importantly these authors describe the political logic of growth in autocratic nations. There are manifest growth advantages for many autocratic governments. Seemingly these nations avoid the anti-coordination problems of newly democratizing, middle-tier nations. Another important aspect of this work is its effort to explain the so-called Dutch disease – why strong natural resource endowments (such as petroleum reserves) so often lead to stagnant growth and poor public goods provision. Further explanatory power would be granted by a discussion of low-level equilibria or growth traps; a phenomena may emerge naturally out of endogeneous growth theories (Fagerberg 1988).

A final topic of research concerns the geographical underpinning of complex social networks. These networks seemingly enable thick social ties at the local level, while spanning continents at the global network (Marrocu et al. 2013). The fact that these networks encompass many kinds of proximity does not lessen the importance of geography in their formation and their persistence. A richer understanding of such epistemic networks is needed for technology policy, as well as for economic geography.

This chapter provided a brief overview of the field of nanotechnology. Once a buzzword of popular science, and then a rallying cry for strategic research, nanotechnology is now a maturing field of technology. A particular, and even somewhat idiosyncratic, selection of geographical ideas are presented in this chapter in support of current thinking in nanotechnology policy. These ideas are used as part of an implicit theory of technological systems which seems to underlie science and technology policy research.

Progress in using these selective geographical concepts in analyzing the incidence of nanotechnology research remains quite limited. There is an increasing amount of empirical evidence about where the nanotechnology research and development is being performed, but there is limited theoretical understanding of locational choice in nanotechnology. This includes critical understanding of where research laboratories and production facilities are being located. Furthermore the degree to which these research facilities are transforming and revitalizing economic districts still remains unclear. Regardless, increasing evidence suggests that nanotechnology developments are incremental, and occurring inside the laboratories of multinational firms.

REFERENCES

Anderson, B., M. Kearnes and R. Doubleday. 2007. Geographies of nano-technoscience. *Area* 39(2): 139–142.

Autant-Bernard, C. 2009. Spatial econometrics of innovation: recent contributions and research perspectives. *Spatial Economic Analysis* 74: 403–419.

Avenel, E., A.V. Favier, S. Ma and V. Mangematin. 2007. Diversification and hybridization in firm knowledge bases in nanotechnologies. *Research Policy* 36: 864–870.

Bell, G. and M. Callon. 1994. Techno-economic networks and science and technology policy. *STI Review* 14: 59–118.

Bertrand, N., J. Wu, X. Xu, N. Kamaly and O. Farokhzad. 2014. Cancer nanotechnology: the impact of passive and active targeting in the era of modern cancer biology. *Advanced Drug Delivery Reviews* 66: 2–25.

Binnig, G., H. Rohrer, C. Gerbe and E. Weibe. 1982. Surface studies by scanning tunneling microscopy. *Physical Review Letters* 49(1): 57–61.

Breshnahan, T. and M. Trajtenberg. 1995. General purpose technologies: engines of growth? *Journal of Econometrics* 65: 83–108.

Bueno de Mesquita, B., A. Smith, R.M. Siverson and J. Morrow. 2004. *The Logic of Political Survival*. Cambridge, MA: MIT Press.

Calignano, G. 2013. Italian organisations within the European nanotechnology network: presence, dynamics and effects. *Die Erde* 145(4): 241–259.

Carlsson, B., S. Jacobsson, M. Holmen and A. Rickne. 2002. Innovation systems: analytical and methodological issues. *Research Policy* 31: 233–245.

Clark, J., H.-I. Huang and J. Walsh. 2009. A typology of innovation districts: what it means for regional resilience. *Cambridge Journal of Regions, Economy and Society* 31: 121–137.

Coccia, M., U. Finardi and D. Margon. 2012. Current trends in nanotechnology research across worldwide geo-economic players. *Journal of Technology Transfer* 37(5): 777–787.

Cunningham, S. and C. Werker. 2012. Proximity and collaboration in European nanotechnology. *Papers in Regional Science* 91(4): 723–742.

Drexler, K.E. 1986. *Engines of Creation*. New York: Doubleday.

European Commission. 2010. Europe 2020, accessed 22 December 2013 from http://ec.europa.eu/europe2020/index_en.htm.

Fagerberg, J. 1988. International competitiveness. *Economic Journal* 98(391): 355–374.

Feldman, M. and R. Florida. 1994. The geographic sources of innovation: technological infrastructure and product innovation in the United States. *Annals of the Association of American Geographers* 84(2): 210–229.

Feynman, R. 1959. *Plenty of Room at the Bottom*. Pasadena, CA: American Physical Society.

Frenken, K., R. Ponds and F. van Oort. 2010. The citation impact of research collaboration in science-based industries: a spatial-institutional analysis. *Papers in Regional Science* 89(2): 351–372.

Goud, P. 2006. Nanomaterials face control measures. *Nano Today* 12: 34–39.

Haggard, S. 1990. *Pathways from the Periphery, The Politics of Growth in the Newly Industrializing Countries*. Ithaca, NY: Cornell University Press.

Harper, T. 2015. Nanotechnology funding: a global perspective, accessed from http://www.nano.gov/sites/default/files/pub_resource/global_funding_rsl_harper.pdf.

Heinze, T. and S. Kuhlman. 2008. Across institutional boundaries? Research collaboration in German public sector nanoscience. *Research Policy* 37: 888–899.

Howard, W. and R. Spotnitz. 2007. Theoretical evaluation of high-energy lithium metal phosphate batteries. *Journal of Power Sources* 165(2): 887–891.

Huang, C., A. Notten and N. Rasters 2011. Nanoscience and technology publications and patents: a review of social science studies and search strategies. *Journal of Technology Transfer* 36: 145–172.

Hullman, A. 2006. Who is winning the global nanorace? *Nature Nanotechnology* 1: 81–83.

Islam, N. and K. Miyazaki 2006. Nanotechnology innovation system: understanding hidden dynamics of nanoscience fusion trajectories. *Technological Forecasting and Social Change* 76: 128–140.

Jia, L., Y. Zhao and X.-J. Liang. 2011. Fast evolving nanotechnology and relevant programs and entities in China. *Nano Today* 6: 6–11.

Malerba, F. and L. Orsenigo. 1996. Schumpeterian patterns of innovation are technology-specific. *Research Policy* 25: 451–478.

Markusen, A. 1996. Sticky places in slippery space: a typology of industrial districts. *Economic Geography* 72(3): 293–313.

Marrocu, E., R. Paci and S. Usai. 2013. Proximity, networking and knowledge production in Europe: what lessons for innovation policy? *Technological Forecasting and Social Change* 80: 1484–1498.

Meyer, M. 2007. What do we know about innovation in nanotechnology? Some propositions about an emerging field between hype and path-dependency. *Scientometrics* 70(3): 779–810.

Miyazaki, K. and N. Islam. 2007. Nanotechnology systems of innovation – an analysis of industry and academia research activities. *Technovation* 27: 661–675.

National Nanotechnology Initiative. 2016. What's so special about the nanoscale?, accessed from http://www.nano.gov/nanotech-101/special.

Nelson, R. (ed.). 1993. *National Systems of Innovation: A Comparative Analysis*. Oxford: Oxford University Press.

Pavitt, K. 1984. Sectoral patterns of technical change: towards a taxonomy and a theory. *Research Policy* 13(6): 343–373.

Pendergast, M. and E. Hoek. 2011. A review of water treatment membrane nanotechnologies. *Energy and Environmental Science* 46: 1946–1971.

Porter, A. and J. Youtie. 2009. How interdisciplinary is nanotechnology? *Journal of Nanoparticle Research* 11(5): 1023–1041.

Powell, W., D. White, K. Koput and J. Owen-Smith. 2005. Network dynamics and field evolution: the growth of interorganizational collaboration in the life sciences. *American Journal of Sociology* 110: 1132–1205.

Rafols, I., P. van Zwanenberg, M. Morgan, P. Nightingale and A. Smith. 2011. Missing links in nanomaterials governance: bringing industrial dynamics and downstream policies into view. *Journal of Technology Transfer* 36: 624–639.

Rao, C.N.R. and A. Cheetham. 2001. Science and technology of nanomaterials: current status and future prospects. *Journal of Materials Chemistry* 11(12): 2887–2894.

Roco, M. 2011. The long view of nanotechnology development: the National Nanotechnology Initiative at 10 years. *Journal of Nanoparticle Research* 13: 427–445.

Roco, M. and W. Bainbridge. 2005. Societal implications of nanoscience and nanotechnology: maximizing human benefit. *Journal of Nanoparticle Research* 71: 1–13.

Rogers, J. and R. Nuzzo. 2005. Recent progress in soft lithography. *Materials Today* 82: 50–56.

Rothaermel, F. and M. Thursby. 2007. The nanotech versus the biotech revolution: sources of productivity in incumbent firm research. *Research Policy* 36: 832–849.

Shapira, P. and J. Youtie. 2008. Emergence of nanodistricts in the United States. *Economic Development Quarterly* 22(3): 187–199.

Shapira, P., J. Youtie and L. Kay. 2011. National innovation systems and the globalization of nanotechnology innovation. *Journal of Technology Transfer* 36: 587–604.

Storper, M. 1995. Territorial development in the global learning economy: the challenge to developing countries. *Revue Région and Dévelopment* 1: 1–37.

Storper, M. 1997. *The Regional World: Territorial Development in a Global Economy*. New York: Guilford Press.

Takeda, Y., S. Mae, Y. Kajikawa and K. Matsushima. 2009. Nanobiotechnology as an emerging research domain from nanotechnology: a bibliometric approach. *Scientometrics* 80(1): 23–38.

Varga, A. 2000. Local academic knowledge transfers and the concentration of economic activity. *Journal of Regional Science* 40(2): 289–309.

Woolcock, M. 1998. Social capital and economic development: toward a theoretical synthesis and policy framework. *Theory and Society* 27: 151–208.

Yeung, H.W. 2009. Regional development and the competitive dynamics of global production networks: an East Asian perspective. *Regional Studies* 43(3): 325–351.

Youtie, J., A. Porter, P. Shapira, L. Tang and T. Benn. 2011. The use of environmental, health and safety research in nanotechnology research. *Journal of Nanoscience and Nanotechnology* 11(1): 158–166.

Zucker, L. and M. Darby. 2007. Virtuous circles in science and commerce. *Papers in Regional Science* 86(3): 445–471.

Zucker, L., M. Darby, J. Furner, R.C. Liu and H. Ma. 2007. Minerva unbound: knowledge stocks, knowledge flows and new knowledge production. *Research Policy* 36: 850–863.

PART VII

LIFE SCIENCE
TECHNOLOGIES

29. Biotechnology: commodifying life
Barney Warf

Biotechnology may be defined as the application of molecular and cellular processes to solve problems, develop products and services, or modify living organisms to carry desired traits. Essentially it means creating or changing organisms to generate profit. Arising in the 1970s, biotechnology has been a rapidly growing industry worldwide, with extensive linkages to agriculture, health care, energy and environmental sciences. Rapid recent advances in genomics (e.g. the ability to decode genomes) have propelled the industry to new heights. Conversely, the biotechnology industry has helped to fuel insights in molecular biology.

This chapter approaches the biotechnology industry from four angles. It opens with a brief survey of its historical trajectory, during which it changed from being a means of enhancing beer production to a technology that facilitated the rise of cloning. Second, it focuses on the impacts of biotech products, which are largely agricultural and medicinal in nature. Third, it turns to the regulation of the industry at the national and international scales. The fourth part examines the locational dynamics of biotechnology firms, particularly their propensity to cluster in distinct districts. The conclusion summarizes the major observations.

A BRIEF HISTORY OF BIOTECHNOLOGY

Human beings have modified the genetic composition of plants and animals for millennia, if one counts selective breeding that began during the Neolithic Revolution. The perfection of the process of fermentation might be considered another step in this process. Biotechnology arose from the field of zymotechnology or zymurgy, which began as a search for a better understanding of industrial fermentation, particularly beer (Bud and Cantley 1994).

However, it was during the early 20th century that modern biotechnology came into its own (Bud 1991; Peacock 2010). In 1917, Chaim Weizmann was the first to use a microbiological culture commercially by manufacturing corn starch using *Clostridium acetobutylicum* to produce acetone, a product in great demand and short supply in Britain during the First World War. The Hungarian Károly Ereky coined the word 'biotechnology' in 1919. Biotechnology was central to the invention of antibiotics; in 1928, Alexander Fleming discovered penicillin, and purified its antibiotic compounds (Colwell 2002; Fiechter and Beppu 2000).

The late 20th century brought a slew of new innovations in this field. Watson and Crick's discovery of the molecular structure of DNA in 1953 was obviously a major advance (Goujon 2001). Not surprisingly, each new scientific advance became a media event designed to capture public support and potential investors. As it became

433

increasingly well established, biotechnology generated new firms such as Biogen and Genentech, and in 1980, the Biotechnology Industry Organization (Rasmussen 2014).

In 1971, Paul Berg at Stanford University pioneered gene splicing, a moment often considered to be the birth of modern biotechnology. In 1972, Herbert Boyer at the University of California at San Francisco and Stanley Cohen at Stanford began transferring genetic material into a bacterium. The discovery in 1973 of recombinant DNA accelerated this process. In 1978 Genentech produced recombinant human insulin for the first time, allowing it to be created in mass quantities, followed by artificial human growth hormone in 1979 and a vaccine for hepatitis B. The first genetically modified organism was patented in 1980, a bacterium designed by Ananda Chakrabarty, working for General Electric, to consume petroleum leaked during spills, a process legalized by a Supreme Court decision. In 1982, the first biotech drug, human insulin, produced in genetically modified bacteria by Genentech and Eli Lilly, was approved by the US Food and Drink Administration (FDA). In 1986, Interferon became the first anticancer drug produced through biotechnology processes. In 1990, the first successful gene therapy was performed on a four-year-old girl suffering from an immune disorder. In 1995, gene therapy, immune-system modulation and recombinantly produced antibodies entered the war against cancer. And in 1997, a sheep famously named Dolly, in Scotland, became the first animal cloned from an adult cell.

In the early 21st century, biotechnology has again experienced multiple rounds of innovation and growth. In 2001, the FDA approved Gleevec, a gene-targeted drug for patients with chronic myeloid leukemia, the first gene-targeted drug to receive FDA approval. In 2003, the Human Genome Project completed sequencing of the human genome, opening the door to a wide array of new possibilities in gene therapy.

APPLICATIONS OF BIOTECHNOLOGY

Biotechnology products have a wide range of industrial and health care uses. In agriculture ('green biotech'), biotechnology was central to the development of genetically modified organisms (GMOs) (Mannion and Morse 2012). GMO crops are those with genetic compositions modified to resist disease, predators or drought, reduce resistance to herbicides, or enhance nutrient quality (e.g. higher protein levels). The first such crop was an antibiotic-resistant tobacco created in 1982. The first GMO crop approved for sale in the United States, in 1994, was the *FlavrSavr* tomato (no longer on the market), which had a longer shelf life because it took longer to soften after ripening (Peacock 2010). In 1995, the Environmental Protection Agency (EPA) approved the BT potato, the first pesticide-producing crop, which was followed by pesticide-producing cotton, canola oil, maize and soybeans. These were followed by virus-resistant squash and vitamin A-enriched golden rice, which is useful in minimizing vitamin A deficiency. Most GMO crops are planted to resist herbicides. Roundup-ready potatoes and soybeans resist pests better than non-GMO equivalents. In 2014, a Chinese researcher filed patents for mildew-resistant wheat. In 2015, the US Department of Agriculture and the EPA approved Arctic apples, which do not turn brown upon opening. Much GM crop output is used for animal feed, and rising global demand for meat has led to higher demand for GM crops. The first GMO animal approved for sale was AquAdvantage salmon, which went on the market in 2015.

In 2003, the first genetically modified pet, GloFish, was sold commercially. Major corporate actors behind these efforts include Monsanto, Calgene and Ciba-Geigy, which jealously guard the intellectual property rights of their innovations (Juma 2014).

GMO crops have long been the subject of enormous debate (Dronamraju 1998). Critics allege that the health risks have not been adequately assessed and that they pose significant risks of environmental or ecological damage (Vallero 2010). Another controversy is whether or not GMO foods should be labeled as such; the FDA currently does not require labeling. The scientific consensus is that GMO crops pose no undue health risks. Nonetheless, opposition to GMOs has included widespread protests and attempts to disrupt research in this area. In 2000, 300 foods, including Starlink corn, which had not been approved for human consumption, were recalled, the first-ever recall of a GMO crop.

Despite heated controversies about the potential health risks of GMO crops, farmers adopted them enthusiastically. Between 1996 and 2013, the world's acreage in GMO-planted crops increased by 10,000%, and today includes more than 430 million acres. Globally, only five countries produce 95% of the world's GMOs, including the United States, Canada, Brazil, Argentina and China (Figure 29.1). In the United States, GMO crops account for 94% of soybeans, 96% of cotton and 92% of corn. Klümper and Qaim (2014) conclude that GMOs have reduced chemical pesticide use by 37%, increased crop yields by 22% and increased farmer profits by 68%.

Biofuels, which use crops as inputs, are another application of biotechnology. During the 1970s, the United States promoted 'gasohol', gasoline with 10% alcohol, as a means to reduce petroleum consumption. Today, the most common biofuel is bioethanol, produced from bioengineered crops such as corn, potatoes, sorghum, soybeans and sugar cane. Bioethanol is often touted as a substitute for gasoline, although there are widespread concerns about its impact on food production and prices, as well as the energy needed to produce it. Global production of ethanol quadrupled between 2000 and 2010. More than 87% of the world's ethanol is produced by the United States and Brazil. Brazil relies on it heavily for engine fuel; 20% of the country's cars rely solely on ethanol. Gasoline there contains at least 25% ethanol. In the United States, it is produced in corn-growing areas such as Iowa.

Industrial applications of biotechnology ('white biotech') include the production of food (e.g. chymosin in cheese making), enzymes that act as catalysts in the production of expensive chemicals, bacteria that detect arsenic in drinking water and genetically modified microbes that devour petroleum or extract heavy metals (notably mercury and selenium) from their environment to make them more recoverable, a process known as biomining or biorecovery. Engineered organisms are used in making paper, textiles and detergents. A genetically modified virus has been used to create a more environmentally friendly lithium-ion battery.

In health care ('red biotech'), biotechnology has led to new pharmaceuticals such as mass-produced insulin, human growth hormones, follistim (to treat infertility), albumin, vaccines and monoclonal antibodies. Transgenic carrots have been used to produce the drug Taliglucerase alfa, which is used to treat Gaucher's disease. Biotech processes include genetic engineering, the modification of the genetic composition of cells. Gene targeting can add or remove genes from cells and allow them to reproduce. The first such organisms were bacteria, such as the ones designed to consume petroleum, although

Figure 29.1 World's major producers of GMOs

insulin-producing mice followed soon thereafter. This process includes the transfer of genes within and across species. Genetic testing perfected by biotechnology methods allows for early identification of inherited disorders. Human treatments often find parallel applications in veterinary care.

Gene therapy involves replacing defective genes (mutants) with effective ones, usually through the delivery of nucleic acid polymers into a patient's cells as a drug to treat disease. The history of this approach dates back to Martin Cline's unsuccessful attempts to modify human DNA in 1980, but the first successful transfer occurred in 1989. Therapeutic gene therapy began in 1990 with the work of French Anderson on patient Ashi DeSilva. A genetically engineered baby was born in 1996, when doctors took a donor's egg with healthy mitochondria, removing its nuclear DNA and filling it with the nuclear DNA of the biological mother. Despite a long early history of multiple failures, including several patient deaths, gene therapy began to enjoy successes in the 2000s with treatments for the retinal disease Leber's congenital amaurosis, adrenoleukodystrophy, chronic lymphocytic leukemia, multiple myeloma (cancer of the plasma cells), hemophilia and Parkinson's disease. China approved the first commercial genetic therapy, Gendicine, in 2003. In 2011, Russia approved Neovasculgen for the treatment of arterial disease. One hope of such approaches is 'personalized medicine' tailored to each individual's unique genetic composition.

THE REGULATION OF BIOTECHNOLOGY

Because it potentially poses significant risks to human health and to the environment, biotechnology is closely monitored and regulated by the state. The invention of recombinant DNA (rDNA) led the National Institutes of Health to form a rDNA advisory committee. In 1982, concerned about the release of GMOs into the environment, the OECD voiced concerns over regulation. In the United States, responsibility for approving GMOs belongs to the Department of Agriculture, the FDA and the EPA (Pew Charitable Trust 2001). GMOs proposed for release into the environment must be assessed under the Plant Protection Act (McHughen and Smyth 2008). In 2000, the Cartagena Protocol, signed by 157 states, was adopted to govern the international transfer, handling and use of genetically modified GMOs. All shipments containing GMOs that are intended to be used as feed, food or for processing must be identified.

The survival and success of biotech firms is heavily affected by federal research funds, primarily through institutions such as the National Science Foundation and the National Institutes of Health as well as the Small Business Technology Transfer program, Small Business Innovation Research, Environmental Protection Agency and the Food and Drug Administration. Federal policies regarding patents and intellectual property rights, subsidies for medical research and national health care programs are all important. State-level determinants are also critical, including regulatory policies, educational systems, university medical facilities, taxation and subsidies (Goetz and Morgan 1995).

Internationally, there are wide variations in the regulation of the industry, most notably between the United States and the European Union, in which consumers have notably different views on the subject (Gaskell et al. 1999). In 1997, the EU began requiring labeling of GMO crops for human consumption, and they are subject to intensive, case-by-case analysis by the European Food Safety Authority (Davison 2010). The EU differentiates between approval for cultivation and approval for the import and processing of biotech products; while only a few GMO crops have been approved for cultivation, many others have been approved for import. Individual EU members may permit or ban a given GMO as they see fit.

THE ECONOMIC GEOGRAPHY OF BIOTECHNOLOGY

In 2014, the world's biotechnology industry consisted of roughly 600 firms that employed 170,000 people and generated $90 billion in output. There is a wide range in the size of firms in this industry, including single proprietorships and firms of more than 500 employees. Salaries in the biotech industry tend to be well above the national average. Because pharmaceutical companies are so heavily involved in funding research for biotechnology products, they are, not surprisingly, the largest firms in this field (Table 29.1); the majority are based in the United States. Among 'pure' biotechnology firms, Amgen, based in Thousand Oaks, California, is the largest, earning $17 billion in 2014. In the United States, the industry employs roughly 100,000 people.

Venture capital is critical to making basic research in biotechnology commercially viable. Most small biotech firms lose money, given the high costs and enormous amounts of research necessary to generate their output and the long lag between R&D and commercial deployment. Only one in 1000 patented biotechnology innovations leads to a successful commercial product, and it may take 15 years and as much as $800 million to reach that point (Baum and Silverman 2004). The vast majority of start-ups go bankrupt within a few years as venture capital runs out, indicating the need for constant supply of new business ventures to keep the sector vibrant. In 2015, the US biotech sector attracted more than $8 billion in venture capital, which comprised 20% of all venture capital in the United States; the bulk of these funds both originated in and went to established firms in California and Boston. Venture capitalists may invest in many different biotech firms, and one biotech firm may receive funding from several venture capitalists. Venture capitalists look for an experienced management team when deciding in which companies they will invest. Venture

Table 29.1   World's 10 largest biotechnology firms, 2015

Rank	Company	Country	Capitalization ($ billions)
1	Johnson & Johnson	United States	284.2
2	Roche	Switzerland	234.7
3	Novartis	Switzerland	206.1
4	Pfizer	United States	199.3
5	Novo Nordisk	Denmark	153.9
6	Merck	United States	147.6
7	Gilead Sciences	United States	145.8
8	Allergan	United States	123.2
9	Amgen	United States	122.5
10	Bristol-Myers	United States	114.8

capitalists provide advice and professional contacts, and serve on the boards of directors of young biotech firms (Jaffe et al. 1993; Zucker et al. 1998). As a biotech firm survives and prospers, venture capitalists typically withdraw from day-to-day management. Many investors sell their shares (stocks) when the firm offers its Initial Public Offering.

Biotech firms commonly tend to cluster in distinct districts, and place-based characteristics are essential for the industry's success in innovation (Prevezer 1997; Hall and Bagchi-Sen 2001; Bagchi-Sen and Scully 2004). Europe, for example, hosts the BioValley Network situated between France, Germany and Switzerland (Claassens 2004; Mytelka 2004; Ter Wal 2014). The Aachen region in Germany has emerged as a significant center in its own right (Plum and Hassink 2011). In the UK, Cambridge has assumed this role (Smith 2004), as has Oxford (Smith 2004). Similarly, Denmark and Sweden formed Medicon Valley (Coenen et al. 2004). Biotech clusters have also been studied in less-established areas not usually associated with the industry, such as Austria (Trippl and Tödtling 2007) and Shanghai (Zhang and Wu 2012).

Geographically, the US biotechnology industry is dominated by a small handful of cities (Cortright and Mayer 2002), particularly Boston, San Diego, Los Angeles, San Francisco, New York, Philadelphia, Seattle, Raleigh (NC) and Washington, DC (Figure 29.2), which combined account for three-quarters of the nation's biotech firms and employment. These cities have excellent universities with medical schools and state-of-the-art infrastructures, and offer an array of social and recreational environments. However, Ferguson (2004) questions whether biotech parks indeed offer a significant advantage not found elsewhere.

Biotechnology firms agglomerate for several reasons. In a highly competitive environment in which the key to success is the rate of new product formation, and in which patent protections lead to a 'winner take all' scenario, the success of biotech firms is closely related to their strategic alliances with universities and large pharmaceutical corporations (Delaney 1993; Deeds and Hill 1996). Although many biotechnology firms engage in long-distance partnering, these tend to be complements to, not substitutes for, co-location in clusters where tacit knowledge is produced and circulated face-to-face, both on and off the job (Arora and Gambardella 1990; Powell et al. 1996, 2005). The flows of information within and between biotechnology clusters, as well as their links to worldwide networks, have been subjects of considerable analysis (Gertler and Levitte 2005; Gittelman 2007).

Because pools of specialized skills and a scientifically talented workforce are essential to the long process of research and development, an essential element defining the locational needs of biotech firms is the location of research universities and institutions and the associated supply of research scientists (Zucker et al. 1998). Most founders of biotech firms are research scientists with university positions. Because knowledge is generated and shared most efficiently within close loops of contact, the creation of localized pools of technical knowledge is highly dependent on the detailed divisions of labor and constant interactions among colleagues in different but related firms. Successful biotechnology firms often revolve around the presence of highly accomplished academic or scientific 'stars' with the requisite technical skills but also the vision and personality to market them.

Work in the biotech industry is characterized by a high degree of uncertainty (Eaton and Bailyn 1999), and its employees often work long hours. Formal linkages among biotech employees may occur over long distances using the Internet, but the informal ones essential to the creation of synergistic economies of scale necessitate geographic proximity (Audretsch and Stephan 1996). In this respect, biotechnology centers offer

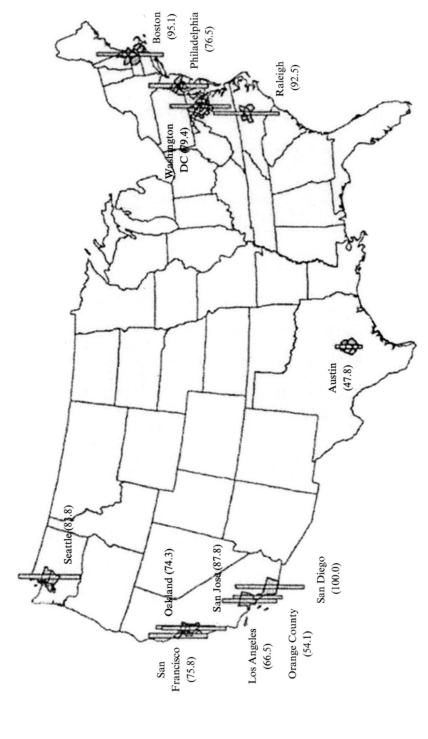

Source: Author.

Figure 29.2 Major biotechnology clusters in the United States

classic examples of clusters that generate innovation and autocatalytic growth (Cooke 2004). Thus, issues of employment in biotechnology cannot be considered separately from those of housing and lifestyle, as Florida's (2004) famed 'creative class' thesis asserts. For a well-educated labor pool in a rapidly growing industry with considerable locational mobility, quality-of-life issues cannot be divorced from those pertaining to work and productivity.

CONCLUSIONS

Human modification of plants and animals is not new, although contemporary biotechnology has taken this process to new levels. Biotechnology innovations grew by leaps and bounds in the late 20th century, and came to include genetically modified organisms. The industry has found a wide range of applications in agriculture, pharmaceuticals and health care. Widespread debates about GMO crops continue, as do real and potential biotech applications such as cloning, stem cell research and gene therapy.

Biotechnology epitomizes the entrepreneurial, knowledge-intensive and networked firms that characterize much of the world's economy today. Critical variables that underpin the location of biotech firms include skilled scientific labor, progressive public policies and access to venture capital. Biotech firms are largely concentrated in a handful of metropolitan areas that offer ready access to well-educated labor, specialized ancillary services, large universities and venture capital firms. Geographic proximity, both on and off the job, is important to the generation of knowledge spillovers and the synergistic effects that these complexes generate. Because many of the contacts among scientists, researchers and other works are informal, local environments conducive to the exchange of tacit knowledge are key to successful biotech regions.

REFERENCES

Arora, A. and A. Gambardella. 1990. Complementary and external linkages: the strategies of the large firms in biotechnology. *Journal of Industrial Economics* 38: 361–379.

Audretsch, D. and P. Stephan. 1996. Company-scientist locational links: the case of biotechnology. *American Economic Review* 86: 641–652.

Bagchi-Sen, S. and J. Scully. 2004. The Canadian environment for innovation and business development in the biotechnology industry: a firm-level analysis. *European Planning Studies* 12: 961–983.

Baum, J. and B. Silverman. 2004. Picking winners or building them? Alliance, intellectual, and human capital as selection criteria in venture financing and performance of biotechnology startups. *Journal of Business Venturing* 19: 411–436.

Bud, R. 1991. Biotechnology in the twentieth century. *Social Studies of Science* 21(3): 415–457.

Bud, R. and M. Cantley. 1994. *The Uses of Life: A History of Biotechnology.* Cambridge: Cambridge University Press.

Claassens, M. 2004. Life sciences cluster in the centre of Europe. *Chimia* 58: 769–770.

Coenen, L., J. Moodysson and J. Asheim. 2004. Nodes, networks and proximities: on the knowledge dynamics of the Medicon Valley biotech cluster. *European Planning Studies* 12: 1003–1018.

Colwell, R. 2002. Fulfilling the promise of biotechnology. *Biotechnology Advances* 20(3–4): 215–228.

Cooke, P. 2004. Regional knowledge capabilities, embeddedness of firms and industry organisation: bioscience megacentres and economic geography. *European Planning Studies* 12: 625–641.

Cortright, J. and H. Mayer. 2002. *Signs of Life: The Growth of Biotechnology Centers in the U.S.* Washington, DC: Brookings Institution.

Davison, J. 2010. GM plants: science, politics and EU regulations. *Plant Science* 178(2): 94–98.

Deeds, D. and C. Hill. 1996. Strategic alliances and the rate of new product development: an empirical study of entrepreneurial biotechnology firms. *Journal of Business Venturing* 11: 41–55.

Delaney, E. 1993. Technology search and firm bounds in biotechnology – new firms as agents of change. *Growth and Change* 24: 206–228.

Dronamraju, K. 1998. *Biological and Social Issues in Biotechnology Sharing*. Brookfield: Ashgate.

Eaton, S. and L. Bailyn. 1999. Work and life strategies of professionals in biotechnology firms. *Annals of the American Academy of Political and Social Science* 562: 159–173.

Ferguson, R. 2004. Why firms in science parks should not be expected to show better performance – the story of twelve biotechnology firms. *International Journal of Technology Management* 28: 470–482.

Fiechter, A. and T. Beppu. 2000. *History of Modern Biotechnology*. Berlin: Springer.

Florida, R. 2004. *The Rise of the Creative Class*. New York: Basic Books.

Gaskell, G., M. Bauer, J. Durant and N. Allum. 1999. Worlds apart? The reception of genetically modified foods in Europe and the U.S. *Science* 285(5426): 384–387.

Gertler, M. and Y. Levitte. 2005. Local nodes in global networks: the geography of knowledge flows in biotechnology innovation. *Industry and Innovation* 12(4): 487–507.

Gittelman, M. 2007. Does geography matter for science-based firms? Epistemic communities and the geography of research and patenting in biotechnology. *Organization Science* 18(4): 724–741.

Goetz, S. and R. Morgan. 1995. State level locational determinants of biotechnology firms. *Economic Development Quarterly* 9: 174–185.

Goujon, P. 2001. *From Biotechnology to Genomes: The Meaning of the Double Helix*. Singapore: World Scientific.

Hall, L. and S. Bagchi-Sen. 2001. An analysis of R&D, innovation, and business performance in the U.S. biotechnology industry. *International Journal of Biotechnology* 3(3): 1–10.

Jaffe, A., M. Trajtenberg and R. Hendersen. 1993. Geographic localization of knowledge spillover as evidenced by patent citations. *Quarterly Journal of Economics* 108: 577–598.

Juma, C. 2014. *The Gene Hunters: Biotechnology and the Scramble for Seeds*. Princeton, NJ: Princeton University Press.

Klümper, W. and M. Qaim, M. 2014. A meta-analysis of the impacts of genetically modified crops. *PLoS One* 9(11): e111629.

Mannion, A. and S. Morse. 2012. Biotechnology in agriculture: agronomic and environmental considerations and reflections based on 15 years of GM crops. *Progress in Physical Geography* 36(6): 747–763.

McHughen, A. and S. Smyth. 2008. US regulatory system for genetically modified [genetically modified organism (GMO), rDNA or transgenic] crop cultivars. *Plant Biotechnology Journal* 6(1): 2–12.

Mytelka, L. 2004. Clustering, long distance partnerships and the SME: a study of the French biotechnology sector. *International Journal of Technology Management* 27: 791–808.

Peacock, K. 2010. *Biotechnology and Genetic Engineering*. New York: Infobase.

Pew Charitable Trust. 2001. *Guide to U.S. Regulation of Genetically Modified Food and Agricultural Biotechnology Products*. Washington, DC: The Pew Charitable Trusts.

Plum, O. and R. Hassink. 2011. On the nature and geography of innovation and interactive learning: a case study of the biotechnology industry in the Aachen technology region, Germany. *European Planning Studies* 19(7): 1141–1163.

Powell, K., K. Koput and L. Smith-Doerr. 1996. Interorganizational collaboration and the locus of innovation: networks of learning in biotechnology. *Administrative Science Quarterly* 41: 116–145.

Powell, W., D. White, K. Koput and J. Owen-Smith. 2005. Network dynamics and field evolution: the growth of interorganizational collaboration in the life sciences. *American Journal of Sociology* 110(4): 1132–1205.

Prevezer, M. 1997. The dynamics of industrial clustering in biotechnology. *Small Business Economics* 9: 255–271.

Rasmussen, N. 2014. *Gene Jockeys: Life Science and the Rise of Biotech Enterprise*. Baltimore, MD: Johns Hopkins University Press.

Smith, H. 2004. The biotechnology industry in Oxfordshire: enterprise and innovation. *European Planning Studies* 12: 985–1001.

Ter Wal, A. 2014. The dynamics of the inventor network in German biotechnology: geographic proximity versus triadic closure. *Journal of Economic Geography* 14(3): 589–620.

Trippl, M. and F. Tödtling. 2007. Developing biotechnology clusters in non-high technology regions – the case of Austria. *Industry and Innovation* 14(1): 47–67.

Vallero, D. 2010. *Environmental Biotechnology: A Biosystems Approach*. Amsterdam: Academic Press.

Zhang, F. and F. Wu. 2012. Fostering indigenous innovation capacities: the development of biotechnology in Shanghai's Zhangjiang high-tech park. *Urban Geography* 33(5): 728–755.

Zucker, L., M. Darby and M. Brewer. 1998. Intellectual human capital and the birth of U.S. biotechnology enterprises. *American Economic Review* 88: 290–306.

30. Creating new geographies of health and health care through technology

Mark W. Rosenberg and Natalie Waldbrook

There are two perspectives that guide the discussion of health and technology in this chapter. The first perspective is how health geographers have incorporated new technologies into their research and their understanding of health and health care issues.[1] The second perspective is how the introduction and adoption of new technologies has created, is creating and will create new geographies of health and health care as the 21st century progresses.

The chapter is divided in two main parts. In the first part, the development and role of geographic information systems (GIS) are highlighted as a critical new technology for geographic research on health and health care. Within this discussion, the use of GIS in mapping infectious and non-infectious diseases, how GIS is being used to untangle compositional and contextual factors in population health, the role of GIS in health care planning and how GIS is being incorporated into disease surveillance, tracking and emergency responses as part of the new public health are reviewed. In the second part, the focus is on how new technologies are creating new geographies of health and health care. Innovations such as telemedicine, 'virtual spaces' of health care and the emerging role of technologies among the older population are highlighted.

GIS APPLICATIONS IN HEALTH GEOGRAPHY

In the latter half of the 20th century, developments in computer technology and statistical databases profoundly affected the way human geographers mapped and visually represented spatial phenomena in their studies. As described by other authors in this volume, the benefits of computer mapping techniques were recognized by geographers concerned with the planning of the physical and built environment from the 1950s and onward. The increase in computer processing power in the decades that followed allowed for the emergence of GIS in the 1990s (Parker and Campbell 1998; Andrews and Evans 2008). The global diffusion of the World Wide Web and the advancements in geocoding by the end of the 20th century further improved GIS techniques. GIS has been defined in other chapters of this book, but we find the definition by Parker and Campbell (1998) to be appropriate for this chapter. Parker and Campbell (1998, 2) define GIS as 'powerful database and display functions that allow for the integration of data from numerous sources and the performance of detailed analysis taking into consideration the location of variables in question'.

Geographers have been using mapping techniques since the late 18th century (Rican and Salem 2010), but the benefits of GIS were not widely recognized in health geography until the end of the 20th century (Parker and Campbell 1998; see Rican and Salem 2010 for a comprehensive review of the history of mapping techniques in discipline of health geography). Considering their long history of interest in mapping techniques, why did it take health geographers so long to incorporate GIS technology into their research methodologies? There has been some reflection on this question in the literature.

Rosenberg (1998) suggests that one reason is the national health datasets that health geographers rely on today were not widely available until the 1990s. The collection of geocoded national health information data from population surveys and censuses has presented health geographers with new methodological opportunities for their research. The rise of the World Wide Web 2.0 at the start of the 2000s further encouraged the use of computer and Internet mapping by health geographers (Rican and Salem 2010). The emergence of Web 2.0 has allowed for more direct user-engagement, in contrast to how users passively took advantage of earlier generations of the Web. Web applications used in combination with GIS have reduced the complexity and time it takes to generate maps, and this has increased its attractiveness to health geographers (Rican and Salem 2010).

Prior to the 2000s, the primary application of GIS in health geography was to produce sophisticated mappings and visual spatial representations of the site and diffusion of illness and disease. Advancements in statistical software (e.g. personal computer-based versions of SPSS and other statistical packages) now allow researchers to layer datasets from different sources to conduct multivariate analyses. Mayer (2010) contends that GIS is not only a tool for visual display of data, but it is considered a method itself for the spatial analysis of health and health care systems phenomena. The reliability and trustworthiness of data are also improving with increased accuracy of geo-referencing and global positioning systems (GPS) (Sui 2007; Rican and Salem 2010; Sabel et al. 2010; Shoval et al. 2014). Developments in geocoding software and Internet mapping systems like Google Earth have reduced barriers for researchers to obtain detailed, accurate spatial and visual information (Rican and Salem 2010; Shoval et al. 2014).

It is important to establish too that not only health geographers are interested in the concept of a spatial scale, but most geographers are inherently concerned with how patterns and phenomena occur over time (Andrews and Kitchen 2005; Rainham et al. 2010; Tanser et al. 2010). Rican and Salem (2010) argue that certain changes in illness and disease can be best understood by analyzing patterns across a 'spatio-temporal scale' (p. 107). Recent innovations have improved the ability of health geographers to conduct spatio-temporal analyses, including dynamic and interactive maps, animation techniques and time-lapse video-recording (Rican and Salem 2010). Overall, the continued advancements made in information and communication technology have facilitated the uptake of GIS into non-traditional fields of application, such as those related to medicine and health research (Rican and Salem 2010; Sabel et al. 2010).

Mapping Infectious and Non-infectious Diseases

One of the first applications of GIS by health geographers was for mapping spatial phenomena related to human and disease ecology (Parker and Campbell 1998). Disease ecology is the study of the interactions between pathogens or parasites and their human and non-human hosts and environments (Oppong and Harold 2010). A traditional focus of disease ecologists is studying human health outcomes in relation to natural environmental factors, such as climate, bodies of water, latitude and elevation (Oppong and Harold 2010).

Health geographers who study the spread of infectious illness and disease share interests with researchers in the field of epidemiology (Parker and Campbell 1998; Rosenberg 1998; Mayer 2010). Infectious diseases are those that are transmittable or communicable from organisms or animals to humans and from humans to humans. Health geographers and epidemiologists have applied GIS techniques to visually represent and analyze the spatial diffusion and distribution of infectious disease (Sui 2007). GIS has been adopted as an effective tool for identifying 'hotspots', or locations where there is a higher than expected concentration of a disease (Castree et al. 2013).

GIS has also been applied by health geographers to map and analyze patterns of non-infectious diseases (those that are non-transmissible between humans). A popular objective in this research has been to identify areas, or populations in which there is a higher than expected occurrence of a non-infectious disease, and to identify risk factors for individuals and populations developing non-transmittable diseases in some regions compared with others. Rican and Salem (2010) explain that GIS can define a 'buffer zone' or 'catchment area' around a site of pollution, for example, to predict the risk of asthma based on exposure. Similar approaches with GIS have been used to explore health outcomes and illness at the neighborhood scale. This has included studies to analyze spatial patterns of overweight/obesity for identifying risk-factors related to 'obesogenic environments' at the local scale (Duncan et al. 2014). Another important application of GIS and GPS has been to analyze health outcomes and rates of non-infectious diseases in areas of higher concentrated poverty (Oppong and Harold 2010).

Untangling Contextual and Compositional Factors in Population Health

A long-standing debate in health geography has been about whether contextual or compositional factors are stronger determinants of the health differences found between different populations in different places (Rosenberg 1998; McLafferty 2003; Sui 2007; Cromley 2009; Rican and Salem 2010). Compositional factors are defined as the characteristics of the individuals who live within a particular region or place, while contextual factors refer to the features of that location or place (Macintyre et al. 2002; Mitchell et al. 2002; Sui 2007). Mitchell et al. (2002) argue that where compositional factors dominate are '*places* with apparently higher levels of sickness or death [and] are those in which a higher proportion of the residents are at a higher risk of death' (p. 15). Exposures (contextual factors) in any given area are expected to influence the risk of illness and death. Mitchell et al. (2002) go on to argue that the influence of exposures can be stronger determinants of population health than the compositional factors of the people who live there (p. 15). Contextual factors often include the presence or absence of clean drinking water, pollution and proximity to sites of waste disposal and contamination (Fernández-Maldonado 2008; Oppong and Harold 2010).

GIS is also commonly used to examine relationships among illness, disease, urbanization and the built environment. Health geographers consider aspects of the built environment that might influence health to include the spatial organization of housing, commercial buildings, industry, road networks, sewage systems, parks and the quality of indoor environments (e.g. air quality, lighting, mold, building age); (Fernández-Maldonado 2008; Oppong and Harold 2010).

The recent advancements in GIS now allow researchers to layer various datasets to conduct multivariate analyses. For example, the role of compositional factors in shaping population health have been explored using GIS by overlaying characteristic data (age, sex, race, ethnicity and socioeconomic status) with environmental and remote sensing imagery (Mitchell et al. 2002; McLafferty 2003; Cromley 2009; Oppong and Harold 2010). This type of multivariate analysis though requires access by the researcher to disaggregated data representing various socio-spatial dimensions at the local scale. In the past, disaggregated data have been difficult to obtain in many countries because of privacy issues, and a lack of government resources to collect detailed spatial data. Recently though, technological improvements in satellite imagery and remote sensing capabilities have presented new ways to collect spatial information in areas that have been difficult to capture in the past (Rican and Salem 2010, 108).

Using GIS, health geographers have also conducted studies on the inequalities in health and the increased risk of illness that is measured in certain groups, despite living in a developed, wealthy country. These concerns by health geographers are shared with researchers in fields of public and community health (Andrews and Evans 2008). GIS technology and more detailed compositional health datasets have improved methods to analyze health and disease inequalities based on social status and the social environment. This area of research has contributed evidence that social factors are important determinants of population health, and in addition to the built environmental and physical factors (McLafferty 2003; Cromley 2009; Oppong and Harold 2010).

Health Care Planning

Aside from mapping disease and illness, GIS has been applied to the planning of health care services. In the early 2000s, McLafferty (2003) published an extensive review in which she identified several areas of health services research that have been conducted with GIS. She argues that GIS has provided a set of tools for illustrating the changing spatial organization of health care systems and delivery. Other geographers have similarly described the usefulness of GIS for 'location-allocation modeling' (Tanser et al. 2010). Location-allocation planning refers to the organization and management of health care services and the distribution of resources across geography and populations (Parker and Campbell 1998, 2). The application of GIS for planning the distribution of medical resources and to identify gaps in health services gained momentum in the 1990s in parallel to the exploitation of GIS in mapping infectious and non-infectious diseases and the analysis of compositional and contextual factors.

Based on McLafferty's (2003) review, a second common application of GIS in health services research has been to examine the relationships between access to care and population health. There has been discussion within the literature regarding how the concept of 'access' should be defined and measured (Tanser et al. 2010). The term 'access' has traditionally signified the availability and proximity to medical treatment and health services within a particular region, or the travel distance to reach a service (McLafferty 2003; Ricketts 2010). McLafferty (2003) makes the important point that, when measuring health outcomes in a population based on access to care, socio-structural factors must be taken into account as well. In studies informed by a social model of health, GIS can be a method for identifying inequalities in access to care and population health outcomes as divided along lines of socioeconomic status, gender, race and ethnicity. McLafferty

(2003) says a third and related objective of using GIS in health services research has been to address barriers to receiving health care across physical space and along lines of social stratification. This area of GIS research can be valuable for the planning and evaluation of health policies.

Surveillance, Tracking and Emergency Response

A relatively new application of GIS and GPS technology has been for surveillance for national and state security (McLafferty 2003; Eccles et al. 2013). The accurate, up-to-date spatial data provided by GPS has made information easier to access in real-time. These GIS applications have been appealing to government bodies tasked with the responsibility of monitoring and making homeland security decisions in their respective countries (McLafferty 2003; Eccles et al. 2013). These same ways of thinking have been incorporated in disease surveillance systems to model and understand how an outbreak of a highly infectious disease can spread in time and over space. Early detection can then be used to intervene at either the local level or to plan interventions to slow or stop the diffusion of a disease to other regions within a country or internationally.

Another related application of GIS and GPS is to improve the response and delivery times for emergency services on the ground, such as ambulance, police, fire and disaster relief (Eccles et al. 2013). The increasing availability of detailed spatial data and integrated geographical information can help to dispatch emergency services in a timelier manner (McLafferty 2003; Eccles et al. 2013). GPS data can improve dispatchers' abilities to predict the arrival time of emergency responders, choosing the quickest route to their destination, and proximity to hospitals. GIS has become an important tool for communicating spatial information and pattern detection to improve speed and accuracy of first-responders to sites of emergency.

CRITICAL GEOGRAPHIES OF TECHNOLOGY, HEALTH AND HEALTH CARE

Alongside the broader postmodern movement and 'cultural turn' turn in the social sciences, the field of health geography has undergone philosophical changes over the past 20 years (Kearns 1993, 1995; Mayer 1996; Kearns and Gesler 1998; Kearns and Moon 2002; Rosenberg 1998; Parr 2004; Andrews and Evans 2008). *A Companion to Health and Medical Geography* (Brown et al. 2010) is an anthology of research and theory capturing the influence of interpretivist philosophy in medical geography, a traditionally objectivist field of study. This philosophical revolution led to the emergence of what is now known as 'geographies of health and health care'. Health geography research is distinguished by its emphasis on critical social theory and the application of social models to the study of health and well-being (Andrews and Kitchen 2005; Andrews and Evans 2008).

Related to technology, a core interest of interpretivist health geographers is to explore how technology, particularly information and communication technologies (ICT), influences interactions between people and places (Graham 1998; Parr 2002; Andrews and Kitchen 2005; Andrews and Evans 2008). At the present time, ICT is regarded in the general population as computers, Internet, interactive web applications, GPS, 'smart' phones and other wireless devices. Technologies though constantly evolve and new applications and devices emerge. For interpretivist health geographers, advancements made in technology are viewed as an ongoing, fluid progression of innovation and not a static process (Graham 1998; Parr 2002; Andrews and Kitchen 2005; Andrews and Evans 2008).

In technology discourse before the mid–late 1990s, cyberspace and the Web tended to be viewed as determining interactions between humans and their relationships with their environments. In some philosophical perspectives, the early discourse surrounding the World Wide Web reflected an objectivist tone. Theorists of the early 1990s often framed cyberspace as existing 'out there', separate from humans yet deterministic of their reality. By the end of the 20th century, this 'technological determinist' view, however, was already losing favor and is now considered passé in the contemporary study of human geography. Graham's (1998) paper marked a turning point in the discipline's view of technology and society. Graham (1998) was writing to the wider audience of the human geography discipline, but his work has been influential and informative for health geographers in their considerations of technology (Cutchin 2002; Andrews and Kitchen 2005). Graham (1998) introduced new ideas about 'space' in light of the spreading adoption of the Internet and information and communication technologies into mainstream society. As ICT becomes more deeply infused in everyday social, economic and cultural life, he suggested that 'virtual' and 'cyberspace' create 'new kinds of space'. Graham (1998, 625–616) writes 'Cyberspace is invisible to our senses, a space which might become more important than physical space itself and which is layered on top, within and between the fabric of traditional geographic space'.

Most interpretivist geographers now widely accept that humans are responsible for the invention of technology and the adaptation of it to fit their needs and suit their environment. Rather than viewing technological progress as something that happens *to* humans, technologies are recognized as the creation and innovation *by* people, *for* people. This theoretical framework largely informs health geographers' work on the application of ICT to health care delivery, which we discuss in the next section.

Telemedicine and 'Virtual' Spaces of Care

Telemedicine was one of the earliest interests of health geographers in ICT. Cutchin (2002, 25) defines telemedicine as an 'interactive video-consultation between medical specialists and local primary care providers'. Discussions about telemedicine can be observed in two major areas of health geography literature. First, telemedicine is described as a method of delivering health care to populations in underserviced and remote areas. The second wave of discussion is about the potential for telemedicine to address gaps in health services created government reforms in many Western economies (Milligan et al. 2011).

Health geographers have explored the opportunities for telemedicine to provide medical treatment and specialist care to people in remote and rural regions, where there are traditionally fewer physicians and services. In studying telemedicine, the systems are typically characterized as involving a doctor or specialist consulting over the Internet with another care provider and their patient, who are in a remote location (Cutchin 2002; Parr 2002; Stowe and Harding 2010; Eccles et al. 2013). The core interest of health geographers has been to assess telemedicine as a modality to overcome the geographical barriers of distance and time to deliver medical treatment and health services to underserviced populations (May et al. 2011; Kilpeläinen and Seppänen 2014).

Health geographers who have mainly focused on rural populations' access to services and proximity and the distance to travel to access services (Skinner and Joseph 2007, 2011; Joseph and Skinner 2011; Keating et al. 2011) have layered into the discussion the added implications that public health care restructuring has had on access to services in already-underserviced areas (Cloutier-Fisher 2006; Skinner and Joseph 2007, 2011; Skinner 2008; Keating et al. 2011). They have pointed to the closure of health service centers and small clinics in rural areas, the relocation of these services to more regional and urban centers, and the critical examination of whether technological fixes can overcome the loss of local services (Cloutier-Fisher 2006; Skinner and Joseph 2007, 2011; Skinner 2008; Joseph and Skinner 2011; Skinner and Power 2011).

Much like in the broader field of human geography, the concept of place is critical to the study of technology by health geographers. Both medical and health geographers agree that physical location is a determinant of access to health care. Health geographers emphasize the importance of social, cultural, political and economic context when analyzing access to care and population health outcomes (Skinner and Joseph 2007, 2011; Skinner and Power 2011; Joseph and Skinner 2011). Some health geographers have theorized that telemedicine and interactive web platforms will allow new networks to form between patients, doctors and other care providers in virtual, or 'cyberspace' (Cutchin 2002; Stowe and Harding 2010; Milligan et al. 2011; Kilpeläinen and Seppänen 2014). This is leading to new 'geographies of care' that have the potential to improve health care delivery to remote populations and in underserviced regions in need (Cutchin 2002; Gilbert et al. 2008; Milligan et al. 2011; Callan and O'Shea 2015).

Situating Technology within Geographies of Aging

A significant amount of the literature written about technology in the field of health geography is in the area of population aging. 'Population aging' is the demographic trend when the median age of population in a country increases. Population aging is occurring at a global scale owing to increased life expectancy, largely resulting from advancements in medical treatment, preventative health care and higher standards of living resulting in good health (Sixsmith 2013). Fertility rates are declining globally, although this is unevenly distributed across world regions, with wealthy developed countries having very low fertility rates and developing countries having relatively high rates. In countries that underwent a 'baby boom' in population growth following the Second World War, the first wave of that cohort is now entering their senior years. This trend will continue over the next two decades and, by 2030, 25% of the population will be 65 years and older in developed countries, such as Canada and the United States, and in Western Europe (Milligan et al. 2011).

Over the last 25 years, health geographers have applied mapping techniques to predict the spatial distribution of the older population across countries and regions. This area of health geography has been concerned with the carrying capacity of a country to support a larger number of older persons. The discourse around population aging has focused on the fiscal implications and challenges for governments to provide health care and long-term care, build accessible infrastructure and support housing needs of large older populations (Gee and Gutman 2000). In the past 10 years though, there has been a turn towards thinking about the older population more positively, rather than fixating on the economic ramifications (Sixsmith 2013; Skinner and Hanlon 2015). This began as a broader dialogue with the World Health Organization's campaign in the early 2000s, urging researchers and government leaders to envision ways of creating 'age-friendly communities' or 'age-friendly environments' at the local scale (Menec et al. 2011; Beringer and Sixsmith 2013; Hwang et al. 2013; Sixsmith 2013; Skinner and Hanlon 2015). As part of this conversation, the concept of aging has moved away from only a biomedical model of health and disease to promote a more well-rounded experience of 'active aging' and 'aging well' (Keating et al. 2011; Sixsmith 2013). Current conceptualizations of what it means to age well are informed by a broader social determinants model, which emphasizes not only health and medical care but also financial security, social support and opportunities for productive participation and social activity (Keating et al. 2011; Menec et al. 2011; Beringer and Sixsmith 2013; Hwang et al. 2013; Sixsmith 2013). The underlying philosophy is that, if older persons are properly supported in their homes and communities, they have a better opportunity for good health, independent living and satisfaction with life.

The promotion of aging-in-place in policy has been one solution by which governments hope to cope with the anticipated pressures on health and long-term care systems (Milligan 2000, 2001, 2003, 2007; Milligan et al. 2011; Sixsmith and Sixsmith 2008). At the same time, policy-makers see aging-in-place as supporting the preferences of older people to live and receive care in their own home as opposed to an institutional setting (Sixsmith and Sixsmith 2008; Milligan et al. 2011). Yantzi (2006) mentions that reorganizing the delivery of long-term care from institution- to home-based settings requires a complex schedule of professional care providers (nurses, personal home support workers, therapists). However, a complicating factor is that health care restructuring and cutbacks have reduced the eligibility and hours for home care services from government-funded sources (Milligan 2000, 2001, 2003; Williams 2002, 2006; Milligan et al. 2011). The direction of health and long-term care policy in most Western economies today is towards encouraging private sector development in seniors' housing, care and support services. The area of research known as 'geographies of voluntarism' reveals that increased pressure on voluntary sector organizations to fill in gaps in home and community-based services for older people is one outcome of government reforms (Milligan 2000, 2007; Milligan and Wiles 2010; Joseph and Skinner 2011; Skinner 2008, 2014; Skinner and Joseph 2007, 2011; Skinner and Power 2011; Skinner and Hanlon 2015). This literature points to the greater pressure being on older people's informal networks, such as family members, neighbors and friends, to act as caregivers (Wiles 2003a, b; Skinner and Joseph 2007, 2011; Skinner and Hanlon 2015).

Academic researchers, government leaders and private industries are looking to innovative technologies to overcome the various challenges for older persons to support aging-in-place. For older adults with chronic disease and physical impairments, home technology can assist with completing day-to-day activities and personal care in the home environment (Czaja et al. 2013; Sixsmith 2013). Traditionally, home technologies have been conceptualized as assistive devices or mobility aids, such as stair lifts, grab bars and automated controls for lights, indoor climate, doors and windows (Milligan et al. 2011;

Stowe and Harding 2010; Peeters et al. 2013). Security alarms and fall detection systems have been available for purchase for several decades, relying initially on radio signals and telephone lines.

Most recently, passive infrared movement sensors and webcams are used in monitoring systems to transfer information about a resident over the Internet to a remote surveillance or control center (Sixsmith and Sixsmith 2008; Sixsmith and Johnson 2004; Blaschke et al. 2009; Chan et al. 2009; Czaja et al. 2013; Peeters et al. 2013). Telemedicine was discussed earlier in this chapter in the context of remote health care delivery. The terms 'tele-care', 'e-health' and 'tele-monitoring' typically refer to innovative systems for installation in the home environment (Sixsmith et al. 2013). These tele-care systems rely on broadband and wireless Internet signals to detect falls, abnormal behavior and the need for emergency response services (Blaschke et al. 2009; Milligan et al. 2011; Bowman et al. 2013; Czaja et al. 2013; Eccles et al. 2013; Peek et al. 2015). Using similar health monitoring technology, devices are available that can be worn by a person in their home to monitor their vital signs, body movement and daily activity (Chan et al. 2009; Stowe and Harding 2010; Milligan et al. 2011).

The most recent advancements in home technology are in the area of artificial intelligence, which has led to the development of 'Smart Home' systems (Sixsmith and Sixsmith 2000; Demiris and Hensel 2009; Chan et al. 2009; Balta-Ozkan et al. 2013; Ehrenhard et al. 2014; Peek et al. 2015). Again, there are varying definitions of what is included in a Smart Home and how it differs from past technology systems designed for the home (Table 30.1). A Smart Home is distinguished by its ambient intelligent network, meaning all the technological components in the system are embedded and integrated directly into the housing infrastructure and furnishings (Balta-Ozkan et al. 2013; Sixsmith 2013; Sixsmith et al. 2013). Each technological component is connected and coordinated to share integrated information about the resident with a remote service agent to monitor health status and activities occurring in the Smart Home (Balta-Ozkan et al. 2013). Other devices that can be 'wired into' the home include intelligent bathtubs that fill and control water temperature based on pre-settings and smart refrigerators. 'Smart Fridges' can be purchased to monitor diet, food intake, nutritional requirements, meal ideas and reminders to replenish groceries. One of the most common features of a Smart Home system is its ability to track energy consumption and to automatically adjust heating, cooling and lighting with preprogrammed controls.

Sixsmith's (2013) edited volume *Technologies for Active Aging* provides the most comprehensive coverage of the research on the various ways technology, particularly ICT, is being applied to support active, healthy aging. Research on intelligent home systems to date is optimistic and shows the potential to improve older people's independence, mobility, daily function and quality of life, as well as alleviate caregiver burden (see Sixsmith 2013). ICT in the home environment can also increase older people's connectivity and engagement with the outside world. This has been found to decrease the risk of social isolation for those aging-in-place (Beringer and Sixsmith 2013; Czaja et al. 2013; Kilpeläinen and Seppänen 2014). Home technologies are also proving to be effective for persons with Alzheimer's disease and dementia (Bierhoff et al. 2013; Mountain 2013). This is in terms of not only safety monitoring systems, but also assistive devices to improve independence (Mountain 2013) and technologies for cognitive exercises (Sixsmith et al. 2010; Hardill and Olphert 2012; Bierhoff et al. 2013). Computer games and online activities are being

Table 30.1   Examples of terminology and definitions of home technologies

---

Assistive and health monitoring device

Blaschke et al. (2009): Sensors and warning systems to alert caregivers of care recipients in need (p. 644).

Chan et al. (2009): Devices worn by user or embedded in the house. Connected through wireless networks to a service center with environmental and diagnostic facilities. Devices can assess sound, images, body motion, ambient parameters (light, temperature, humidity), vital signs (blood pressure, respiration, body temperature, heart rate, body/weight, blood oxygen generation, ECG), sleep patterns, daily activities, social interactions (p. 92).

Peeters et al. (2013): Assistive technology applications can range from specific alarm and monitoring devices to ambient living technology (p. 5543).

Stowe and Harding (2010): Assistive technology includes equipment ranging from a simple grab rail to accessible transport and electronic fall detectors (p. 193).

Home automation devices

Milligan et al. (2011): Home automation technologies can include lights, climate control, control of doors and windows, security/surveillance systems and care/diagnostic systems using sensors and webcam technologies (p. 349).

Telemedicine, tele-care, e-health

Blaschke et al. (2009): Passive monitoring systems for fall and movement sensors, remote exchange of health status data between patients and health-care professionals, and videoconferencing systems that allow patients to interact with social network and health professionals (p. 645).

Peek et al. (2015): E-health encompasses a broad range of technologies including online tools to support self-management of chronic disease (p. 2).

Peeters et al. (2013): Home tele-monitoring is an automated process for data transmission about a patient's health status from the patient's home to the respective health-care setting (p. 5543).

Stowe and Harding (2010): Telecare is a tool for professionals to deliver health care and support directly to the user (patient) (p. 193).

Smart Home systems

Chan et al. (2009): A residence equipped with technology that allows monitoring of inhabitants and equipped with sensors, actuators and/or biomedical monitors. The devices operate in a network connected to a remote center for data collection and processing (p. 91).

Demiris and Hensel (2009): Residential settings equipped with sensors or devices that use motion sensors, radio frequency sensors, video cameras or wearable devices to enable the monitoring of residents (p. 33).

Ehrenhard et al. (2014): Wireless connected and integrated health monitoring systems, security and safety monitoring services for security, fall detection, immobility identification, activity tracking and intelligent energy consumption (p. 307).

Peek et al. (2015): Smart Homes include emergency help systems, vital sign monitoring devices and fall detection systems (p. 2).

---

explored as cognitive exercises for persons with dementia and for social interaction among older persons in general (Astell 2013).

As mentioned previously in this chapter, one of the major challenges to studying health and social implications of technology is that new innovations are constantly emerging and replacing current generations of devices. Chan et al. (2009) argues that the wide range of terminology and definitions of what constitutes home technology for older persons has led to ambiguity and confusion in the literature. To illustrate this point, Table 30.1 provides examples of terminology and definitions of home technologies that are intended for the older population from recently published literature. Sixsmith et al. (2013) provides a useful categorization to summarize the wide range of home technologies that are available to older people into three 'generations'. The first two generations include emergency response alarms and smart devices, respectively. The third generation, Sixsmith et al. (2013) explains, is the broader study and development of information and communication technologies and the application of these advancements to assist older persons.

A common theme in the literature has been to identify the factors that determine or influence older people's decisions to adopt technology in the home. Studies have generally concluded that older adults' individual characteristics and the perceived need for home technology are among the strongest determinants of adoption (Sixsmith and Sixsmith 2000; Bowman et al. 2013; Czaja et al. 2013; Lee and Coughlin 2014; Peek et al. 2015). Other factors that have been found to influence the decision to adopt technology in the older population include: the design and functionality of the specific device or system; consumer awareness, previous experience and exposure to technology; and the manner through which technology is advertised, distributed and delivered to this demographic (Bowman et al. 2013; Czaja et al. 2013; Morris et al. 2013; Sixsmith 2013; Lee and Coughlin 2014). For a more in-depth discussion of the barriers and facilities of technology adoption found among the older population, see the work of Czaja et al. (2013).

The barriers that prevent older people from adopting advanced technologies in the home environment have been identified as well. These include the high cost of purchasing and maintaining advanced technological systems and Internet services; concerns about accessing technical support when needed; and having the personal 'know how' to troubleshoot problems with technology (Bowman et al. 2013; Czaja et al. 2013; Morris et al. 2013; Lee and Coughlin 2014).

Health geographers have also contributed a discussion of how the adoption of home health technology can be viewed as 'medicalizing' the home environment, not just for older people but for all age groups (Yantzi and Rosenberg 2008; Demiris and Hensel 2009). A concern is that outfitting a home with medical equipment and advanced technologies alters the subjective experience of the home, and this potentially interferes with the benefits to aging-in-place. The medicalization of the home environment also blurs the boundaries between the private sphere of the home and health care systems with the presence of workers (Dyck et al. 2005; Demiris and Hensel 2009; Milligan and Wiles 2010; Eccles et al. 2013). Another concern raised is whether medical equipment and a hospital-like setting in the home causes stigmatization or essentialist views about aging, illness and health by visitors to the home (Demiris and Hensel 2009). The underlying question is whether these various issues arising from the adoption of home technology might actually have negative implications for perceived satisfaction, quality of life and mental well-being. The emerging discourse about home technology and older people is moving in the direction of promoting an active, fulfilling lifestyle and de-emphasizing technology for the purposes of medical treatment (Bierhoff et al. 2013; Eccles et al. 2013; Sixsmith 2013).

Governments' encouragement of older persons to age-in-place and adopt home technology is intended to reduce demand on public health and long-term care resources.

One major issue that continues to be overlooked in research and policy is that a significant share of the older population are without the financial resources to purchase, install and maintain advanced home technology systems. The issue of inequality in access to technology based on socioeconomic and sociocultural characteristics requires greater attention in research and policy that promotes aging-in-place (Gilbert et al. 2008; Sixsmith 2013). The field of health geography and policy would benefit from further research into the regions in a country where older populations have unequal access to technology and technological infrastructure to support aging-in-place.

CONCLUSION: EMERGING GEOGRAPHIES OF TECHNOLOGIES AND HEALTH

Throughout this chapter, the emphasis has been on how health geographers have incorporated new technologies into their research and their understanding of health and health care issues and how the introduction and adoption of new technologies has created, is creating and will create new geographies of health and health care as the 21st century progresses. There are many reasons to be optimistic that new technologies will continue to open up new areas of investigation and provide new opportunities for health geographers to contribute to the improvement of the health and well-being of people at geographical scales from the local to the global.

We conclude, however, with one caveat. Nowhere do issues of privacy and technology intertwine more than at the interface of the health of the individual and the health of a community. There can be no doubt that geo-referenced data, GIS, smart technologies and the 'app' revolution have the potential to help people remain healthy or overcome their health problems and to create new geographies of health and health care. It is also increasingly obvious, however, that the protections of individual privacy lag behind the introduction of new technologies and the potential to stigmatize, discriminate or even isolate people and places based on their health profiles demands the greater attention of health geographers specifically and geographers in general. Health geographers and geographers in general need to be actively involved in both the debates and protecting human rights in their use of new technologies, not passive bystanders, if we are to ensure that the emerging geographies of technologies and health are to benefit everyone.

NOTE

1. Although terms like 'medical geography' and 'medical geographers' were commonly used up until the 1990s and continue to be used by some geographers, terms like 'health geography' and 'health geographers' are now more commonly used. To avoid confusion, we use the current terminology except where someone is being directly quoted.

REFERENCES

Andrews, G. and J. Evans. 2008. Understanding the reproduction of health care: towards geographies in health care work. *Progress in Human Geography* 32(6): 759–780.

Andrews, G. and R. Kitchin. 2005. Geography and nursing: convergence in cyberspace? *Nursing Inquiry* 12(4): 316–324.

Astell, A. 2013. Technology and fun for a happy old age. In A. Sixsmith and G. Gutman (eds), *Technologies for Active Aging*, vol. 9. pp. 169–188. New York: Springer Science and Business Media.

Balta-Ozkan, N., R. Davidson, M. Bicket and L. Whitmarsh. 2013. Social barriers to the adoption of smart homes. *Energy Policy* 63: 363–374.

Beringer, R. and A. Sixsmith. 2013. Videoconferencing and social engagement for older adults. In A. Sixsmith and G. Gutman (eds), *Technologies for Active Aging*, vol. 9. pp. 189–200. New York: Springer Science and Business Media.

Bierhoff, I., S. Müller, S. Schoenrade-Sproll, S. Delaney, P. Byrne, V. Dolničar, B. Magoutas, Y. Verginadis, E. Avatangelou and C. Huijnen. 2013. Ambient assisted living systems in real-life situations: experiences from the SOPRANO Project. In A. Sixsmith and G. Gutman (eds), *Technologies for Active Aging*, vol. 9. pp. 123–153. New York: Springer Science and Business Media.

Blaschke, C., P. Freddolino and E. Mullen. 2009. Ageing and technology: a review of the research literature. *British Journal of Social Work* 39(4): 641–656.

Bowman, S., K. Hooker, C. Steggell and J. Brandt. 2013. Perceptions of communication and monitoring technologies among older rural women: problem or panacea? *Journal of Housing for the Elderly* 27(1–2): 48–60.

Brown, T., S. McLafferty and G. Moon. 2010. *A Companion to Health and Medical Geography*. New York: Wiley.

Callan, A. and E. O'Shea. 2015. Willingness to pay for telecare programmes to support independent living: results from a contingent valuation study. *Social Science and Medicine* 124: 94–102.

Castree, N., R. Kitchen and R. Alisdair. 2013. Medical geography. In *A Dictionary of Human Geography*. Oxford: Oxford University Press.

Chan, M., E. Campo, D. Estève and J. Fourniols. 2009. Smart homes – current features and future perspectives. *Maturitas* 64(2): 90–97.

Cloutier-Fisher, D. 2006. Levelling the playing field? Exploring the implications of managed competition for voluntary sector providers of long-term care in small town Ontario. *Health and Place* 12(1): 97–109.

Cromley, E. 2009. Breadth and depth in research on health disparities: commentary on the work of Nancy Krieger. *GeoJournal* 74(2): 115–121.

Cutchin, M. 2002. Virtual medical geographies: conceptualizing telemedicine and regionalization. *Progress in Human Geography* 26(1): 19–39.

Czaja, S., S. Beach, N. Charness and R. Schulz. 2013. Older adults and the adoption of healthcare technology: opportunities and challenges. In A. Sixsmith and G. Gutman (eds), *Technologies for Active Aging*, vol. 9. pp. 27–46. New York: Springer Science and Business Media.

Demiris, G. and B. Hensel. 2009. Technologies for an aging society: a systematic review of 'smart home' applications. *Yearbook of Medical Informatics* 24(1): 33–40.

Duncan, D., S. Regan, D. Shelley, K. Day, R. Ruff, M. Al-bayan and B. Elbel. 2014. Application of global positioning system methods for the study of obesity and hypertension risk among low-income housing residents in New York City: a spatial feasibility study. *Geospatial Health* 9(1): 57–70.

Dyck, I., P. Kontos, J. Angus and P. McKeever. 2005. The home as a site for long-term care: meanings and management of bodies and spaces. *Health and Place* 11(2): 173–185.

Eccles, A., L. Damodaran, W. Olphert, I. Hardill and M. Gilhooly. 2013. Assistive technologies: ethical practice, ethical research, and quality of life. In A. Sixsmith and G. Gutman (eds), *Technologies for Active Aging*, vol. 9. New York: Springer Science and Business Media.

Ehrenhard, M., B. Kijl and L. Nieuwenhuis. 2014. Market adoption barriers of multi-stakeholder technology: smart homes for the aging population. *Technological Forecasting and Social Change* 89: 306–315.

Fernández-Maldonado, M. 2008. Expanding networks for the urban poor: water and telecommunications services in Lima, Peru. *Geoforum* 39(6): 1884–1896.

Gee, E. and G. Gutman. 2000. *The Overselling of Population Aging: Apocalyptic Demography, Intergenerational Challenges, and Social Policy*. Don Mills, ON: Oxford University Press.

Gilbert, M.R., M. Masucci, C. Homko and A. Bove. 2008. Theorizing the digital divide: information and communication technology use frameworks among poor women using a telemedicine system. *Geoforum* 39(2): 912–925.

Graham, S. 1998. The end of geography or the explosion of place? Conceptualizing space, place and information technology. *Progress in Human Geography* 22(2): 165–185.

Hardill, I. and C.W. Olphert. 2012. Staying connected: exploring mobile phone use amongst older adults in the UK. *Geoforum* 43(6): 1306–1312.

Hwang, E., A. Park, A. Sixsmith and G. Gutman. 2013. The virtual environment in communication of age-friendly design. In A. Sixsmith and G. Gutman (eds), *Technologies for Active Aging*, vol. 9. pp. 155–168. New York: Springer Science and Business Media.

Joseph, A. and M. Skinner. 2011. Voluntarism as a mediator of the experiences of growing old in evolving rural spaces and changing rural places. *Journal of Rural Studies* 28(4): 380–388.

Kearns, R. 1993. Place and health: towards a reformed medical geography. *Professional Geographer* 45(2): 139–147.

Kearns, R. 1995. Medical geography: making space for difference. *Progress in Human Geography* 19(2): 251–259.

Kearns, R. and W. Gesler. 1998. *Putting Health into Place: Landscape, Identity, and Well-being*. Syracuse, NY: Syracuse University Press.

Kearns, R. and G. Moon. 2002. From medical to health geography: novelty, place and theory after a decade of change. *Progress in Human Geography* 5: 605–625.

Keating, N., J. Swindle and S. Fletcher. 2011. Aging in rural Canada: a retrospective and review. *Canadian Journal on Aging/La Revue Canadienne Du Vieillissement* 30(3): 323–338.

Kilpeläinen, A. and M. Seppänen. 2014. Information technology and everyday life in ageing rural villages. *Journal of Rural Studies* 33: 1–8.

Lee, C. and J. Coughlin. 2014. Older adults' adoption of technology: an integrated approach to identifying determinants and barriers. *Journal of Product Innovation Management* 32(5): 747–759.

Macintyre, S., A. Ellaway and S. Cummins. 2002. Place effects on health: how can we conceptualise, operationalise and measure them? *Social Science and Medicine* 55: 125–139.

May, C., T. Finch, J. Cornford, C. Exley, C. Gately, S. Kirk and F. Mair. 2011. Integrating telecare for chronic disease management in the community: what needs to be done? *BMC Health Services Research* 11(1): 131.

Mayer, J. 1996. The political ecology of disease as a new focus of medical geography. *Progress in Human Geography* 20: 441–456.

Mayer, J. 2010. Medical geography. In T. Brown, S. McLafferty and G. Moon (eds), *A Companion to Health and Medical Geography*. pp. 33–54. Malden, MA: Wiley-Blackwell.

McLafferty, S. 2003. GIS and health care. *Annual Review of Public Health* 24(1): 25–42.

Menec, V., R. Means, N. Keating, G. Parkhurst and J. Eales. 2011. Conceptualizing age-friendly communities. *Canadian Journal on Aging/La Revue canadienne du vieillissement* 30(3): 479–493.

Milligan, C. 2000. Bearing the burden: towards a restructured geography of caring. *Area* 32: 49–58.

Milligan, C. 2001. *Geographies of Care: Space, Place, and the Voluntary Sector*. Aldershot: Ashgate.

Milligan, C. 2003. Location or dislocation: from community to long-term care: the caring experience. *Journal of Social and Cultural Geography* 4(4): 455–470.

Milligan, C. 2007. Geographies of voluntarism: mapping the terrain. *Geography Compass* 1–2: 183–199.

Milligan, C. and J. Wiles. 2010. Landscapes of care. *Progress in Human Geography* 34(6): 736–754.

Milligan, C., C. Roberts and M. Mort. 2011. Telecare and older people: who cares where? *Social Science and Medicine* 72(3): 347–354.

Mitchell, R., D. Dorling and M. Shaw. 2002. Population production and modelling mortality – an application of geographic information systems in health inequalities research. *Health and Place* 8(1): 15–24.

Morris, M.E., B. Adair, K. Miller, E. Ozanne, R. Hansen, A.J.C. Pearce and C.M. Said. 2013. Smart-home technologies to assist older people to live well at home. *Journal of Aging Science* 1(1): 1–9.

Mountain, G. 2013. Using technology to support people with dementia. In A. Sixsmith and G. Gutman (eds), *Technologies for Active Aging*, vol. 9. pp. 105–122. New York: Springer Science and Business Media.

Oppong, J. and A. Harold. 2010. Disease, ecology, and environment. In T. Brown, S. McLafferty and G. Moon (eds), *A Companion to Health and Medical Geography*. pp. 85–95. Malden, MA: Wiley-Blackwell.

Parker, E. and J. Campbell. 1998. Measuring access to primary medical care: some examples of the use of geographical information systems. *Health and Place* 4(2): 183–193.

Parr, H. 2002. New body-geographies: the embodied space of health and medical information on the Internet. *Environment and Planning D: Society and Space* 20: 73–95.

Parr, H. 2004. Medical geography: critical medical and health geography? *Progress in Human Geography* 28(2): 246–257.

Peek, S., K. Luijkx, M. Rijnaard, M. Nieboer, S. van der Voort, S. Aarts, J. van Hoof, H. Vrijhoef and E. Wouters. 2015. Older adults' reasons for using technology while aging in place. *Gerontology* 62(2): 226–237.

Peeters, J., T. Wiegers and R. Friele. 2013. How technology in care at home affects patient self-care and self-management: a scoping review. *International Journal of Environmental Research and Public Health* 10(11): 5541–5564.

Rainham, D., I. McDowell, D. Krewski and M. Sawada 2010. Conceptualizing the healthscape: contributions of time geography, location technologies and spatial ecology to place and health research. *Social Science and Medicine* 70(5): 668–676.

Rican, S. and G. Salem. 2010. Mapping disease. In T. Brown, S. McLafferty and G. Moon (eds), *A Companion to Health and Medical Geography*. pp. 96–110. Malden, MA: Wiley-Blackwell.

Ricketts, T. 2010. Accessing health care. In T. Brown, S. McLafferty and G. Moon (eds), *A Companion to Health and Medical Geography*. pp. 521–539. Malden, MA: Wiley-Blackwell.

Rosenberg, M. 1998. Medical or health geography? Populations, peoples and places. *International Journal of Population Geography* 4(3): 211–226.

Sabel, C., D. Pringle and A. Schaerström. 2010. Infectious disease diffusion. In T. Brown, S. McLafferty and G. Moon (eds), *A Companion to Health and Medical Geography*. pp. 111–132. Malden, MA: Wiley-Blackwell.

Shoval, N., M.-P. Kwan, K. Reinau and H. Harder. 2014. The shoemaker's son always goes barefoot: implementations of GPS and other tracking technologies for geographic research. *Geoforum* 51: 1–5.

Sixsmith, A. 2013. Technology and the challenge of aging. In A. Sixsmith and G. Gutman (eds), *Technologies for Active Aging*, vol. 9. pp. 7–26. New York: Springer Science and Business Media.

Sixsmith, A. and N. Johnson. 2004. A smart sensor to detect the falls of the elderly. *Pervasive Computing, IEEE* 3(2): 42–47.

Sixsmith, A. and J. Sixsmith. 2000. Smart care technologies: meeting whose needs? *Journal of Telemedicine and Telecare* 6(Supplement 1): 190–192.

Sixsmith, A. and J. Sixsmith. 2008. Ageing in place in the United Kingdom. *Ageing International* 32(3): 219–235.

Sixsmith, A.J., R.D. Orpwood and J.M. Torrington. 2010. Developing a music player for people with dementia. *Gerontechnology* 9(3): 421–427.

Sixsmith, A., M. Carrillo, D. Phillips, P. Lansley and R. Woolrych. 2013. International initiatives in technology and aging. In A. Sixsmith and G. Gutman (eds), *Technologies for Active Aging*, vol. 9. pp. 201–222. New York: Springer Science and Business Media.

Skinner, M. 2008. Voluntarism and long-term care in the countryside: the paradox of a threadbare sector. *Canadian Geographer* 52(2): 188–203.

Skinner, M. 2014. Ageing, place and voluntarism: towards a geographical perspective on third sector organisations and volunteers in ageing communities. *Voluntary Sector Review* 5(2): 165–176.

Skinner, M. and N. Hanlon. 2015. *Ageing Resource Communities: New Frontiers of Rural Population Change, Community Development, and Voluntarism*. London: Routledge.

Skinner, M. and A. Joseph. 2007. The evolving role of voluntarism in ageing rural communities. *New Zealand Geographer* 63: 119–129.

Skinner, M. and A. Joseph. 2011. Placing voluntarism within evolving spaces of care in ageing rural communities. *GeoJournal* 76: 151–162.

Skinner, M. and A. Power. 2011. Voluntarism, health and place: bringing an emerging field into focus. *Health and Place* 17(1): 1–6.

Stowe, S. and S. Harding. 2010. Technology applied to geriatric medicine: telecare, telehealth, and telemedicine. *European Geriatric Medicine* 1: 193–197.

Sui, D. 2007. Geographic information systems and medical geography: toward a new synergy. *Geography Compass* 1(3): 556–582.

Tanser, F., P. Gething and P. Atkins. 2010. Location-allocation planning. In T. Brown, S. McLafferty and G. Moon (eds), *A Companion to Health and Medical Geography*. pp. 540–566. Malden, MA: Wiley-Blackwell.

Wiles, J. 2003a. Daily geographies of caregivers: mobility, routine, scale. *Social Science and Medicine* 57(7): 1307–1325.

Wiles, J. 2003b. Informal caregivers' experiences of formal support in a changing context. *Health and Social Care in the Community* 11(3): 189–207.

Williams, A. 2002. Changing geographies of care: employing the concept of therapeutic landscapes as a framework in examining home space. *Social Science and Medicine* 55(1): 141–154.

Williams, A. 2006. Restructuring home care in the 1990s: geographical differentiation in Ontario, Canada. *Health and Place* 12(2): 222–238.

Yantzi, N. and M. Rosenberg. 2008. The contested meanings of home for women caring for children with long-term care needs in Ontario, Canada. *Gender, Place and Culture* 15(3): 301–315.

31. Biometric technologies and the automation of identity and space
Gabriel Popescu

Biometrics are the latest major addition to the lexicon of the digital era, joining other terms such as 'big data', 'cloud computing' and 'geo-location'. Generally, biometrics are defined as measurements of a person's unique physical characteristics to verify or establish their identity. Until recently, biometric technologies were surrounded by an aura of science fiction in the popular imagination, and their applications remained largely confined to police and security work. Currently, they are expanding rapidly, experiencing substantial rates of adoption spurred on to a significant extent by their integration in various mainstream consumer applications. Apple's adoption of fingerprint-based sign-in for their iconic iPhone in 2013, followed by Wells Fargo's introduction of iris recognition-based personal banking sign-in in 2016, constitute momentous steps in bringing biometrics to the masses. If we also consider the ballooning of the security markets in the context of the purported fight against terrorism, where biometrics have long played a role, as well as the burgeoning biometric applications in the healthcare and entertainment industries, the picture that we see emerging is that of a technology on its way to permeating daily life and playing key roles in shaping its future.

At the same time, biometrics are also surrounded by much controversy and suspicion, like few other digital technologies before. The scale of the opposition has been considerable, spanning the political spectrum, social divisions and countries. There have been street protests against the introduction of biometric passports and national IDs in various parts of the world, and employee backlash against US companies adopting healthcare insurance plans that include biometrics-dependent variable rates. The impact of this opposition has created an image issue for biometrics that is taking a long time to turn around. Their initial association with law enforcement and criminality has been difficult to shake off and has certainly been a factor in the resistance they encounter. However, there is more to the resistance than an image problem. At the heart of the matter is the fact that biometrics are related to certain intimate aspects of human beings such as our bodies and our sense of identity, and thus raise issues of privacy rights and civil liberties. This situation opens up an entire new set of ethical and political questions regarding their regulation and their costs and benefits for society.

Biometric technologies relate to space in several fundamental ways. Their primary purpose is to collect data about the smallest and most personal of spaces, which is the human body. From this perspective, they touch on issues regarding the spatial and embodied nature of human experience. Another essential connection to space emerges from biometrics' role of regulating access to a variety of spaces, both material and virtual. Although their immediate goal is to confirm the identity of a person, this is often done in view of providing access to the location of certain resources and services. Additionally, biometrics have a close association with mobility and flows, if we consider the essential mobile nature of human bodies. In essence, biometrics represent a nexus between digital technology, identity, body and space. In these circumstances, the scarce attention geographic scholarship has paid to biometric technologies leaves a lot to be desired, as it foregoes an opportunity to explore important avenues for the production of space and the influence of space on social relations in the digital era.

SITUATING BIOMETRICS

Biometrics, as a practice of measuring people's distinctive body features in order to identify them, are not new (Lockie 2009). Historical records trace them back to antiquity, in places like China, where palm prints were used to sign legal documents, and Egypt, where detailed descriptions of the appearances of workers building the Pyramids were kept to assure the rightful distribution of their compensation. The first study credited with examining the unique patterns of fingerprints dates from 1684, and the first paper to associate the use of fingerprints with the identification of criminals was published in 1880 in the UK by Dr Henry Faulds. By the end of the nineteenth century, police departments in Argentina and the United States started using fingerprints to solve crimes. During the twentieth century, biometrics used in documents such as pre-digital passports and national IDs were descriptive in nature, and usually included pictures, signatures, eye color and height (Salter 2003).

Modern biometric technologies use digital sensor devices such as cameras and scanners to automatically acquire unique body features that are then algorithmically encrypted into data and stored for retrieval in centralized databases or on chips inserted into personal e-documents and on mobile devices such as smartphones (Das 2015; van der Ploeg 1999b). This process is called enrollment because it is at this stage that people's biometrics are recorded in a system for the first time. At a checkpoint, a person has to present the enrolled body part together with the e-document to a digital reader device in order to be identified. The person's identity will then be digitally read from his or her body and checked against the data stored in databases or on chips. If a match is found, this verifies their identity and grants them the access to spaces, services or resources they are seeking. This process is known as authentication, because at this stage the identity of an individual is confirmed. Accordingly, computer-aided digitization and automation emerge as key elements of modern biometrics that differentiate them from the past and shape their development.

Today's biometrics can be derived from reading diverse body parts such as hands, fingers, eyes and face, as well as from interpreting diverse embodied activities such as typing, writing, speaking and walking. These human characteristics fall under two general categories: physiological biometrics and behavioral biometrics (Das 2015; Epstein 2008; Lodge 2007; Lockie 2009). In the first category, there are personal attributes that we are born with. The most common types of such biometrics used today are fingerprints, hand geometry, vein pattern, face shape, iris patterns and retina patterns. In the second category, there are personal traits that develop over time and that once formed are believed to remain constant throughout our lives. The most common examples here include signature, keystroke and voice recognition. Several other types of biometrics are under development such as gait recognition, heartbeat, body odor and

DNA. Undoubtedly, DNA code is the top prize biometric as it is the most unique and information rich. At the same time, this is also the most controversial. Nonetheless, the technology to automate DNA analysis is not yet viable for large-scale applications.

An additional definition of biometrics is emerging in the field of healthcare and in some commercial applications such as wearable devices that view biometrics in terms of an individual's vital signs, such as blood pressure, pulse, body mass index and other health-related metrics. In this paper we will retain the initial meaning of biometrics as identification technologies.

Underlining biometrics' appeal is the assumption of infallible personal identification as summarized by the expression 'the body does not lie' (Aas 2006). The belief is that the more accurately bodily characteristics can be technologically captured and interpreted, the more human subjectivity can be eliminated from the process and the more unfaltering personal identification becomes. At the core of this assumption there are several biometrics attributes (Das 2015). The most important is uniqueness. Biometrics are unique to each person, which means that they are accurate and difficult to falsify or reproduce. Second is their universality. Everybody has them at all times, and as such they can be used in any circumstance or place. Third, they are permanent. They are not changing (too much) over time, thus they are highly consistent. Last is convenience, as they are easy to capture and automate, and they cannot be forgotten or misplaced.

In consequence, among identification technologies biometrics are considered to provide the highest level of security possible (Jain et al. 2006; Lockie 2009). Along the authentication continuum, metal keys or magnetic IDs, used for a long time to unlock doors, are devices we own or have. We need to carry them with us to gain access. One step further along the continuum, passwords and PINs are something we need to know to obtain access. Fingerprints and irises, however, represent something we are, and we do not need to carry or remember them. In other words, the body works much like a password, providing personal identification on the move to gain access to diverse spaces and services. From this vantage point, biometrics are simply updated versions of previous identification technologies. They are quicker, more efficient and more precise than their predecessors. They are also believed to be virtually tamper-proof.

Modern biometric technologies were initially used on a small scale during the 1990s, mainly in banking to grant secure access to money, in surveillance operations and crime fighting, and in security establishments to control access to buildings and other spaces. During the early 2000s they were incorporated into border securitization practices in the United States and Europe to control illegal immigration and to manage trusted traveler schemes (Sparke 2006; van der Ploeg 1999a). By the late 2000s, biometrics had become mainstays of border securitization regimes from Thailand and Australia to Nigeria and the United States. Today, they are being used in large-scale applications affecting hundreds of millions of people, most notably through government-issued biometric passports and IDs, as well as through the setup of numerous border-related databases. In parallel, the use of biometrics in large-scale surveillance and other security operations has also exploded during the 2000s, aided, among others, by the incorporation of facial recognition systems in CCTV cameras and by advances in miniaturization to allow the incorporation of biometrics in mobile devices.

Despite the use of biometrics in security practices in various countries before 11 September 2001, there is little doubt that the attacks on New York and Washington,

DC, provided critical impetus for their widescale adoption. First officially heralded by the US Patriot Act of 2001, biometric technologies became generalized with the US Enhanced Border Security and Visa Entry Reform Act of 2002, which mandated the creation of inter-operational biometric immigration databases, as well as the introduction of biometric e-passports for US citizens. One particular provision of this law stipulated that the Visa Waiver countries, whose citizens could visit the United States without a visa for up to three months, would have to introduce biometric e-passports as well if they wished to maintain these privileges. These legal provisions reaching outside the borders of the United States effectively worked to globalize the use of biometrics in border securitization practices. The process was bolstered in 2003 when the International Civil Aviation Organization, a UN body comprising 190 states that regulates air traffic, recommended the adoption of biometric e-passports including digitized photographs as the new standard for international travel documents. The governments of the European Union soon introduced their own biometric e-passports based not on one but on two sets of biometrics, photos and fingerprints (Epstein 2007). These developments left many governments in the developing world little choice but to follow suit, investing in biometric technologies in order to meet the passport standards demanded by developed countries. Among other examples of large-scale biometrics adoption, the most notable is India's ambitious national ID program to include 10 fingerprints and iris scans for 1.2 billion people. In these circumstances, governments remain the largest customers of the biometrics industry, despite the booming commercial applications in the private sector.

IN THE BELLY OF THE BEAST: UNDERSTANDING THE BIOMETRIC MACHINE

The most advanced biometric systems can identify a person in less than a second. Speed is a key factor behind the power of biometrics because it addresses issues of convenience and dependability. Behind the speed, however, there are relatively complex technical processes at work that include several stages and are heavily reliant on mathematical algorithms (Jain et al. 2006). Generally, sensors in digital devices first have to capture an image of the chosen body part in order to store it in a database. Nonetheless, the image itself is of secondary importance. Instead, the emphasis is on extracting unique features and patterns from the image that are then assembled into a template. In effect, a biometric template is a digital representation of the features of the image captured from the body part (Das 2015). For example, from fingerprint images, the shape of the ridges and the depth of valleys are recorded, while from facial images the recorded elements are the position and distance between eyes, cheekbones, chin, etc. In the case of behavioral biometrics, such as keystroke, for example, a series of recordings need to be made that are then assembled into a pattern with the help of statistical modeling. Contrary to popular belief, there is no picture stored in the database but an algorithmically encrypted binary file. These files have the advantage of adding an extra layer of security because, if biometric templates are stolen, one also needs to have the encoding algorithms in order to make full use of them.

The next stage of identification involves comparing the original template stored in the database with a temporary one, generated each time a person presents a body part for authentication at a terminal. If there is a positive match, then the identity of the person is confirmed and access authorized. Algorithmic equations and statistical modeling are heavily at work in this stage. For reasons that have to do with enrollment circumstances and variation in device specifications, there are no two identical biometric templates despite the fact that our biometric features are unique. The positive match, then, represents a statistical probability of the two templates matching, and not an exact match. This situation presents an opportunity for reflection about taking biometrics for granted as infallible identification technologies and about how bias is built into biometric technologies.

Biometric identification is used in two major ways: to verify someone's identity and to establish someone's identity (Lyon 2008; van der Ploeg 1999b). In the first instance, the use of biometrics is justified by the need to verify a person's identity to confirm that they are who they say they are. This is the most common use of biometrics to date and also the one that can potentially raise fewer concerns if properly regulated. From a technological perspective, verification boils down to as an in-and-out computerized task. A claim is made to a biometric system and a data sample is submitted for verification. The system has to confirm the identity claim by verifying that the sample data are the same as the data on file in a database. This is a one-to-one operation. Once the task is performed, the process ends and temporary data are discarded. However, verification is also the least useful from a law enforcement and national security perspective. The point is that biometrically verifying Osama bin Laden's identity would only have confirmed that he was who he said he was. Biometrics stored on bin Laden's passport chip would not have said that he was a terrorist.

The second major use of biometrics is aimed at addressing the question, 'Who is this person?' Answering this question requires establishing a person's identity. In this case, biometric systems have to perform sophisticated investigative work to identify one person from a group of people. It is not known if the person's biometrics already exist in the database. The systems have to perform a one-to-many operation to compare someone's biometric template with those of a multitude of other individuals. In other words, biometric technologies have to assign an identity to a person. This task necessitates the establishment and maintenance of huge databases to store biometrics from a large number of people. This is how the database emerges as a key element of biometric technologies. Biometric passports and IDs that a person carries with them, for example, have to be linked to databases that are administered separately by authorities in each country or by each security company operating the biometric system. Ideally, for biometric technologies to work best, there should be one global database that contains several types of biometrics for each person, enrolls the entire population of the planet and is stored in cyberspace to be easily accessible.

Nonetheless, vast volumes of biometric data alone are still not sufficient to meaningfully determine someone's identity (van der Ploeg 1999b). For this to happen, databases have to contain additional information about peoples' everyday lives that can be electronically mined to reveal patterns of behavior and association. Once someone's biometrics are recorded into a database, personal information can be collected every time there is a hit on them. For example, every time someone's biometric passport is checked, this leaves a trace in the US border security database that is stored under that person's profile according to predetermined criteria such as name, address, country of origin, how many times the person crossed a border, which borders were crossed and in what places, means of transportation used, form of payment for the trip, duration of stay, driving history, type of meal consumed in flight, seating preference and more (Department of Homeland Security 2006). This is how a person's digital identity takes shape in a database without the person even being aware of it. The belief is that these security systems will prevent the repetition of situations like 11 September 2001, by allowing law enforcement authorities to 'connect the dots' between apparently unrelated bits of information. Such systems have been functioning in secret at US borders since 2002 under the name of Automated Targeting Systems. They integrate several separate governmental and private databases and have collected data on hundreds of millions of US visitors and US citizens alike. The European Union has implemented similar systems dubbed Automated Border Control.

Other widespread law enforcement biometric databases in existence in various countries are known as automated fingerprint identification systems and contain hundreds of millions of fingerprint records. In addition, numerous private companies maintain smaller and more targeted customer biometric databases for commercial purposes, such as identifying purchasing patterns or spending habits that are then used to calculate risk scores to establish insurance premiums or to personalize advertising. Under these circumstances, database fraud and data theft has become a major criminal activity both between governments and between private actors.

LOOKING BEHIND THE FAÇADE: A CRITICAL ASSESSMENT

A critical examination of the assumptions behind the deployment of biometric technologies reveals major issues concerning their logic of identity management and indicates a wide gap between expectations and what these technologies are actually able to deliver. To be sure, biometrics can play a role in managing twenty-first century personal identification challenges. The problem is with the uncritical adoption of biometric technologies and the resulting societal impacts. It is important to weigh the potential costs and benefits the widespread use of biometrics brings to society in order to make informed decisions about their future.

One major issue is the conflation of identification with identity (van der Ploeg 1999b). This means that identifying a person's body is the same as knowing a person's identity. An associated assumption is that a person's identity makes a good risk predictor (Amoore 2009). Therefore, the identity of an individual provides the basis for identifying the risk they pose to society or to various kinds of profit maximization schemes.

Another major issue resides in the technologically deterministic thinking surrounding the adoption and implementation of biometrics. This works to obscure their inherently biased nature and results in biometric technologies being taken for granted as panacea for identification and security. Moreover, these attitudes promote a view of identification and security as simply technical problems that require a technical fix, and overlook the fact that the social use of technology is a two-way street (Popescu 2011). Technology alone cannot provide exclusive benefits to security agencies and law-abiding citizens. Societies evolve with their technology, including peoples' capacity to defeat the system.

Last, there are issues related to the regulation of biometric technologies. Given the power of biometric technologies, their growing use in daily life situations, and their potential impact on privacy and civil liberties, the lack of laws to govern the use and abuse of biometric data is surprising. Presently, there is no effective regulation of biometric technologies anywhere in the world. It is not clear what data are being collected, for how long, to what purposes, where and how they should be stored, what level of protection they should have, and if they are used to other ends beyond their formally stated purpose. Moreover, many major biometric systems lack systematic provisions and back up procedures to govern instances when people are wrongly identified. Although most biometric technologies claim a rate of error of around 1 percent or less, this is not as insignificant as it first seems when we think that numerous biometric systems enroll millions of people. Depending on circumstances, misidentifying someone can create major problems in their lives, for example, when there are no dependable alternatives to using a certain biometric system, or when one is left by themselves to prove their identity credentials to the system. These situations demonstrate why, despite the financial appeal of automatization, human intervention should always be part of biometric systems.

The implementation of biometric technologies raises two broad sets of problems related to security and privacy issues. The first problem relates to the limitations of the biometric technology itself, while the second relates to how biometrics are used and to what ends. Particularly questionable is the belief that biometrics are infallible identity management technologies (Lodge 2007). The aura of foolproof technology often arises from the fact that most people only see the end result of biometric technology when their body parts are mysteriously recognized with the authority of an impersonal machine. The granular process of identification is hidden from view inside computerized devices. Such experiences tend to lead many people to take the expertise of the computer for granted and to believe that its qualities are superior to those of human beings or to pre-digital identification technologies. What is overlooked here is that the individuals who seek to steal or compromise biometric data can have the same level of technical knowledge as the designers and programmers of biometric systems.

In reality, fingerprint readers can be successfully fooled by attaching silicon patches to a person's fingertips, as well as by surgically grafting patches of skin from other parts of the body to the fingertips (Heussner 2009). Most notoriously, Apple's fingerprint sign-in system was hacked shortly after its official launch using a fake fingerprint generated from a clean fingerprint lifted off a surface. Face and iris recognition systems can be relatively easy to deceive as well. The answer to these issues has been to demand the use of even more biometrics. The latest generation of e-documents normally include two or more types of biometrics to minimize the chance of fraud. Even better, current biometric systems can determine whether the body part to be identified belongs to a live body or not to avoid easy spoofing. However, these responses are still not deterring potential fraud. Rather, they push hackers to use more sophisticated methods.

Another problem arises from the multiplication of mobile biometric devices ranging from scanners to e-passports that use wireless communication to transfer data back and forth in a biometric system. These can be particularly vulnerable to hacking via their wireless capabilities that offer avenues for remote access to their data. Hackers can break the existing encryption codes and copy their contents without the owner's knowledge

(Albrecht 2008). The data can then be cloned into a new e-document or device to impersonate the owner (Koscher et al. 2009). These issues are addressed by continuously upgrading wireless security features to close all foreseeable loopholes (Liersch 2009).

A significant shortcoming of biometrics is that they do not offer 100 percent identification accuracy in all cases. Rather, they offer a high statistical probability of a positive match between a body part and a previously submitted sample of the body part. Certain categories of people, like the elderly, Asian women and the disabled, present physical features that are difficult to enroll in biometric programs like fingerprinting. Biometric systems flag members of such groups more often, thus discriminating against them.

Additional concerns involve the securing of immense biometric databases that would have to be accessed in thousands of locations worldwide. If they are lost or hacked into, the damage to the lives of people affected can be incalculable. What is important to realize about this kind of personal data is that once it is lost, the victims will bear the consequences for the rest of their lives. Biometric data are not cancellable. Since these are bodily data, one cannot just declare them lost and have new ones issued. It is somewhat ironic that biometrics' most appealing characteristic – uniqueness – is at the same time its biggest weakness.

At the same time, biometric technologies push the issues of identification and security further inside society instead of resolving them. The fact that peoples' biometrics have to be prerecorded in e-documents and databases in order for biometric systems to work means that biometric systems can only be as secure as the process of biometric enrollment. Instead of physically tampering with an e-passport or with somebody's fingertips, it becomes more effective to forge the much less sophisticated feeder documents required to establish a biometric identity, such as birth certificates, proof of residence and other documents. In this way, a known terrorist or a spy can acquire a perfectly authentic biometric identity under another name at a passport-issuing office somewhere in the world. Provided that government biometric databases do not contain the terrorist's or spy's photograph, fingerprints or iris scans, he or she can sail smoothly through the biometric systems.

BIOMETRICS AND THE AUTOMATIC PRODUCTION OF SPACE AND IDENTITY

Biometrics are bringing the connections between space and identity into the spotlight via the medium of the human body. From a spatial perspective, biometric technologies are about providing differentiated access to space. With daily lives becoming more and more connected and networked because of people living and working across multiple places, the need for speedy and reliable identification becomes more critical. In this sense, the production of space and biometrics work hand-in-hand in the digital era.

Embedding identification technologies into the human body presents obvious advantages since it allows constant and accurate monitoring of movement at the smallest spatial scale (Popescu 2011). The body makes the ideal ID, as it is always at hand, ready to be performed whenever circumstances require. In this way, mobile risks can be estimated from mobile bodies and efficiently eliminated along the way so that traffic flows are not disrupted. Space can be governed following a logic of inclusion through exclusion, where all people have to be enrolled in biometric databases in order then to have separate access to places. Everyone can move but but not everyone can reach all destinations. What this logic achieves is a more fine-grained and automated control over space that allows the targeting of individuals or groups in society to ban them from certain places or to keep them moving only along designated routes. In addition, access to space becomes not only differentiated but also code-like given the binary calculations behind biometrics. This is how it is possible that people who find themselves in close territorial proximity in a city or a country might have little interaction with each other, while they might be better connected with people located in remote places. Such geographies can best be made sense of by applying a topological analytical lens where inside can be outside and close can be far away.

The problem of biometric identification goes well beyond technical concerns with national security and personal privacy, as significant as these may be. At stake here is redefining the relationships between social life, power and space. The work of Michel Foucault (1977, 1978, 2007, 2008) has provided productive analytical grounding for scholarship seeking to understand the impacts that current biometric identity management practices have on social relations (Amoore 2009; Epstein 2007; Salter 2006). Foucault's analyses of strategies of power that unearthed the connections between power, knowledge and space remain highly relevant in the current context (Dobson and Fisher 2007). Particularly insightful are his concepts of governmentality and biopolitics that illustrate how in modern states providing security has become a form of governing populations, and how such governing is performed through the management of populations' biological characteristics, behaviors and movements (Dillon and Lobo-Guerrero 2008; Elden 2007).

In the twenty-first century, power is increasingly derived by securing populations. Making biometric technologies part of everyday life as a means to securing it means acquiring power to order everyday lives. Biometrics technologies, together with the entire technological arsenal available for law enforcement and corporate decision-makers, are contemporary tools that can be used to render populations knowable by statistically classifying them along pre-established criteria, calculating their behavior in terms of risk assessment, and tracking their movements. Biometric technologies' preoccupation with acquiring comprehensive knowledge about large numbers of mobile bodies is a power strategy for controlling the fluid and networked territoriality of movement. Classifying and ranking bodies in terms of good versus bad mobility or their economic worth creates categories that are then amenable to risk contingency calculus. In this way, knowledge of the body results in power over the body. This is, at the same time, power over the most intimate and mobile of spaces. The goal of biometric technologies is not to differentiate between political subjects, but to get the data from the body. The identity of the body really does not matter beyond its flesh, blood and bone existence. The body in this techno-logic of power is the kind of body as a living organism, to be kept alive to be governed, not the political body, with rights and duties, to be the subject of government (Agamben 1998; Epstein 2007).

While not being able to actually say who one really is, biometric technologies are in fact producing one's identity to say who one must be (Amoore 2006). This is essentially an imagined identity that few understand how it was arrived at but that everybody has to trust. The identity that emerges out of algorithmic calculations of personal risk profiles is not synonymous with a person's identity, which emerges over time from processes like interpersonal communication and self-reflection (van der Ploeg 1999b). This is not the identity of a person as a social being, but the identity of an object that has been rendered knowable. Instead of verifying that 'you are who you say you are', the purpose changes so that 'you are who we say you are'. Implied here is the promise that these technologies can predict how one will behave in the future based on past patterns of behavior. The question that arises now, however, is who decides what constitutes good and bad behavior? How does one know which are good friendships to have or what is the right food to eat, and which ones will increase their risk score at the border or for getting a loan, or increase their health insurance premium? The fact is that the biometric data in a database can be made to tell multiple stories about a person according to what type of information the algorithmic software is programmed to look for. The algorithms define good and bad citizens according to how they are written. As the criteria by which the algorithms works are kept secret and can change with the powers that be, the politics of biometrics become crucially important going forward (van der Ploeg 1999b). The concern is that, without an understanding of what these new technologies of power can and cannot do, identity management practices may incorporate racial, class, ethnic or gender stereotypes and prejudices that perpetuate existing inequalities and create new ones.

CONCLUSION

Biometrics are powerful technologies of the digital age, with transformative impacts on society's relationship to space. They bring together issues of identity and space, redefining these categories as well as the relationship between them. During the span of one decade they have transitioned from being marginal to being central identification management tools, and they have been widely adopted in government, law enforcement, security, healthcare, finance and commerce.

The main appeal of biometric technologies consists in their bodily embedded nature and automation, which open up new possibilities for the organization of interaction in space. At the same time, these characteristics are also the most polarizing. Some people and interest groups enthusiastically embrace them while others actively oppose them. Biometrics' intimate association with the human body raises legitimate privacy concerns about their use, and elicits a sense of inescapable and overwhelming personal control that leads to mistrust and makes them feared. The uniqueness of a person's bodily data gives new meaning to the phrases like 'identity theft' and 'identity crisis', given the current lack of alternatives to replace such data. Unlike passwords and keys, having one's biometrics stolen is similar to having one's own body stolen forever.

The technologically deterministic discourse that often accompanies the implementation of large-scale biometric systems has perpetuated popular beliefs in their supernatural qualities, while the absence of meaningful regulation had done little to address existing concerns. Presently, there is an acute need for thorough and overarching regulation if biometrics are to make a positive contribution to social life in the digital age instead of harming it. It is important to understand that at the end of the day biometrics are just a technology, and thus they will not be able to solve the world's structural problems such as insecurity, underfunded healthcare systems and bank fraud. They can make suitable tools to address social problems but they cannot be the answers.

From a spatial perspective, when our bodies are passwords we have to use them to gain access to spaces. This also means that we as persons are under suspicion when encountering biometric decision-making systems, until the proof of our right to access is extracted from our bodies. In other words, our identities are treated as potentially culpable until we prove ourselves innocent. As biometric technologies are becoming ubiquitous in everyday life, our bodies will have to clear such encounters time after time and place after place as we go about our daily routines. This is why getting the politics of biometrics right and avoiding fetishizing these technologies is essential to make sure we are not creating more oppressive tools instead of using them to enhance democratic liberties.

REFERENCES

Aas, K. 2006. The body does not lie: identity, risk and trust in technoculture. *Crime, Media, Culture* 2: 143–158.

Agamben, G. 1998. *Homo Sacer: Sovereign Power and Bare Life*. Stanford, CA: Stanford University Press.

Albrecht, K. 2008. RFID tag – You're it. *Scientific American* 299: 72–77.

Amoore, L. 2006. Biometric borders: governing mobilities in the war on terror. *Political Geography* 25: 336–351.

Amoore, L. 2009. Algorithmic war: everyday geographies of the war on terror. *Antipode* 41: 49–69.

Das, R. 2015. *Biometric Technology: Authentication, Biocryptography, and Cloud-based Architecture*. Boca Raton, FL: CRC Press.

Department of Homeland Security. 2006. *Privacy Impact Assessment for the Automated Targeting System*, 22 November.

Dillon, M. and L. Lobo-Guerrero. 2008. Biopolitics of security in the 21st century: an introduction. *Review of International Studies* 34: 265–292.

Dobson, J. and P. Fisher. 2007. The panopticon's changing geography. *Geographical Review* 97: 307–323.

Elden, S. 2007. Governmentality, calculation, territory. *Environment and Planning D* 25: 562–580.

Epstein, C. 2007. Guilty bodies, productive bodies, destructive bodies: crossing the biometric borders. *International Political Sociology* 1(2): 149–164.

Epstein, C. 2008. Embodying risk: using biometrics to protect the borders. In L. Amoore and M. de Goede (eds), *Risk and the War on Terror*. pp. 178–193. London: Routledge.

Foucault, M. 1977. *Discipline and Punish: The Birth of the Prison*. London: Penguin.

Foucault, M. 1978. *The History of Sexuality: An Introduction*. New York: Vintage.

Foucault, M. 2007. *Security, Territory, Population: Lectures at the College de France, 1977– 1978*. New York: Palgrave Macmillan.

Foucault, M. 2008. *The Birth of Biopolitics: Lectures at the College de France, 1978–1979*. New York: Palgrave Macmillan.

Heussner, K.M. 2009. Surgically altered fingerprints help woman evade immigration. ABC News, 11 December, accessed from http://abcnews.go.com/Technology/GadgetGuide/surgically-altered-fingerprints-woman-evade-immigration/story?id=9302505&page=3.

Jain, A., R. Bolle and S. Pankanti (eds). 2006. *Biometrics: Personal Identification in Networked Society*. New York: Springer.

Koscher, K., A. Juels, V. Brajkovic and T. Kohno. 2009. EPC RFID tag security weaknesses and defenses: passport cards, enhanced drivers licenses, and beyond. In *Proceedings of the 16th ACM Conference on Computer and Communications Security*. pp. 33–42.

Liersch, I. 2009. Electronic passports – from secure specifications to secure implementations. *Information Security Technical Report* 14: 96–100.

Lockie, M. 2009. *Biometric Technology*. Chicago, IL: Heinemann.

Lodge, J. (ed.). 2007. *Are You Who You Say You Are? The EU and Biometric Borders*. Nijmegen: Wolf Legal.

Lyon, D. 2008. Biometrics, identification and surveillance. *Bioethics* 22(9): 499–508.

Popescu, G. 2011. *Bordering and Ordering the Twenty-First Century: Understanding Borders*. Lanham, MD: Rowman & Littlefield.

Salter, M. 2003. *Rights of Passage: The Passport in International Relations*. Boulder, CO: Lynne Reinner.

Salter, M. 2006. The global visa regime and the political technologies of the international self. *Alternatives* 31: 167–189.

Sparke, M. 2006. A neoliberal nexus: economy, security and the biopolitics of citizenship on the border. *Political Geography* 25(2): 151–180.

van der Ploeg, I. 1999a. The illegal body: 'Eurodac' and the politics of biometric identification. *Ethics and Information Technology* 1(4): 295–302.

van der Ploeg, I. 1999b. Written on the body: biometrics and identity. *Computers and Society* 29(1): 37–44.

Index